INDUSTRIAL FERMENTATIONS

EDITED BY

LELAND A. UNDERKOFLER, PH.D., D.SC.
Professor, Chemistry Department
Iowa State College, Ames, Iowa

AND

RICHARD J. HICKEY, PH.D.
Research Microbiological Chemist
Commercial Solvents Corporation
Terre Haute, Indiana

VOLUME II

1954
CHEMICAL PUBLISHING CO., INC.
212 Fifth Avenue, New York, N. Y.

Printed in the United States of America

CONTRIBUTORS TO VOLUME II

A. M. Buswell, Chief, Illinois State Water Survey, Urbana, Ill.

R. D. Coghill, Director of Research, Abbott Laboratories, North Chicago, Ill.

R. J. Hickey, Research Microbiological Chemist, Commercial Solvents Corporation, Terre Haute, Ind.*

J. C. Hoogerheide, Research Department, Pabst Brewing Company, Milwaukee, Wisc.

G. A. Ledingham, Director, Prairie Regional Laboratory, National Research Council of Canada, Saskatoon, Sask.

C. C. Lindegren, Chairman, Department of Microbiology, and Director, Biological Research Laboratory, Southern Illinois University, Carbondale, Ill.

L. B. Lockwood, Microbiologist, Miles-Ames Research Laboratory, Elkhart, Ind.

Elizabeth F. McCoy, Professor of Agricultural Bacteriology, University of Wisconsin, Madison, Wisc.

L. E. McDaniel, Microbiologist, Merck & Co., Inc., Rahway, N. J.

A. C. Neish, Senior Fermentation Chemist, Prairie Regional Laboratory, National Research Council of Canada, Saskatoon, Sask.

* Present address, University of Pennsylvania, Philadelphia 4, Pa.

Mary S. Peterson, University of Wisconsin, Madison, Wisc.

W. H. Peterson, Professor of Biochemistry, University of Wisconsin, Madison, Wisc.

A. R. Stanley, Microbiologist, The Ralph M. Parsons Company, Facilities Operation Division, Braddock Heights, Md.

J. C. Sylvester, Manager, Fermentation Research, Abbott Laboratories, North Chicago, Ill.

L. A. Underkofler, Professor of Chemistry, Iowa State College, Ames, Iowa.

J. M. Van Lanen, Assistant Director of Research, Hiram Walker & Sons, Inc., Peoria, Ill.

R. H. Vaughn, Associate Professor, Food Technology Division, University of California, Davis, Calif.

H. B. Woodruff, Assistant Director and Head of Research Section, Department of Microbiology, Merck & Co., Inc., Rahway, N. J.

FOREWORD

Plans for an authoritative compilation of modern industrial practices in the fermentation industries were originally made about 1948. Because the field is so diversified, assistance of individuals with special experience was solicited to contribute chapters relating to the different fermentations. Many unavoidable delays occurred between inception of the idea and collection of all the chapter manuscripts. When all the chapters had been assembled, the bulk of the manuscript was such that it became necessary to publish in two volumes.

All the contributors have generously given most valuable assistance, some in the face of overwork and ill health. They have felt with the editors the disappointments attendant on the various delays that have occurred between first writing and actual publication. In those fermentation fields where rapid progress is being made, contributors have laboriously revised and rewritten their chapters several times so that they will be as nearly up-to-date as possible. The editors wish again to express their sincere appreciation to these contributors who have been so generous with their time and effort in this cooperative undertaking.

Over the space of five years, the difficulties of writing, editing, collaborating with distant workers, and producing these volumes covering such a vast and rapidly expanding field introduced possibilities of omissions, errors, repetition, and even minor contradictions. In these matters, the editors ask the indulgence of readers. Despite imperfections, we believe that these volumes are unique in their authoritative treatment of one of the most improtant branches of chemical industry. If this compilation of the current knowledge and industrial practices is found of value by industrial workers and students interested in fermentations, the arduous task has not been completed in vain.

L. A. UNDERKOFLER
R. J. HICKEY

TABLE OF CONTENTS

CONTENTS OF VOLUME I

Part I. THE MICROBIOLOGICAL KETOGENIC PROCESSES

CHAPTER 1

KETOGENIC FERMENTATION PROCESSES

Lewis B. Lockwood

The ketogenic fermentation processess discussed in this chapter are principally those dehydrogenation or oxidation reactions by which polyhydric alcohols, sugars, or polyhydroxy acids are converted respectively into ketoses or keto-sugar acids. Several additional fermentations, including the production of α-keto-glutaric acid and kojic acid will be considered briefly. The most important processes are the oxidation of sorbitol to L-sorbose, and of D-glucose or D-gluconic acid to 2-keto-D-gluconic acid or 5-keto-D-gluconic acid. Numerous other analogous fermentation reactions have been studied on a laboratory scale, but have not been conducted in pilot-plant or factory-size operations. One important fermentation ketone, acetone, is not covered in this chapter on ketogenic fermentations because it is associated with anoxidative processes yielding *n*-butyl alcohol or *n*-butyric acid, and is treated elsewhere (see Chapter 11 of volume I). Acetone is the product of biochemical and physical systems greatly different from those yielding the highly oxidized products discussed here.

The type reaction for the bacterial oxidative ketogenic processes is:

$$\mathrm{H{-}\overset{|}{\underset{|}{C}}{-}OH} \longrightarrow \mathrm{\overset{|}{\underset{|}{C}}{=}O}$$

The product of the reaction is governed by the choice of substrate and by the specificity of the enzyme system of the fermenting organism. The fundamental requirements of high oxygen tension in the solution and adequate nutrient supply are the same for all the bacterial ketogenic processes considered here.

KETOGENIC PROCESSES OF THE ACETOBACTERS

Many species of the genus *Acetobacter* produce ketogenic dehydrogenations. Butlin[3] has extensively reviewed the biochemical activities of the *Acetobacters.* The first example of this type of biological process was described by Bertrand,[4] who isolated the "sorbose bacterium" from fermenting juice of the berries of the mountain ash *(Sorbus aucuparia L.)* and identified the L-sorbose produced when the sorbitol in the juice was oxidized by the bacteria. Later Bertrand's "sorbose bacterium" was identified with *Bacterium xylinum* Brown, which is now classified as *Acetobacter xylinum.*

Bertrand[5] subjected a considerable number of sugar alcohols and sugars to the action of his organism and, on the basis of his findings, postulated a rule regarding compounds which were and were not oxidized by the organism. Bertrand's rule will be considered in a subsequent section.

The Sorbose Fermentation

The sorbose fermentation is presented first for several reasons. It was the first bacterial ketogenic fermentation discovered, it is one of the simplest of such processes, and it is the process of this type having greatest industrial significance. The operative conditions for all these processes are similar, but some variations, such as the addition of a neutralizing agent or a difference in concentration of substrate may be found in the operation of the different processes.

L-Sorbose is produced from the polyhydric alcohol, sorbitol, by the action of several species of bacteria of the genus *Acetobacter*. *Acetobacter suboxydans* appears to be the species most widely used industrially, although many other species, including *A. xylinum,*

A. gluconicum, A. xylinoides, A. pasteurianum, A. kutzingianum, A. orleanse, and *A. hoshigaki,* also produce sorbose readily. *A. aceti,* the common vinegar bacterium, does not bring about this fermentation or produces sorbose in yields too poor to be practicable.

As noted previously, the oxidation of sorbitol to sorbose was first described by Bertrand.[4] There was only academic interest in this fermentation until it was discovered that L-sorbose was the logical intermediate in the commercial manufacture of ascorbic acid.

Early studies on the production of sorbose from sorbitol were conducted, using unagitated media, with bacterial pellicles floating on the surface of the solutions. Bertrand's organism, *A. xylinum,* formed very tough, cellulosic pellicles, was slow in its action, and gave relatively poor yields of sorbose. If the pellicles became submerged the fermentation stopped. The fermentation by *A. xylinum* was completed in about 6 weeks and gave sorbose yields of 40 to 60%.

Another organism, *A. suboxydans,* discovered by Kluyver and de Leeuw,[22] formed very light, fragile pellicles, acted much more rapidly than *A. xylinum,* and gave better yields of sorbose. Fulmer, Dunning, Guymon, and Underkofler[14] found that surface fermentations by this organism were complete in about 7 days and gave yields of 80 to 90%. Wells, Stubbs, Lockwood, and Roe[50] grew *A. suboxydans* in vigorously aerated media, under 30 lb gage pressure of air, and obtained rapid oxidation of sorbitol to sorbose. This work was extended by Wells, Lockwood, Stubbs, Roe, Porges, and Gastrock[49] through the semiplant scale. Yields of approximately 98% of theory, based on the sorbitol supplied, were obtained in 14 hours when 10% solutions of sorbitol were fermented. Solutions containing 20% sorbitol required 24 hours and 29% sorbitol solutions required 40 hours.

The starting material in the sorbose fermentation is sorbitol (D-glucitol), a polyhydric alcohol, obtained when the aldehydic terminal group of the D-glucose molecule is reduced to a primary alcohol. This reduction is brought about by either electrolytic or catalytic methods. In the first process, some mannitol is also formed. A nickel catalyst is used for the hydrogenation reaction in the catalytic method. It is essential to use sorbitol free of nickel for the fermentation, since *A. suboxydans* is very sensitive to the nickel ion.

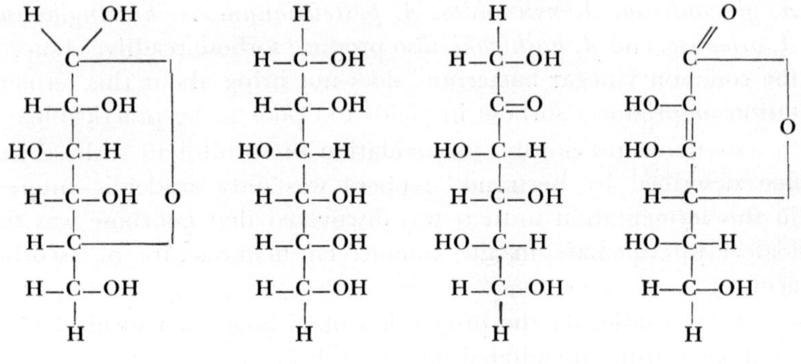

D-Glucose Sorbitol (D-Glucitol) L-Sorbose L-Ascorbic acid

Culture Isolation and Care

Acetic acid bacteria suitable for the production of sorbose are readily isolated from fruits and berries undergoing natural alcoholic fermentation by yeasts. They are characterized by their ability to convert ethyl alcohol into acetic acid and to metabolize inorganic nitrogen compounds, such as ammonium salts. They do not produce the water-soluble, yellow-green, fluorescent pigments typical of the genus *Pseudomonas.* Some species of *Acetobacter* require the addition of vitamins to their culture media. For example, Underkofler, Bantz, and Peterson[44] have found that *p*-aminobenzoic acid, pantothenic acid, and nicotinic acid are required as growth factors by *A. suboxydans,* which has been recommended by Landy and Dicken[23] as an assay organism for *p*-aminobenzoic acid. Cultures of acetic acid bacteria, including those of *A. suboxydans,* may be preserved by lyophilization in blood serum or by frequent transfer on agar slants. The semisolid medium should contain 3% glycerol, 0.5% yeast extract, and 1.5 to 2% agar. Sorbitol or glucose may be substituted for the glycerol. However, if glucose is used in the culture medium, calcium carbonate sufficient to neutralize the equivalent gluconic acid must be added to maintain the pH at a suitable value.

Laboratory cultures in liquid medium are prepared for inoculation of plant fermentations. Media which contain 5 ml corn steep liquor (50% solids basis) or 5 g yeast extract, or 10 g killed yeast per liter, in addition to sorbitol, are suitable for the preliminary phases of the culture propagation prior to full-size plant use. The ratio of inoculum for the preparation of successively larger cultures should be about 3:100. Each culture should be aerated with ⅓ to 2 volumes of air per volume of culture liquor per minute, the rate depending on the efficiency of dispersion of the air through the medium, the pressure, and the efficiency of agitation. The acetic acid bacteria may be readily overwhelmed in culture if contaminating organisms, such as *Pseudomonas fluorescens, Bacillus cereus, B. subtilis, B. megatherium,* or *Leuconostoc mesenteroides,* are present. Consequently, rigid sanitary precautions are essential in developing the cultures to plant-scale operations. Bacteriophage has not been a problem in *Acetobacter* fermentations and specific bacteriophages for most species of *Acetobacter* are unknown.

Plant Process

The plant fermentation is conducted in a conventional-type vat fermentor equipped with porous stones, perforated pipes, or other means for the dispersion of air. The fermentor should be constructed of nickel-free metal. High purity aluminum or nickel-free stainless steel is satisfactory for this process. Mechanical agitation may be advantageous, but is not essential, since aeration of the liquor serves also as a means of stirring.

A mash composition suitable for plant-scale fermentation contains:

	per liter	*per gallon*
Sorbitol	100-250 g	0.8-2.0 lb
Corn steep liquor (50% solids)	5 ml	18.9 ml
Antifoam	1 g	3.8 g

Octadecanol, lard oil, or soybean oil are suitable antifoam agents. When the higher sorbitol concentrations are employed, the initial lag period in the fermentation is prolonged. This can be avoided by starting at a 10 or 15% level and feeding additional sorbitol during the logarithmic phase of growth.

The mash is prepared and sterilized by heating under pressure. After cooling, the sterile mash in the fermentor is inoculated with about 3% by volume of active plant culture of *A. suboxydans,* produced in smaller fermentation vessels under aeration. During fermentation, vigorous aeration with sterile air must be provided. Fermentation under an air pressure of 2 atmospheres (30 lb gage pressure) results in a more rapid fermentation, but yields are little different from those obtained when the process is operated at atmospheric pressure. The volume of air supplied will depend on the efficiency with which the solution is oxygenated. The air requirement will be reduced if mechanical agitation and operation under super atmospheric pressure are used. With vigorous stirring at 30 lb gage pressure, ⅓ volume of air per minute results in rapid fermentation. A higher aeration rate is not beneficial. The preferred temperature range is between 30° and 35°C. Considerable heat is evolved during the fermentation and cooling equipment is necessary in large-scale operations. Coils, jackets, or sprinkler systems have been found satisfactory for temperature control.

Analytical Methods

The initial sorbitol content may be conveniently determined by measuring the refractive index of the mash and reading the concentrations from a graph on which the concentration of sorbitol is plotted against refractive index. The progress of the fermentation may be followed by the use of any of the well-known copper-reduction methods for the determination of reducing sugars. A special graph is prepared, on which milligrams of cuprous oxide are plotted against the experimentally determined cuprous oxide-sorbose ratios, or cuprous oxide may be related to sorbose in a table.

Recovery of Product

Filter aid is added to the fermented liquor and the mash is filtered to remove bacteria and suspended solids. An ordinary filter press is suitable for this step. The clear liquor is concentrated under reduced pressure and crystallized according to the conventional methods for handling sugars. The temperature of evaporation should not exceed 50°C. Analytical yields of 95 to 98% of the sorbitol supplied should be obtained in the fermentation. Actual recovery of 80 to 90% of the sugar should be possible. An activated-carbon treatment is necessary in order to obtain a product of good quality. Wash water from the centrifugation of the crystallized sugar is returned to the filtered-beer line.

There are no valuable recoverable by-products of the fermentation. The only residues presenting disposal problems are a small amount of sorbose molasses and the filter cake.

Uses, Competitive Processes, Markets

The sole commercial use of sorbose is in the manufacture of ascorbic acid. The nonfermentative steps in the conversion of sorbose to ascorbic acid involve the prepartion of the diacetone derivative which is then oxidized, the acetone groups are removed, and the resultant 2-keto-L-gulonic acid is isomerized to the ene-diol, with ring closure. Any cheaper method of making ascorbic acid, not involving sorbose as an intermediate, would tend to remove the sorbose fermentation from commercial practice. A process involving the fermentation of L-idonate to 2-keto-L-gulonate by *Pseudomonas mildenbergii* has been patented. This is, however, not competitive at present, since the L-idonate is prepared by hydrogenation of 5-ketogluconic acid, which is itself the product of an *Acetobacter* fermentation.

Production figures for sorbose are not available, but some idea of the quantity produced can be obtained from consideration of the ascorbic acid manufactered. In 1948, 830,528 lb of synthetic ascorbic acid were produced in the United States.

Stereochemical Requirements for Acetobacter Oxidations

The conversion of sorbitol to sorbose is a simple dehydrogenation reaction, typical of those produced by the *Acetobacters* which conform to Bertrand's rule. After the study of the oxidation of numerous alcohols, Bertrand[5] concluded that *Acetobacter xylinum* oxidizes a secondary alcoholic group of a polyhydric alcohol to a ketone when its position is between a primary and a secondary alcohol group and is *cis* to the secondary alcohol:

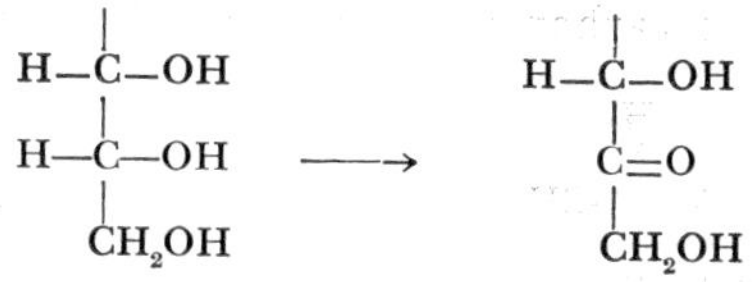

but

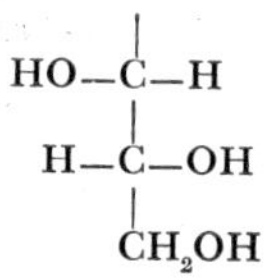

is not dehydrogenated.

The stereochemical relationship of polyhydric alcohols oxidized by *Acetobacter suboxydans* has been reviewed in considerable detail by Fulmer and Underkofler.[15] They classified the polyhydric alcohols which they studied into four groups, depending on the action of *A. suboxydans* on them: (1) those oxidized in solutions of 25% or greater concentration (sorbitol, mannitol); (2) those not oxidized in solutions at high concentrations, but oxidized readily in solutions of low concentration (glycerol, erythritol); (3) those which require an assimilable substrate for continuous subculture (*meso*-inositol, *meso*-2,3-butanediol, D-(−)-2,3-butanediol); and (4) those alcohols not oxidized even in the presence of an assimilable substrate. The first and second groups are suitable carbon sources for continuous subculture and are oxidized in accordance with Bertrand's rule. The third group can serve as source of energy, but its members are not adequate for the production of cellular substance.

In accordance with Bertrand's rule, mannitol is oxidized to D-fructose, D-perseitol to L-perseulose, D-arabitol to D-xylulose, erythritol to L-erythrulose, and D-gluconates to 5-ketogluconates by *A. suboxydans*. Compounds of a structure not conforming to Bertrand's rule, in which the secondary alcohol is dehydrogenated, include glycerol which yields dihydroxyacetone, 2,3-butanediol which yields acetoin, and *meso*-hexanediol which yields D-hexane-3-one-4-ol. It was shown by van Risseghem[36] that only the D-carbon atom was oxidized when both D- and L-carbons were present in the same molecule, as in *meso*-hexanediol. Whistler and Underkofler[51] demonstrated this with erythritol. Grivsky[17] and Underkofler, Fulmer, Bantz, and Kooi[46] demonstrated the same specificity for D-carbon atoms using isomers of 2,3-butanediol. Here the *meso*-isomer yielded L-(+)-acetoin, the D-(−)-isomer yielded D-(−)-acetoin, but the L-(+)-butanediol which contains no D-carbon atoms was not dehydrogenated.

$$\underset{meso\text{-2,3-Butanediol}}{CH_3-\underset{\underset{OH}{|}}{\overset{\overset{H}{|}}{C}}-\underset{\underset{OH}{|}}{\overset{\overset{H}{|}}{C}}-CH_3} \longrightarrow \underset{L\text{-}(+)\text{-Acetoin}}{CH_3-CO-\underset{\underset{OH}{|}}{\overset{\overset{H}{|}}{C}}-CH_3}$$

$$\underset{D\text{-}(-)\text{-2,3-Butanediol}}{CH_3-\underset{\underset{OH}{|}}{\overset{\overset{H}{|}}{C}}-\underset{\underset{H}{|}}{\overset{\overset{OH}{|}}{C}}-CH_3} \longrightarrow \underset{D\text{-}(-)\text{-Acetoin}}{CH_3-CO-\underset{\underset{H}{|}}{\overset{\overset{OH}{|}}{C}}-CH_3}$$

$$\underset{L\text{-}(+)\text{-2,3-Butanediol}}{CH_3-\underset{\underset{H}{|}}{\overset{\overset{OH}{|}}{C}}-\underset{\underset{OH}{|}}{\overset{\overset{H}{|}}{C}}-CH_3} \longrightarrow \text{not fermented}$$

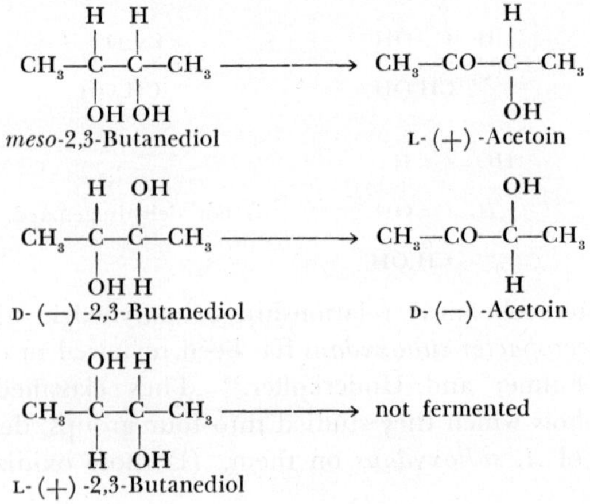

Ketogenic Fermentation

The specificity for D-carbon atoms may explain the failure of the acetic acid bacteria to ferment L-rhamnitol, which contains only L-carbon atoms, but conforms to Bertrand's rule. The enantiomorphic D-rhamnitol is readily fermented.

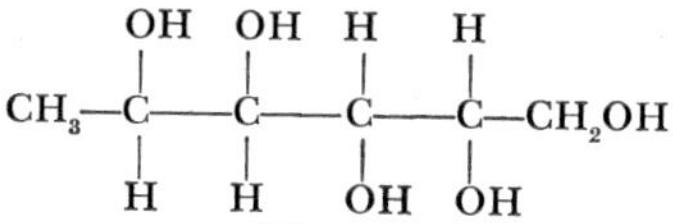

$$CH_3-\underset{H}{\overset{OH}{C}}-\underset{H}{\overset{OH}{C}}-\underset{OH}{\overset{H}{C}}-\underset{OH}{\overset{H}{C}}-CH_2OH$$
L-Rhamnitol

Hann, Tilden, and Hudson,[18] studied the action of *A. suboxydans* on a considerable number of sugar alcohols and extended Bertrand's rule as applied to this organism, stating that the pair of *cis* secondary alcohol groups must have the D-configuration.

One apparently anomalous result of the work of Hann, Tilden, and Hudson was their discovery that L-fucitol

$$CH_3-\underset{H}{\overset{OH}{C}}-\underset{OH}{\overset{H}{C}}-\underset{OH}{\overset{H}{C}}-\underset{H}{\overset{OH}{C}}-CH_2OH$$

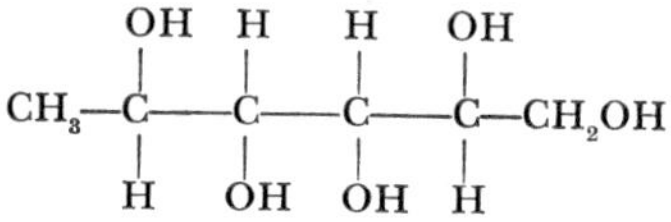

was oxidized by *A. suboxydans*. Bollenback and Underkofler[6] confirmed the oxidation of L-fucitol and studied the action of *A. suboxydans* on a number of other ω-desoxy sugar alcohols. They found a number of them, including D-rhamnitol, were oxidized readily by *A. suboxydans* while several others were not attacked. Richtmyer, Stewart, and Hudson[34] identified the oxidation product of L-fucitol as L-fuco-4-ketose. They[35] extended the work of Bollenback and Underkofler to additional ω-desoxy sugar alcohols and found that the configurations

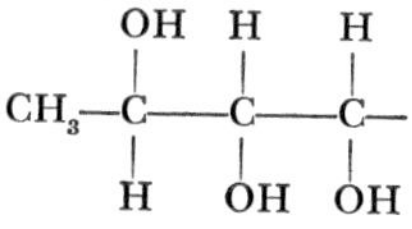

$$CH_3-\underset{H}{\overset{OH}{C}}-\underset{OH}{\overset{H}{C}}-\underset{OH}{\overset{H}{C}}-\quad\text{and}\quad CH_3-\underset{OH}{\overset{H}{C}}-\underset{OH}{\overset{H}{C}}-\underset{OH}{\overset{H}{C}}-$$

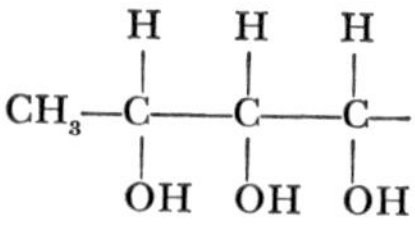

are oxidized by *A. suboxydans*, whereas other types of ω-desoxy sugar alcohols so far tested are not attacked. This permits extension of Bertrand's rule to predict the action of *A. suboxydans* on the ω-desoxy sugar alcohols by considering the CH_3-CHOH group as an elongated CH_2OH group where one of the hydrogen atoms of the primary alcohol group is replaced by a methyl group.

The 5-Ketogluconic Acid Fermentation

The identification of 5-ketogluconic acid or 5-ketogluconate as a product of the action of acetic acid bacteria was first made by Boutroux[7] in 1880. Hermann and Neuschul[19] later investigated this biochemical reaction. 5-Ketogluconic acid is produced by many species of *Acetobacter* when glucose is fermented, or when calcium gluconate is fermented by special strains. It represents 60 to 90% of the acid formed by various strains of *Acetobacter suboxydans.*

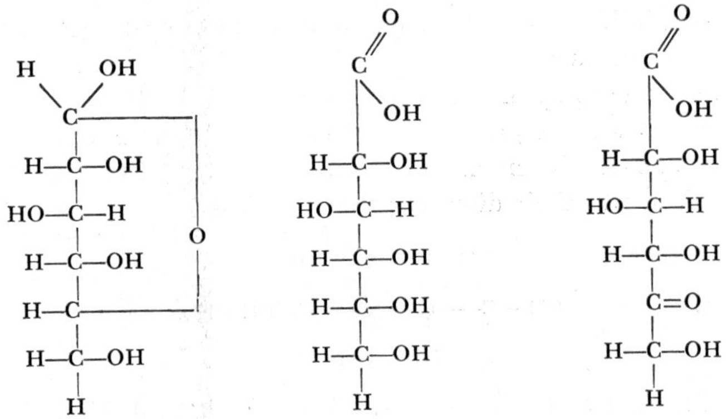

D-Glucose D-Gluconic acid 5-Keto-D-gluconic acid

Stubbs, Lockwood, Roe, Tabenkin, and Ward[39] studied the production of 5-ketogluconic acid by *A. suboxydans.* They found that gluconic acid is an intermediate in the oxidation of glucose to 5-ketogluconic acid; indeed, no 5-ketogluconic acid is formed until nearly all the glucose in the culture has been oxidized to gluconic acid. All cell multiplication occurs during the period of gluconic acid production, and the further oxidation of gluconic acid to 5-ketogluconic acid is accomplished by already grown non-proliferating cells. The latter oxidation step is accomplished by enzymes already present in the cells, and is much slower than the first step. The initial growth and gluconic acid production step requires 16 to 20 hours to oxidize a 10% glucose solution; and 20 to 40 hours are required for the second step, the over-all time being 35 to 60 hours. Stubbs, Lockwood, Roe, and Ward[40] patented the process for fermentative production of 5-ketogluconic acid. The cultures used in its manufacture are the same as those used for the production of sorbose. The cultural and operating requirements are, consequently, the same, except that it is necessary to add to the medium, at the time of inoculation, 27 g of calcium carbonate for each 100 g of glucose supplied, to neutralize the gluconic acid as it is formed. The calcium carbonate is sterilized dry. The acidity of the glucose solution should be not less than pH 4.5 during sterilization. At slightly alkaline pH values, the Lobry de Bruyn-van Eckenstein reaction occurs, with the consequent rearrangement of a considerable portion of the glucose molecules to a configuration unsuitable for the fermentation.

Analytical Methods

Since 5-ketogluconic acid is formed in fermentation liquors only after all the glucose is consumed, an analytical procedure based on the reduction of alkaline copper sulfate can be used to follow the entire fermentation. Glucose is determined by any of the conventional sugar methods during the first phase of the fermentation. The slight solubility of calcium 5-ketogluconate, about 0.25%, makes necessary considerable care in sampling during the later phase of the fermentation. The mash must be well agitated during sampling and a measured volume of sample taken. A measured quantity of concentrated hydrochloric acid is added to the entire sample and the final volume recorded after solution of the calcium 5-ketogluconate is complete. The 5-ketogluconic acid in the acidified sample is determined by the method of Shaffer and Hartmann,[38] using previously determined cuprous oxide-ketogluconic acid ratios. A final correction is made for the dilution of the sample on acidification.

Recovery and Purification

Calcium 5-ketogluconate precipitates during the later stages of the fermentation, and is filtered off at harvest. The crude substance is dissolved in dilute hydrochloric acid, treated with activated carbon, and refiltered with filter aid to remove bacteria. the calcium 5-ketogluconate is reprecipitated in the filtrate by slow addition of ammonia while stirring rapidly. The recrystallized salt is filtered and washed free of ammonium chloride with water. The yield of 5-ketogluconic acid should be 90 to 95 g from 100 g of glucose. A very little 2-ketogluconic acid is also formed in this process, and remains in the original filtrate.

Uses and Competitive Products

5-Ketogluconic acid may be used as an intermediate in the

production of D-tartaric acid. It is oxidized to this acid by air, in the presence of a vanadium pentoxide catalyst.[1] The economy of the fermentation depends on the demand for tartaric acid, which is a by-product of the wine industry. Recent improvements in the processing of wines, lees, and argols would indicate that tartaric acid from the wine industry will continue to be cheaper than the acid made from 5-ketogluconic acid. An additional process for the production of tartaric acid by the oxidation of starch is also competitive.

Other Acetobacter Fermentations of Practical Importance

Although the production of sorbose, and possibly of 5-ketogluconic acid, are the only large-scale *Acetobacter* fermentations of industrial importance, the action of *Acetobacter suboxydans* on sugar alcohols of suitable structure, conforming to Bertrand's rule, furnishes a convenient method for obtaining certain otherwise unobtainable ketose sugars. Several of these sugars have been produced by laboratory investigators by this method for different purposes. For example, Whistler and Underkofler[51] produced L-erythrulose from erythritol, Tilden[43] obtained L-perseulose from D-perseitol, and Underkofler and Fulmer[45] developed an efficient method for producing dihydroxyacetone from glycerol. Dihydroxyacetone has considerable importance as a fine chemical and is produced commercially in large-scale laboratory apparatus by this fermentation process to supply the limited market. The *Acetobacters* oxidize glucose to gluconic acid, galactose to galactonic acid, etc., and selected strains may be utilized for these fermentations (Currie and Carter;[12] Takahashi;[41] Takahashi and and Asai[42]). However, mold fermentations or chemical oxidation of glucose supply the market for gluconic acid and calcium gluconate. These processes have advantages over the *Acetobacter* fermentation for producing gluconic acid, since this compound is further converted to ketogluconic acids by most *Acetobacter* species (see Chapter 14 of volume I).

With the strains of *A. suboxydans* employed by investigators in the United States, the principal oxidation product obtained by action of this organism on glucose in the presence of calcium carbonate or on calcium gluconate has been identified as calcium 5-ketogluconate. However, there is apparently considerable strain difference with various *Acetobacters.* Bernhauer and Görlich[3] identified 2-ketogluconate formed, along with much larger amounts of 5-ketogluconate, by *A. gluconicum.* Recently Riedl-Tumova and Bernhauer[33a] reported that a strain identified as *A. suboxydans muciparum* produced mainly 5-ketogluconate, a strain of *A. suboxydans* produced principally 2-ketogluconate, while *A. melanogenum* produced another as yet unidentified ketogluconic acid salt along with 5-ketogluconate.

Inositol is oxidized to keto-inositol by *Acetobacter suboxydans.* Dunning, Fulmer, Guymon, and Underkofler[13] first prepared a keto-inositol by oxidation of *meso*-inositol with *A. suboxydans.* Kluyver and Boezaardt[20] obtained a mono-keto-inositol identical with the inosose of Posternak.[33] Carter, Belinskey, Clark, Flynn, Lytle, McCasland, and Robbins[11] also obtained inosose in good yields by the oxidation of *meso*-inositol by *A. suboxydans.* Magasanik and Chargaff[28] studied the oxidation of various inositols by *A. suboxydans* and found that *meso*-inositol gave *meso*-inosose (monoketo), *epi*-inositol gave *l-epi*-inosose and probably *d-epi*-inosose (monoketo), while *d*- and *l*-inositol gave α-diketoinositols.

KETOGENIC PROCESSES OF THE PSEUDOMONAS

The 2-Ketogluconic Acid Fermentation

2-Ketogluconic acid has been recognized as a keto acid produced by *Acetobacter* species. Bernhauer and Görlich[3] were apparently the first workers to recognize this acid in the fermentation products from *A. gluconicum.* However, the 2-ketogluconic acid is produced in relatively minor amounts by most *Acetobacters,* and along with the 5-ketogluconic acid. It may represent as much as 10 to 40% of the total final acid yield in 5-ketogluconic acid fermentations. The actual ratio of 2-ketogluconic acid to 5-ketogluconic acid depends on the strain of *Acetobacter* chosen.

Practical means for the production of 2-ketogluconic acid were first presented by Stubbs, Lockwood, Roe, Tabenkin, and Ward[39] who described the production of this acid in submerged cultures by an organism isolated from aberrant 5-ketogluconic acid fermentation liquors. The culture was later identified by Lockwood, Tabenkin, and Ward[26] as a strain of *Pseudomonas fluorescens* and a survey of other species revealed that the property of oxidizing glucose to 2-ketogluconic acid was widespread in the genus *Pseudomonas.* The acid is produced by *P. aeruginosa, P. fragi, P. mildenbergii, P. ovalis, P. pavonacea, P. putida, and P. schuylkilliensis.*

2-Ketogluconic acid is produced by *Pseudomonas* species when glucose or gluconate is the substratum in vigorously aerated cultures. Gluconic acid is an intermediate in the oxidation of glucose to 2-ketogluconic acid. 2-Ketogluconic acid is structurally related to both gluconic acid and glucosone and may be derived from either by chemical oxidation. Glucosone is not known as a fermentation product and the ability of bacteria to oxidize glucosone to 2-ketogluconic acid has not been reported. The enzyme system responsible for the *Pseudomonas* ketonization differs from that of the Bertrand oxidation by *Acetobacter* in that one of a *trans* pair of secondary alcoholic groups is oxidized, while the Bertrand enzyme system requires the *cis* configuration.

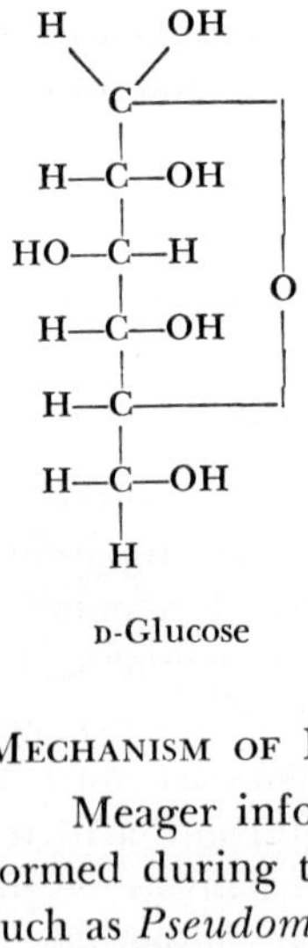

D-Glucose

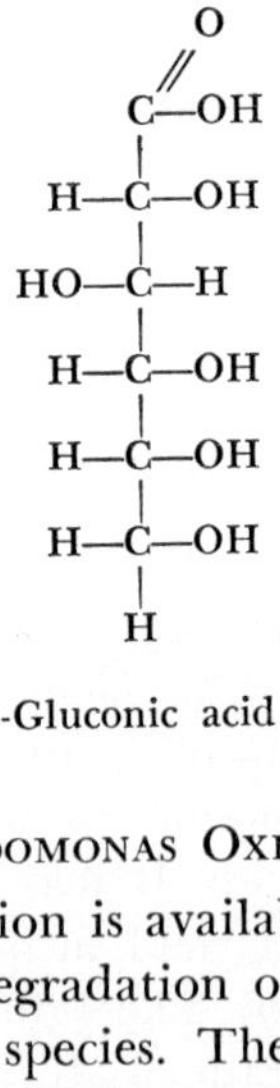

D-Gluconic acid

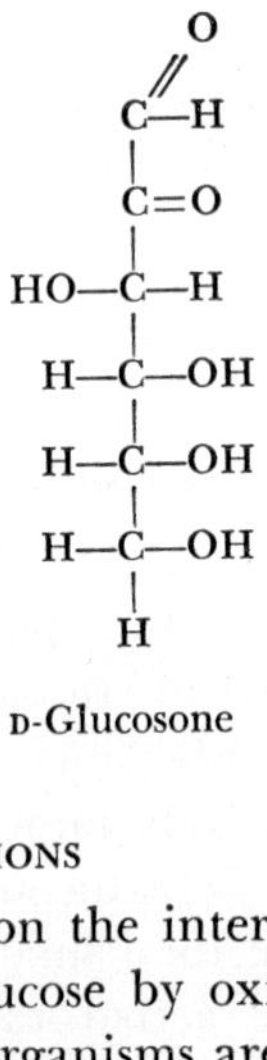

D-Glucosone

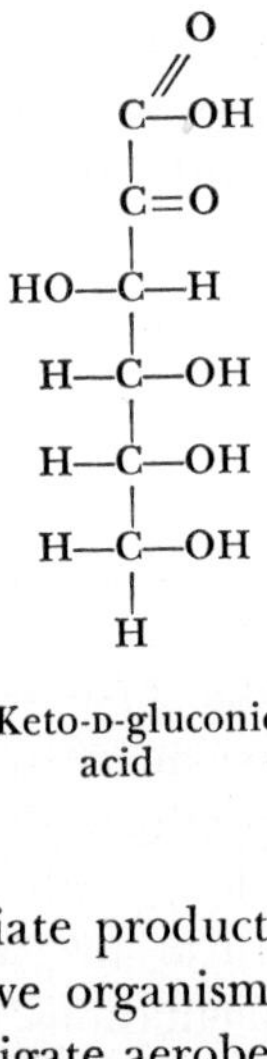

2-Keto-D-gluconic acid

MECHANISM OF PSEUDOMONAS OXIDATIONS

Meager information is available on the intermediate products formed during the degradation of glucose by oxidative organisms such as *Pseudomonas* species. These organisms are obligate aerobes and without intense aeration, the end products of glucose metabolism are carbon dioxide and water. Under conditions of vigorous aeration, Lockwood, Tabenkin, and Ward[26] found a number of cultures of *Phytomonas* and *Pseudomonas* which yielded 58 to 96% gluconic acid or better than 70% 2-ketogluconic acid from glucose. With intense aeration, these acids appear as the end products of the oxidation of glucose by *Pseudomonas* species.

Recently, workers at the University of British Columbia have been publishing a series of papers on their investigations of the intermediate metabolism of *Pseudomonas aeruginosa.* This organism has been shown to produce appreciable quantities of 2-ketogluconic acid from glucose under conditions of intense aeration,[31] but in still cultures, it metabolizes glucose to carbon dioxide and water. By paper chromatography, Norris and Campbell[32] showed that both gluconic and 2-ketogluconic acids were present in detectable amounts in 16- and 24-hour cultures of *P. aeruginosa* grown in still culture in glucose-ammonium phosphate liquid medium. After 38 hours, all glucose had disappeared and only the faintest trace of gluconic or 2-ketogluconic acids could be detected chromatographically. Gluconic and 2-ketogluconic acids as substrates were found to be oxidized rapidly by *P. aeruginosa.* It was concluded that gluconic and 2-ketogluconic acids were being formed at a continuous rate and being removed by continuous oxidation and that, therefore, these compounds were part of the system through which most if not all of the glucose was oxidized. Gluconic and 2-ketogluconic acids are evidently intermediates in the oxidation of glucose by *P. aeruginosa.*

Campbell and Norris[9] showed that neither the fermentation liquor nor the cells of the organism contained any of the hexose phospate members of the Embden-Meyerhof scheme of glucose dissimilation. Resting cells of the organism failed to remove any glucose or fructose or to take up phosphate. It was concluded that *P. aeruginosa* did not initiate dissimilation of glucose by the Embden-Meyerhof pathway. Rather, under normal physiological conditions, the organism oxidizes glucose by a pathway not previously reported in any type of tissue or bacterial cell, through formation of gluconic and 2-ketogluconic acids as intermediate products.

Further work by Warburton, Eagles, and Campbell[47] identified pyruvate as a later intermediate in the oxidation of glucose by *P. aeruginosa* and Campbell and Stokes[10] detected the presence of a tricarboxylic acid cycle in the metabolism of this organism. Much further work will be necessary to fully elucidate the complete mechanism of the oxidative dissimilation of glucose by *P. aeruginosa.* It will be of interest to ascertain whether a similar pathway through gluconic acid may be employed by other oxidative organisms, such as some of the *Acetobacters* and fungi.

Cultures and Culture Care

The *Pseudomonas* organisms responsible for the oxidation of glucose to 2-ketogluconic acid are extremely widespread and can be readily isolated from soil. In fact, they are among the most common soil organisms. Suitable strains occur in temperate climates in almost all fertile soils except those which are very acid. Some of the *Pseudomonas* cultures which give good yields of 2-ketogluconic acid are plant or animal pathogens, but suitable non-pathogenic strains are also easily isolated.

Cultures may be stored in dry loam according to the method of Greene and Fred,[16] or as slants on vegetable or liver decoction agar media.

The Fermentation of Glucose to 2-Ketogluconic Acid

2-Ketogluconic acid has not been manufactured on a commercial scale. However, equipment used for the manufacture of sorbose should be entirely suitable for the production of this acid. Rigid exclusion of nickel is not necessary, since *Pseudomonas* are remarkably insensitive to this metallic ion. The temperature, aeration, and agitation optima are the same for the three fermentations: sorbose, 5-ketogluconic acid, and 2-ketogluconic acid. In addition to the 100 g of glucose, 5 ml of corn steep liquor, and 27 g of calcium carbonate per liter required in the 5-ketogluconate fermentation, it is necessary to supply 0.25 g of $MgSO_4 \cdot 7H_20$, 0.6 g of KH_2PO_4, and 2 g of urea per liter. The urea must be sterilized separately from the glucose; this is usually done by autoclaving a 20% solution. The calcium carbonate is sterilized dry, and added at the time of inoculation. A sterile, aqueous slurry could also be used. The composition of the media for the preliminary propagative steps and the final fermentation is the same. The ratio of inoculum to inoculated mash is about 3:100. The fermentation is conducted under 30 lb gage pressure, at 30°C. It is complete in 32 to 35 hours.

Contamination has not been a serious problem in pilot-plant studies. Several bacteriophages for *Pseudomonas aeruginosa* are known. Consequently, it appears that bacteriophages may occur and cause trouble in industrial operation of this process. Immunization of cultures may become necessary.

Analytical Methods

2-Ketogluconic acid is produced in culture media while a

considerable quantity of glucose remains in the mash. The analysis of these liquors is ordinarily based on the reduction of alkaline copper sulfate, as in the determination of glucose, sorbose, 5-keto-gluconic acid, or other compounds having the configurational group:

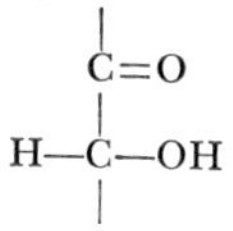

Cuprous oxide : 2-ketogluconic acid ratios have been determined and may be used in either tabular or graphic form.

Since both glucose and 2-ketogluconic acid reduce alkaline copper reagents, such as Benedict's solution, it is necessary to differentiate between the copper reduction of each of these two compounds. This is done on the basis of optical rotation. The rotation of 1% glucose solution is +0.55°, whereas that of a 1% solution of 2-ketogluconic acid is —0.88°. An algebraic equation has been worked out to facilitate the calculation of the 2-keto-gluconic acid and glucose contents of such mixtures:

$$CuK = 0.353CU - 13.28\alpha$$

CuK is the cuprous oxide produced by the reducing action of 2-keto-gluconic acid on cupric ions; *CU* is the total cuprous oxide, and α is the observed rotation of the sample in a 1 dm tube. The total cuprous oxide less the cuprous oxide due to the 2-ketogluconic acid equals the cuprous oxide, *CuG*, produced by reduction of cupric ions by glucose:

$$CU - CuK = CuG$$

These cuprous oxide values may be readily converted to glucose or 2-ketogluconic acid by use of appropriate tables or graphs.

Recovery and Uses of 2-Ketogluconic Acid

The 2-ketogluconic acid is recovered as the calcium salt on crystallization from the filtered, evaporated culture liquor. The concentration is done under reduced pressure, so that the temperature does not exceed 50°C. No by-products of this fermentation are to be expected, since yields as great as 90% of theory can be obtained. Some calcium salt of gluconic acid, or of acids which form on further oxidation of 2-ketogluconic acid, may be present.

The principal use for 2-ketogluconic acid appears to be as an intermediate in the preparation of D-araboascorbic acid.

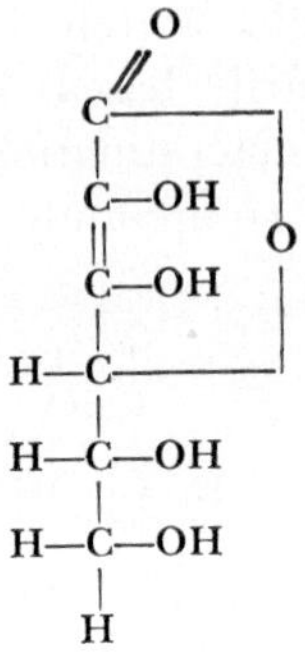

D-Araboascorbic acid

This compound has much less vitamin C potency than does L-ascorbic acid, but exerts a vitamin C sparing antioxidant action in preserving fruits and other food products. Some fatty acid esters of D-araboascorbic acid may find use as antioxidants in the prevention of rancidity in oils and fats of vegetable or animal origin. The leading competitor of D-araboascorbic acid will undoubtedly be L-ascorbic acid.

α-Ketoglutaric Acid Fermentation

Lockwood and Stodola[24] found that α-ketoglutaric acid, $HOOC \cdot CH_2 \cdot CH_2 \cdot CO \cdot COOH$, is produced when a 2-ketogluconic acid fermentation by certain strains of *Pseudomonas fluorescens* is allowed to proceed beyond the point of maximum 2-ketogluconic acid content. At this time, the aeration rate is reduced to approximately half the initial rate. At the point at which all the 2-ketogluconic acid has been destroyed, about 25 g of α-ketoglutaric acid is present for each 100 g of glucose initially supplied. A patent on this process has been granted to Lockwood and Stodola.[25]

More recently, Koepsell, Stodola, and Sharpe[19a] have reported yields of about 41 g of α-ketoglutaric acid per 100 g of glucose supplied in shake-flask cultures. They employed the same organism, *Ps. fluorescens* NRRL B-6, and a medium containing, per 100 ml, 9 g glucose, 0.32 g $(NH_4)_2SO_4$ (urea, sterilized separately, could be substituted on an equimolar basis), O.11 g KH_2PO_4, 0.024 g $MgSO_4$, 1 ppm Fe supplied as ferrous ammonium sulfate, and 3.75 g $CaCO_3$ (sterilized separately). In the fermentations, conversion of glucose to gluconate and 2-ketogluconate had occurred at 40 hours, but appreciable α-ketoglutarate was not detectable until after 60 hours. About 11 days were required to achieve maximum yields of α-ketoglutarate. In many cases, calcium α-ketoglutarate separated during fermentation from the medium, which became a thick slurry by the end of fermentation. Sometimes supersaturation occurred and seeding was necessary to start precipitation. For recovery of the free acid, the whole culture was acidified with hydrochloric acid and extracted with ethyl acetate. Crude crystalline α-ketoglutaric acid was obtained by concentrating the extract and the pure acid by recrystallization from ethyl acetate—petroleum ether. The authors stated that the ease of recovery would enhance the possibility of producing the substance on an industrial scale by fermentation methods.

Pseudomonas Oxidation of Inositol

Kluyver, Hof, and Boeezaardt[21] have reported that inositol is oxidized by *Pseudomonas beijerinckii* to a triketo-inositol which oxidizes to tetrahydroxybenzoquinone.

KOJIC ACID FERMENTATION

Kojic acid was first identified as a mold metabolic product and its structure was worked out by Yabuta.[52] Saito[37] had previously observed the compound when ferric chloride was added to the culture filtrates of *Aspergillus oryzae.* It is a pyrone compound, the acidic properties of which are due to the enolic hydroxyl group. No carboxyl group is present.

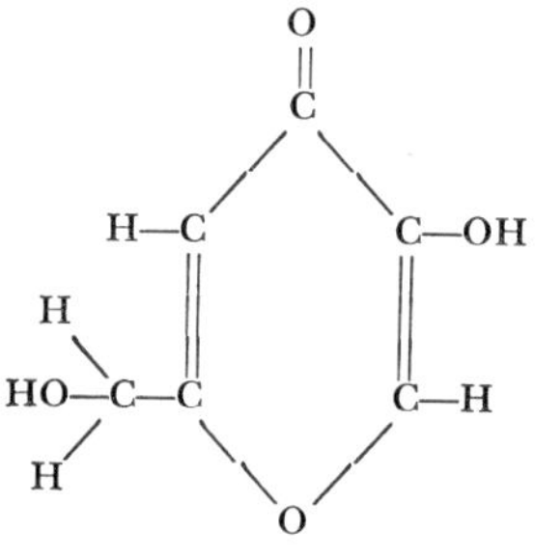

Kojic acid

Pioneering work on the production of kojic acid by mold fermentation was reported by May, Moyer, Wells, and Herrick.[29]

The acid is produced by members of the *Aspergillus flavus-oryzae* and *A. tamarii* groups, growing on the surface of the culture solution. There is considerable variation from strain to strain in the ability to produce kojic acid.

A suitable nutrient composition for the production of kojic acid contains 200 g glucose, 0.054 g H_3PO_4, 0.1 g KCl, 0.5 g $MgSO_4 \cdot 7H_2O$, and 1.125 g NH_4NO_3 per 1. The addition of ethylene chlorohydrin, up to 0.1 g per 1 of culture solution, appears to result in considerable increase in yield.[30]

Kojic acid is synthesized by the mold from glucose, sucrose, xylose, or dihydroxyacetone. Consequently, it appears probable that the larger sugar units are broken down into smaller fragments, from which the acid is synthesized.

Barham and Smits[2] reported an extensive investigation of conditions required for the maximum production of kojic acid from xylose by surface cultures of *A. flavus.*

Since kojic acid contains a carbonyl group adjacent to a secondary alcoholic group, the compound reduces alkaline copper reagents, such as Fehling's or Benedict's solution. The kojic acid reduction values published by May, Moyer, Wells, and Herrick[29] are:

Kojic acid, mg	5	10	20	30
Cuprous oxide, mg	9.5	17	35	52

Correction for copper reduced by kojic acid is made in the determination of sugar in kojic acid culture liquors by copper reduction methods. The amount of correction to be applied depends on the quantity of kojic acid present. This is determined by titration, using alizarin orange as indicator.

The yield of kojic acid should be about 45 g per 100 g of glucose supplied, after a 12-day fermentation period. The product crystallizes out readily on evaporation of the culture liquor. To obtain a colorless product, iron must be carefully excluded, since minute traces of this metal will give a deep-red color with kojic acid.

Although kojic acid is easily produced by fermentation, no uses for the compound have been developed. Should uses be found, undoubtedly submerged processes would be developed to replace the surface process.

OTHER KETOGENIC FERMENTATION PROCESSES

Several ketogenic processes besides those previously mentioned have been described, in addition to the many *Acetobacter* oxidations in accordance with Bertrand's rule. Among these should be mentioned the production of acetoin by yeast and by *Bacillus polymyxa, Aerobacter aerogenes, A. cloacae,* and various species of *Proteus* (see Chapter 2). The metabolism of these bacteria is essentially of the butanediol-producing type, but the diol may be oxidized to acetoin if the cultures are aerated. Bacteria of the genus *Pseudomonas* have been reported to oxidize the hydroxyl group on the side chain of a steroid associated with adrenal cortical hormone to the corresponding ketone, thus increasing the sex-hormonal potency of the extracts of the adrenal cortex.

BIBLIOGRAPHY

1. Barch, W. E., *J. Am. Chem. Soc.,* **55**, 3653 (1932).
2. Barham, H. N., and B. L. Smits, *Ind. Eng. Chem.,* **28**, 567 (1936).
3. Bernhauer, K., and B. Görlich, *Biochem. Z.,* **280**, 367, 375 (1935).
4. Bertrand, G., *Compt. rend.,* **122**, 900 (1896).
5. Bertrand, G., *Ann. Chim. Phys.,* **(8) 3**, 181 (1904).
6. Bollenback, G. N., and L. A. Underkofler, *J. Am. Chem. Soc.,* **72**, 741 (1950).
7. Boutroux, L., *Compt. rend.,* **91**, 236 (1880).
8. Butlin, K. R., *The Biochemical Activities of the Acetic Acid Bacteria,* Gr. Br. Dept. Sci. Ind. Research, Spec. Rept. No. 2, London, H. M. Stationery Office, 1936.
9. Campbell, J. J. R., and F. C. Norris, *Can. J. Research,* **28C**, 203 (1950).
10. Campbell, J. J. R., and F. N. Stokes, *J. Biol. Chem.,* **190**, 853 (1951).
11. Carter, H. E., C. Belinskey, R. K. Clark, E. H. Flynn, B. Lytle, G. E. McCasland, and M. Robbins, *J. Biol. Chem.,* **174**, 415 (1948).
12. Currie, J. N., and R. H. Carter, U. S. Patent 1,896,811 (1933).
13. Dunning, J. W., E. I. Fulmer, J. F. Guymon, and L. A. Underkofler, *Science,* **87**, 72 (1938).
14. Fulmer, E. I., J. W. Dunning, J. F. Guymon, and L. A. Underkofler, *J. Am. Chem. Soc.,* **58**, 1012 (1936).

15. Fulmer, E. I., and L. A. Underkofler, *Iowa State Coll. J. Sci.*, **21**, 251 (1947).
16. Greene, H. C., and E. B. Fred, *Ind. Eng. Chem.*, **26**, 1297 (1934).
17. Grivsky, E., *Bull. soc. chim. Belg.*, **51**, 63 (1942).
18. Hann, R. M., E. B. Tilden, and C. S. Hudson, *J. Am. Chem. Soc.*, **60**, 1201 (1938).
19. Hermann, S., and P. Neuschul, *Zentr. Bakt. Parasitenk., Abt. II*, **93**, 25 (1935).
19a. Koepsell, H. J., F. H. Stodola, and E. S. Sharpe, *J. Am. Chem. Soc.*, **74**, 5142 (1952).
20. Kluyver, A. J., and A. G. J. Boezaardt, *Rec. trav. chim.*, **58**, 956 (1939).
21. Kluyver, A. J., T. Hof, and A. G. J. Boezaardt, *Enzymologia*, **7**, 257 (1939).
22. Kluyver, A. J., and F. J. de Leeuw, *Tijdschr. verg. Geneesk.*, **10**, 170 (1924).
23. Landy, M., and D. M. Dicken, *J. Biol. Chem.*, **146**, 109 (1942).
24. Lockwood, L. B., and F. H. Stodola, *J. Biol. Chem.*, **164**, 81 (1946).
25. Lockwood, L. B., and F. H. Stodola, U. S. Patent 2,443,919 (1948).
26. Lockwood, L. B., B. Tabenkin, and G. E. Ward, *J. Bact.*, **42**, 51 (1941).
27. Lockwood, L. B., G. E. Ward, J. J. Stubbs, E. T. Roe, and B. Tabenkin, U. S. Patent 2,277,716 (1942).
28. Magasanik, B., and E. Chargaff, *J. Biol. Chem.*, **174**, 173 (1948).
29. May, O. E., A. J. Moyer, P. A. Wells, and H. T. Herrick, *J. Am. Chem. Soc.*, **53**, 774 (1931).
30. May, O. E., G. E. Ward, and H. T. Herrick, *Zentr. Bakt. Parasitenk., Abt. II*, **86**, 129 (1932).
31. Ney, P. W., Thesis, University of British Columbia, Vancouver, B. C., (1948).
32. Norris, F. C., and J. J. R. Campbell, *Can. J. Research*, **27C**, 253 (1949).
33. Posternak, T., *Helv. Chim. Acta*, **24**, 1045 (1941).
33a. Riedl-Tumova, E., and K. Bernhauer, *Biochem. Z.*, **320**, 472 (1950).
34. Richtmyer, N. K., L. C. Stewart, and C. S. Hudson, *J. Am. Chem. Soc.*, **72**, 4934 (1950).
35. Richtmyer, N. K., L. C. Stewart, and C. S. Hudson, *Abstracts 118th Meeting, Am. Chem. Soc.*, Chicago, September 1950.

36. Risseghem, H. van, *Bull. soc. chim. Belg.,* **45,** 21 (1936).
37. Saito, K., *Botan. Mag.* (Tokyo), **21,** No. 249 (1907).
38. Shaffer, P. A., and A. F. Hartmann, *J. Biol. Chem.,* **45,** 365 (1920).
39. Stubbs, J. J., L. B. Lockwood, E. T. Roe, B. Tabenkin, and G. E. Ward, *Ind. Eng. Chem.,* **32,** 1626 (1940).
40. Stubbs, J. J., L. B. Lockwood, E. T. Roe, and G. E. Ward, U. S. Patent 2,318,641 (1943).
41. Takahashi, T., U. S. Patent 1,953,694 (1934).
42. Takahashi, T., and T. Asai, *Zentr. Bakt. Parasitenk., Abt. II,* **93,** 248 (1936).
43. Tilden, E. B., *J. Bact.,* **37,** 629 (1939).
44. Underkofler, L. A., A. C. Bantz, and W. H. Peterson, *J. Bact.,* **45,** 183 (1943).
45. Underkofler, L. A., and E. I. Fulmer, *J. Am. Chem. Soc.,* **59,** 301 (1937).
46. Underkofler, L. A., E. I. Fulmer, A. C. Bantz, and E. R. Kooi, *Iowa State Coll. J. Sci.,* **18,** 377 (1944).
47. Warburton, R. H., B. A. Eagles, and J. J. R. Campbell, *Can. J. Botany,* **29,** 143 (1951).
48. Wells, P. A., L. B. Lockwood, and J. J. Stubbs, U. S. Patent 2,121,533 (1938).
49. Wells, P. A., L. B. Lockwood, J. J. Stubbs, E. T. Roe, N. Porges, and E. A. Gastrock, *Ind. Eng. Chem.,* **31,** 1518 (1939).
50. Wells, P. A., J. J. Stubbs, L. B. Lockwood, and E. T. Roe, *Ind. Eng. Chem.,* **29,** 1385 (1937).
51. Whistler, R. L., and L. A. Underkofler, *J. Am. Chem. Soc.,* **60,** 2507 (1938).
52. Yabuta, T., *J. Coll. Agr. Tokyo,* **5,** 51 (1912).

PART II. THE FERMENTATIVE PRODUCTION OF 2,3-BUTANEDIOL

CHAPTER 2

FERMENTATIVE PRODUCTION OF 2,3-BUTANEDIOL

G. A. Ledingham and A. C. Neish

INTRODUCTION

Interest in four-carbon compounds as possible precursors for the manufacture of 1,3-butadiene during World War II led to many studies on the production of 2,3-butanediol, also called 2,3-butylene glycol, by government, university, and industrial laboratories in the United States and Canada. As a result of these studies, knowledge available early in the century on the production of this compound by different becteria has been greatly expanded and several new fermentations have been described. None of these processes has actually been applied commercially, but large-scale pilot-plant operations have been conducted. Although not utilized industrially at present, an extensive treatment of the development work connected with the 2,3-butanediol fermentation is justified, not only because of its commercial potential, but also to illustrate the progress which can result from intensive research on a fermentation process. Since the work of the various laboratories was correlated through conferences and free exchange of information, there has been little hesitancy in publishing results. Consequently, the literature relating to this fermentation is voluminous. Some five hundred references, more or less closely connected with the microbiology, production, and analysis of 2,3-butanediol and related compounds, are in the files of the authors. Even by limiting references to those directly relating to the discussion, the bibliography at the end of this chapter has become quite lengthy.

One of the earliest references to the natural occurrence of butanediol is possibly that of Henninger,[91] who isolated what he thought was isobutylene glycol from wine in 1882. Later it was found that pure-yeast fermentations produced the same compound.[41] It is now known that yeast produces small quantities of 2,3-butanediol and it is very likely that this compound was isolated by Henninger. The first oxidation product of 2,3-butanediol, acetylmethylcarbinol or acetoin, has also been recognized as a product of bacterial fermentations since 1901.[79] In 1906, Harden and Walpole[89] first proved that bacteria produced 2,3-butanediol in appreciable quantities, using cultures which would now be classified as *Aerobacter aerogenes.* Walpole[263] reported further studies in 1911 and Scheffer[216] in 1928. Donker[61] published the first important contribution to the *Bacillus polymyxa* fermentation in 1926. The initial work of Donker and Scheffer led to a number of patents typical of which were those of Verhave,[259,260,261] Kluyver and Scheffer,[110,111,112] and Scheffer.[217] In 1933, Fulmer, Christensen, and Kendall,[73] at Iowa State College, showed that the *Aerobacter aerogenes* fermentation had industrial possibilities, because they were able to obtain good yields in much shorter time than previously. Many other contributions were later made by Werkman, Fulmer, Underkofler, and their students from this College.

The wartime investigations on the production of 2,3-butanediol were begun early in 1942 in both the United States and Canada. The Northern Regional Research Laboratory of the United States Department of Agriculture in Peoria, Ill., held a number of conferences which were attended by representatives from Iowa State College, University of Wisconsin, University of Alberta, Commercial Solvents Corporation, Schenley Distillers, Joseph E. Seagram and Sons, Inc., Lucidol Corporation, Heyden Chemical Corporation, Merck and Company, Columbia Brewing Company, and the National Research Council of Canada. Progress on the problem was greatly expedited through the pooling of reports and exchange of cultures. Laboratory studies were followed by pilot-plant investigations in the Northern Regional Research Laboratory, Joseph E. Seagram and Sons, Inc., Louisville, Ky., Schenley Distillers, Lawrenceburg, Ind., Commercial Solvents Corporation, Terre Haute, Ind., and the National Research Laboratories, Ottawa, Canada. Most of this wartime work was concentrated on the *Aerobacter* and the *Bacillus polymyxa* fermentations and on efforts to convert the 2,3-butanediol to 1,3-butadiene.

PROPERTIES OF 2,3-BUTANEDIOL IN RELATION TO POSSIBLE USES

Physical Properties

2,3-Butanediol is a colorless, odorless, viscous liquid. It exists in three stereoisomeric forms, which are depicted by the conventional planar projection formulas as:

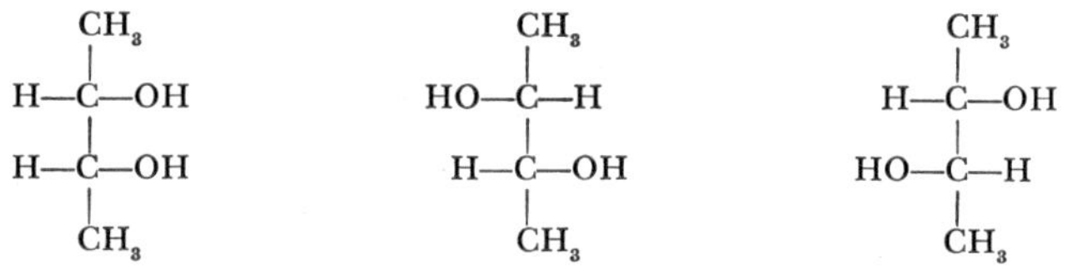

meso-2,3-Butanediol D-(—)-2,3-Butanediol L-(+)-2,3-Butanediol

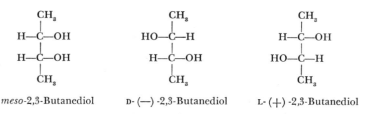

The D and L forms are mirror images and thus are identical in all common physical properties except the direction of rotation of polarized light. All three stereoisomers are found in nature in varying proportions, depending on the organism which makes them. This is illustrated in Table 1. Most bacteria give predominantly *meso*-2,3-butanediol, but the dextro-rotatory and levo-rotatory isomers are also found, particularly the second which is formed to the practical exclusion of all the other isomers by *Bacillus polymyxa*. The values given in Table 1 for the composition of the diol mixture are rather approximate. *A. aerogenes* is reputed to give a small amount of *racemic*-2,3-butanediol[26,249] and some of the other species may produce this also.

Table 1. OPTICAL ROTATION OF 2,3-BUTANEDIOL[a] PRODUCED BY VARIOUS BACTERIA

Organism	$[\alpha]_D^{23-26°C}$	Approximate Composition of mixture	References
Aerobacter aerogenes	+0.8 to + 1.8	5-14% L-(+), remainder *meso*	26, 70, 89, 152, 263
Bacillus polymyxa	—13.0 to —13.34	At least 98% D-(—)	115, 132, 156 209, 265
Pseudomonas hydrophila	—1.25	About 8% D-(—), remainder *meso*	43, 231
Bacillus subtilis (Ford's type)	—3.86 to —5.0	Up to 40% D-(—), remainder *meso*	21,166
Serratia marcescens	+0.07 to +0.35	Predominantly *meso*	167
Serratia plymuthicum	—0.18	Predominantly *meso*	168
Serratia anolium	+0.03	Predominantly *meso*	168

[a] Diol samples purified by distillation before measurement of the optical rotation.

Some of the physical properties of the stereoisomers of 2,3-butanediol are summarized in Table 2. The pure dextro-rotatory isomer has not been obtained, but its properties should correspond to those of the levo-rotatory isomer except for the sign of rotation. *Racemic*-2,3-butanediol acts as a fourth compound with somewhat different physical properties from those of the pure isomers and its identity as an equimolecular mixture of the D and L forms can be shown only by resolving it with optically active agents. It has been shown by biochemical[80] and chemical[152] studies that

Table 2. PHYSICAL PROPERTIES OF THE STEREOISOMERIC 2,3-BUTANEDIOLS

Constant	*Meso*-2,3-butanediol	D-(—)-2,3-butanediol	*Racemic*-2,3-butanediol	References
Melting point, °C	34.4	19.0	7.6	265, 279
Boiling point, °C at 745 mm	181-182	179-180	177	132, 152, 279
Specific rotation at 26°C (D line)	0.00°	—13.34°	0.00°	156
Density, g per ml at 25°C	0.9939	0.9869	—	115
Refractive index at 25°C	1.4366	1.4308	—	42, 115
Surface tension, dynes per cm at 25°C	—	30.61	—	115
Viscosity, centipoises at 35°C	65.6	21.8	—	115
Specific viscosity at 30°C	15.72	5.34	—	132
Melting point of di-*p*-nitrobenzoate, °C	193	143	128	209
Specific rotation of di-*p*-nitrobenzoate in chloroform at 25°C	—	51.0°	—	209

2,3-butanediol and acetoin have the configuration D-(—) and L-(+). In the balance of this chapter, the symbols (—) and (+) will be used for levo- and dextro-rotatory 2,3-butanediols, respectively.

It is obviously of practial importance to determine the ratio of the isomers in the 2,3-butanediol produced by any given organism. There are many strains of bacteria for which this has not been done. Methods have been suggested based on measurements of the specific viscosity, or the specific conductivity, or pH in boric acid solutions.[132] Some use might be made of the observation that the rotation of 2,3-butanediol is six hundred times as great in cuprammonium solution as in water.[202] A quantitative method might be based on specific oxidation of 2,3-butanediol isomers by different bacteria.[233] It is necessary to interpret with caution the results of physical measurements on purified samples, since 2,3-butanediol is extremely hygroscopic[132] and a small amount of water produces a relatively large change in optical rotation and some other properties.[42]

It is evident from the data of Table 2 that *meso*-2,3-butanediol has quite different physical properties from those of the other isomers. In addition, differences have been noted in the Raman spectra.[245]

One of the greatest differences, which is of considerable practical importance, is in the freezing point–composition diagrams of 2,3-butanediol-water mixtures (see Figure 1). Both (—) and *meso* isomers form hydrates as shown by the effect of water on the density and specific rotation, but the hydrate formed by the *meso* isomer crystallizes readily while that formed by the (—) isomer does not. As a result of this difference the *meso* isomer is useless as an antifreeze while careful studies of the (—) isomer in binary [43] and ternary[44] mixtures have shown that it may be considered as a substitute for ethylene glycol in permanent type antifreezes. This use has been patented in both Canada[44] and the United States.[46]

In addition to its use as an antifreeze, 2,3-butanediol might find other applications based on its low vapor pressure and hygroscopicity. One that has been suggested[143] is for the moistening and softening of glue, gelatin, casein, tobacco, composition cork, textile fibers, etc. The *meso* isomer would probably be as good as the (—) isomer for these uses.

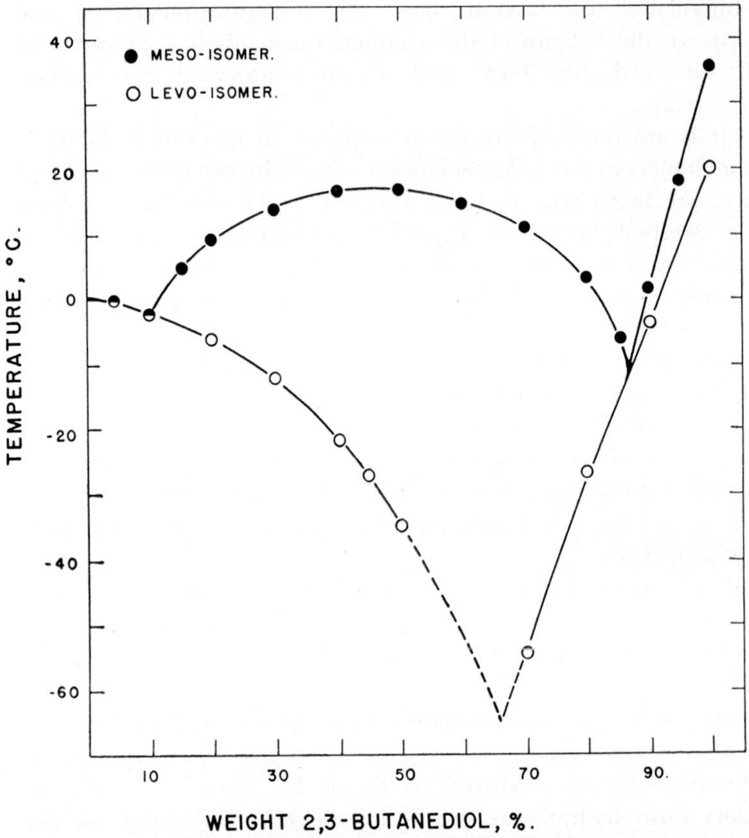

FIGURE 1. *Freezing-Point Diagram of Aqueous Solutions of* Levo- *and* Meso-*2,3-Butanediols*

Information is available on the vapor pressure of 2,3-butanediol and detailed studies have been made of vapor-liquid equilibria of the system 2,3-butanediol-water for both of these isomers.[24,115,189,218,254] Measurements have been made of ultrasonic velocity in 2,3-butanediol[276,277] and also of its heat of formation.[155]

Chemical Properties

Certain potential uses of 2,3-butanediol are based on its conversion to other chemicals. Some of the products which can be formed in good yields by processes which might be commercially feasible are shown in Table 3. Space here permits brief discussion of only a few of these reactions. Acetoin and diacetyl are formed by oxidation of 2,3-butanediol:

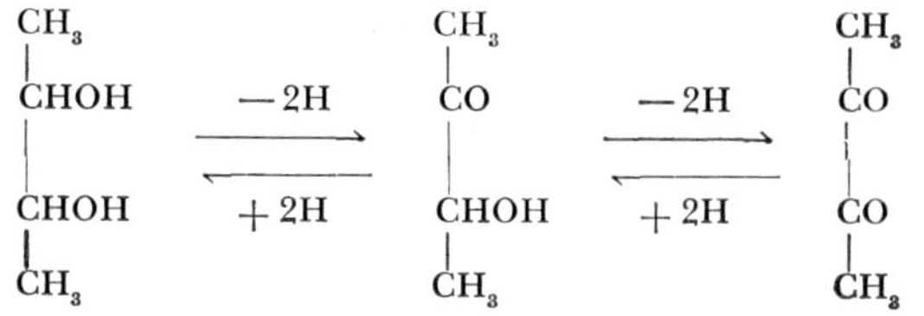

2,3-Butanediol   Acetoin   Diacetyl

This reaction can be carried out by bacteria in the presence of air and it is possible to build up appreciable concentrations of acetoin in the medium. Thus acetoin can be produced directly by fermentation of carbohydrates under aerobic conditions. Alternatively, 2,3-butanediol can be converted to acetoin and diacetyl by catalytic oxidation of dehydrogenation over copper catalysts.[23,53,116,145,157]

TABLE 3. COMPOUNDS READILY PREPARED FROM 2,3-BUTANEDIOL

Compound	Method of preparation	References
$CH_3 \cdot CHOH \cdot CO \cdot CH_3$ Acetoin	Formed during fermentation when strongly aerated	111, 135, 191, 261
	Vapor-phase oxidation or dehydrogenation of 2,3-butanediol over copper catalysts	23, 53, 145, 157
$CH_3 \cdot CO \cdot CO \cdot CH_3$ Diacetyl	Vapor-phase oxidation or dehydrogenation of 2,3-butanediol over copper catalysts	53, 116, 145, 157
$CH_3 \cdot CO \cdot C_2H_5$ Methyl ethyl ketone	2,3-Butanediol heated with sulfuric acid, the diol being added continuously as volatile products distill	6, 169
	2,3-Butanediol vapor passed over dehydration catalysts, such as alumina, bentonite, or phosphorus pentoxide in a heated tube	10, 27, 50, 84, 153, 185
$CH_3 \cdot CH(O \cdot CO \cdot CH_3) \cdot CH(O \cdot CO \cdot CH_3) \cdot CH_3$ 2,3-Butanediol diacetate	2,3-Butanediol and sulfuric acid passed countercurrent to acetic acid in a heated column	82, 223
	2,3-Butanediol heated with acetic and sulfuric acids in distillation column, using an entrainer to remove water	151, 226

Compound	Method of preparation	References
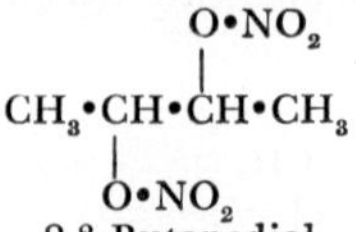 2,3-Butanediol dinitrate	D-(—)-2,3-Butanediol esterified by nitric acid, using sulfuric acid as catalyst	146
Polyester of 2,3-butanediol and phthalic acid	2,3-Butanediol heated with *o*-phthalic anhydride	28, 268
$CH_3 \cdot CHOH \cdot CH{=}CH_2$ Methyl vinyl carbinol	2,3-Butanediol vapor passed over thorium dioxide under reduced pressure at 350°C	280
	Vapor of 2,3-butanediol, mixed with methyl ethyl ketone, water, and triethylamine, passed over a mixed catalyst at 235°C	85
$CH_3 \cdot CO \cdot CH{=}CH_2$ Methyl vinyl ketone	Vapor of acetoin, mixed with water and triethylamine, passed over alumina-silica-tungstic oxide catalyst	85
$CH_2{=}CH \cdot CH{=}CH_2$ 1,3-Butadiene	2,3-Butanediol vapor passed over thorium dioxide at 350°C under reduced pressure	280
	Vapor of a mixture of 2,3-butanediol, water, methyl ethyl ketone, and triethylamine passed over mixed catalyst at 235°C	85
	2,3-Butanediol diacetate vapor passed through tube at 585° to 595°C	93, 151, 153, 222
	2,3-Butanediol diacetate vapor passed through heated tube packed with copper phosphate catalyst	196
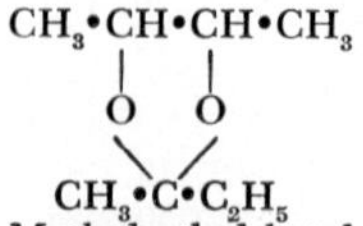 Methyl ethyl ketal of 2,3-butanediol	Continuous distillation of 2,3-butanediol from sulfuric acid under reduced pressure	169
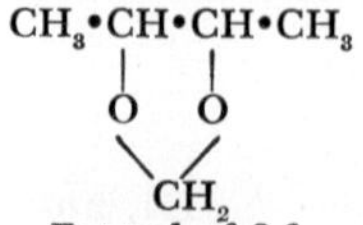 Formal of 2,3-butanediol	Distills when acidified fermentation solution is heated with formaldehyde	225

Compound	Method of preparation	References
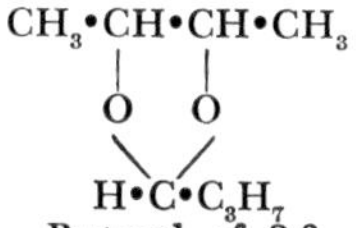 Butyral of 2,3-butanediol	Acidified fermentation solution extracted with butyraldehyde; the butyral is nearly all in solvent phase	248
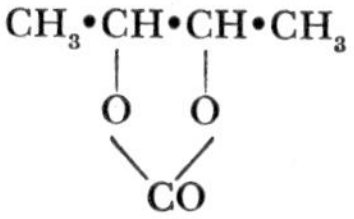 Cyclic carbonate of 2,3-butanediol	Reaction of 2,3-butanediol and phosgene	117
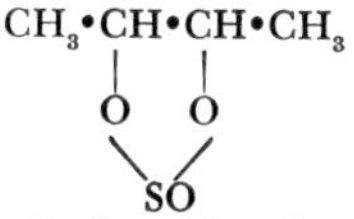 Cyclic sulfite of 2,3-butanediol	Reaction of 2,3-butanediol and thionyl chloride	50, 208

The catalytic dehydration of 2,3-butanediol can lead to a variety of compounds. This type of reaction has received considerable attention because the chief initial impetus for the development of 2,3-butanediol fermentations was the production of 1,3-butadiene, the ultimate dehydration product of this diol. It is rather difficult to convert 2,3-butanediol to compounds with ethylenic double bonds by catalytic dehydration due to the strong tendency toward formation of methyl ethyl ketone instead, presumably by the reaction:

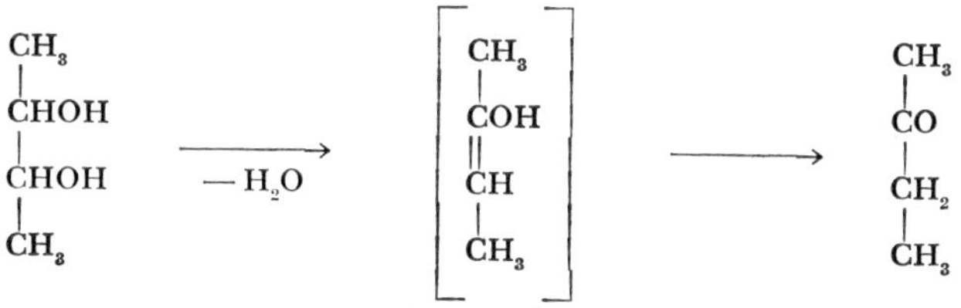

The majority of catalysts tested, acting in liquid or vapor phase, produce the ketone as the main product of the dehydration and this reaction could be used for the manufacture of methyl ethyl ketone if 2,3-butanediol were cheap enough to warrant it. However, conditions have been found which yield the more desirable ethylenic compounds, particularly methylvinylcarbinol and 1,3-butadiene.

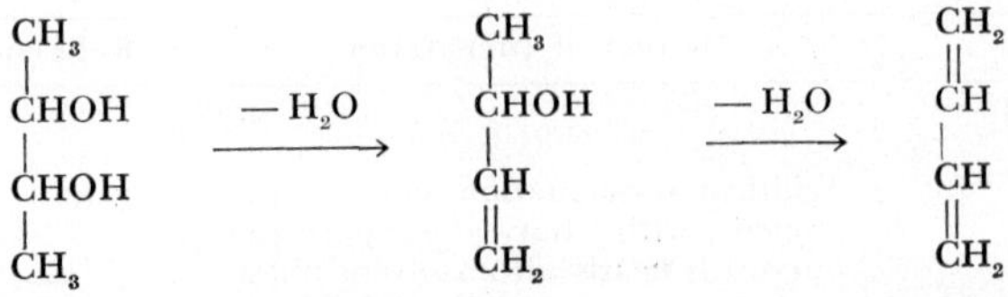

This can be accomplished by using a thorium oxide catalyst at a pressure of about 65 mm and a temperature of 350°C. Single-pass conversions of 60% to butadiene or 80% to methylvinylcarbinol plus butadiene have been obtained with cumulative yields of the carbinol up to 90% of the theoretical value.[280] Some methyl ethyl ketone is obtained as a side reaction. Good yields of ethylenic compounds have also been claimed, using an oxide catalyst containing aluminum, tungsten, and silicon oxides.[85] The 2,3-butanediol is mixed with 10% methyl ethyl ketone, 50% triethylamine, and 20% water, the mixture vaporized and passed over the catalyst at 225° to 235°C. Butadiene and methylvinylcarbinol are the main products with little or no methyl ethyl ketone being formed. If acetoin is treated similarly (without addition of methyl ethyl ketone) a good yield of methyl vinyl ketone is obtained with 72% conversion in a single pass.

The most successful process for the manufacture of butadiene from 2,3-butanediol is pyrolysis of the diacetate:[93,151,153,222]

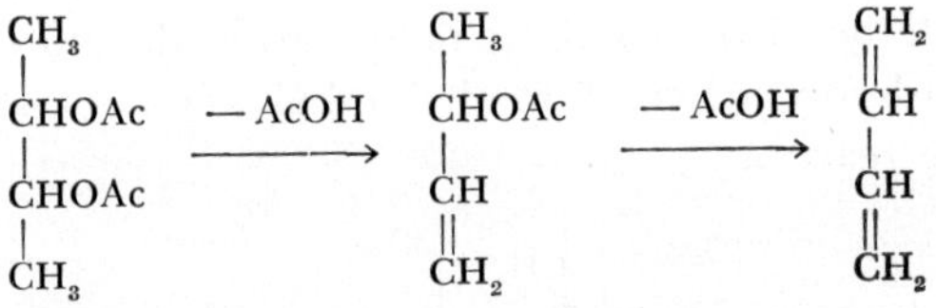

This process is known to work equally well with the *meso* and *levo* isomers and gives butadiene of over 99% purity. The reaction is usually accomplished by passing the vapor of the diacetate through an unpacked tube at 585° to 595°C. The acetic acid is recovered and used again. Yields of 82% are obtained on one pass with a cumulative yield of about 87% of theory.

2,3-Butanediol reacts readily with a number of compounds to give products which contain a five-membered ring with oxygen atoms in 1,3 positions. The three general types of reaction giving this ring are:

(a) Cyclic acetal formation[10,160,172] catalyzed by strong acids:

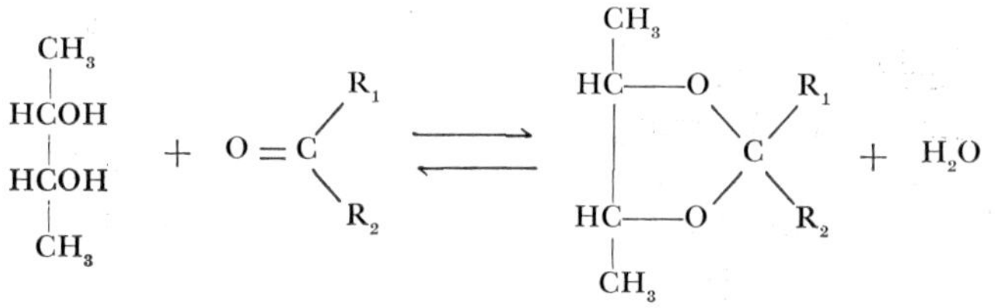

(b) Condensation with acid dihalides such as phosgene,[117] thionyl chloride,[50,208] or phenyl phosphorus oxydichloride:[255]

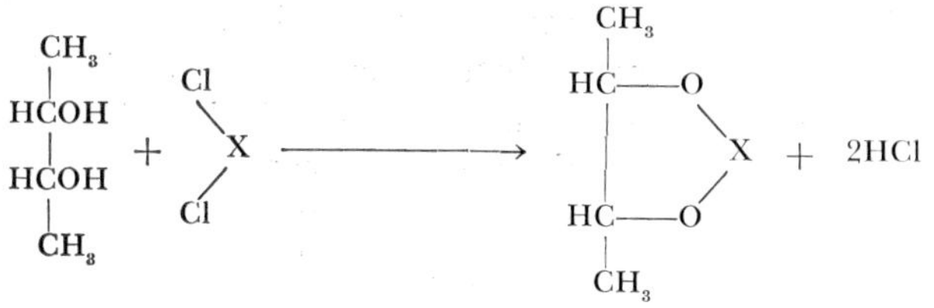

Good yields are obtained in these reactions, which occur spontaneously at room temperature.

(c) Condensation with dihydroxy acids such as arsenoacetic acid[66] and boric acid:[154,206]

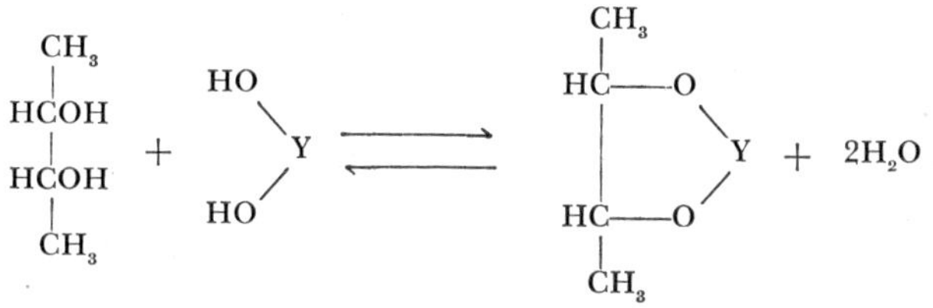

At least two borates can be obtained from *meso*-2,3-butanediol.

There are no definite uses for these compounds. The cyclic acetals are of interest because of their ease of formation, which can be utilized in recovery of 2,3-butanediol from dilute aqueous solutions. They are volatile liquids not miscible with water which form diphasic, low-boiling azeotropes with the butanediol. If a fermentation solution is treated with formaldehyde and boiled, the formal of 2,3-butanediol distills as the azeotrope which separates into two layers.[225] The 2,3-butanediol is recovered by methanolysis of this compound. Alternatively, if the acidified fermentation solution (2 to 4% of 2,3-butanediol) is mixed with one-half volume of butyraldehyde, more than 98% of the 2,3-butanediol is extracted into the butyraldehyde layer as the butyral which is easily purified by distillation.[248] Hydrolysis of the butyral with dilute acid gives the 2,3-butanediol in high yields. These recovery methods would result in a cheaper production of the formal and butyral of 2,3-butanediol than of 2,3-butanediol itself and would be most attractive if direct uses could be found for these cyclic acetals. Catalytic cracking of these cyclic compounds gives chiefly methyl ethyl ketone from the part of the molecule contributed by 2,3-butanediol,[50,170] so that they do not appear promising as precursors of butadiene.

Both isomers of 2,3-butanediol form cyclic acetals readily but the (—) isomer reacts much better than the *meso* isomer with methyl ketones. If (—)-2,3-butanediol is dehydrated by heating with sulfuric acid, 95% of the theoretical amount of the methyl ethyl ketal can be obtained by continuous distillation.[169] Presumably, the methyl ethyl ketone forms and then reacts so rapidly with the excess diol that very little escapes in the free state, if the conditions are right.

These 1,3-dioxacyclopentane derivatives are good solvents, but like other ethers, they show a strong tendency to auto-oxidation. Antioxidants, such as hydroquinone, will control this tendency. Some of the lower members might serve as fuels for internal-combustion engines and some of the more viscous ones might be useful as plasticizers.

In addition to being a possible source for butadiene, 2,3-butanediol might be a source of monomers, such as methylvinylcarbinol and methyl vinyl ketone, for the plastics industry. It might also be used in place of glycerol in making resins of the polyester type. Some interesting studies have been reported on the polyphthalate[28,268] and this use might prove to be an important outlet for 2,3-butanediol. Perhaps these esters could be made from the butyral.

Physiological Properties

Neish[158] has reviewed the possible uses of 2,3-butanediol with reference to the pharmaceutical and food industries. Although, it is less toxic than ethanediol, it appears to be more toxic than 1,2-propanediol and is thus probably less desirable for pharmaceutical uses.

Acetoin and 2,3-butanediol are tolerated by rabbits in doses of 1 to 2 g per kg when administered subcutaneously.[173] Rats will tolerate about 3 g per kg, will recover from the effects of 6 g per kg, but are usually killed by 13 g per kg.[272] Acetoin disappears from the blood at a rate proportional to its concentration and at about the same rate as ethanol.[78] Only a small percentage is excreted unchanged in the urine,[78,173,272] but 5 to 25% may be excreted as 2,3-butanediol. Twelve to 14% of administered 2,3-butanediol is excreted in the urine unchanged, or in combination with glucuronic acid.

Administration of acetoin and 2,3-butanediol to rabbits significantly increased the biological acetylation of *p*-aminobenzoic acid; sodium acetate was without effect under the same condition.[59] This suggests that acetoin plays a role in metabolic acetylations. This is the only physiological use so far suggested to account for the presence of acetoin and acetoin-forming enzymes in animal tissues.

Diacetyl has antiseptic properties and a 0.1% solution will kill most pathogenic bacteria.[142] It will also inhibit the growth of the tubercle baccillus[14,16] and many nonpathogenic organisms[124] more effectively than benzoic acid.

ANALYTICAL METHODS FOR THE DETERMINATION OF 2,3-BUTANEDIOL, ACETOIN, AND DIACETYL

Lack of simple quantitative analytical methods has undoubtedly deterred bacteriologists from investigating the production of 2,3-butanediol, acetoin, and diacetyl by microorganisms. However, most of these difficulties have now been overcome and the investigator has a wide choice of methods available for use in analyzing fermentation solutions. Other compounds, such as ethanol, glycerol, acetic, formic, lactic, and succinic acids, are usually present and must be taken into consideration. Detailed methods for these need not be listed here, but have been published.[163] The authors recommend the use of microdiffusion[281] for ethanol, partition chromatography[161] for the acids, and colorimetric estimation of the formaldehyde formed by periodate oxidation[125] for glycerol. The specificity of the glycerol determinations can be greatly increased if it is first separated from interfering substances by chromatography.[162]

Methods for the Determination of Diacetyl and Acetoin

Diacetyl has never been reported in very large amounts in fermentation solutions, but its determination merits consideration since many methods proposed for the estimation of acetoin and 2,3-butanediol depend on their conversion to diacetyl. Many of the methods for determining diacetyl depend on converting it to dimethylglyoxime by treating with an excess of hydroxylamine. There are numerous ways of estimating dimethylglyoxime.[11,55,94,96,183,199,220,240,246,274] There are also three direct colorimetric methods for determining diacetyl,[48,64,119,195,270] and there is a polarographic method.[78]

Several methods have been proposed for the determination of acetoin. It can be readily oxidized to diacetyl in high yields by heating with ferric chloride and the yields are almost quantitative if a mixture of ferric and ferrous salts is used.[113,134,183,270] This reaction can be conducted in a tightly stoppered tube.[270] The diacetyl is then distilled from the oxidation mixture and may be estimated by any of the available methods.

Other methods for the determination of acetoin depend on its separation from interfering substances, such as sugars, by distillation of the dilute aqueous solution. Acetoin behaves somewhat like the volatile fatty acids, a definite fraction of it distilling along with a definite fraction of the water.[128] Most of the important methods utilize this property in order to attain some degree of specificity. The acetoin in the distillate may be determined iodimetrically by oxidation with alkaline iodine,[128] acidimetrically by titration of acetic acid formed on oxidation by periodate,[33,159] by measuring the reducing power with alkaline copper sugar reagents,[33,133,263] by titration of the hydrochloric acid released on reaction of acetoin with hydroxylamine hydrochloride,[190] and gravimetrically as *p*-nitrophenylosazone.[180] Acetoin and diacetyl can be determined by analyzing the first half of the distillate colorimetrically, using the creatine-α-naphthol reaction.[12,163,187,270] If the color is allowed to develop for 1 hour under prescribed conditions,[270] then equal weights of diacetyl and acetoin give the same intensity of color.[163] The diacetyl can be estimated separately, using the hydroxylamine-urea color reaction[272] and the acetoin obtained by difference. This is the method used by the authors.[163] Acetoin is difficult to recover quantitatively in an ether extract.

However, it may be quantitatively separated from many other substances by partition chromatography and determined by the creatine-α-naphthol color reaction.[163]

Methods for the Determination of 2,3-Butanediol

2,3-Butanediol is usually the major product of the fermentations discussed in this chapter. The numerous procedures for its determination fall into two general classes.

(a) Methods based on oxidation to diacetyl by bromine and ferric chloride belong to the first group.[83,94,95,96,109,113,134,192,221,252] The diacetyl formed is estimated by previously outlined procedures. Methods based on the measurement of diacetyl are specific, but not very accurate since the yields are not quantitative. Furthermore, these methods are difficult to develop into rapid, precise procedures such as are useful in routine analysis.

(b) The second group includes those methods which are based on oxidation to acetaldehyde by periodic acid:

$$\begin{array}{c} CH_3 \\ | \\ CHOH \\ | \\ CHOH \\ | \\ CH_3 \end{array} + HIO_4 \longrightarrow 2 \begin{array}{c} CH_3 \\ | \\ CHO \end{array} + HIO_3 + H_2O$$

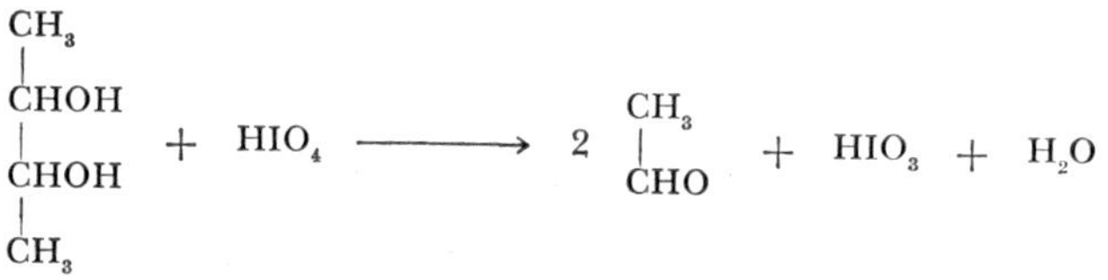

In these methods, either the acetaldehyde formed or the periodate consumed may be measured. Obviously, it is more specific to measure the acetaldehyde, but the titration of periodate is very accurate and may be preferable in the absence of interfering substances.[68,200] If the diol is first extracted from a neutral aqueous solution by ether[18,100,163] or butanol,[194] the interference due to sugars is eliminated and the specificity increased. The diol in *A. aerogenes* and *B. polymyxa* fermentations can be determined in this fashion, but the glycerol present in *B. subtilis* fermentations causes a + 10% error unless correction is made.[163]

The most useful methods for the estimation of 2,3-butanediol are probably those based on the measurement of the acetaldehyde formed by periodate oxidation. They are fairly specific though not as specific as those based on the measurement of diacetyl. However, it is possible to determine 2,3-butanediol in the presence of equal or greater amounts of sugar or glycerol by some of these procedures.

In order to do this, it is necessary to separate the acetaldehyde formed by periodate oxidation from the formaldehyde formed at the same time or to use a colorimetric method in which formaldehyde does not interfere. The blue color developed on treating dilute acetaldehyde solutions with piperazine and sodium nitroprusside permits rapid estimation of 2,3-butanediol, since it can be applied in the presence of formaldehyde and excess periodate.[54] Determinations accurate to within 2 to 3% can be made, in spite of the instability of the color, if a photoelectric colorimeter is used. This method can be applied to diluted fermentation solutions, ether extracts, or butanol extracts[163] with no appreciable interference from equal amounts of glycerol or glucose. If the glycerol concentration is much higher than the diol concentration the reaction can be carried out in a microdiffusion unit and in this way, 2,3-butanediol can be estimated in the presence of fifty times as much glycerol.[163]

The volumetric determination of acetaldehyde absorbed in hydroxylamine hydrochloride[71] or bisulfite[100,186,205,272,282] solutions may yield accurate results, but precautions must be taken to exclude formaldehyde. This makes a distillation necessary[100,186,203,205,227,239,272] in the procedure, if much glycerol is present. However, microdiffusion[163,280] into bisulfite can be used for the volumetric determination of 2,3-butanediol in the presence of an equal amount of glycerol.

The specificity of methods based on periodate oxidation is increased by introducing some procedure for the separation of 2,3-butanediol from other substances that react with periodate. This is often done by steam distillation of the solution after addition of sodium carbonate or by continuous extraction with ether. The extraction with ether may be accomplished quantitatively in 2 hours, if small extractors of 5 ml capacity are used[163] and the organic acids can then be extracted from the same sample after acidification with hydrochloric acid. For this reason, ether extraction is preferred over distillation or butanol extraction if a complete analysis is to be made.

These techniques do not give a sharp separation of the 2,3-butanediol from glycerol, and acetoin is still present, though not necessarily in quantitative amounts. A method based on fractional steam distillation[186] has been proposed which gives an almost complete separation of acetoin and 2,3-butanediol and appears to have a high degree of specificity. Perhaps the most specific method is based on the use of partition chromatography on celite-water columns.[162,163] This gives a sharp, quantitative separation of acetoin, 2,3-butanediol and glycerol, and permits the quantitative determination of each in separate fractions.

The selection of the method to be used for the determination of 2,3-butanediol or acetoin depends on the sample being analyzed. If very little glycerol is present, as in the *A. aerogenes* and *B. polymyxa* fermentations, the diol may be extracted with ether or butanol and determined by titration of the periodate consumed. The most rapid methods for analysis of large numbers of cultures[163] are the colorimetric procedures based on the use of creatine-α-naphthol for acetoin,[270] and piperazine-sodium nitroprusside for 2,3-butanediol.[54] If acetoin and 2,3-butanediol are the major products, these reactions can be run directly on the diluted fermentation solutions at the rate of about one hundred determinations a day per worker. However, if they are minor components of the mixture being analyzed, or if interfering substances are present, they must be separated from the mixture by distillation or ether extraction respectively before developing the color, since substances may be present which modify the intensity of the color. The 2,3-butanediol value must be corrected for the acetoin since 1 mol of acetoin gives 1 mol of acetaldehyde on periodate oxidation.

NATURAL OCCURRENCE OF 2,3-BUTANEDIOL, ACETOIN, AND DIACETYL

Acetoin and 2,3-butanediol are found in small amounts in many common materials, particularly foodstuffs (see Table 4). Usually, they are formed as a result of the action of microorganisms, but it is interesting to note that both are also present in the tissues of higher plants and animals. Foods and beverages which have been subjected to the action of yeasts during their manufacture invariably contain 2,3-butanediol and often acetoin as well. The amount found in wines is quite variable and different types of wines can be distinguished by measurement of their 2,3-butanediol contents.[114] Dry wines, to which no spirits have been added, contain about one hundred to two hundred times as much ethanol as 2,3-butanediol, while fortified sweet wines contain five hundred to twenty-five hundred times as much. Normally, there is not much acetoin in wines, although some yeasts in pure culture form quite appreciable amounts.[8,55] The apiculate yeasts form acetoin, but cannot reduce it to 2,3-butanediol, while the elliptical yeasts, which are always present, reduce acetoin readily.[7,8,9] Therefore, acetoin is usually found in very small amounts. When vinegar is made from wine, the acetoin content increases without much change in the 2,3-butanediol content, indicating a possible synthesis of acetoin from ethanol by the acetic acid bacteria.[150]

TABLE 4. NATURAL OCCURRENCE OF 2,3-BUTANEDIOL, ACETOIN AND DIACETYL

Source	Quantity	References
Beer	18 to 26 ppm acetoin; 10 to 51 ppm 2,3-butanediol	205
Wines	Up to 1,350 ppm 2,3-butanediol and up to 20 ppm acetoin. Ciders may have 300 to 400 ppm acetoin	55, 75, 91, 92, 114, 150, 200, 205
Vinegar	Up to 500 ppm 2,3-butanediol; up to 800 ppm acetoin	75, 150
Bread	Up to 4 ppm acetoin; less than 10 ppm 2,3-butanediol	205, 262
Butter	Up to 4 ppm acetoin and about 1.7 ppm diacetyl	184, 194
Cheese	Diacetyl up to 3.4 ppm, but usually less than 0.5	36, 37
Coffee	10 to 15 ppm of both acetoin and diacetyl in freshly roasted coffee. Diacetyl changes to acetoin as coffee stales	98
Blood and urine of higher animals	About 1 ppm acetoin and 4 ppm 2,3-butanediol in blood—less in urine	140, 221
Seedlings of higher plants	Up to 91 ppm acetoin and 28 ppm 2,3-butanediol	139

Diacetyl has a strong odor and is believed to be the chief component of butter aroma[121a,121b,184,194a] and to contribute to the aroma of other foods. It contributes to the aroma of tobacco smoke,[218a] roasted coffee,[98] honey, and dark beer.[219] Acetoin is also found in these products, usually in higher concentration than diacetyl. Since it is easily oxidized to diacetyl, it may be regarded as a less volatile reserve supply of aroma. A great many papers have been published on the relation of diacetyl to the aroma of butter. It is normally found in a concentration of 1 to 2 ppm,[49] although a concentration of 5 ppm is considered more palatable. It may be used for flavoring butter or margarine and has also preservative value. For example, 10 to 12 ppm of diacetyl will preserve maragarine against development of "off flavors" better than 0.2% benzoic acid.[221] In addition, it gives a butterlike aroma when added in the recommended quantity of 6 ppm.

Bacterial cultures are used in the manufacture of butters of high aroma. Milk is acidified with citric acid and bacteria, such as *Streptococcus citrovorus* and *S. lactis*, are allowed to develop. These ferment the sugar and citric acid, producing 2,3-butanediol, acetoin, and diacetyl.[194] When a properly ripened culture is added to the pasteurized cream, it enables the manufacturer to produce a butter with better aroma and keeping qualities than untreated butter.[67] Aeration favors production of diacetyl and acetoin rather than the nonaromatic 2,3-butanediol,[32,86] which is usually present also in butter cultures.

A great many species of bacteria produce acetoin and/or 2,3-butanediol. The tests for acetoin are widely used by bacteriologists and are of particular value in classifying the Enterobacteriaceae and Bacillaceae.[17,137] In an investigation on one hundred seventy-four strains of anaerobic bacteria, belonging to sixty-four species, it was found that 72.4% formed acetoin[198] and this was considered to be a valuable feature to use in their classification.

Many of these organisms probably give quite small amounts of acetoin and 2,3-butanediol. This is true of all the species of *Clostridium* which have been analyzed carefully. *Cl. felsineum* and *Cl. butyricum* produce acetoin, but are unable to reduce it to 2,3-butanediol.[35] In this respect, they are similar to the apiculate yeasts mentioned before.

Other organisms, such as the *Aerobacter* and *Serratia* species, *B. polymyxa*, *B. subtilis* and *Pseudomonas hydrophila* will ferment more than half of the sugar to 2,3-butanediol and acetoin. It is interesting that these 2,3-butanediol bacteria should be found in three families, showing such marked morphological differences.

Many of the organisms known to produce acetoin have not been tested very thoroughly and under optimum conditions might form substantial amounts of acetoin and 2,3-butanediol. Most cultures of acetoin-producing bacteria normally contain 2,3-butanediol, often in higher concentrations than the acetoin. There are probably many strains of bacteria which are of potential value for 2,3-butanediol production. Any organism which shows a positive test for acetoin is worth testing further, provided it will ferment at a reasonably rapid rate. In the following pages, only a few species, which have been investigated rather extensively, will be considered.

BACTERIOLOGICAL METHODS

The most intensive studies for 2,3-butanediol production have been on species of *Aerobacter* and *Serratia* of the Enterobacteriaceae and *Bacillus polymyxa* and *B. subtilis* of the Bacillaceae. No attempt will be made to deal with problems of nomenclature, nor to list the many synonyms of the species mentioned. The system of nomenclature recommended by Bergey's *Manual of Determinative Bacteriology,* Sixth Edition,[17] is followed here.

Aerobacter aerogenes (Kruse) Beijerinck

Organisms of this genus are universally distributed in nature, being very abundant in soil and plant products and also occurring in the intestines of man and animals. Fresh vegetables, such as celery, spinach, carrots, lettuce, yams, onions, and potatoes, have been found most useful for seeding enrichment cultures. Cultures are also readily available from many culture collections.

Aerobacter cultures may be maintained by transfer on slants of the medium suggested by the Northern Regional Research Laboratory which contains, per liter, 5.0 g tryptone, 1.0 g glucose, 5.0 g yeast extract, and 15.0 g agar. Starter cultures for *Aerobacter* fermentations were developed[264] by growing the organism for 8 to 20 hours at 30°C in an unagitated, unaerated medium which contained, per liter, 50 g glucose, 0.25 g $MgSO_4 \cdot 7H_2O$, 0.60 g KH_2PO_4, 5.0 g $CaCO_3$, and 0.20 g urea. All ingredients were mixed together, except the urea, and the medium was sterilized at 121°C for 30 minutes; then 1 ml of sterile 20% urea solution was added for each 100 ml of basic medium.

Serratia marcescens Bizio

Species of this genus are small, aerobic rods, usually producing a bright-red or pink pigment on agar and gelatin. Cultures have been isolated from water, soil, milk, foods, silk worms and other insects, but no dependable enrichment techniques are available for obtaining them. Most fermentation studies have been made on cultures available from collections.

Bacillus polymyxa (Prazmowski) Migula

The classification, morphological and biochemical properties of *B. polymyxa* have been described[61,197,229] and also procedures for isolation of strains having active fermentation properties.[131] Rich garden soil is a very dependable source of inoculum for enrichment cultures, though other natural sources, such as soil water, and fresh or decaying vegetables, may also be used. It was found that pasteurization of the original inoculum in water at 80°C for 10 minutes is very important. The pasteurized suspension may be inoculated in 1 ml amounts into tubes of lactose or 1.5% starch media and incubated at 30°C. Streak plates are made in 2 or 3 days from cultures showing gas production. Neutral red agar, plain starch-peptone, or starch yeast-extract agar are all satisfactory media for this purpose. Presence of neutral red, which stains *B. polymyxa* colonies pink or red, aids in their selection from other species. Slime formation and the presence of gas bubbles, as well as a pleasant fruity odor, are additional characteristics which aid in the selection of *B. polymyxa*. However, great differences in colony form, consistency, surface, and elevation occur between isolates and the final selection depends on biochemical characteristics.

Stock cultures in active use may be carried on agar slants, containing 5% whole wheat, or 2% soluble starch, containing 0.5% yeast extract and 0.5% calcium carbonate.

Bacillus subtilis (Ford strain)

There is still much uncertainty as to the taxonomic position of the Ford strain of *B. subtilis*. Gibson[76] regards it as a distinct species, *B. licheniformis*, but Smith, Gordon, and Clark[229] and Bergey's Manual[17] do not distinguish between it and the recognized type culture, the so-called Marburg strain. It has been found[22,165] that under anaerobic conditions, the Ford strain produces glycerol and 2,3-butanediol whereas the Marburg strain is a strict aerobe and does not produce glycerol.

B. subtilis is very widely distributed in the soil and decomposing organic matter. Enrichment cultures of dried hay, straw, or soils rich in organic matter may be pasteurized at 90 to 100°C for 10 to 30 minutes and samples used to inoculate tubes of nutrient media. After incubation at 35°C for 24 hours, cultures of *B. subtilis* and related species may be picked from streak plates. These cultures are then differentiated by biochemical tests as outlined by

Smith et al.[229] After this, it is necessary to carry out strictly anaerobic fermentations on all *B. subtilis* types to differentiate Ford and Marburg strains.

FERMENTATION OF CARBOHYDRATES BY 2,3-BUTANEDIOL BACTERIA

Types of Fermentation

The 2,3-butanediol bacteria may be defined as those in which 2,3-butanediol is the major fermentation product (i.e., usually 20% or more based on the weight of the sugar fermented). The first species to be placed in this group was *Aerobacter aerogenes* which was found, in 1906, to give good yields of 2,3-butanediol.[89] This organism grows well anaerobically in a calcium carbonate-buffered medium, containing sugars, and forms 2,3-butanediol according to the equation:

$$C_6H_{12}O_6 \longrightarrow CH_3{\bullet}CHOHCHOH{\bullet}CH_3 + 2\,CO_2 + H_2$$

Other 2,3-butanediol bacteria, such as *Bacillus polymyxa,*[61] *Aerobacter cloacae,*[247] *Aerobacter faeni,*[30] *Aerobacter indologenes,*[149] *Pseudomonas hydrophila,*[231] and *Serratia plymuthicum,*[167,192] also ferment carbohydrates according to this general equation under appropriate conditions. However, it is evident that 2,3-butanediol bacteria which do not form molecular hydrogen must have some other way of disposing of the two hydrogen atoms left over when 2,3-butanediol is made from glucose. Some bacteria of the genus *Serratia* accomplish this, under anaerobic conditions, by production of formic acid:[167,192]

$$C_6H_{12}O_6 \longrightarrow CH_3{\bullet}CHOH{\bullet}CHOH{\bullet}CH_3 + CO_2 + HCOOH$$

However, the group of bacteria represented by Ford's type of *Bacillus subtilis* form glycerol:[21,166]

$$3\,C_6H_{12}O_6 \rightarrow 2\,CH_3{\bullet}CHOH{\bullet}CHOH{\bullet}CH_3 + 2\,CH_2OH{\bullet}CHOH{\bullet}CH_2OH + 4\,CO_2$$

All these bacteria are heterofermentative and carry out the well-known lactic acid or alcoholic fermentations as well as 2,3-butanediol fermentations:

$$C_6H_{12}O_6 \longrightarrow 2\,C_2H_5OH + 2\,CO_2$$
$$C_6H_{12}O_6 \longrightarrow 2\,CH_3{\bullet}CHOH{\bullet}COOH$$

The relative importance of these reactions in the different

species of 2,3-butanediol bacteria is shown in Table 5. These figures were obtained by culturing the bacteria anaerobically in a calcium carbonate-buffered medium, containing 5% glucose with accessory nutrients, usually supplied by 0.5% yeast extract. The actual weight of products obtained under these conditions is shown in Table 6 for the four most interesting species. These organisms should not be compared solely under these conditions, since all 2,3-butanediol fermentations are sensitive to changes in conditions. For example, *Aerobacter aerogenes* gives the lowest yield of useful products anaerobically, but it is probably the best organism to use for production of 2,3-butanediol, since it grows and ferments more vigorously than the other organisms and the yield of 2,3-butanediol can be increased to 40 lb of 2,3-butanediol per 100 lb of glucose fermented by moderate aeration.[70,186] Thus under readily attainable conditions, it will produce more 2,3-butanediol than the other organisms whose yield is not raised by aeration as much. *B. polymyxa* is of interest since it will ferment starchy matter directly and produces pure (—)-2,3-butanediol. *S. marcescens* produces considerable amounts of formic acid, but it is doubtful if this could be economically recovered on a commercial scale. *B. subtilis* is interesting in that it gives good yields of glycerol as well as of 2,3-butanediol and the sum of these two products is considerably greater than the theoretical amount of diol produced by the other species.

Table 5. RELATIVE IMPORTANCE OF THE VARIOUS FERMENTATION REACTIONS IN SPECIES OF BACTERIA PRODUCING 2,3-BUTANEDIOL

Organism	Percentage of sugar dissimilated by fermentations producing				
	Diol-hydrogen	Diol-formic acid	Diol-glycerol	Ethanol	Lactic acid
Aerobacter aerogenes	52	2	—	23	2
Pseudomonas hydrophila	56	—	—	26	12
Bacillus polymyxa	68	—	—	33	—
Serratia plymuthicum	49	2	—	23	14
Serratia marcescens	1	55	1	27	6
Serratia indica	1	50	1	22	5
Serratia anolium	1	48	1	21	4
Bacillus subtilis (Ford's Type)	—	—	84	3	12

TABLE 6. MAJOR PRODUCTS OF ANAEROBIC 2,3-BUTANEDIOL FERMENTATIONS

Product	Pounds per 100 lb glucose fermented[a]			
	Bacillus polymyxa	*B. subtilis* (Ford's Type)	*Aerobacter aerogenes*	*Serratia marcescens*
2,3-Butanediol	32.6	27.5	24.0	32.0
Ethanol	17.1	2.0	17.2	11.8
Glycerol	—	29.4	—	—
Lactic acid	—	11.6	2.4	5.0
Formic acid	—	0.3	0.8	12.3
Carbon dioxide	48.2	26.9	43.8	29.6
Hydrogen	0.78	—	0.47	—
	98.68	97.7	88.67	90.7

[a] Acetic acid, succinic acid, and assimilated carbon account for the remainder of the glucose used.

Effects of Aeration

Aeration of the cultures during the fermentations has a marked effect. In the presence of oxygen, all these organisms tend to ferment according to the equation:

$$C_6H_{12}O_6 + \frac{1}{2}\,O_2 \longrightarrow CH_3{\cdot}CHOH{\cdot}CHOH{\cdot}CH_3 + 2\,CO_2 + H_2O$$

Thus under aerobic conditions *S. marcescens* gives very little formic acid, *B. subtilis* yields hardly any glycerol, and hydrogen production by the other species is lowered to varying extents, depending on the organism.

In considering the effect of aerating these fermentations, it must be realized that the over-all effect on a complex system is being observed. A suitable medium is inoculated with a comparatively small number of cells. These cells multiply, reach a peak population, and eventually die off. During this period, the sugar is fermented. Aeration may affect both the rate of growth and the rate of fermentation per cell, as well as the relative concentration or activities of the enzymes produced by the cells. If air or oxygen is passed through a fermenting medium, it can act by agitating the medium, thus aiding neutralization of acids by the calcium carbonate, or it may affect the fermentation by removing the carbon dioxide or by acting as a hydrogen acceptor. Some experiments have been designed to separate these variables and it is possible to make certain inferences concerning the results observed when a stream of air is passed through a 2,3-butanediol fermentation.

In general, aeration increases the yield of the C_4 compounds (2,3-butanediol and acetoin) as compared with the other products.

[2,70,186,211,263] Experiments with *B. polymyxa* have shown that there is a gradual decline in the proportion of the carbohydrate-forming C_4 compounds during anaerobic fermentation and that aeration raises the yield and prevents this decline.[211] Oxygen retards the ethanol formation by *B. polymyxa* and it is possible to change the diol to ethanol ratio from 1.4 to 3.6 if it is used.[2] *A. aerogenes* acts similarly and gives a greatly increased yield of 2,3-butanediol when aerated.

In addition to changing the yields of the products, aeration also affects the over-all rate of fermentation. Oxygen is not necessary to obtain these increased rates if *A. aerogenes* and *B. polymyxa* are used; gases, such as nitrogen or hydrogen, passed through fermenting solutions also cause an increased rate.[2,217] Carbon dioxide is not suitable for this purpose and it is possible that the beneficial effects of the other gases is due to the removal of carbon dioxide. This view is supported by the fact that more rapid fermentations occur with a high surface to volume ratio or when fermentations are run under reduced pressure.[4,102]

It is safe to say that aeration should be beneficial to most 2,3-butanediol fermentations, except *B. subtilis* fermentations, where it greatly reduces the yield of glycerol.[21,166] The exact amount of aeration required for any fermentation has to be worked out on a pilot-plant scale for the organism and medium to be used.

Effect of pH on the Fermentation

Due to the sensitivity of bacteria to variations in the environment, there is still much work to be done in determining optimum conditions for the various fermentations. A good example of the need for close control of conditions may be seen in the effects of the hydrogen-ion concentration of the medium on the rate and course of the fermentation. This has been shown in experiments on fermentations controlled to within 0.05 pH unit by automatic addition of ammonium hydroxide.[171] The effect of pH on the rate of carbohydrate breakdown by the four most interesting species of bacteria are shown in Figure 2. They all show an optimum rate near pH 7, but *B. subtilis* and *S. marcescens* are more sensitive to acid. These four species of bacteria produce 2,3-butanediol well at pH 6.2 or below, but above this pH, the yield drops rapidly until at pH 7.6 little or none is formed (see Figure 3). Earlier work on *B. polymyxa*[3] and *A. indologenes*,[148] using manual control, had indicated that lower diol yields were obtained at high pH values and that a critical point occurred at pH 6.3. This seems to be true for the five species tested so far. The reason for this rapid decrease in the yield is not known, but it seems to be at least partly due to the effect of pH in altering the enzymic constitution of the cells which are grown in the medium.[31] In addition, it is probable that pH affects the mature cells as well.

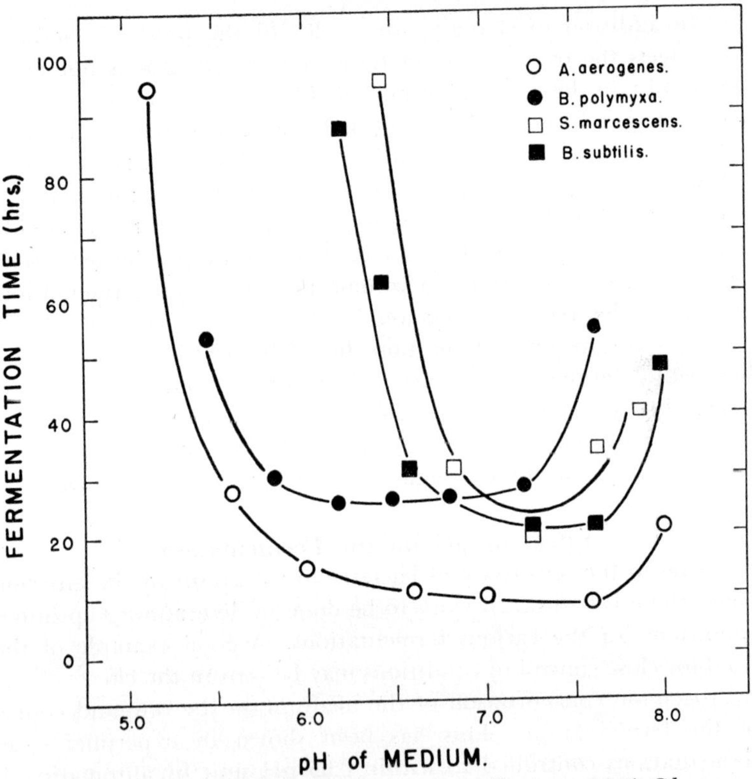

FIGURE 2. *Effect of pH on Rate of Dissimilation of 5% Glucose*

The failure of these bacteria to form much carbon dioxide or hydrogen in alkaline media is remarkable. Formic acid is formed instead and the diol is replaced largely by acetic acid. There seems to be a definite relation between the yields of acetic acid and 2,3-butanediol. When acetate is added to a fermenting *Aerobacter* culture at a pH favorable for 2,3-butanediol fermentation, the acetate is reduced and the 2,3-butanediol yield is increased enough to account for all the acetate disappearing (diol increased from 64 mols to 88 mols per 100 mols of glucose fermented). This may result in almost complete suppression of the hydrogen formed.[148,204] Similar results have been obtained with *B. polymyxa*.[207,230] Propionate acts similarly to acetate in increasing the diol and decreasing the hydro-

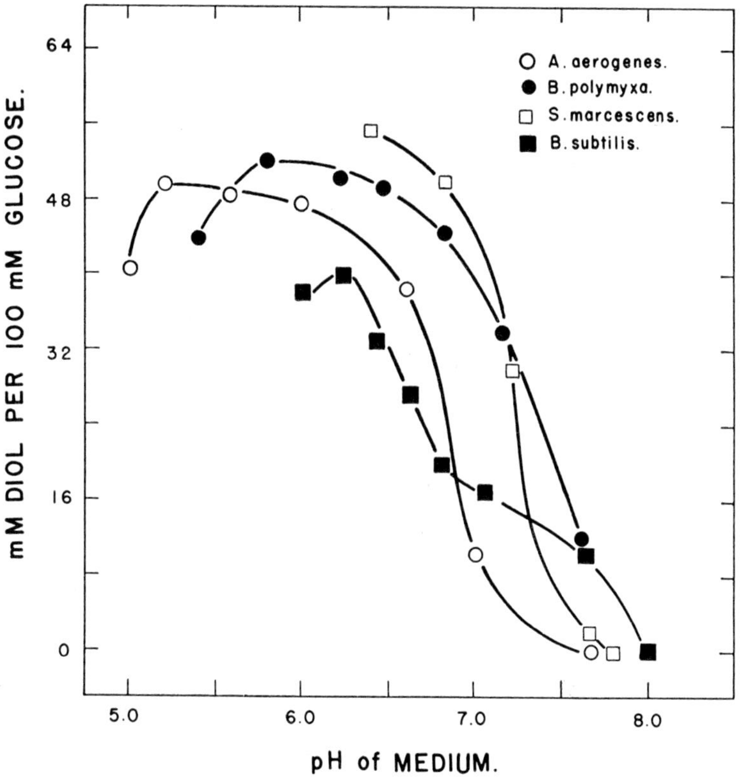

FIGURE 3. *Effect of pH on Production of 2,3-Butanediol by Four Species of Bacteria*

gen yield while succinate increases the diol and formic acid yields, with decreasing hydrogen production. Since propionate is reduced to propanol, this suggests that these acids act by a mechanism similar to that of other hydrogen acceptors in raising the yield of 2,3-

butanediol. However, ferric hydroxide is reduced by *B. polymyxa,* without increasing the diol yield.[207] Since reduction of acetates tends to make the medium alkaline, they might be useful buffering agents for controlling the pH of 2,3-butanediol fermentations on an industrial scale. The neutralizing agent consumed would then be recovered as the product. This possibility seems to have been overlooked by most investigators who have studied these fermentations from an industrial point of view.

If the acids formed as by-products by most of these bacteria are not neutralized, the pH falls too low for vigorous growth and fermentation. Calcium carbonate has been the most widely used buffering agent; it can be added in excess and goes into solution as it is required. It usually buffers in the range of pH 5.6 to 5.9 which is favorable for 2,3-butanediol formation but low enough to limit the rate of the fermentations. The best pH for carrying out 2,3-butanediol fermentations is about 6.2, since good diol yields are still obtained and the reaction is considerably faster than in calcium carbonate fermentations.[171,269] It is necessary to use an automatic pH-control system to achieve these conditions. In this case, ammonia is a good neutralizing agent. *B. polymyxa* fermentations on a sucrose medium are five times as fast when ammonium hydroxide is used as when sodium or potassium hydroxide is used as a neutralizing agent. The optimum conditions for anaerobic fermentation of sucrose by *B. polymyxa* are believed to be in a medium containing 8% sucrose at 33°C, controlled at pH 6.2 ± 0.05 by automatic addition of ammonium hydroxide. The fermentation is complete in 30 hours giving 130 mols of diol and 122 mols of ethanol per 100 mols of sucrose fermented.[269] It is likely that the same conditions would be suitable for *Aerobacter aerogenes,* although a temperature of 35°C is probably better.[70] Aeration under these conditions will raise the yield of 2,3-butanediol but decrease the rate of the fermentation as far as *B. polymyxa* is concerned.

Although most large-scale work on 2,3-butanediol fermentations has been done with calcium carbonate as a buffering agent, it appears that ammonia has the advantages of lower cost, inherent sterility, greater range of control, ease of addition in the liquid or gaseous state, greater nutrient value, and acceleration of fermentation.[3,269]

Effect of Sugar Concentration

Increasing the initial sucrose concentration increases the yield of 2,3-butanediol in *B. polymyxa* fermentations.[73,193] Under aerobic conditions on a medium containing 10% sucrose with pH controlled at 6.2 a diol + acetoin to ethanol ratio of 11 to 1 may be obtained with *B. polymyxa*.[269] A 6% solution is almost completely fermented in 71 hours, giving 104 mols of 2,3-butanediol, 48.6 mols of acetoin and 40 mols of ethanol per 100 mols of sucrose.

Aerobacter aerogenes is capable of fermenting fairly high concentrations of sugars, if they are added gradually. Experiments have been performed in which a concentrated solution of glucose was fed into a fermentation at such a rate as to keep the sugar concentration in the fermentor close to 3%. In this way, sugar concentrations corresponding to 26.5% were fermented in 108 hours, giving 2,3-butanediol concentrations up to 9.8%.[62,186]

Nutrient Requirements

Organisms of the *Aerogenes* group *(Aerobacter, Serratia)* have fairly simple nutrient requirements and will ferment well on a synthetic medium containing only sugars and mineral salts. *B. subtilis* and *B. polymyxa* need a complex nitrogen source in order to grow and ferment well. However, *B. polymyxa* is able to ferment starchy substrates, such as whole wheat or corn mashes,[102,121,130,131,266] directly, whereas a preliminary hydrolysis is necessary before they can be fermented by *Aerobacter* strains.

It is more desirable from a chemurgic and engineering viewpoint to separate the starch from the other components of grain before fermentation. Consequently, investigations have been conducted on the fermentation of starch by *B. polymyxa*.[69,103,104,121] This organism requires biotin and complex nitrogen sources to ferment well.[105]

Good fermentations of cornstarch were obtained by Kooi, Fulmer, and Underkofler[121] with a medium containing 0.5% of corn gluten, 7.5% of cornstarch, and 0.0069% of potassium permanganate. Addition of malt sprouts or brewers' yeast as nutrients raised the yield of 2,3-butanediol. The resulting fermentation solution is more readily filtered than a whole-corn mash fermentation and valuable by-products are recovered in preparing the cornstarch.

Wheat starch is a poor medium for fermentation and supplements have to be added. In this case, the gluten is not stimulatory, but the bran and shorts are. A 1% addition of malt sprouts was the most effective supplement for wheat starch with shorts, bran, Cerogras, alfalfa, soybeans, yeast extract, and corn-steep liquor in decreasing order of effectiveness. It is possible to ferment purified wheat starch by some strains of this organism provided isoleucine, tyrosine, glycine, methionine, and asparagine are added. The nutritional requirements of different strains vary somewhat and some amino acids are inhibitory in concentrations above 0.02%. Potassium stimulates 2,3-butanediol production from starch and increases the diol to ethanol ratio. Phosphate also stimulates diol production, especially in the presence of potassium.[104] The production of 2,3-butanediol by *A. aerogenes* is also stimulated by phosphate as well as by peptone.[215]

In addition to fermenting glucose, fructose, and sucrose, the 2,3-butanediol bacteria will also ferment xylose and mannitol.[2,5,30,231,247,263] On each of these substrates, 2,3-butanediol is formed, although usually in lower yields than from glucose. *Aerobacter* strains will also produce 2,3-butanediol from mannose, galactose, arabinose, isodulcite, adonitol, and glycerol.[87,88] *Aerobacter* will ferment a much wider range of substrates than yeast and, for this reason, may be useful for fermenting hydrolysates of fibrous materials which usually contain xylose. It has been shown that wood hydrolysates can be successfully fermented by this organism, producing a yield of 35% of 2,3-butanediol based on the fermented sugar.[193]

MECHANISM OF THE BIOLOGICAL SYNTHESIS OF ACETOIN AND 2,3-BUTANEDIOL

Biological Synthesis of Acetoin

Pyruvic acid is readily fermented by 2,3-butanediol bacteria and is probably an important intermediate in the synthesis of acetoin and 2,3-butanediol from sugars. The synthesis of acetoin seems to be closely linked to the decarboxylation of pyruvic acid; in fact, a model experiment, using light energy to decarboxylate pyruvic acid, will give higher yields of acetoin[56] than can be obtained by irradiation of acetaldehyde.

The first clue to the mechanism of the biological synthesis of acetoin was the observation that acetylphenylcarbinol was formed when benzaldehyde was added to a sugar solution being fermented by yeast.[174,178,179] Since the compound obtained was optically active, it was obviously a result of biological synthesis. Similar experiments, using acetaldehyde in place of benzaldehyde, have yielded optically active acetoin.[180] This suggested the presence of an enzyme, called carboligase, capable of catalyzing the reaction:

$$2CH_3{\bullet}CHO \longrightarrow CH_3{\bullet}CHOH{\bullet}CO{\bullet}CH_3 \qquad (I)$$

The high yields obtained made it seem likely that some of the acetoin was formed from the sugar. Further experiment showed that this condensation occurred only in the presence of fermenting carbohydrates or pyruvate.[57,58,182] Acetoin can be formed from pyruvate alone, although the yield may be increased by addition of acetaldehyde.[181] Since the acetoin-forming enzyme has been found by most workers to have the same properties as the decarboxylating enzyme found in the same material, it is considered likely that they are identical.[58,126,228,243] These conclusions have been questioned by workers who have evidence that there are actually two enzymes concerned since they obtained acetoin from acetaldehyde, without simultaenous carbon dioxide formation.[122]

Since the formation of acetoin seems to require at least 1 mol of "nascent acetaldehyde," formed from pyruvate by decarboxylation, the reaction may be written:

$$CH_3CHO + CH_3{\bullet}CO{\bullet}COOH \longrightarrow CH_3{\bullet}CHOH{\bullet}CO{\bullet}CH_3 + CO_2 \qquad (II)$$

Acetoin-forming enzymes are widely distributed in nature, being also found in yeasts of genera *Zygosaccharomyces* and *Schizosaccharomyces*,[250] in bacteria,[107,138,228,251] in the seeds[243,244,252] and leaves [238] of higher plants, and in various animal tissues, especially muscle.[77,252,271]

These acetoin-forming enzymes are probably not all identical, since the optical activity of the acetoin obtained varies considerably, depending on the source of the enzyme preparation. Yeasts, bacteria, and animal-tissue enzymes all give levo-rotatory acetoin, while the enzymes of the higher plants give dextro-rotatory acetoin.[9,77,138,180,243,244,252,271] The acetoin produced by muscle tissue has $[\alpha]_D = -71°$ to $-81°$ in dilute aqueous solution as compared with $[\alpha]_D = +37°$ to $+40°$ for acetoin formed by leguminous seed meals. There must be at least two acetoin-producing enzymes, one found in plants and another found in animals and microorganisms. The highest rotations reported for dilute aqueous solutions of acetoin are $[\alpha]_D = -98°$ for acetoin formed by *Bacillus subtilis (B. natto),*[250] and $[\alpha]_D = 103°$ for an enzyme preparation from yeast maceration juice.[58] *A. aerogenes* maceration juice also gives acetoin with a high *levo* rotation.[228] It is obvious that most enzyme preparations form both isomers of acetoin in unequal proportions or else they contain a racemase.[251] The specific rotation of optically pure acetoin is not known, so that it is not certain what proportions of isomers are produced by the various preparations.

The enzymes of yeast and muscle, which have been studied the most extensively, are quite similar in their properties. They both carry out the general reaction:

$$CH_3CO{\bullet}COOH + RCHO \longrightarrow RCHOH{\bullet}CO{\bullet}CH_3 + CO_2 \qquad \text{(III)}$$

and it is possible to form a variety of compounds similar to acetoin. The muscle enzyme has the properties of a carboxylase. It requires diphosphothiamin and magnesium ions for coenzymes. It will decarboxylate other α-keto acids, such as α-ketobutyric and α-ketoglutaric acid, thus catalyzing reactions, such as:[15,77]

$$2C_2H_5{\bullet}CO{\bullet}COOH \longrightarrow C_2H_5{\bullet}CO{\bullet}CHOH{\bullet}C_2H_5 + 2CO_2 \qquad \text{(IV)}$$
$$C_2H_5{\bullet}CO{\bullet}COOH + CH_3CHO \longrightarrow C_2H_5CHOH{\bullet}CO{\bullet}CH_3 + CO_2 \qquad \text{(V)}$$
$$COOH{\bullet}CO{\bullet}CH_2{\bullet}CH_2{\bullet}COOH \longrightarrow CHO{\bullet}CH_2{\bullet}CH_2{\bullet}COOH + CO_2 \qquad \text{(VI)}$$

The compound formed in reaction III appears to derive the $CH_3CO{\bullet}$ group from the pyruvate, since addition of benzaldehyde to yeast fermentations[52,178,179] and propionaldehyde plus pyruvate to muscle preparation[15] gives as products the compounds $C_6H_5CHOH{\bullet}CO{\bullet}CH_3$ and $C_2H_5{\bullet}CHOH{\bullet}CO{\bullet}CH_3$, rather than the isomers containing a $CH_3CHOH{\bullet}$ group. By analogy, one would expect that acetoin formed by reaction (II) would derive the $CH_3CO{\bullet}$ end of the molecule from pyruvate and the ${\bullet}CHOH{\bullet}CH_3$ end from the acetaldehyde added. To check this point, heavy-carbon acetaldehyde was added to yeast juice containing pyruvate, the acetoin formed was isolated, and the position of the heavy carbon determined.[81] The isotopic carbon was found in both ends of the molecule, although there was a somewhat greater excess in the carbinol end, as expected. This means that the acetoin must have been partly formed through a symmetrical molecule or else racemization occurred after formation. As pointed out previously, yeast preparations often yield acetoin which is certainly not optically pure; furthermore racemization might occur during isolation, particularly if alkaline conditions are used.[244] Racemization probably occurs through the symmetrical dienol form whether catalyzed by alkali or enzymes.

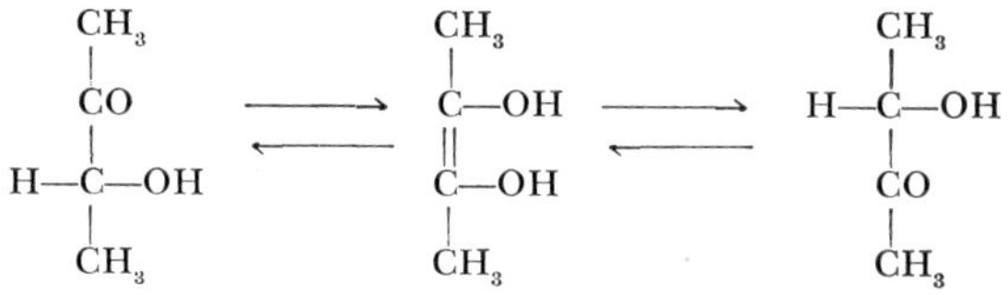

$$\begin{array}{c} CH_3 \\ | \\ CO \\ | \\ H-C-OH \\ | \\ CH_3 \end{array} \quad \rightleftarrows \quad \begin{array}{c} CH_3 \\ | \\ C-OH \\ || \\ C-OH \\ | \\ CH_3 \end{array} \quad \rightleftarrows \quad \begin{array}{c} CH_3 \\ | \\ H-C-OH \\ | \\ CO \\ | \\ CH_3 \end{array} \quad (VII)$$

As a consequence of racemization, the labeled carbon will be found in both ends of the molecule. Experiments of this type should include measurements of the optical rotation, since samples of acetoin with the highest optical rotation would be expected to have the greatest concentration of isotopic carbon in the $CH_3CHOH\cdot$ end of the molecule. This type of experiment is rather important since another mechanism suggested for acetoin formation involves diacetyl as an intermediate:[147,242]

$$CH_3 \cdot CO \cdot COOH \longrightarrow CH_2=C=O + CO_2 + 2H \quad (VIII)$$

$$CH_2=C=O + CH_3CHO \longrightarrow$$
$$CH_3 \cdot CO \cdot CO \cdot CH_3 \xrightarrow{+2H} CH_3 \cdot CO \cdot CHOH \cdot CH_3 \quad (IX)$$

If acetoin is formed by these reactions, rather than by reaction (II), the isotopic carbon should be equally distributed between both ends of the molecule.

The acetoin-forming enzymes of bacteria differ from those of yeast and animal tissues in not being able to utilize acetaldehyde. These enzymes form 2 mols of carbon dioxide and 1 mol of acetoin from 2 mols of pyruvate:

$$2CH_3 \cdot CO \cdot COOH \longrightarrow CH_3 \cdot CO \cdot CHOH \cdot CH_3 + 2CO_2 \quad (X)$$

This has been found to hold for *B. subtilis*,[123,138] *A. aerogenes*,[13,228] and *Clostridium acetobutylicum*.[101] The enzyme system of *A. aerogenes* has been studied on cell-free preparations.[228] It differs from the yeast enzyme in being more sensitive to heat and alkalies and in not being active toward acetaldehyde in the presence of pyruvate. Although it is a carboxylase, using the same coenzyme, trapping experiments with bisulfite and dimedon failed to show intermediate formation of acetaldehyde which can be readily demonstrated with yeast carboxylase and even with bacteria growing on sugars.[177,230] The only products formed by decarboxylation of pyruvate were carbon dioxide and acetoin. Good yields of enzyme are obtained from cells grown in acid but not in alkaline media. Phosphate is necessary for the action of this enzyme and although no phosphorylated intermediates have yet been demonstrated, it is probable they exist.

Very recently significant advances have been made in studies on the mechanism of acetoin formation. It was shown by Juni[101a] that α-acetolactic acid is an intermediate in the formation of acetoin from pyruvate by bacteria but apparently is not in yeast or mammalian tissue. These conclusions are supported by the results of other workers.[56a,59a,86a] An enzyme extract of *Aerobacter aerogenes* was resolved into components catalyzing the following reactions separately:

$$2CH_3COCOOH \longrightarrow CH_3COH(COCH_3)COOH + CO_2$$
$$CH_3COH(COCH_3)COOH \longrightarrow CH_3CHOHCOCH_3 + CO_2$$

The dextro-rotatory acetolactic acid formed is decarboxylated to levo-rotatory acetoin; the other isomer of acetolactic acid is not attacked at any appreciable rate. Yeast and mammalian tissue will not decarboxylate either isomer of this acid. However, tracer experiments indicate an asymmetric intermediate since the acetoin formed by pig-heart enzymes from pyruvate-2-C_{14} and acetaldehyde is labeled preponderantly in the carbonyl group, while that formed by yeast enzymes from acetaldehyde-1-C_{14} and pyruvate is labeled mainly in the carbinol group. There must be at least two mechanisms for the biosynthesis of acetoin and, in view of the known relation of acetoin to biological acetylation,[59] it would not be surprising if coenzyme A is involved in some systems.

Biological Synthesis of 2,3-Butanediol

2,3-Butanediol is probably formed by reduction of acetoin. Its more reduced state and the fact that some organisms form acetoin but no 2,3-butanediol,[7,8,9,35] both suggest that acetoin is formed before 2,3-butanediol rather than *vice versa.* These compounds are in equilibrium with each other and with diacetyl in fermentation solutions:

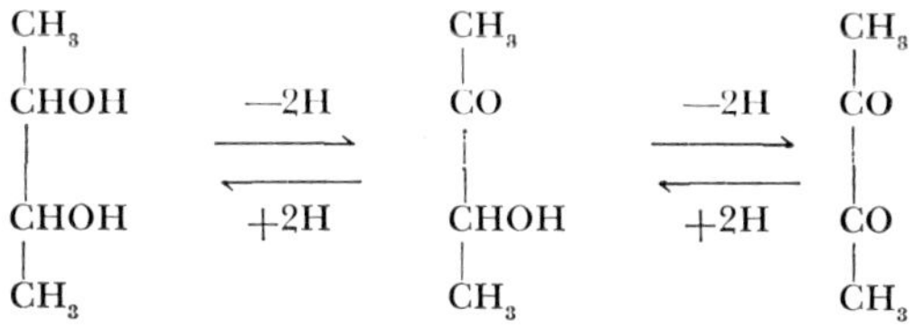

(XI)

Under aerobic conditions, acetoin may be the major product of a 2,3-butanediol fermentation since all bacteria making 2,3-butanediol can oxidize it to acetoin under aerobic conditions.[111,135,191,259] Diacetyl is not usually formed in large amounts, but is present in strongly aerated cultures.[32,97] Acetoin is quite commonly formed in aerobic cultures. When the fermentation is vigorous, the redox potential falls and 2,3-butanediol is the main product, but when the sugar is used up, the redox potential rises and acetoin is formed.[230,263] Addition of a hydrogen acceptor, such as methylene blue, at a concentration of about 33 ppm, catalyzes the reaction and permits more effective use of air.[190] Removal of acetoin by oxime formation causes the reaction to go to completion.[191] Some 2,3-butanediol bacteria, such as *Aerobacter aerogenes,* will oxidize acetoin further, yielding such products as acetic acid if the aeration is prolonged.[233,278]

Diacetyl[176] is reduced by fermenting yeast, giving 2,3-butanediol with $[\alpha]_D = -2.4°$, while racemic acetoin[175] is reduced under the same condition to give 2,3-butanediol with $[\alpha]_D = -5.5°$. Since the 2,3-butanediol naturally formed from yeast[164] has a specific rotation of at least $-8.5°$, it is difficult to see how it could be formed by reduction of diacetyl, although it could be formed from acetoin, since the acetoin produced in a natural yeast fermentation is optically active. Bacteria will also reduce acetoin and diacetyl to 2,3-butanediol.[86]

Bacteria which oxidize 2,3-butanediol show stereochemical specificity and usually only oxidize an atom with either a D(—) or L(+) configuration, depending on the species. *Aerobacter aerogenes* and *Pseudomonas hydrophila* do not oxidize atoms with the D(—) configuration and thus will not oxidize the levo-rotatory (—,—) diol, but will rapidly oxidize the *meso* (+,—) diol. *B. polymyxa* can oxidize D(—) atoms and consequently will oxidize both the (—,—) and the (+,—) diols.[233] *Acetobacter suboxydans,*[74] *Acetobacter xylinum,* and *Acetobacter aceti*[80] also oxidize these diols specifically, all attacking the D(—) atom preferentially (see Chapter 1, Volume II).

The mechanism of the formation of 2,3-butanediol by microorganisms is not definitely established. However, from the work reviewed in the preceding pages, it appears likely that acetoin is first formed from pyruvate, or a closely related compound, and it is then reduced to 2,3-butanediol. Only one acetoin-forming enzyme may be present in bacteria since all those tested form (—) acetoin from pyruvate. Although acetoin has not been definitely isolated in an optically pure form, this may be due to enzymic racemization. However, it is necessary to postulate two acetoin reducing enzymes since, both (—,—) and (—,+) diols are commonly formed by bacteria. The simplest way to understand this is to postulate a 2,3-butanediol dehydrogenase reacting reversibly with D(—) atoms and another which acts reversibly with L(+) atoms. The formation of (+) acetoin by enzymes of higher plants makes it appear that at least four acetoin enzymes are found in nature; two which form acetoin and two which reduce it. However, it is possible that an enzyme exists which is capable of changing (—) acetoin into (+) acetoin, but which is incapable of synthesizing it from pyruvate. There is an interesting field here for further enzyme studies of fundamental importance in connection with the metabolism of pyruvic acid.

LABORATORY AND PILOT-PLANT STUDIES ON THE PRODUCTION OF 2,3-BUTANEDIOL

Many laboratory investigations have been conducted on the production of 2,3-butanediol from various substrates, some of which have been referred to previously. Several firms have carried out pilot-plant-scale fermentations for making of 2,3-butanediol, but few of their results have been recorded in the literature. Patent claims filed in Europe, Great Britain, the United States and Canada all give examples which indicate the use of pilot-plant equipment.

Fermentation Equipment

In order to obtain satisfactory results, butanediol fermentation must be carried out in modern, closed fermentation equipment. Owing to the sensitivity of most bacteria to traces of copper in the media, it is recommended to use only mild steel or stainless steel fermentors. The plant should be so designed that all pipes and vessels are readily sterilized by means of live steam under pressure. Valves and flange connections should be protected against contamination by steam pressure when not in use. Since all raw materials must be completely sterile, the use of a continuous high-pressure cooker[258] is desirable. If the proper alloy is used in the manufacture of the continuous cooker, it may also be employed for acid hydrolysis of grain mashes, such as was carried out by Joseph E. Seagram and Sons, Inc.[38,234,257] Installation of agitators in the fermentors is advisable for maintaining a constant and uniform temperature and to assist in proper aeration. Aerators should be mounted so that they can readily be removed for cleaning between fermentations as whole grain mashes have a tendency to plug them. Propagator equipment may be relatively small and must be so ar-

FIGURE 4. *Battery of 200-gal Stainless-Steel Experimental Fermentors in a Fermentation Pilot Plant* (Courtesy—Prairie Regional Laboratory, Saskatoon, Saskatchewan)

ranged as to eliminate possibilities of contamination during transfers. In general, much less inoculum is satisfactory in these fermentations than is necessary with yeasts in the production of industrial alcohol; inoculation with 0.1% by volume of active culture gives very satisfactory fermentations. A battery of three 200-gal stainless-steel fermentors in a pilot plant is shown in Figure 4.

Raw Materials

Most raw materials suitable for the production of industrial alcohol, as well as those containing pentose sugars, may be used for the manufacture of 2,3-butanediol. These may be classified into three main groups: (1) starchy substrates, including corn, wheat, barley, sweet potatoes, and potatoes; (2) sugars and sugar residues, such as beet and cane molasses; (3) wood hydrolyzates and pulp mill residues, including sulfite waste liquor. Bacteria used in the diol fermentation are much less tolerant of toxic substances than yeasts. Therefore, the use of very complex mixtures, such as sulfite waste liquor, or even cane or beet molasses, presents special problems. The most intensive studies to date have been made with whole ground cereal grains or degerminated corn and most pilot-plant data are based on the use of these materials. *Bacillus polymyxa* produces its own amylolytic enzymes which hydrolyze whole-grain mashes during fermentation. But if *Aerobacter aerogenes, Bacillus subtilis, Pseudomonas hydrophila,* or species of *Serratia* are used in the fermentation, the starch must first be converted to sugars. Conventional malting processes are not likely to be useful in saccharifying starchy substrates for these fermentations, since the introduction of contaminants is difficult to avoid. Acid hydrolysis has been employed most frequently, and studies have been carried out at Iowa State College using mold enzymes for the saccharification procedure.

Fermentation Processes

The 2,3-butanediol fermentations, with literature references to the substrates on which they were grown and conditions used, have been compiled[256] and are shown in Table 7. Relatively few of these publications refer to pilot-plant work, however.

TABLE 7. ORGANISMS PRODUCING 2,3-BUTANEDIOL

Organism	Substrate	Cultural condition	Literature references
Bacillus polymyxa	dextrose	anaerobic	5, 60, 61, 103, 210, 269
	sucrose		
(*Aerobacillus polymyxa*)	xylose	anaerobic	5
	mannitol	anaerobic	5
	grain mashes	anaerobic	1, 2, 3, 4, 20, 22, 102, 120, 121, 129, 130, 131, 212, 232, 253, 266, 273
	grain mashes	aerobic	2, 22, 102, 131, 253, 273
	wheat starch	anaerobic	2, 69, 102, 103, 104
	cornstarch	anaerobic	120, 121
Aerobacter aerogenes including synonyms	dextrose	anaerobic	2, 4, 39, 89, 177, 186, 216, 263
A. faeni	dextrose	aerobic	2, 90, 120, 186
A. pectinovorum	xylose	anaerobic	29, 30, 213
	sucrose	anaerobic	29, 30, 70, 72, 73, 106, 213
	corn mash[a]	aerobic	90
	corn mash[b]	aerobic	120
	grain mash[c]	aerobic	234, 235, 236, 237
	starch[c]	aerobic	264
	starch[c]	anaerobic	264
	beet and cane molasses		72
	wood sugars[c]	aerobic	193
Aerobacter cloacae	dextrose	anaerobic	87, 214, 247
	xylose	anaerobic	213
	sucrose	anaerobic	213
Aeromonas hydrophila	dextrose	anaerobic	231
	xylose	anaerobic	231
Bacillus subtilis	dextrose	anaerobic	21, 165, 166
	dextrose	aerobic	21, 166
Serratia marcescens and other species	dextrose	anaerobic	168, 192
	dextrose	aerobic	168

[a] Mashes saccharified by fungus.
[b] Mashes saccharified by malt.
[c] Materials hydrolyzed by acid.

Source: Underkofler and Fulmer.[256]

FERMENTATIONS ON UNTREATED WHOLE GRAIN OR STARCH MASHES

Since whole grain mashes contain their own nitrogen and mineral sources with sufficient accessory nutrient factors to maintain a vigorous fermentation, their use has been intensively studied, employing *Bacillus polymyxa,* without preliminary acid hydrolysis or enzymic saccharification. Examples have been given[266] of an anaerobic fermentation, using *Bacillus polymyxa* in a wheat mash, a ground-corn mash, and a sweet-potato mash. Fermentation was carried out at 30°C, the mash was neutralized with 1% calcium carbonate, and fermentation was completed after 48 hours. The yields were approximately 10 lb of diol and 6 to 8 lb of alcohol for a 56-lb bushel of corn. With wheat, approximately 9.4 lb diol and 8.6 lb alcohol per 60-lb bushel were obtained. Schenley Distillers[143] at Lawrenceburg, Ind., conducted extensive pilot experiments on the fermentation of whole-grain mashes with *Bacillus polymyxa*. They carried out fermentations, using several strains of *Bacillus polymyxa*, obtained from the Northern Regional Research Laboratory, National Research Council of Canada, and others isolated in their own laboratory. In a 200-bushel per day capacity plant, they found that approximately 9 lb of 2,3-butanediol per bushel of wheat could be produced. They also used 10 to 12% corn mashes, obtaining an optimum yield of 10 lb per bushel with an average of approximately 9 lb per bushel of 2,3-butanediol. The National Research Council of Canada[273] recently published results of pilot-plant investigations on the *Bacillus polymyxa* fermentation. A yield of 8.9 lb of (—)-2,3-butanediol, and 5.9 lb of ethanol, per 60-lb bushel of wheat, was obtained, which is a fermentation efficiency of 90%. The best fermentations were obtained at 32°C, using a 15% mash buffered with calcium carbonate. Agitation increased the fermentation efficiency at 96 hours, but decreased the diol to ethanol ratio. Reduced pressure did not change the efficiency but markedly reduced the ratio of diol to ethanol. Aeration had little effect in the large fermentors. In these pilot-plant experiments, it was found that adding inoculum at levels of 0.2, 0.1 and 0.05% gave completely satisfactory results. The use of this low amount of inoculum is of importance in plant design, as it eliminates large propagator equipment. On the above basis, only 10 to 20 gal of inoculum would be required for a 20,000 gal fermentor. In these pilot-plant studies, the inoculum was prepared from agar slants by three 24-hour stages. Transfers were made from slant to 10 ml of medium. This was employed to inoculate 200 ml of medium and the 200 ml culture to inoculate 2,000 ml, which was finally used to inoculate the 600 gal fermentation. The inoculum medium consisted of 1% ground wheat, 3% starch, 1.5% yeast extract and 1% calcium carbonate. On this medium, 500 million organisms per milliliter, as estimated by direct count with a

Petroff-Hauser plate, were obtained after an incubation period of 24 hours. Malt sprouts or shorts offer a good commercial substitute for yeast extract as the added nutrient. Where no nutrient is added, growth is slower and the numbers are much lower at 24 hours. The calcium carbonate may be omitted if the incubation period is not longer than 24 hours. Because the effect of aeration in large fermentors is less marked than on a laboratory scale, it was concluded that the beneficial effects of aeration in commercial practice would not justify the initial cost of installing air compressors and aeration equipment. This also applies to the application of reduced pressure. The installation of agitators would be advisable to facilitate removing the mash and for maintaining a constant and uniform temperature. If the cooked mash were to be cooled in the fermentors, agitation would also be necessary.

Barley mashes were also successfully fermented by *Bacillus polymyxa* on the pilot-plant scale.[253] It was found possible to ferment mashes up to 14.5% of concentration. The highest efficiency, 94%, was obtained with a 14% mash. The yield from the 48 lb bushel of barley was 6.9 lb of diol and 5 lb of alcohol. Comparison of the product yields of wheat and barley fermentations are given in Table 8. Estimates showed that on the basis of the long-term average grain prices in western Canada, a 10% cost reduction could be expected if barley, rather than wheat, was used as a raw material for this fermentation.

TABLE 8. YIELDS FROM WHEAT AND BARLEY FERMENTATIONS[a]

Products	Wheat	Barley
Butanediol recovered, gal	846	781
Alcohol recovered, gal 95%	675	610
Bran or hulls, 13.5% moisture, tons	1.89	5.12
Residue, 13.5% moisture, tons	12.20	9.53
Total residues, tons	14.09	14.65

[a] Basis: 30 tons a day—1,000 bu wheat or 1,250 bu barley at 13.5% moisture.
Source: Tomkins, Scott, and Simpson.[253]

While whole-grain mashes do not present any serious fermentation problems and have certain advantages in that they contain growth factors, nitrogen, and minerals, the large percentage of nonfermentable matter complicates subsequent diol-recovery operations. Furthermore, the nonstarch portions of the grain may be seriously degraded as a feed by the recovery treatments. For these reasons, many laboratories carried out experimental work on purified corn or wheat starch and also on granular wheat flour or "alcomeal," used for industrial-alcohol production. The laboratory investigations of additional nutrient requirements to obtain good fermentations of purified starch by *Bacillus polymyxa* have already been mentioned. Apparently no pilot-plant studies have been made with purified starch, although laboratory studies have indicated that cornstarch could be fermented satisfactorily with *B. polymyxa.* The best medium of Kooi, Fulmer, and Underkofler[121] contained 7.5 g commercial cornstarch, 0.5 g corn gluten, 0.006 g potassium permanganate and 0.5 g calcium carbonate per 100 ml. Yields from 200 ml fermentations on this medium were 26.8% diol and 14.8% ethanol, based on the weight of starch present. The efficiency of this fermentation was slightly lower than the best fermentations which may be obtained on whole grains. However, the total solids were reduced by 50%, resulting in greater ease of filtration.

Acid-Hydrolyzed Whole-Grain or Purified-Starch Mashes

In 1942 and 1943 quite extensive studies were made on the acid hydrolysis of whole grains for the 2,3-butanediol fermentation produced by *Aerobacter aerogenes.* A number of industrial firms, together with the Northern Regional Research Laboratory, the University of Wisconsin, and the University of Alberta in Canada, carried out laboratory and in some cases pilot-plant studies. Descriptions of the pilot-plant work, conducted by Joseph E. Seagram and Sons, Inc., have been published.[38,234,235,236] Many factors affecting the diol yields, mash preparations, bacterial cultures, nitrogen supplements, pH, temperature, aeration, sugar concentrations, and time of fermentation were investigated in laboratory studies and most of these confirmed by later pilot-plant work on corn and wheat mashes. The grain mash was acid hydrolyzed by a continuous process developed by the same firm.[257] Two different cultures were compared for their fermentation ability, No. 199 obtained from the Northern Regional Research Laboratory and No. 766 from the University of Wisconsin. The second gave higher yields and was not as sensitive to many factors as culture 199. In Table 9, the fermentation yields obtained with culture 766 on a number of pilot-plant runs are shown.

TABLE 9. PILOT-PLANT FERMENTATION YIELDS OBTAINED WITH CULTURE 766[a]

Run No.	Fermentor 1		Fermentor 2	
	Diol, lb per bu	Ethanol, lb per bu	Diol, lb per bu	Ethanol, lb per bu
91	13.1	2.69	13.8	1.51
92	13.3	2.69	13.1	1.51
93	14.9	1.56	13.4	0.57
94	13.0	2.39	13.1	1.05
95	14.2	2.35	15.7	1.24

[a] Fermentation medium: Acid-hydrolyzed wheat mash.
Source: Stark, Adams, Kolachov, and Willkie.[235]

The grain mash was acid hydrolyzed, neutralized to pH 6, and 0.5% calcium carbonate and 0.2% urea were added. Fermentations were conducted at 86 to 90°F, with agitation, and aerated at the rate of 5 cu ft per minute per 1,000 gal of mash. The average pilot-plant yields, as shown in Table 9 were approximately 14.2 lb of butanediol per bushel of grain; the maximum yield obtained was 15.7 lb. Small quantities of ethanol (approximately 0.75 to 1.5 lb per bushel) and also traces of acetoin (0.6 to 1.2 lb per bushel) were produced. The Seagram investigators concluded that a practical fermentation for the production of 2,3-butanediol from grain mash could be developed and that additional work ought to raise the yield to 15 or 16 lb per bushel. Wheat and corn were satisfactory raw materials.

The production of 2,3-butanediol by *Aerobacter aerogenes*[200] has been studied on a large laboratory scale, using apparatus designed for the acid hydrolysis of wheat mashes with agitation and another type of apparatus for hydrolysis without agitation.[186] This work was carried out in glass vessels. It was found that the stirring of wheat mashes reduced the time necessary for hydrolysis to one third that required with unstirred mashes. Yields on acid-hydrolyzed wheat were approximately 28 g of diol per 100 g of grain.

The Northern Regional Research Laboratory carried out work on the fermentation of acid-hydrolyzed whole grain (corn and wheat) mashes and also of acid-hydrolyzed purified corn and wheat starch.[264] The results on the last are the only ones which have so far been published, although it is mentioned that acid-hydrolyzed whole-grain mashes were found to yield 13 to 14 lb of butanediol per bushel of grain processed. From work with whole grain mashes the metallic ion content was not an important factor in determining the outcome of the fermentation. It was found, however, where purified starches were used, that certain heavy metals were toxic. This is probably due to the lack of proteins in the mash which would protect the organism against such substances. Traces of some metals play an important role in the production of 2,3-butanediol from acid-hydrolyzed starch. The presence of certain ions, particularly those of manganese, cobalt, or molybdenum, was reflected in a decreased production of diol and an increased production of ethanol under aerobic conditions and they also caused an increased rate of fermentation under anaerobic conditions. The balance between these trace elements seems to be the important factor. It was possible to treat the mashes either by an ion-exchange agent to remove certain of these elements and to bring them into balance or by the addition of traces of other metallic ions, such as copper and zinc. The second method is only applicable if the amounts of trace elements present in the starch are known so that they can be balanced. It is somewhat easier to treat the starch hydrolyzate with an ion-exchange column and remove certain ions. Acid-hydrolyzed starches fermented by *Aerobacter aerogenes* have given 13.5 to 14 lb of 2,3-butanediol per 34 lb of pure starch (equivalent of 1 bushel corn).[264] It has been found in anaerobic fermentations that all mashes produced large quantities of ethanol, the ratio butanediol to ethanol being approximately 2:1, i.e., very similar to that obtained with *Bacillus polymyxa* under the same conditions. With proper aeration, however, the yield of ethanol was reduced and much higher ratios were obtained. The effect of various growth factors, such as biotin, nicotinic acid, *p*-aminobenzoic acid, pyridoxin, inositol, thiamin, riboflavin, and sodium pantothenate were studied. No difference in the proportion of ethanol and diol were observed when aerated glucose cultures were supplied. Each liter of the saccharified starch medium contained 0.25 g $MgSO_4 \cdot 7H_2O$, 0.60 g KH_2PO_4, 2.0 g urea, and 5 to 10 g $CaCO_3$. The first two salts were added to the medium before cooking and the urea and calcium carbonate were each sterilized separately and added before inoculation. The quantity of calcium carbonate employed was sufficient to neutralize at pH 5.5 to 6 and also to act as a means of pH control during the fermentation. Each 3-l lot of medium was inoculated with 100 ml of an 8- to 20-hour inoculum culture. The fermentations were conducted in rotating aluminum drum fermentors, which were operated at 30°C,

10 to 12 rpm, 5 psi gage pressure, and an air flow of 50 to 100 ml per minute which had been found near optimal in previous studies of diol production from glucose.

ENZYME-HYDROLYZED WHOLE-WHEAT AND STARCH MASHES

Although many investigators undoubtedly have used malt to hydrolyze starch for butanediol fermentations, relatively few reports have been published on this process. Possibly this is due to the fact that use of the conventional malting process necessarily introduces many other organisms which interfere with the subsequent diol fermentation. If the partially converted mash is autoclaved prior to fermentation, extra equipment, steam, and water is required and there is a 20% reduction in yield as well.[236] The production of 2,3-butanediol by *Aerobacter* from corn has been studied, employing malt saccharification to determine whether brewery equipment and technique could be adapted to the fermentation.[120] The ground corn was mashed with water, cooked, and saccharified by means of barley malt. The solids were removed by filtration to obtain a clear wort which contained the equivalent in reducing sugars of about 90% of the starch present in the corn and the malt employed. Urea and calcium carbonate slurries were sterilized separately and added to the wort before fermentation. These experiments showed that in a wort containing up to 12% reducing sugar as maltose, fermentation was satisfactory, but with higher sugar concentrations decreased yields were obtained. It was necessary to use aeration for rapid fermentation and yields of the diol obtained were about 20% of the weight of maltose present in the wort. By the addition of freshly prepared malt extract to the medium after an active fermentation had started, it was possible to obtain better fermentations with yields as high as 38% of the weight of the initial maltose in the wort. In pilot-plant fermentations, at least 35% was obtained.

In work carried out in the Department of Soils at the University of Alberta, in 1943, malt was used to hydrolyze whole wheat and starch mashes.[19] Low sugar concentrations were used in most of the experimental work and excellent fermentations were obtained in a number of cases. However, the use of a malt-hydrolyzed medium did not appear to be feasible in larger-scale work, owing to the difficulty of obtaining high sugar concentrations and good sugar utilization by the organism.

In an investigation on starch hydrolysis by *Endomycopsis fibuliger* at the Northern Regional Research Laboratory,[275] mixed cultures of this yeast and *Aerobacter aerogenes* were used for the production of 2,3-butanediol from 15% wheat mash. A tube, containing 100 ml of 15% wheat mash, was inoculated with 2 ml of the cell suspension of a strain of *E. fibuliger* and simultaneously with 5 ml of a 20-hour culture of *A. aerogenes,* grown in a 5% malt-converted wheat mash. Before inoculation, each tube received 1 ml of sterile 20% urea solution to provide the necessary nitrogen for *A. aerogenes.* All cultures were incubated at 38°C for 72 hours. Two cultures received 10 ml of air per minute for 72 hours and the other two cultures were not aerated. It was found that the fermentation in unaerated cultures was much better than usual. Apparently, the presence of the yeast, or the liberation of some nutrient factor, stimulated the development and action of *Aerobacter,* especially in regard to the production of ethanol. The best yields obtained were 27% of 2,3-butanediol and 13% of ethanol, based on the weight of the starch present.

Corn mashes can be fermented satisfactorily by *A. aerogenes,* following saccharification by *Aspergillus oryzae.*[90] From a medium containing 15% corn meal, 0.4% dipotassium phosphate, and 1% calcium carbonate, up to 38% of 2,3-butanediol was obtained, based on the weight of the starch present. To date, no work has been reported with enzymes produced by submerged cultures of *Aspergillus niger* for saccharification of mashes prior to fermentation with any of the butanediol-producing bacteria. Investigations are now under way in the Prairie Regional Laboratory at Saskatoon, Saskatchewan, to determine the usefulness of these enzymes in the *Bacillus subtilis* fermentation for 2,3-butanediol and glycerol production.

Studies on Purified Sugar and Beet and Cane Molasses

Laboratory investigations at many centers have shown that high yields of butanediol can be obtained with aerated cultures of *Aerobacter aerogenes* grown on dextrose solutions. Sugar solutions up to 10% concentration can be fermented in less than 48 hours with a yield of 80 millimols of glycol per 100 millimols of glucose under optimum conditions.[186] If anaerobic cultures of this organism are used, the yield is approximately half this amount. No pilot-plant experiments on the fermentation of dextrose have been reported, though undoubtedly diol was produced for recovery and chemical studies by this method. Several laboratory studies have been made on the fermentation of sucrose, largely as preliminary work to the utilization of beet or cane molasses. The most detailed studies reported appear to be those from the Imperial Chemical Industries in Britain.[70,71] Detailed investigations were made on the conversion of sucrose to 2,3-butanediol with *A. aerogenes,* investigating pH, sucrose concentration, optimal temperatures, aeration, and the addition of organic nutrients. The pH optimum range for diol production was found to be 5 to 5.5 and the optimum temperature 35°C. Rapid fermentation and high diol yields were obtained with an initial sucrose concentration in the range of 5 to 20%. A maximum yield of 87.4% of theory was obtained with a sucrose concentration of 15%. In aerated, stirred cultures, containing 10% sucrose with 1.2% calcium carbonate, rapid fermentation began after a lag period of 5 to 7 hours, and was complete in 24 hours. It was found that with unaerated cultures, high yields of ethanol, amounting to 12 to 14% of the sugar fermented, and of lactic acid (5 to 10%) were obtained. However, under optimum conditions approximately 85% of theory of 2,3-butanediol and relatively small concentrations of ethanol were obtained. Different types of molasses were investigated, including high-test molasses, crude-beet molasses, South African blackstrap molasses, West Indian blackstrap molasses, home-grown beet molasses, and crude commercial glucose, manufactured from corn. A rapid, complete fermentation of blackstrap molasses took place under suitable conditions, leading to 75 to 80% of theoretical yield of diol. Rates of fermentation compared unfavorably with purer substrates, such as sucrose or high-test molasses, but by using acclimatized strains of the organism, both fermentation time and yields could be markedly improved. Fermentations on a 12.5 l scale were carried out successfully and there is good reason to expect that this fermentation can be put into large-scale practice. In addition to the diol, considerable amounts of ethanol, lactic acid, and acetoin were formed under certain conditions. The relative amounts of these products, particularly the diol to ethanol ratios, depended on aeration and on protein and trace-metal concentrations of the medium.

The National Research Council of Canada has recently carried out pilot-plant studies on the production of 2,3-butanediol from Ontario sugarbeet molasses. Some twelve or fifteen 800-gal fermentations have now been run, with automatic pH control in most of them. It was found that the addition of phosphate increased the rate and, in common with all rate increasing factors, decreased the diol to alcohol ratio. By decreasing the amount of phosphate and increasing the aeration rate, it was possible to get a ratio greater than 3:1 and still complete the fermentation in less than 72 hours. Many of these fermentations were completed within 36 to 48 hours, 95 to 98% of the sugar having been utilized with a total fermentation efficiency of 90 to 95%. It is apparent that there would be little difficulty in obtaining a commercially feasible 2,3-butanediol molasses fermentation using *Aerobacter aerogenes.*

If it is desirable to produce *levo*-2,3-butanediol, it will be necessary to use *Bacillus polymyxa.* So far only preliminary pilot-plant work has been carried out with this organism on Ontario beet molasses, but fermentation of pure sucrose under controlled pH conditions has been studied recently.[269] It was found that the use of 8% sucrose gave the most efficient fermentation in 30 hours at pH 6, yielding 65 millimols of diol per 100 millimols of invert sugar fermented. This laboratory investigation has been discussed before. Ammonium hydroxide was found to be much superior to either sodium or potassium hydroxide as neutralizing agent, because of its effect on the rate of sugar dissimilation.

Recently, Simpson and Stranks[228a] have reported a laboratory study on some factors affecting the fermentation of beet molasses by *B. polymyxa.* Strains selected on the basis of their ability to ferment glucose to 2,3-butanediol gave better diol yields from beet molasses when acclimatized to beet molasses medium. Best yields were obtained by adjusting the molasses media to pH 5.6 prior to sterilization since this gave the optimum pH of 6.2, after autoclaving at 15 psi gage pressure for 15 minutes. Most rapid fermentation and best diol yields were obtained by agitation and aeration of the medium. Addition of organic stimulants to molasses medium resulted in more rapid fermentations and increased final diol yields in several cases. Yeast extract, malt sprouts, whole wheat and wheat bran were the most effective. Wheat bran was adopted as the cheapest adjuvant, and was tested at several levels. With the molasses sample employed, fermentation of 10% molasses medium containing 0.11% wheat bran at 35°C, with agitation and aeration, was complete in 60 to 72 hours and yielded 2.04 g diol plus 0.88 g ethanol per 100 ml of medium.

Subsequently Murphy, Stranks, and Harmsen[154b] reported that *B. polymyxa, A. aerogenes, B. subtilis,* and *Serratia marcescens,* when cultivated on molasses media, were stimulated in growth and butanediol production by adding organic substances, such as wheat bran, yeast extract or corn-steep liquor. It was found that the principal beneficial factor in all these substances was the phosphorus content. Beet molasses was found to be an excellent substrate for butanediol production by all these organisms when sufficient soluble phosphate salts were added. It is necessary to add the phosphorus as orthophosphate, or in a form which readily yields orthophosphate. However, the natural organic substances contained, in addition to phosphate, some minor unidentified factors. Addition of wheat bran to a molasses medium containing added phosphate resulted in a small but noticeable improvement in 2,3-butanediol yield when fermented by *B. polymyxa.*

Other studies have also been made in the Canadian Research Council on the fermentation of molasses with *Bacillus subtilis* and *Pseudomonas hydrophila* for making butanediol. These investigations have not yet been carried out on the pilot-plant scale, but there is every reason to expect that most of the butanediol-producing bacteria can be satisfactorily acclimatized to the utilization of beet and cane molasses.

Most recently, Murphy, Watson, Muirhead, and Barnwell[154c] have reported extensive study of beet-molasses fermentation by two strains of *Pseudomonas hydrophila.* These strains differed from those employed by Stanier and Adams[231] in that they produced 2,3-butanediol containing a high percentage of racemic mixture, gave higher yields of diol, and fermented sugars more rapidly and completely. Dissimilation of molasses sugars was practically complete in 24-hour fermentations in molasses media containing 5 and 7.5% sugar, and much slower and incomplete when 10% sugar was present. The media contained, in addition to molasses, ammonium phosphate and wheat bran to supply minor nutrients. Yields of products from numerous fermentations, conducted with agitation and aeration at 32°C, and without pH control, averaged 87.8% of theoretical diol plus acetoin, 13.1% ethanol, and 2.2% organic acids. Fermentations by high-yielding strains of *A. aerogenes* on media containing the same molasses averaged 82.9% of theoretical diol plus acetoin, 19.8% ethanol and 4.5% organic acids. Aeration had less tendency to convert diol to acetoin in the *P. hydrophila* fermentation. It, therefore, appears that *P. hydrophila* produces a higher yield of diol plus acetoin than *A. aerogenes*. In two fermentations, with the initial pH adjusted to 6.2, approximately 95% of theoretical diol and no ethanol were produced. The mixture of stereoisomeric 2,3-butanediols produced by *P. hydrophila* by physical methods was calculated to contain 50% *racemic,* 48% *meso* and 2% *levo.*

Fermentations on Hydrolyzates of Wood or Agricultural Residues

Since the 2,3-butanediol-producing bacteria ferment xylose and mannitol[119,136,139,144] as well as glucose, fructose, and sucrose, the sugars derived from different cellulosic materials, such as wood waste and agricultural residues, ought to be considered as possible raw materials. Workers in the Northern Regional Research Laboratory[63] have used *Aerobacter aerogenes* to ferment saccharified agricultural residues, such as corncobs, bagasse, oat hulls, and flax shives, but no details of these experiments have as yet been published.

Hydrolyzates of both hard and soft woods, including southern red oak, Douglas fir, white spruce, and southern yellow pine, have been used as substrates for the production of 2,3-butanediol.[193] Two types of treatment of the wood sugars, prior to fermentation, were used. In one, the solutions were neutralized to pH 6 by addition of a slurry of lime or alkali, after which the precipitated calcium sulfate was removed by filtration. The second method involved the addition of excess alkali up to pH 10 or 11, filtration, and then readjustment with sulfuric acid to pH 6 and filtering. Dilute solutions from the first treatment fermented well, but addition of malt sprouts or yeast extract was necessary on more concentrated solutions. The second method of treatment gave good fermentations in solutions containing up to 17.8% sugars. Approximately 35% of the fermented sugar was converted to 2,3-butanediol. It was possible to acclimatize the organism to wood-hydrolyzate media by repeated transfers. With acclimatized cultures, improved yields were obtained on wood-hydrolyzate media, without addition of auxiliary substances or treatment involving more than direct adjustment of the pH to 6.0.

Recent Fermentations

Since this chapter was first written, a number of papers have appeared confirming and extending previous work on the production of 2,3-butanediol from various substrates. These include work on the fermentation of waste sulfite liquor,[154a] cellulosic materials,[58a] and wheat,[48a] as well as further work on the fermentation of glucose by *Aerobacter aerogenes*[187a] and *Bacillus mesentericus*.[9a]

CONTAMINANTS AFFECTING 2,3-BUTANEDIOL FERMENTATIONS

Bacteriophage and foreign spore-forming bacteria are the most important contaminants likely to affect the 2,3-butanediol fermentation. The most frequent cause of contamination in pilot-plant fermentations of wheat arises from insufficient sterilization of the whole-wheat mash, rather than from the entrance of foreign organisms into the system after sterilization.[21] Most of the isolated contaminating organisms were spore formers. It has been found that No. 1 Manitoba Northern Wheat has the lowest bacterial count, but in lower grades, counts ranged from 280,000 to 164,000,000 for bacteria, 420 to 1,870 for fungi, and 6,200 to 64,000 for yeasts per gram of grain.[99] Experiments on the use of ultraviolet germicidal lamps made it possible to reduce cooking time by 15 minutes, but recontamination makes this process unlikely to be of much commercial value. The effect of bacteriophage has been studied in relation to diol production,[105] but no relationship has been found between the ability of strains of *B. polymyxa* to produce diol and their susceptibility to bacteriophage. It was found that bacteriophage could be carried through the spores of *B. polymyxa* and that it was quite heat resistant, since a temperature of 80°C for 60 minutes did not inactivate it.

Workers at Joseph E. Seagram and Sons, Inc., have also investigated the bacteriophage present in *Aerobacter aerogenes* fermentations. Standard methods of phage isolation and culture failed to demonstrate the presence of phage conclusively, despite the fact that positive indications of this type of infection were obtained. It was found that the addition of mash and calcium carbonate to the tubes cleared by phage activity of the original filtrate before filtering enabled active cultures of the phage to be obtained.

RECOVERY PROCESSES AND EQUIPMENT

Because of its high boiling point, 2,3-butanediol is not easily recovered from fermented mashes. The initial steps in all suggested recovery methods are similar and consist of removal of ethanol, filtration or screening, and concentration of the liquor by evaporation.[241] Ethanol, which is a by-product of the fermentation, can be recovered either by a separate operation or during evaporation. Methods which have been suggested for the recovery of 2,3-butanediol from evaporator sirup are: spray drying,[143] drum drying,[127] chemical conversion,[78,225] dialysis,[47] solvent extraction,[94,127,137] ethanol precipitation,[34] kerosene distillation,[127] and steam stripping.[113,246] Only those considered most promising will be described here.

Spray Drying

A recovery process for fermentation residues, containing high-boiling constituents, such as glycerol, butylene glycols, ethylene and propylene glycols, was patented[51] in 1942. The method consists of spraying the fermentation residue into a stream of inert gas at a temperature of approximately 900°F, to provide vaporized volatile products and separating the solids from the gas stream while the temperature is above the point at which condensation of volatile products can occur, then cooling the stream, and condensing the products to be recovered. In spray-drying of 2,3-butanediol the beer is concentrated by evaporation to about one fourth to one fifth its original volume and the heated concentrate is flashed into a vacuum chamber.[143] The vapors consisting of steam and diol are directly fractionated and the diol collected separately. The residual solids from this process are used as animal feed.

Chemical Conversion

A chemical process for the recovery of 2,3-butanediol from fermentation beers has been described.[225] The method consisted of distilling an acidified mixture of whole beer and formaldehyde into a decanter which permitted return of the lower aqueous layer to the distilling vessel. The top oil layer, which was collected, contained, as the formal, 98 to 100% of the butanediol present in the beer. The 2,3-butanediol formal was then resolved into two geometrical isomers. The formal was allowed to react with acid methanol and converted to 2,3-butanediol and methylal in 95% conversion. A method was developed for the recovery of formaldehyde and methanol from the methylal. A similar process, using butyraldehyde as the solvent, has recently been developed by the National Research Council of Canada.[248] In this process, butyral is formed, which is then separated by distillation, and later 2,3-butanediol can be recovered by acid treatment.

Solvent Extraction

In the laboratory, 2,3-butanediol can be readily recovered from fermentation beers by continuous extraction with diethyl ether.[18,69,100,118,159] Extraction by means of *n*-butyl alcohol has been employed in several pilot-plant installations. Several ternary liquid systems have been investigated, including diol, water, and individual solvents which might be suitable, such as butyl alcohol, butyl acetate, butylene glycol diacetate, and methylvinylcarbinol acetate.[188] This experimental work allowed the investigation of respective advantages of the solvents and the desirable operating temperatures.

Steam Stripping

Possibly, the most practical method of recovering 2,3-butanediol from fermentation sirups is through vapor-phase steam extraction. This method has been successfully employed on pilot-plant scale by both the Northern Regional Research Laboratory and the National Research Council of Canada. A process for the recovery of *meso*-2,3-butanediol has been described in which the concentrated liquor is steam stripped in a packed column at elevated temperatures.[25] A similar process has been successful, using a whole-wheat mash fermented by *Bacillus polymyxa*.[273] The mash was stripped of ethanol, screened and concentrated, and the resulting sirup was then passed through a steam-stripping column and the *levo*-2,3-butanediol in the concentrate was recovered by rectification. The composition of intermediate products, performance of pilot-plant equipment, and data and recommendations for the design of a large-scale plant were worked out.

RESIDUES AND WASTES

It is unlikely that the fermentation of whole-grain mashes for the production of 2,3-butanediol will be able to compete with a molasses fermentation, unless the residues can be recovered in the form of a high-grade protein feed. Very little experimental work has been carried out to determine whether the organisms employed in the fermentation produce sufficient riboflavin or other vitamins to enhance the feeding qualities of the nonfermentable portions of the grain. Analysis of the residues obtained from fermentation of whole-wheat mashes by *Bacillus polymyxa* in the National Research Council of Canada[273] are shown in Table 10, with data for other fermentation residues given for comparison. The ash content of of the stripper residue was high, because it contained all the calcium carbonate added to the mash. The butanediol content of these residues is higher than necessary, since it could be reduced by more efficient stripping of the stripper residue or by more complete washing of the bran prior to drying. No thiamin was present in any of the solids, but the riboflavin content of the stripper residue represents approximately 100%, and the nicotinic acid content

TABLE 10. ANALYSES OF PILOT-PLANT BUTANEDIOL WHEAT MASH AND COMMERCIAL RESIDUES

	Dried stripper residue		Dried screen solids	Distillers' grains from rye	Brewers' grains over 25% protein	Strained distillery slop[a]
	No. 1	No. 2				
Butanediol, %	1.27	1.58	0.53			
Moisture, %	8.42	11.8	10.7	7.2	7.5	10.0
Soluble solids, %	59.4	55.6	0.0			
Ash, %	11.7	13.5	1.65	3.9	3.5	6.6
Total nitrogen, %	5.59	6.02	1.21	3.7	4.24	4.91
Water-soluble nitrogen, %	4.05	4.40	0.21			
Total P_2O_5, %	2.33	2.96	1.33	0.83	0.99	
Available P_2O_5, %	2.21	2.82	1.13			
Total K_2O, %	0.11	0.10	0.07	0.24	0.09	
Total CaO, %	5.4	5.5	0.74		0.16	
Crude protein, % $N \times 6.25$	34.0	37.6	7.56	23.1	26.5	30.7
Fat, %	5.1	5.0	3.4	7.8	6.9	15.4
Fiber, %	2.0	2.3	27.0	10.9	14.6	4.4
Nitrogen-free extract, %	37.9	29.8	49.5	47.1	41.0	32.9
Thiamine, ppm	0.0		0.0			
Riboflavin, ppm	1.64		1.3			
Nicotinic acid, ppm	88		6.4			

[a] Calculated on a 10% moisture basis.

about 50%, of that present in the wheat, although the residue was only 30% of the original weight of wheat used. The bran contained about 50% of the riboflavin and 100% of the nicotinic acid usually present in untreated wheat brans.

The stripper residue appears to have possibilities as a livestock feed, though feeding tests were not carried out. It is higher in protein and minerals than other commercial residues with which comparsons were made, but somewhat lower in fat and fiber content. The dried bran would be of little value as a feed other than roughage. Neither the stripper residue nor the bran would be useful as a fertilizer without being enriched with phosphorus or potash.

COST OF PRODUCTION

Because none of the 2,3-butanediol fermentations has ever been applied on an industrial scale, very little detailed information is available on which to base costs of production. An estimate, based on pilot-plant operations by Schenley Distillers,[143] indicates that butanediol may be produced at a price of 10 to 12¢ per pound. Calculations from the Northern Regional Research Lavoratory,[127] based on grain at 70¢ per bushel and a diol yield of 9.86 lb per bushel in an 8,500 bushels per day plant gave a cost of 10.62¢ per pound. Both estimates are appreciably lower than those based on pilot-plant studies of the *Bacillus polymyxa* fermentation carried out in the national research Council of Canada.[253,273] An analysis of costs and credit distribution using wheat is shown in Table 11

TABLE 11. COST AND CREDIT DISTRIBUTION FOR BUTANEDIOL PRODUCTION

Distribution	Percentage of total	Cents per imperial gal
Costs		
Fixed costs	14.30	39.10
Wheat	47.70	130.10
Other raw materials	5.46	14.89
Steam	8.15	22.20
Wages	13.00	35.50
Other production costs	11.39	31.13
	100.00	272.83
Credits		
Butanediol	63.8	173.15
Ethanol	17.2	47.88
Dried solids	19.0	51.80
	100.00	272.83

Source: Wheat, Leslie, Tomkins, Milton, Scott, and Ledingham.[273]

and a comparison of costs of wheat and barley fermentations in Table 12. With present prices of cereals, cost of production from grains is unduly high and there is little prospect of commercial development until cheaper sources of raw materials become available. Undoubtedly *(meso-dextro)*-2,3-butanediol could be produced by *Aerobacter aerogenes* on molasses for considerably less per pound but no cost analyses have been published.

TABLE 12. COSTS FOR WHEAT AND BARLEY FERMENTATIONS[a]

Production costs and credits	Wheat		Barley	
	Present	Long-term average	Present	Long-term average
Costs	dollars	dollars	dollars	dollars
Grain[b]	1,270.00	870.00	1,025.00	637.50
Utilities (steam, power, water)	265.00	265.00	261.00	261.00
Other costs (wages, fixed costs, etc.)	941.50	941.50	941.50	941.50
Total daily cost	2,476.50	2,076.50	2,227.50	1,840.50
Credits	dollars	dollars	dollars	dollars
Ethanol at $0.60 per gal	405.00	405.00	366.00	366.00
Residues at $30.00 per ton	422.70	422.70	439.50	439.50
Butanediol	1,648.80	1,248.80	1,422.00	1,035.00
	cents	cents	cents	cents
Cost butanediol per lb	19.5	14.8	18.2	13.3

[a] Basis: 30 tons a day—1,000 bu wheat or 1,250 bu barley at 13.5% moisture.
[b] Wheat—present price, $1.27 per bu; long-term average, $0.87 per bu.
Barley—present price, $0.82 per bu; long-term average, $0.51 per bu.
Source: Tomkins, Scott, and Simpson.[253]

BIBLIOGRAPHY

1. Adams, G. A., *Can. Chem. Process Inds.,* **27**, 390 (1943).
2. Adams, G. A., *Can. J. Research,* **24F**, 1 (1946).
3. Adams, G. A., and J. D. Leslie, *Can. J. Research,* **24F**, 12 (1946).
4. Adams, G. A., and J. D. Leslie, *Can J. Research,* **24F**, 107 (1946).
5. Adams, G. A., and R. Y. Stanier, *Can J. Research,* **23B**, 1 (1945).
6. Akabori, S., *J. Chem. Soc. Japan,* **59**, 1132 (1938).
7. Antoniani, C., and S. Gugnoni, *Biochim. terap. sper.,* **28**, 7 (1941).
8. Antoniani, C., and S. Gugnoni, *Ann. chim. applicata,* **31**, 417 (1941).
9. Antoniani, C., S. Gugnoni, and P. Scrivani, *Biochim. terap. sper.,* **28**, 119, 143 (1941).

9a. Asai, T., Y. Ikeda, and T. Sasaki, *J. Agr. Chem. Soc. Japan,* **24**, 101 (1951).
10. Backer, H. J., *Rec. trav. chim.,* **55**, 1036 (1936).
11. Barnicoat, C. R., *Analyst,* **60**, 653 (1935).
12. Barritt, M. M., *J. Path. Bact.,* **42**, 441 (1936).
13. Barritt, M. M., *J. Path. Bact.,* **44**, 679 (1937).
14. Baumann, E., *Klin. Wochschr.,* **17**, 382 (1938).
15. Berg, R. L., and W. W. Westerfeld, *J. Biol. Chem.,* **152**, 113 (1944).
16. Berger, J., *Physiol. expt. pharmakol.,* **101**, 633 (1937).
17. Bergey's *Manual of Determinative Bacteriology,* 6th ed., Baltimore, Williams and Wilkins, 1948.
18. Birkinshaw, J. H., J. H. V. Charles, and P. W. Clutterbuck, *Biochem. J.,* **25**, 1522 (1931).
19. Blackwood, A. C., M. S. Thesis, University of Alberta, 1944.
20. Blackwood, A. C., and G. A. Ledingham, *Can. J. Research,* **25F**, 180 (1947).
21. Blackwood, A. C., A. C. Neish, W. E. Brown, and G. A. Ledingham, *Can, J. Research,* **25B**, 56 (1947).
22. Blackwood, A. C., J. A. Wheat, J. D. Leslie, G. A. Ledingham, and F. J. Simpson, *Can. J. Research,* **27F**, 199 (1949).
23. Blom, R. H., and A. Efron, *Ind. Eng. Chem.,* **37**, 1237 (1945).
24. Blom, R. H., G. C. Mustakas, A. Efron, and D. L. Reed, *Ind. Eng. Chem.,* **37**, 870 (1945).
25. Blom, R. H., D. L. Reed, A. Efron, and G. C. Mustakas, *Ind. Eng. Chem.,* **37**, 865 (1945).
26. Böeseken, J., and R. Cohen, *Rec. trav. chim.,* **47**, 839 (1928).
27. Bourns, A. N., and R. V. V. Nicholls, *Can. J. Research,* **25B**, 80 (1947).
28. Bradley, T. F., U. S. Patent 1,890,668 (1932).
29. Breden, C. R., Ph.D. Thesis, Iowa State College, 1930.
30. Breden, C. R., and E. I. Fulmer, *Iowa State Coll. J. Sci.,* **5**, 133 (1931).
31. Brewer, C. R., M. N. Mickelson, and C. H. Werkman, *Arch. Biochem.,* **15**, 379 (1947).
32. Brewer, C. R., C. H. Werkman, M. B. Michaelian, and B. W. Hammer, *Agr. Expt. Sta., Research Bull. No. 233* (1938).
33. Brockmann, M. C., L. Manna, and W. H. Stark, Unpublished work, Joseph E. Seagram and Sons, Inc., Louisville, Ky.
34. Brockmann, M. C., and C. H. Werkman, *Ind. Eng. Chem., Anal. Ed.,* **5**, 206 (1933).
35. Brown, R. W., G. L. Stahly, and C. H. Werkman, *Iowa State Coll. J. Sci.,* **12**, 245 (1938).

36. Calbert, H. E., and W. V. Price, *J. Dairy Sci.*, **32**, 515 (1949).
37. Calbert, H. E., and W. V. Price, *J. Dairy Sci.*, **32**, 521 (1949).
38. Callaham, J. R., *Chem. Met. Eng.*, **51**, No. 11, 94 (1944).
39. Canepa, D. R., and C. S. de la Serna, *Folio Biol.*, **1**, 238 (1935).
40. Chappell, C. H., Ph.D. Thesis, Iowa State College, 1935.
41. Claudon, E., and E. C. Morin, *Compt. rend.*, **104**, 1109 (1887).
42. Clendenning, K. A., *Can. J. Research*, **24B**, 269 (1946).
43. Clendenning, K. A., *Can. J. Research*, **24F**, 249 (1946).
44. Clendenning, K. A., Canadian Patent 446, 755 (1948).
45. Clendenning, K. A., and D. E. Wright, *Can. J. Research*, **24F**, 287 (1946).
46. Coghill, R. D., and R. T. Milner, U. S. Patent 2,369,435 (1945).
47. Cornwell, R. T. K., U. S. Patent 2,390,779 (1945).
48. Cox, G. A., and W. J. Wiley, *J. Council Sci. Ind. Research*, **12**, 227 (1939).
48a. Crewther, W. G., *Australian J. Applied Sci.*, **1**, 437, 447, 468 (1950).
49. Davies, W. L., *Dairy Ind.*, **1**, 165 (1936).
50. Denivelle, L., *Compt. rend.*, **208**, 1024 (1939).
51. Dennis, W., U. S. Patent 2,375,288 (1945).
52. Dentice di Accadia, F., *Ann. Inst. Pasteur*, **73**, 1114 (1947).
53. De Simo, M., and S. H. McAllister, U. S. Patent 2,043,950 (1936).
54. Desnulle, P., and M. Naudet, *Bull. soc. chim.*, **12**, 871 (1945).
55. Diemair, W., and J. Kleber, *Z. Untersuch. Lebensm.*, **81**, 386.
56. Dirscherl, W., *Z. physiol. Chem.*, **188**, 225 (1930).
56a. Dirscherl, W., and H. Hofermann, *Biochem. Z.*, **322**, 237 (1951).
57. Dirscherl, W., and A. Schöllig, *Z physiol. Chem.*, **252**, 53 (1938).
58. Dirscherl, W., and A. Schöllig, *Z. physiol. Chem.*, **252**, 70 (1938).
58a. Doi, S., *J. Agr. Chem. Soc. Japan*, **24**, 215 (1951).
59. Doisy, E. A., Jr., and W. W. Westerfeld, *J. Biol. Chem.*, **149**, 229 (1943).
59a. Dolin, M. I., and I. C. Gunsalus, *J. Bact.*, **62**, 199 (1951).
60. Donker, H. J. L., *Tijdschr. Vergel. Geneeskunde*, **11**, 78 (1924).

61. Donker, H. J. L., Ph.D. Dissertation, Delft, Netherlands, 1926.
62. Dudykina, N. V., *Mikrobiologiya,* **18**, 181 (1949).
63. Dunning, J. W., and E. C. Lathrop, *Ind. Eng. Chem.,* **37**, 24 (1945).
64. Eggleton, P., S. R. Elsden, and N. Gough, *Biochem, J.,* **37**, 526 (1943).
65. Elion, L., *Biochem. Z.,* **169**, 471 (1926).
66. Englund, B., *J. prakt. Chem.,* **129**, 1 (1931).
67. Fabricius, N. E., *Dairy Inds.,* **5**, 159 (1940).
68. Fleury, P., and J. Lange, *J. pharm. chim.,* **17**, 107, 196 (1933).
69. Fratkin, S. B., and G. A. Adams, *Can. J. Research,* **24F**, 29 (1946).
70. Freeman, G. G., *Biochem. J.,* **41**, 389 (1947).
71. Freeman, G. G., and R. I. Morrison, *Analyst,* **71**, 511 (1946).
72. Freeman, G. G., and R. I. Morrison, *J. Soc. Chem. Ind.,* **66**, 216 (1947).
73. Fulmer, E. I., L. M. Christensen, and A. R. Kendall, *Ind. Eng. Chem.,* **25**, 798 (1933).
74. Fulmer, E. I., and L. A. Underkofler, *Iowa State Coll. J. Sci.,* **21**, 251 (1947).
75. Garino-Canina, E., *Ann. chim. applicata,* **23**, 14 (1933).
76. Gibson, T., *J. Dairy Research,* **13**, 248 (1944).
77. Green, D. E., W. W. Westerfeld, B. Vennesland, and W. E. Knox, *J. Biol. Chem.,* **145**, 69 (1942).
78. Greenberg, L. A., *J. Biol. Chem.,* **147**, 11 (1943).
79. Grimbert, L., *Compt. rend.,* **132**, 706 (1901).
80. Grivsky, E., *Bull. soc. chim. Belg.,* **51**, 63 (1942).
81. Gross, N. H., and C. H. Werkman, *Arch. Biochem.,* **15**, 125 (1947).
82. Grubb, H. W., *Chem. Eng. Progress,* **43**, 437 (1947).
83. Guittonneau, M. G., M. Bejambes, and J. Tavernier, *Ann. fermentations,* **6**, 159 (1941).
84. Gutner, R. A., and D. V. Tishchenko, *J. Gen. Chem. (U.S.S.R.),* **6**, 1729 (1936). (See Universal Oil Products Co. Survey of Foreign Petroleum Literature, Translation S 307A.)
85. Hale, W. J., and H. Miller, U. S. Patent 2,400,409 (1946).
86. Hammer, B. W., G. L. Stahly, C. H. Werkman, and M. B. Michaelian, *Iowa Agr. Expt. Sta., Research Bull.* No. 191 (1935).
86a. Happold, F. C., and C. P. Spencer, *Biochim. et Biophys. Acta,* **8**, 543 (1952).

87. Harden, A., and D. Norris, *Proc. Roy. Soc. (London),* **84B**, 492 (1912).
88. Harden, A., and D. Norris *Proc. Roy. Soc. (London),* **85B**, 73 (1913).
89. Harden, A., and G. S. Walpole, *Proc. Roy. Soc. (London),* **77B**, 399 (1906).
90. Hendlin, D., M. S. Thesis, Iowa State College, 1943.
91. Henninger, A., *Compt. rend.,* **95**, 94 (1882).
92. Higashi, T., *Sci. Papers Inst. Phys. Chem. Research (Tokyo),* **33**, No. 727, 1 (1937).
93. Hill, R., and E. Isaacs, British Patent 483,989 (1938).
94. Hooreman, M., *Compt. rend.,* **222**, 1257 (1946).
95. Hooreman, M., *Compt. rend.,* **225**, 208 (1947).
96. Hooreman, M., Thesis, Paris, 1948.
97. Horovitz-Vlassova, L. M., and E. A. Rodionova, *Zentr. Bakt. Parasitenk.,* II Abt., **87**, 333 (1933).
98. Hughes, E. B., and R. F. Smith, *J. Soc. Chem. Ind.,* **68**, 322 (1949).
99. James, N., J. Wilson, and E. Stark, *Can. J. Research,* **24C**, 224 (1946).
100. Johnson, M. J., *Ind. Eng. Chem., Anal. Ed.,* **16**, 636 (1944).
101. Johnson, M. J., W. H. Peterson, and E. B. Fred, *J. Biol. Chem.,* **101**, 145 (1933).
101a. Juni, E., *J. Biol. Chem.,* **195**, 715, 727 (1952).
102. Katznelson, H., *Can. J. Research,* **22C**, 235 (1944).
103. Katznelson, H., *Can. J. Research,* **24C**, 99 (1946).
104. Katznelson, H., *Can. J. Research,* **25C**, 129 (1947).
105. Katznelson, H., and A. G. Lochhead, *Can. J. Research,* **22C**, 273 (1944).
106. Kendall, A. R., Ph.D. Thesis, Iowa State College, 1934.
107. Kitasato, T., *Biochem. Z.,* **195**, 118 (1928).
108. Kluyver, A. J., and H. J. L. Donker, *Proc. Acad. Sci. Amsterdam,* **28**, 314 (1925); *Verslag Akad. Wetenschappen Amsterdam,* **33**, 915 (1924).
109. Kluyver, A. J., H. J. L. Donker, and F. Visser't Hooft, *Biochem. Z.,* **161**, 361 (1925).
110. Kluyver, A. J., and M. A. Scheffer, Canadian Patent 296,070 (1929).
111. Kluyver, A. J., and M. A. Scheffer, U. S. Patent 1,899,094 (1933).
112. Kluyver, A. J., and M. A. Scheffer, U. S. Patent 1,899,156 (1933).

113. Kniphorst, L. C. E., and C. I. Kruisheer, *Z. Untersuch. Lebensm.*, **73**, 1 (1937).
114. Kniphorst, L. C. E., and C. I. Kruisheer, *Z. Untersuch. Lebensm.*, **74**, 477 (1937).
115. Knowlton, J. W., N. C. Schieltz, and D. Macmillan, *J. Am. Chem. Soc.*, **68**, 208 (1946).
116. Kolfenbach, J. J., *Iowa State Coll. J. Sci.*, **19**, 35 (1944).
117. Kolfenbach, J. J., E. I. Fulmer, and L. A. Underkofler, *J. Am. Chem. Soc.*, **67**, 502 (1945).
118. Kolfenbach, J. J., E. R. Kooi, E. I. Fulmer, and L. A. Underkofler, *Ind. Eng. Chem., Anal. Ed.*, **16**, 473 (1944).
119. Komm, E., and J. Flügel, *Z. Untersuch. Lebensm.*, **79**, 246 (1940).
120. Kooi, E. R., Ph.D. Thesis, Iowa State College, 1946.
121. Kooi, E. R., E. I. Fulmer, and L. A. Underkofler, *Ind. Eng. Chem.*, **40**, 1440 (1948).
121a. Krenn, J., and D. Valik, *Proc. 12th Intern. Dairy Congr.* (Stockholm) **2**, 516 (1949).
121b. Krishnaswamy, M. A., and F. J. Babel, *J. Dairy Sci.*, **34**, 374 (1951).
122. Kuzin, A. M., and E. V. Budnitskaya, *Biokhimiya*, **5**, 309 (1940).
123. Lafon, M., *Bull. soc. chim. biol.*, **14**, 263 (1932).
124. Lagoni, H., *Zentr. Bakt. Parasitenk., II Abt.*, **103**, 225 (1941).
125. Lambert, M., and A. C. Neish, *Can. J. Research*, **28B**, 83 (1950).
126. Langenbeck, W., H. Wrede, and W. Schlockermann, *Z. physiol. Chem.*, **227**, 263 (1934).
127. Langford, C. T., Eleventh Annual Chemurgic Conference, St. Louis, Mo., *Chemurgic Papers, Series 3, No. 465* (1946).
128. Langlykke, A. F., and W. H. Peterson, *Ind. Eng. Chem., Anal. Ed.*, **9**, 163 (1937).
129. Ledingham, G. A., and G. A. Adams, *Can. Chem. Process Inds.*, **28**, 742 (1944).
130. Ledingham, G. A., and G. A. Adams, Canadian Patent 435,716 (1946).
131. Ledingham, G. A., G. A. Adams, and R. Y. Stanier, *Can. J. Research*, **23F**, 48 (1945).
132. Lees, T. M., E. I. Fulmer, and L. A. Underkofler, *Iowa State Coll. J. Sci.*, **18**, 359 (1944).
133. Lemoigne, M., *Ann. Inst. Pasteur*, **27**, 856 (1913).
134. Lemoigne, M., *Compt. rend.*, **170**, 131 (1920).
135. Lemoigne, M., *Compt. rend. soc. biol.*, **88**, 467 (1923).

136. Lemoigne, M., and M. Croson, *Bull. soc. chim. biol.,* **28,** 110 (1946).
137. Lemoigne, M., B. Delaporte, and M. Croson, *Ann. Inst. Pasteur,* **70,** 65 (1944).
138. Lemoigne, M., M. Hooreman, and M. Croson, *Ann. Inst. Pasteur,* **76,** 303 (1949).
139. Lemoigne, M., and P. Monguillon, *Compt. rend.,* **190,** 1457 (1930).
140. Lemoigne, M., and P. Monguillon, *Compt. rend.,* **191,** 80 (1930).
141. Leslie, J. D., and A. Castagne, *Can. J. Research,* **24F,** 311 (1946).
142. Levy-Bruhl, M., and Y. Cado, *Compt. rend. soc. biol.,* **122,** 373 (1936).
143. Liebmann, A. J., *Oil & Soap,* **22,** 31 (1945).
144. Lubin, M., and W. W. Westerfeld, *J. Biol. Chem.,* **161,** 503 (1945).
145. McAllister, S. H., and M. de Simo, U. S. Patent 2,051,266 (1936).
146. McKay, A. F., R. H. Meen, and G. F. Wright, *J. Am. Chem. Soc.,* **70,** 430 (1948).
147. Martius, C., *Z. physiol. Chem.,* **279,** 96 (1943).
148. Mickelson, M. N., and C. H. Werkman, *J. Bact.,* **36,** 67 (1938).
149. Mickelson, M. N., and C. H. Werkman, *Iowa State Coll. J. Sci.,* **13,** 157 (1939).
150. Mohler, H., and W. Hämmerle, *Mitt. Lebensm. Hyg.,* **29,** 53 (1938).
151. Morell, S. A., U. S. Patent 2,372,221 (1945).
152. Morell, S. A., and A. H. Auernheimer, *J. Am. Chem. Soc.,* **66,** 792 (1944).
153. Morell, S. A., H. H. Geller, and E. C. Lathrop, *Ind. Eng. Chem.,* **37,** 877 (1945).
154. Morell, S. A., and E. C. Lathrop, *J. Am. Chem. Soc.,* **67,** 879 (1945).
154a. Murphy, D., and D. W. Stranks, *Can. J. Technol.,* **29,** 413 (1951).
154b. Murphy, D., D. W. Stranks, and G. W. Harmsen, *Can. J. Technol.,* **29,** 131 (1951).
154c. Murphy, D., R. W. Watson, D. R. Muirhead, and J. L. Barnwell, *Can. J. Technol.,* **29,** 375 (1951).
155. Moureu, H., and M. Dode, *Bull. soc. chim.,* (5) **4,** 637 (1937).
156. Neish, A. C., *Can. J. Research,* **23B,** 10 (1945).

157. Neish, A. C., *Can. J. Research,* **23B**, 49 (1945).
158. Neish, A. C., *Am. Perfumer,* **48**, No. 10, 59 (1946).
159. Neish, A. C., *National Research Council of Canada, Rept. No. 46-8-3,* June 1946.
160. Neish, A. C., *Can. J. Research,* **25B**, 423 (1947).
161. Neish, A. C., *Can. J. Research,* **27B**, 6 (1949).
162. Neish, A. C., *Can. J. Research,* **28B**, 535 (1950).
163. Neish, A. C., *Analytical Methods for Bacterial Fermentations,* Saskatoon, Sask., Prairie Regional Laboratory, National Research Council of Canada, 1950.
164. Neish, A. C., *Can. J. Research,* **28B**, 660 (1950).
165. Neish, A. C., A. C. Blackwood, and G. A. Ledingham, *Science,* **101**, 245 (1945).
166. Neish, A. C., A. C. Blackwood, and G. A. Ledingham, *Can. J. Research,* **23B**, 290 (1945).
167. Neish, A. C., A. C. Blackwood, F. M. Robertson, and G. A. Ledingham, *Can. J. Research,* **25B**, 65 (1947).
168. Neish, A. C., A. C. Blackwood, F. M. Robertson, and G. A. Ledingham, *Can. J. Research,* **26B**, 335 (1948).
169. Neish, A. C., V. C. Haskell, and F. J. Macdonald, *Can. J. Research,* **23B**, 281 (1945).
170. Neish, A. C., V. C. Haskell, and F. J. Macdonald, *Can. J. Research,* **25B**, 266 (1947).
171. Neish, A. C., and G. A. Ledingham, *Can. J. Research,* **27B**, 694 (1949).
172. Neish, A. C., and F. J. Macdonald, *Can. J. Research,* **25B**, 70 (1947).
173. Neuberg, C., and A. Gottschalk, *Biochem. Z.,* **162**, 484 (1925).
174. Neuberg, C., and J. Hirsch, *Biochem. Z.,* **115**, 282 (1921).
175. Neuberg, C., and M. Kobel, *Biochem. Z.,* **160**, 250 (1925).
176. Neuberg, C., and F. F. Nord, *Ber.,* **52B**, 2248 (1919).
177. Neuberg, C., F. F. Nord, and W. Wolff, *Biochem. Z.,* **112**, 144 (1920).
178. Neuberg, C., and H. Ohle, *Biochem. Z.,* **127**, 327 (1922).
179. Neuberg, C., and H. Ohle, *Biochem. Z.,* **128**, 610 (1922).
180. Neuberg, C., and E. Reinfurth, *Biochem. Z.,* **143**, 553 (1923).
181. Neuberg, C., and O. Rosenthal, *Ber.,* **57**, 1436 (1924).
182. Neuberg, C., and E. Simon, *Biochem. Z.,* **156**, 374 (1925).
183. Niel, C. B., van, *Biochem. Z.,* **187**, 472 (1927).
184. Niel, C. B., van, A. J. Kluyver, and H. G. Derx, *Biochem. Z.,* **210**, 234 (1929).
185. Northern Regional Research Laboratory, U.S.D.A., Peoria,

Ill., *Mimeo. Report NM204,* August 3, 1942.
186. Olson, B. H., and M. J. Johnson, *J. Bact.,* **55**, 209 (1948).
187. O'Meara, R. A. Q., *J. Path. Bact.,* **34**, 401 (1931).
187a. Orlova, N. V., *Mikrobiologiya,* **18**, 423 (1949); **19**, 326 (1950).
188. Othmer, D. F., W. S. Bergen, N. Shlechter, and P. F. Bruins, *Ind. Eng. Chem.,* **37**, 890 (1945).
189. Othmer, D. F., N. Shlechter, and W. A. Koszalka, *Ind. Eng. Chem.,* **37**, 895 (1945).
190. Paretsky, D., and C. H. Werkman, *Arch. Biochem.,* **14**, 11 (1947).
191. Paretsky, D., H. G. Wood, and C. H. Werkman, *J. Bact.,* **44**, 257 (1942).
192. Pederson, C. S., and R. S. Breed, *J. Bact.,* **16**, 163 (1928).
193. Perlman, D., *Ind. Eng. Chem.,* **36**, 803 (1944).
194. Peterson, N., *Fette u. Seifen,* **50**, 447 (1943).
194a. Pette, J. W., *Proc. 12th Intern. Dairy Congr. (Stockholm)* **2**, 572 (1949).
195. Pien, J., J. Baisse, and R. Martin, *Lait,* **17**, 673 (1937).
196. Pines, H., and V. N. Ipatieff, U. S. Patent 2,391,508 (1945).
197. Porter, R., C. S. McCleskey, and M. Levine, *J. Bact.,* **33**, 163 (1937).
198. Prevot, A. R., and J. Taffanel, *Rev. can. biol.,* **6**, 797 (1947).
199. Prill, E. A., and B. W. Hammer, *Iowa State Coll. J. Sci.,* **12**, 385 (1938).
200. Pritzker, J., and R. Jungkunz, *Mitt. Lebensm. Hyg.,* **21**, 236 (1930).
201. Rappaport, F., I. Reifer, and H. Weinmann, *Mikrochim. Acta,* **1**, 290 (1937).
202. Reeves, R. E., *J. Am. Chem. Soc.,* **69**, 1836 (1947).
203. Reinke, R. C., and E. N. Luce, *Ind. Eng. Chem., Anal. Ed.,* **18**, 244 (1946).
204. Reynolds, H., B. J. Jacobsson, and C. H. Werkman, *J. Bact.,* **34**, 15 (1937).
205. Ribereau-Gayon, J., and E. Peynaud, *Bull. soc. chim.,* **1947**, 894.
206. Rippere, R. E., and V. K. La Mer, *J. Phys. Chem.,* **47**, 204 (1943).
207. Roberts, J. L., *Soil Sci.,* **63**, 135 (1947).
208. Robertson, F. M., and A. C. Neish, *Can. J. Research,* **25B**, 491 (1947).
209. Robertson, F. M., and A. C. Neish, *Can. J. Research,* **26B**, 737 (1948).

210. Rose, D., *Can. J. Research,* **24F**, 320 (1946).
211. Rose, D., *Can. J. Research,* **25F**, 273 (1947).
212. Rose, D., and W. S. King, *Can. J. Research,* **23F**, 79 (1945).
213. Sakaguti, K., K. Ohara, and S. Kikuti, *J. Agr. Chem. Soc. Japan,* **16**, 1012 (1940); *Bull. Agr. Chem. Soc. Japan,* **16**, 150 (1940).
214. Sakaguti, K., K. Ohara, and S. Kobayasi, *J. Agr. Chem. Soc. Japan,* **15**, 1075 (1939); *Bull. Agr. Chem. Soc. Japan,* **15**, 148 (1939).
215. Salmoiraghi, G. C., *Bull. soc. ital. biol. sper.,* **23**, 30, 31 (1947).
216. Scheffer, M. A., *De suikervergisting door bacterien der coligroep,* Ph.D. Dissertation, Meinema, Delft, Netherlands, 1928.
217. Scheffer, M. A., U. S. Patent 2,064,359 (1936).
218. Schierholtz, O. J., and M. L. Staples, *J. Am. Chem. Soc.,* **57**, 2709 (1935).
218a. Schmalfuss, H., *Rev. intern. tabacs,* **25**, 89 (1950).
219. Schmalfuss, H., and H. Barthmeyer, *Biochem. Z.,* **216**, 330 (1929).
220. Schmalfuss, H., and H. Rethorn, *Z. Untersuch. Lebensm.,* **70**, 233 (1935).
221. Schmalfuss, H., H. Schaake, and H. Barthmeyer, *Z. physiol. Chem.,* **200**, 169 (1931).
222. Schneipp, L. E., J. W. Dunning, H. H. Geller, S. A. Morell, and E. C. Lathrop, *Ind. Eng. Chem.,* **37**, 884 (1943).
223. Schneipp, L. E., J. W. Dunning, and E. C. Lathrop, *Ind. Eng. Chem.,* **37**, 872 (1945).
224. Senkus, M., U. S. Patent 2,382,622 (1945).
225. Senkus, M., *Ind. Eng. Chem.,* **38**, 913 (1946).
226. Shlechter, N., D. F. Othmer, and S. Marshak, *Ind. Eng. Chem.,* **37**, 900 (1945).
227. Shupe, I. S., *J. Assoc. Offic. Agr. Chemists,* **26**, 249 (1943).
228. Silverman, M., and C. H. Werkman, *Proc. Soc. Exptl. Biol. Med.,* **43**, 777 (1940); *J. Biol. Chem.,* **138**, 35 (1941).
228a. Simpson, F. J., and D. W. Stranks, *Can. J. Technology,* **29**, 87 (1951).
229. Smith, N. R., R. E. Gordon, and F. E. Clark, *U.S.D.A. Misc. Publ. No. 559* (1946).
230. Stahly, G. L., and C. H. Werkman, *Biochem. J.,* **36**, 575 (1942).
231. Stanier, R. Y., and G. A. Adams, *Biochem. J.,* **38**, 168 (1944).

232. Stanier, R. Y., G. A. Adams, and G. A. Ledingham, *Can. J. Research,* **23F**, 72 (1945).
233. Stanier, R. Y., and S. B. Fratkin, *Can. J. Research,* **22B**, 140 (1944).
234. Stark, W. H., S. L. Adams, S. S. Block, and R. M. Strain, *Abstracts of 109th Meeting, Am. Chem. Soc.*, p. 9A, Atlantic City, 1946.
235. Stark, W. H., S. L. Adams, P. Kolachov, and H. F. Willkie, *Congr. intern. inds. fermentation, confs. et communs.,* **1947**, 330.
236. Stark, W. H., S. L. Adams, P. Kolachov, and H. F. Willkie, *Rept. Proc. 4th Intern. Congr. Microbiol. 1947,* 549 (1949).
237. Stark, W. H., S. L. Adams, P. Kolachov, and H. F. Willkie, *Rev. fermentations et inds. aliment.,* **3**, 7 (1948).
238. Stepanov, A., and A. Kuzin, *Ber.,* **64**, 1345 (1931).
239. Stotz, E., *J. Biol. Chem.,* **148**, 585 (1943).
240. Stotz, E., and J. Raborg, *J. Biol. Chem.,* **150**, 25 (1943).
241. Strohmaier, A. J., and C. L. Lovell, *Ind. Eng. Chem.,* **38**, 721 (1946).
242. Suomalainen, H., and L. Jännes, *Nature,* **157**, 336 (1946).
243. Tanko, B., and L. Munk, *Z. physiol. Chem.,* **262**, 144 (1939).
244. Tanko, B., L. Munk, and I. Abonyi, *Z. physiol. Chem.,* **264**, 91 (1940).
245. Taufen, H. J., M. J. Murray, and F. F. Cleveland, *J. Am. Chem. Soc.,* **65**, 1130 (1943).
246. Testoni, G., and W. Ciusa, *Ann. chim. applicata,* **21**, 147 (1931).
247. Thompson, J., *Proc. Roy. Soc. (London),* **84B**, 500 (1912).
248. Tink, R. R., M. S. Thesis, University of Saskatchewan, 1950.
249. Tipson, R. S., *J. Am. Chem. Soc.,* **70**, 3610 (1948).
250. Tomiyasu, Y., *Biochem. Z.,* **289**, 97 (1936).
251. Tomiyasu, Y., *Biochem. Z.,* **292**, 234 (1937).
252. Tomiyasu, Y., *Enzymologia,* **3**, 263 (1937).
253. Tomkins, R. V., D. S. Scott, and F. J. Simpson, *Can. J. Research,* **26F**, 497 (1948).
254. Tomkins, R. V., J. A. Wheat, and D. W. Stranks, *Can. J. Research,* **26F**, 168 (1948).
255. Toy, A. D. F., U. S. Patent 2,382,622 (1945).
256. Underkofler, L. A., and E. I. Fulmer, *Wallerstein Lab. Comuns.,* **11**, 41 (1948).
257. Unger, E. D., and H. W. Grubb, Am. Chem. Soc. Meeting, Cleveland (1944).

258. Unger, E. D., H. F. Willkie, and H. C. Blankmeyer, *Trans. Am. Inst. Chem. Engrs.*, **40**, 421 (1944).
259. Verhave, T. H., British Patent 315,263 (1928).
260. Verhave, T. H., British Patent 335,280 (1929).
261. Verhave, T. H., British Patent 337,025 (1929).
262. Visser't Hooft, F., and F. J. G. de Leeuw, *Cereal Chem.*, **12**, 312 (1935).
263. Walpole, G. S., *Proc. Roy. Soc. (London)*, **83B**, 272 (1911).
264. Ward, G. E., O. G. Pettijohn, and R. D. Coghill, *Ind. Eng. Chem.*, **37**, 1189 (1945).
265. Ward, G. E., O. G. Pettijohn, L. B. Lockwood, and R. D. Coghill, *J. Am. Chem. Soc.*, **66**, 541 (1944).
266. Ward, G. E., L. J. Wickerham, O. G. Pettijohn, and L. B. Lockwood, U. S. Patent 2,359,950 (1944).
267. Warshowsky, B., and P. J. Elving, *Ind. Eng. Chem., Anal. Ed.*, **18**, 253 (1946).
268. Watson, R. W., and N. H. Grace, *Can. J. Research*, **26B**, 783 (1948).
269. Watson, R. W., F. Tamboline, and G. W. Harmsen, *Can. J. Research*, **27F**, 457 (1949).
270. Westerfeld, W. W., *J. Biol. Chem.*, **161**, 495 (1945).
271. Westerfeld, W. W., *Proc. Soc. Exptl. Biol. Med.*, **71**, 28 (1949).
272. Westerfeld, W. W., and R. L. Berg, *J. Biol. Chem.*, **148**, 523 (1943).
273. Wheat, J. A., J. D. Leslie, R. V. Tomkins, H. E. Mitton, D. S. Scott, and G. A. Ledingham, *Can. J. Research*, **26F**, 469 (1948).
274. White, A. G., L. O. Krampitz, and C. H. Werkman, *Arch. Biochem.*, **9**, 229 (1946).
275. Wickerham, L. J., L. B. Lockwood, O. G. Pettijohn, and G. E. Ward, *J. Bact.*, **48**, 413 (1944).
276. Willard, G. W., *J. Acoust. Soc. Am.*, **12**, 438 (1941).
277. Willard, G. W., *J. Acoust. Soc. Am.*, **19**, 235 (1947).
278. Williams, O. B., and M. B. Morrow, *J. Bact.*, **16**, 43 (1928).
279. Wilson, C. E., and H. J. Lucas, *J. Am. Chem. Soc.*, **58**, 2396 (1936).
280. Winfield, M. E., *J. Council Sci. Ind. Research*, **18**, 412 (1945).
281. Winnick, T., *Ind. Eng. Chem., Anal. Ed.*, **14**, 523 (1942).
282. Winnick, T., *J. Biol. Chem.*, **142**, 461 (1942).

PART III. THE PRODUCTION OF ENZYMES

CHAPTER 3

FUNGAL AMYLOLYTIC ENZYMES

L. A. Underkofler

Enzymes may be defined as biocatalysts which bring about specific biochemical reactions. In their natural condition, the enzymes perform metabolic processes of the living cell. Enzymes occur in every living organism from the highest-developed animals and plants to the simplest, unicellular forms of life. Although most of the enzymes are endocellular, some are exocellular and are excreted outside the living cell. Enzymes are of special importance in the fermentation industries, since all fermentation processes are the result of the enzymic activities of microorganisms.

Enzymes were used in everyday life and industry to accomplish biochemical reactions apart from the cell long before the nature and function of enzymes were understood. The use of barley malt for starch conversion in brewing and a similar use of crude fungal preparations in oriental countries are examples of such ancient use of enzymes. With the gradual development of knowledge of the nature of enzymes, crude preparations from certain animal tissues, such as pancreas and the mucosa of the stomach, or from plant tissues, such as malt and papaya fruit, were prepared, which readily found technical applications in the textile, leather, brewing, and other industries. Once the favorable results of employing enzyme preparations were established, a search began for better, less expensive and more readily available sources of such enzymes. It was found that certain microorganisms produced useful enzymes and this led to the development of technical processes for making such microbial enzymes on a commercial scale.

Takamine, in 1894, was probably the first to realize the technical possibilities of cultivated enzymes from molds and to introduce such enzymes to industry. At present, several concerns are preparing and marketing fungal amylases, proteases, and pectinases. The discussion in this chapter will be restricted to fungal carbohydrases which act on starch. Other fungal enzymes, as well as bacterial enzymes, will be considered in Chapter 4.

Many species and strains of molds produce a considerable number of enzymes, the amount of the individual enzymes varying widely between species and strains. Tauber[33] has listed twenty-three enzymes which have been identified from the mold *Aspergillus oryzae* and twenty from *Aspergillus niger*.

The fungal carbohydrases seem to be exocellular enzymes. These enzymes belong to the class of hydrolases, which bring about hydrolytic reactions. Most of the known carbohydrases are produced by different species of molds, many species forming several different carbohydrases. A remarkable property of the hydrolytic enzymes is their great specificity of substrate; each enzyme readily brings about hydrolysis of a particular substrate, but not of others. Among the more important fungal carbohydrases are α-amylase, limit dextrinase, cellulase, emulsin, inulase, invertase, maltase, and lactase. Of these enzymes α-amylase, maltase, and probably limit dextrinase are of industrial importance. α-Amylase is frequently termed the dextrinizing enzyme since it hydrolyzes large starch molecules more or less at random with formation of dextrins. Fungal maltase not only converts maltose to glucose, but also hydrolyzes starch and dextrins producing smaller dextrins and glucose. A better designation which is coming into use for this enzyme is amyloglucosidase. Limit dextrinase hydrolyzes malt limit dextrins with the production of fermentable sugars. A more detailed discussion of the nature of fungal amylolytic enzymes will be given later.

The absolute and relative amounts of the various enzymes produced by molds vary markedly between species and even between strains of the same species. For example, Le Mense, Corman, Van Lanen, and Langlykke[14] studied the production of amylases in submerged culture by three hundred sixty-seven cultures of molds which included species of the genera *Penicillium, Aspergillus, Rhizopus, Mucor,* and *Monilia.* Only eight out of eighty cultures of *Penicillium* demonstrated α-amylase dextrinizing activity and this was of very low order. Cultures of *Rhizopus, Mucor,* and *Monilia* produced little or no dextrinizing enzyme. Greatest dextrinizing activity was observed in cultures of the genus *Aspergillus.* Of two hundred seventy-eight strains of *Aspergillus* studied, thirty-four produced the dextrinizing enzyme. Highest concentrations of this enzyme were produced by strains of *A. niger, A. oryzae, A. alliaceus, A. foetidus,* and *A. wentii.* There was considerable variation between strains of the same species in both α-amylase and maltase (amyloglucosidase) activity, according to Corman and Langlykke.[4] More recently, Dingle and Solomons[5a] tested one hundred thirteen strains of molds by both surface and submerged culture for production of α-amylase as well as other enzymes. These authors likewise found great differences in enzyme production between species and between strains of the same species.

Not only the difference in strain but also the medium affects the enzyme production markedly, according to Tsuchiya, Corman, and Koepsell,[34] Shu and Blackwood,[25] and Shu.[24a] For the maximum yield of a desired enzyme, a potent mold strain must be selected and the optimum medium for production of the enzyme by the selected strain must be developed.

NATURE OF FUNGAL AMYLOLYTIC ENZYMES

Several of the carbohydrases produced by fungi act on starch or partially degraded starch and thus may be considered the amylolytic enzymes of these organisms. These include α-amylase, amyloglucosidase (maltase), and limit dextrinase.

According to our present knowledge, α-amylase attacks α-1,4-glucosidic linkages of large starch molecules, with the formation of lower-molecular-weight dextrins. This action results in rapid liquefaction and reduction in viscosity of the starch. If the action is continued long enough, substantial saccharification results, the end product being maltose.

Fungal maltase rapidly converts maltose to glucose, but is also an amylolytic enzyme, since it will hydrolyze starch and dextrins, producing lower-molecular-weight dextrins and glucose. This enzyme is now preferably designated amyloglucosidase. Corman and Langlykke[4] found that the efficiency of alcoholic fermentation of grain was more closely related to the maltase content of the fungal amylase preparation employed as conversion agent than to the α-amylase content. Subsequently, Corman and Tsuchiya[5] confirmed that alcohol yield from grain mashes converted with fungal amylase was positively correlated with the maltase content of the fungal amylase, but only if an adequate amount of α-amylase was also present. Pan, Andreasen, and Kolachov[17] reported that in the slow secondary phase of alcohol fermentation of malt-saccharified grain mashes, the rate of fermentation depended on the rate of hydrolysis of the dextrins. Addition of fungal amylases gave a rapid fermentation of the dextrins which was related to the maltase activity of the enzyme preparation rather than to the contents of α-amylase or limit dextrinase. Pool and Underkofler[19a] could find no direct correlation between the alcohol yields from grain mashes saccharified with fungal amylase preparations and the absolute amounts of α-amylase, amyloglucosidase, and limit dextrinase in either bran or submerged culture preparations of *A. oryzae* or two strains of *A. niger*.

The enzyme produced by molds which hydrolyzes malt limit dextrins with the production of fermentable sugars, has been designated limit dextrinase. Very little is known about limit dextrinase. It is possible that this enzyme may hydrolyze the α-1,6-glucosidic linkages at the branch points in the amylopectin fraction of starch. Tsuchiya, Montgomery, and Corman[35] found that over 90% of isomaltose could be hydrolyzed by a culture filtrate from *A. niger* NRRL 330. Isomaltose differs from maltose in that the chemical linkage between the glucose residues is of the α-1,6 type in isomaltose, whereas it is of the α-1,4 type in maltose. The mold filtrate obviously contains an enzyme capable of hydrolyzing 1,6-glucosidic linkages. Much additional investigation will be needed to ascertain whether or not this enzyme is limit dextrinase and to learn about the properties and function of the enzyme.

RAW MATERIALS

The raw materials for fungal-amylase production depend on

the process employed. Originally, Takamine used steamed rice, which is common practice in oriental countries. He next turned to other cereals[27] and then to wheat bran.[28,29] The bran method proved superior to other known procedures of producing mold enzymes and consequently has found extensive application for the production of commercial fungal enzymes. When bran is used, it is moistened with water or dilute acid,[29,39] sterilized, inoculated with mold spores, and cultivated in relatively thin layers. For the bran process, the raw materials are wheat bran, water, hydrochloric acid, and perhaps traces of salts. After growth of the mold on the substrate, the enzymes are obtained in aqueous solution by percolation systems, followed by precipitation of the enzymes as solid concentrates by means of alcohol.

With the development of submerged-culture methods for the production of antibiotics by molds, similar cultural procedures were successfully developed for making enzymes. In general, different strains of molds may be required for best enzyme production by submerged culture as compared with the bran-culture method. At least one concern has produced enzyme concentrates on a large scale by the submerged culturing of molds. Although undoubtedly the largest proportion of commercial fungal enzymes still comes from the bran process, it is probable that the inherent advantages of the submerged-culture process will in time lead to its eventual replacement of the older method. However, Lockwood[15a] has reported that efforts of industrial concerns to obtain by submerged culturing good, dry, powdered fungal amylase preparations having the same balance of enzymes as from bran cultures have generally been unfruitful.

For the submerged process, several types of media have been suggested and used. The media chosen depend on the uses for which the fungal enzymes are produced. Where fungal amylase is produced in large volumes for saccharification of alcoholic-fermentation mashes, by-products of the alcohol plant are the favored raw materials. In these cases, the principal raw materials are thin stillage, or distillers' dried solubles, and a little ground grain. For the production of commercial enzyme concentrates, undoubtedly purer raw materials are preferred to facilitate the precipitation and recovery of the enzymes. One plant has employed media containing starch and inorganic salts. Recovery of solid enzyme concentrates then required simply filtration and precipitation of the enzymes by alcohol.

CULTURES AND THEIR MAINTENANCE

Many strains of various species of molds produce carbohydrases. As previously mentioned, the kinds and amounts of the carbohydrases produced depend on the species and strain of mold. Strains of the *Rhizopus* species and of the *Aspergillus* species are the principal ones which have been employed on industrial scale for enzyme production, the second being by far more important. *Rhizopus* cultures have been used in the amylo process for alcoholic fermentation of starchy substrates and *Aspergillus* cultures for production of mold bran and of enzyme concentrates. Of the *Aspergilli,* strains of *Aspergillus oryzae* have been employed most commonly. It is probable that strains of this organism are generally employed by all manufacturers of fungal-enzyme concentrates, using the bran system. It is known, however, that certain manufacturers use other species for the production of special-purpose enzyme concentrates. Strains of *A. oryzae* have been employed for commercial production of mold bran.

It may be expected that as the submerged culture system for fungal-enzyme production expands, selection of strains and species will be made for the production of enzyme concentrates best adapted to specific purposes. Le Mense, Corman, Van Lanen, and Langlykke[14] investigated amylase production by three hundred sixty-seven strains of molds of five genera. Species of *Aspergillus* showed highest amylolytic activity. Of these, several strains of *A. niger* showed higher α-amylase activity when cultivated in submerged culture than the strains of *A. oryzae* tested. Moreover, there was a considerable difference in the activity of the various strains of *A. niger.* Some had much higher α-amylase activity than others, whereas other strains had higher maltase activity. Dingle and Solomons[5a] also found that strains of *A. niger* gave highest yields of α-amylase in submerged culture while strains of *A. oryzae* gave highest yields in surface culture. Corman and Langlykke[4] verified the findings of Le Mense *et al.* and also found that certain species of *Rhizopus,* while almost devoid of α-amylase activity, showed fairly good maltase activity. Therefore, it is quite possible that in the future, enzyme producers will select the mold strains and media employed to produce enzyme concentrates especially suited for diverse uses.

Stock cultures of molds are usually maintained in the laboratory on agar slants of a suitable medium. Various media may be employed. For the growth and isolation of molds, a modification of Czapek's medium[26] has been widely used. Malt-extract medium, yeast-extract sugar medium, and various natural media have had application in cultivating molds. Ideally, for laboratory stock cultures, media should be employed which give good growth and heavy sporulation of the mold cultures. The author has kept cultures of *Aspergillus oryzae* and several other molds for many years by transferring every 2 months on slants containing 2.5% glucose, 2.5% glycerol, 0.5% Difco yeast extract and 1.5% Bacto agar, with storage in the refrigerator after incubating for about 5 days at 30°C to obtain heavy sporulation. Strains of *A. oryzae,* acquired in 1939 and carried on this medium, still have their original growth characteristics and capacity for amylase production. If desired stock cultures may be preserved by means of soil cultures or by lyophilization. These procedures are discussed in Chapter 12.

For plant use, laboratory spore cultures are generally grown on bran medium. A typical procedure is to mix wheat bran with water or dilute hydrochloric acid, distribute in small quantities in wide-mouth flasks or Mason jars, sterilize under steam pressure, inoculate with spores of the mold, and incubate with the flasks or jars laid on their sides. Moisture content of the medium is adjusted as experience dictates, so that the bran dries at the proper rate during incubation. If the bran dries too rapidly, poor spore production results; if the bran does not dry rapidly enough after sporulation is complete, autolysis occurs with a rapid loss in spore viability. One commercial plant[39] produced spore cultures on a medium prepared by mixing 10 g ground corn, 100 g wheat bran, and 60 ml of 0.2 *N* hydrochloric acid, containing 0.62 ppm $ZnSO_4 \cdot 7H_2O$, 0.63 ppm $FeSO_4 \cdot 7H_2O$ and 0.08 ppm $CuSO_4 \cdot 5H_2O$, distributing the moist mass in 10 g quantities in 250-ml, wide-mouth Erlenmeyer flasks, or in 40 g quantities in 2-qt Mason jars, and sterilizing under steam pressure. The cooled, sterile mixture was then inoculated with the mold spores, the bran distributed on one side of the container by gentle tapping, and the flask or jar incubated at 30°C while lying on the side. With this procedure the mold culture grows, sporulates and then dries out. The dry bran spore cultures may be left in the incubator undisturbed for several weeks until required for use.

The laboratory spore stocks are used as starter cultures for larger-scale inoculum production in the plant. For the bran process, spore inoculation is preferred and large spore stocks to serve as plant inoculum may be grown on thin layers of bran medium in large bottles or jars, or in trays.

Jeffreys[11] described a method of growing plant-spore cultures in shallow metal trays, equipped with perforated covers. The sterilized and cooled bran is inoculated from laboratory spore stock, the inoculated bran spread in a thin layer in the tray, a piece of newspaper placed over the top of the tray, and the cover put on. The newspaper, according to Jeffreys, prevents contamination and absorbs moisture. The trays are incubated on racks in special incubator rooms, at 78°F, with forced air circulation. After a suitable incubation period, the tray covers are removed and the mold spore culture is dried.

Underkofler, Severson, Goering, and Christensen[39] describe a similar procedure for tray growing of plant spore culture which has been used on a commercial scale. These authors employed metal pans $24 \times 36 \times 4$ in. in size. The moistened, sterilized, and inoculated bran was spread in $\frac{1}{2}$ in. thick layers in the sterile pans and the covers put on the pans. The covers were provided with air spreaders in the center and air outlets in the corners of the covers, the pan being placed in a special incubator cabinet. Humidified sterile air was passed into the pans through the air spreaders at the rate of 1,200 to 1,800 ml per minute. When heavy sporulation had occurred, the covers were removed and the spore cultures dried in the incubation cabinet.

For submerged culturing of molds, mycelial inoculum is most commonly employed. A laboratory spore culture is used to inoculate flasks of liquid medium which are then incubated on a shaker in a laboratory incubator. The flask shake cultures are employed to inoculate larger flasks or bottles of liquid medium provided with aerators. The submerged cultures are incubated with aeration and the bottle mycelial cultures employed to inoculate plant seed tanks.

Use of vegetative inoculum is a time saver since the new culture starts with virtually no lag. As growth starts from each viable particle, the larger the number of viable particles in the inoculum the faster the development. Maximum efficiency may be secured by using a Waring blender for particle-size reduction. The first use of this procedure was in the gluconic acid process, as reported by Moyer, Wells, Stubbs, Herrick, and May.[16] Dorrell and Page[6] recently applied it to other organisms. Savage and Vander Brook[22] reported that "blended" *Penicillium* submerged mycelium, diluted as much as forty thousand times, adequately substituted for unblended submerged inoculum used at a 10% inoculation ratio. One of the most common faults in submerged mold cultivation is the use of too little inoculum. Not only is the development of the culture delayed, but the desired homogeneous, dispersed type of growth, consisting of small individual colonies or clumps of mycelia, fails to develop. Instead, the relatively few viable cells from the inoculum develop into large, individual, usually spherical pellets which may grow as large as marbles or larger. In most cases, an inoculation ratio of 5 to 10% (v/v) of a heavy submerged vegetative culture is most desirable for the plant-scale submerged-culture process. The intermediate cultures must, therefore, be successively worked up in aerated fermentors of increasing size to sufficient volume for the final fermentation.

LABORATORY PRODUCTION

Fungal amylase may be produced in the laboratory by growth of the mold on fibrous substrates, usually wheat bran, or by growth of the mold in submerged liquid culture. Prior to World War II, the first method was almost universally employed and the enzymic product has come to be known as *mold bran.* Since the establishment of submerged-culture fungal processes, this submerged technique has been widely adopted for the laboratory cultivation of molds for enzyme production.

Pioneering investigations on methods for mold-bran production were reported from Iowa State College. Underkofler, Fulmer, and Schoene[36] employed a rotating drum technique and later Hao, Fulmer, and Underkofler[10] adopted a covered aluminum pan with perforated bottom through which air could be passed by alternately applying air pressure and suction as the mold grew on the moistened bran in the container. The laboratory and pilot-plant work which led to a process for the commercial production of mold bran on the tonnage basis has been reviewed by Underkofler, Severson, Goering, and Christensen.[39]

The cultivation of molds by submerged culture received pioneering attention since World War II by workers at the Northern Regional Research Laboratory of the U. S. Department of Agriculture, although the procedure was first proposed by Woolner and Lassloffy[42] as early as 1909. For laboratory cultivation on a small scale, the shake-flask technique is employed and on a larger laboratory scale, the mold is grown in submerged culture in flasks or bottles up to 5 gal capacity with vigorous aeration supplied by means of a tube, connected to a suitable air dispenser, which leads to the bottom of the vessel.

Various media have been used for the submerged cultivation of molds for amylase production. Goering, Le Mense, and Ogden[9] successfully employed a synthetic medium containing starch and inorganic salts for both laboratory and pilot-plant work. In a patent, Corman[3a] claimed the use of synthetic media containing starch, a potassium phosphate buffer, compounds such as nitrites and nitrates, aspartic acid, urea, and asparagine as nitrogen sources, and small amounts of salts. In most of the reported work, media containing thin stillage from alcohol fermentation, or distillers' dried solubles, and corn meal have been used. Earlier workers[8,14] found that the addition of 0.5% calcium carbonate to such media was requisite for obtaining maximum yields of amylase.

Tsuchiya, Corman, and Koepsell[34] found that the yields of α-amylase and of maltase depended on the terminal pH of the submerged cultures. This terminal pH should not be allowed to drop below pH 4.2. In the absence of calcium carbonate, the terminal pH of the cultures of *A. niger* NRRL 337 can be controlled by adjustment of the concentrations of the ingredients of the medium, distillers' thin-stillage solids and corn meal. Increasing the thin-stillage solids content of the medium results in the rise of the terminal pH with a consequent increase in the yield of α-amylase. Increasing the corn-meal content of the medium results in a lowering of the terminal pH and an increase in the yield of maltase. By adjusting the concentrations of distillers' thin stillage and corn meal in the medium, it is possible to control, to some degree, the yields of both α-amylase and maltase. The concentration variation of the ingredients of the medium affected the production of α-amylase and maltase in cultures of *A. niger* NRRL 330 and *A. oryzae* NRRL 458 less markedly than with *A. niger* NRRL 337.

Tsuchiya, Corman, and Koepsell[34] also found that incorporation of calcium carbonate in the medium, in concentrations previously recommended, results in lowering the yield of maltase. Since the terminal pH can be controlled in the absence of calcium carbonate by using the proper amounts of stillage solids, it was recommended to eliminate calcium carbonate from the medium used for making fungal amylase. A medium containing 5% thin-stillage solids and 5% corn meal gives excellent yields of both α-amylase and maltase with *A. niger* NRRL 337.

Recently, Shu and Blackwood[25] have reported a rather extensive investigation on the effect of the variation of carbon and nitrogen sources on the production of fungal amylolytic enzymes by submerged culture. These investigators employed *A. niger* PRL 558 in shake-flask culture. The basal medium contained 0.5% $CaCO_3$, 0.1% KH_2PO_4 and 0.05% $MgSO_4 \cdot 7H_2O$ to which various carbon and nitrogen sources were added. In media containing 1.5% hydrolyzed isoelectric casein several carbon compounds, including polyalcohols, acids and carbohydrates, were tested. The enzyme activities of the culture filtrates varied widely, depending on the carbon source present. Maltose, dextrin, and starch yielded much more α-amylase than other compounds used. Maltase and limit dextrinase production were best with these substrates. However, rather good yields of maltase were obtained with glucose and the production of limit dextrinase was least dependent on the substrate.

With starch as a substrate, four types of nitrogen sources (nitrates, amides, ammonium and amino compounds) were investigated. Casein hydrolyzate gave best enzyme yields. Amides did not support good growth. Single amino acids and inorganic nitrogen compounds gave poor yields of enzymes. However, inorganic nitrogen compounds, both ammonium compounds and nitrates, when added to media containing casein hydrolyzate, significantly stimulated the production of α-amylase and maltase. This effect was much less prominent when the level of casein hydrolyzate in the medium was raised to 6.0%. Whole proteins gave much poorer results than hydrolyzed proteins. Yields of the enzymes were comparable with enzyme-hydrolyzed casein, gelatin, gluten, and edestin. It was concluded that the availability of the nitrogen source to the organism is one of the most important factors to be considered in the production of fungal amylolytic enzymes.

The relative amounts of α-amylase and maltase were controlled by Shu and Blackwood[25] by varying the fermentation time as well as the carbohydrate and protein contents of the media. It was found that an increased amount of either carbohydrate or protein increases the time required to attain the maximum amylase yield and also increases the maximum yields of all three of the amylolytic enzymes. A culture filtrate of low α-amylase, but high maltase and high limit dextrinase content was obtained by cultivating the mold on a medium containing 2% starch and 6% casein hydrolyzate for 6 to 8 days. A low maltase preparation, with fair α-amylase activity, was obtained by culturing for 48 hours on a medium containing 3% starch and 4% casein hydrolyzate. Highest yields of all three amylolytic enzymes were obtained by culturing for 6 days on a medium containing 8% starch and 4% casein hydrolyzate.

Shu[24a] more recently reported further studies on nitrogen sources for the production of amylolytic enzymes by submerged culture of *A. niger*. He found that amino acids or proteinaceous substances are not required for production of high yields of the enzymes. A readily available nitrogen source, preferably a potential alkali donor, such as ammonium acetate, should be used for high yields of α-amylase. Accumulation of maltase and limit dextrinase was suppressed by maintaining a high pH (about 8) during fermentation.

Another noteworthy contribution made by Shu[24b] has been the design of a closed shake-flask fermentor which simulates conditions in shake-flask culture in a cotton-plugged Erlenmeyer flask. In this apparatus, the uptake of oxygen during a fermentation is recorded automatically. In the past, oxygen utilization could not be followed in shake-flask cultures. This was a distinct disadvantage since information on substrate utilization, including that of oxygen, and product formation under various conditions is necessary if optimal fermentation conditions are to be established. Shu's equipment and procedures should be very useful to investigators developing new aerobic submerged-culture fermentations of all types.

Shu investigated the oxygen requirements of three different submerged fermentations with his apparatus, including fermentations producing citric acid, ustilagic acid, and α-amylase. For the α-amylase fermentation he employed *A. niger* PRL 558 and a medium containing 3% soluble starch, 2% yeast extract and 0.5% calcium carbonate. The mycelial growth was filamentous and mushy. The oxygen uptake curves showed that, after the initial lag phase, the rate of oxygen uptake decreased with time. This was more pronounced at reduced oxygen tension. The maximum efficiency of enzyme synthesis by the mycelium occurred at an oxygen uptake rate in the neighborhood of 20 millimols per hour per liter.

THE PLANT FERMENTATION PROPER

Bran Process

Little has been published on the exact procedures employed by the plants producing industrial fungal enzymes. Takamine was the first to use the bran process and secured some nineteen patents on various aspects of the production and use of fungal enzymes. He particularly recommended[32] and patented[31] the use of a rotating horizontal drum. In the drum method, the bran is moistened with an approximately equal weight of water and sterilized in the drum by steaming while the drum is rotated. The bran is then cooled by means of a stream of cool air, inoculated by blowing in dry spores of the mold *Aspergillus oryzae,* and incubated. During the incubation period, a stream of cool, humidified air is passed through the drum as it rotates. The movement and air stream serve to dissipate the large amount of heat formed by oxidation of carbohydrates by the fungus. Water may also be sprayed on the outside of the drum to prevent the temperature from rising to a level inhibitory to the growth of the mold.

It is known that some industrial enzyme plants currently employ drums, but other manufacturers who have used drums have abandoned them in favor of tray or cabinet culturing methods.[15a] In the drum process, it is difficult to control the temperature effectively, there is danger of contamination, and the rubbing action of the bran particles, as they are tumbled in the rotating drum, damages the delicate mold mycelium during the early stages of growth. However, with careful control, excellent enzyme production is achieved in the plants employing the drum system.

Shellenberger[24] has briefly described a typical commercial method for preparing fungal enzymes on wheat bran. The moistened bran is first cooked for about 1 hour in a rotating cooker by direct steam under pressure to gelatinize the starch and sterilize the mash. The cooked bran is then cooled to about 40°C in a water-jacketed cooler and prepared for transfer to the growing chamber by adjusting the pH (lactic acid may be used), adding mineral salts if necessary, and inoculation with mold-spore culture.

The inoculated bran is then spread out in layers in the growing compartments and optimum growth conditions maintained by regulating the temperature. Temperature control is the most important of the process variables, but the prevention of contamination in the incubation chambers is also of major importance. It is usually necessary to warm the molding bran layer at the beginning and to cool when growth becomes rapid. Temperatures may be maintained at 32°C for the first 24 hours, but when the cake begins to heat as a result of active growth, the temperature of the growing compartment is reduced to aboue 26°C. After 4 days, the mold mycelium has matted the medium together and enzyme production has reached a maximum.

Moist culture, on removal from the growing chambers, is dried in either rotary or tray driers. Highest temperature of the mass should not exceed 50°C, because, if subjected to higher temperature, the enzymes would be inactivated. Initial temperature depends on moisture content of the culture, type of drier, and size of charge. After drying, the mass is broken up by grinding and may be marketed as such or processed further for production of enzyme concentrates. For this purpose, the mold bran is extracted with water, filtered, and the enzyme precipitated from the filtrate by successive additions of ethanol. The precipitate is then separated from the liquor by centrifuges, dried in a vacuum drier, being careful not to exceed 55°C, and the dried product ground and blended to obtain a final product of uniform activity. Shellenberger[24] states that in addition to amylase, other enzymes can also be manufactured by this or similar methods. These include protease, protopectinase, and pectinase (Chapter 4).

For the bran process, shallow trays undoubtedly have had widest application. Since details of the processes employed by commercial enzyme concerns are not available, those interested in large-scale production of fungal enzymes, for such uses as saccharifying agent in alcohol production and as new producers of commercial enzymes, have been compelled to develop their own procedures, and some of these have been published. Undoubtedly, the concerns manufacturing commercial fungal enzymes employ similar processes, the variations probably being only in the manner of handling trays, in the types of incubation chambers employed, etc.

Perhaps the most extensive investigation of methods for cultivating the mold, *Aspergillus oryzae,* on bran for large-scale production of mold bran was reported by Underkofler, Severson, Goering, and Christensen.[39] This process proved satisfactory for the plant production of mold bran at the rate of 10 tons a day. The process flow sheet for this plant is shown in Figure 5.

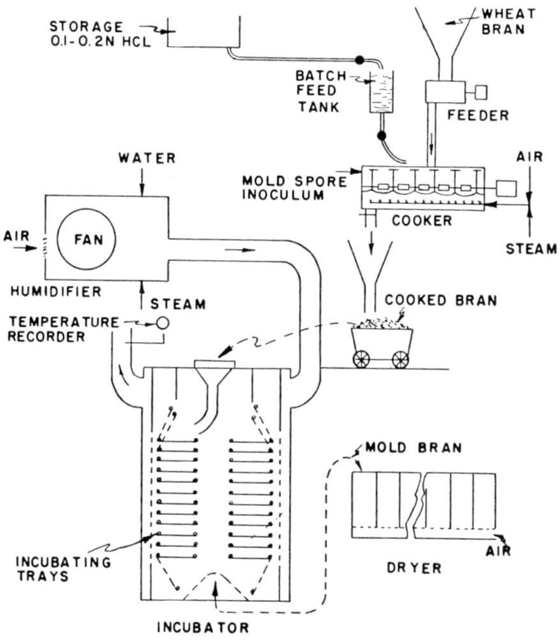

FIGURE 5. *Flow Diagram of Mold-Bran Production at Eagle Grove, Iowa*

In this process, the bran was fed batchwise into the cooker and mixed by means of the cooker agitator with the requisite amount of 0.2 *N* hydrochloric acid, the proportions being in the neighborhood of 2 parts bran to 1 part acid. The moisture content was varied in accordance with experience so that condensation of steam during cooking should yield a mash containing approximately 50% moisture. The moist bran was sterilized in the cooker by injecting steam directly into the mash for 30 minutes with continuous agitation. At the end of the heating period, the steam was shut off and air injected, with the agitators running. This rapidly cooled the mass by evaporation. When the mix had been cooled to about 35°C, the air was shut off and spore inoculum fed into the cooker by means of a dusting atomizer and air stream. When thoroughly mixed, the batch of inoculated bran was conveyed to the incubation rooms. Here the inoculated bran was spread in 2-in. layers on trays having hardware cloth (wire screen) bottoms. The trays were hinged lengthwise from the wall and filled from the bottom trays to the top until all trays in a room were filled. The temperature of the incubating mold bran was controlled by continuous circulation of humidified air over and under the trays by means of air ducts in the walls of the incubation rooms. At the conclusion of the incubation period, the trays were dropped, one at a time, to dislodge the mold bran cake, beginning with the bottom trays. The moist mold bran was conveyed by mechanical conveyors to bins with perforated bottoms. Here it was dried by means of a current of warm air, blown up through the mass.

Jeffreys[11] has described a tray method employed in another plant. In this plant, mechanical handling is provided for as much as is possible. The bran is fed by a percentage feeder on a slow-moving conveyor, where it is mixed with the required amount of water, heated with steam to at least 190°F, and the steaming continued for 15 minutes. The steamed mixture is then dropped on a second conveyor, where it is cooled by means of a current of cold, filtered air. The prepared substrate is inoculated by spraying a suspension of spores in sterile water on it. At the same time, it is also inoculated with dry spores fed by means of a percentage feeder and blower.

The substrate and spores are thoroughly mixed and dropped on a spreader which distributes it uniformly in sterilized trays. The loaded trays are conveyed to a truck which holds twenty trays 4 in. apart. The loaded trucks are pushed into the culturing tunnel on overhead trolleys. Cool, humidified air is circulated in the culturing tunnels to maintain the temperature of the growing mold bran slightly below 37°C. If it is necessary to warm the air at the beginning of the culturing period, this is accomplished by steam jets controlled by thermostat.

During passage of a truck through the culturing tunnel, the mold growth reaches an optimum in 24 to 30 hours and the trucks are then transferred to the drying tunnels. Warm dry air is circulated through these tunnels. The air flows parallelly to the direction in which the trucks are moving in the primary stage and countercurrent to their direction in the secondary stage. During drying, the moisture content is reduced to about 8% and the temperature of the product is not allowed to exceed 110°F. The dried mold bran is ground and bagged. For preparation of enzyme concentrates, the mold bran may be extracted either before or after drying.

After the contents of the trays are dried, they are removed from the truck and dumped. They are then carried inverted on a chain belt over revolving brushes which clean them. They are next passed through an oven at 300°F for 3 minutes. After being automatically turned, they go to the spreader for reloading with a fresh batch of inoculated bran substrate.

Submerged Process

There is no description in the literature of the commercial production of enzymes by submerged culturing of molds. However, several papers have appeared on the submerged culturing of molds for producing diastatic material for the saccharification of alcohol-fermentation mashes. This procedure was first proposed by Woolner and Lassloffy[42] as early as 1909. Erb and Hildebrandt[7] grew submerged cultures of *Rhizopus delemar* in plant-scale quantities for use as adjunct with malt in saccharifying alcohol-fermentation mashes. More recently, a number of investigators[1,4,8,14,15] employed submerged cultures, especially of *Aspergillus niger* NRRL 337, to totally replace malt for the saccharification of grain-fermentation mashes. This work furnished the basis for commercial-scale experimental units for submerged fungal amylase production and successful plant-scale application, in the grain-alcohol distillery of the Grain Processing Corporation, Muscatine, Iowa.[2,3,40]

The author has had close contact with the pilot-plant and semicommercial production of fungal amylase concentrates by submerged culture at one commercial concern. This work was conducted for the development and production of a concentrate high in α-amylase and as low as possible in proteolytic activity. One medium employed[9] was a starch, inorganic salts medium which contained, per liter, 30 g cornstarch, 10 g KNO_3, 5 g $CaCO_3$, 0.8 g KH_2PO_4, 0.4 g $MgSO_4$, 0.4 g KCl, 0.0035 g each of $FeSO_4$, $ZnSO_4$, and $CuSO_4$, and 3 ml oleic acid. The culture employed was a selected strain of *Aspergillus oryzae*. Laboratory spore cultures were carried in the usual manner in flasks on bran medium. From this laboratory culture, spore inoculation was made into flasks of liquid medium which were incubated on a reciprocal shaker at 30°C. After about 72 hours, the shake-flask cultures were employed to inoculate 2 gal of liquid medium in 2.5-gal, tall-form pyrex bottles and incubated at 30°C, with vigorous aeration, sterile air being introduced through a glass tube leading to the bottom of the bottle where it was dispersed in bubbles through a cloth sack of several thicknesses. After incubation for 24 hours, a bottle of culture was used to inoculate 60 gal of medium in a glass-lined seed tank. The contents of the plant seed tank, after 24 hours of incubation with aeration, were used to inoculate the production fermentor. Two types of production fermentors were used. In the glass-lined type, 300 to 600 gal of liquid were employed and in the steel fermentors, 600 to 1,200 gal of liquid.

The seed tanks and production fermentors were equipped with agitators and air spargers in the bottom of the tanks. Air was provided under pressure and was sterilized by passage through carbon filters.

The course of the fermentations was followed by periodic sampling. Samples were inspected microscopically for bacterial contamination and analyzed for α-amylase activity by the method of Sandstedt, Kneen and Blish.[21] When the maximum α-amylase activity had been reached, as shown by the analyses on consecutive samples, the aeration was stopped and the fermentor contents processed to obtain the amylase. Depending on a number of factors, such as medium composition, rate of aeration, and amount of inoculum, the maximum amylase content was reached in 64 to 135 hours in individual fermentations. These were considered successful when α-amylase activity amounted to about 75 SKB units (30°C) per ml. Occasional potencies as high as 140 units per ml were obtained.

The amylase was recovered from the fermented liquors by separating the liquid from mold mycelium by means of filter presses, concentrating the liquid about fourfold by vacuum evaporation, and precipitating the enzyme with 1.6 to 2.3 volumes of isopropyl alcohol per volume of concentrated filtrate at pH 7.5 to 8.5. Under the optimum conditions almost quantitative recovery of the α-amylase was obtained in the precipitates from the alcohol treatment. The solid enzyme concentrate was dried in a vacuum drier, and powdered by grinding. Enzyme concentrates containing more than 6,000 SKB units per g were made. All preparations were standardized at 3,000 SKB units per gram by dilution of the solid enzyme concentrates with inert ingredients. For food products, dilution with starch and/or dextrin was satisfactory. For nonfood uses where an insoluble residue was not objectionable, diatomaceous earth could be the diluent.

INDUSTRIAL APPLICATIONS AND ECONOMICS

No information is available as to the costs of production of commercial fungal amylases. The fungal amylase concentrates marketed are rather expensive in terms of unit weight, but in terms of unit activity are not more costly than competing products obtained from other sources, such as pancreatin, malt, and bacterial preparations.

The largest potential use for fungal amylases is as saccharifying agent for grain-alcohol fermentation mashes. Slightly better alcohol yields result from the use of fungal amylases than from malt for this purpose. Tables 10 and 11 in Chapter 2 of Volume I give a comparison of alcohol yields obtained from grain mashes saccharified by malt, mold bran, and submerged fungal amylase preparations. The use of mold bran on the commercial scale has been shown by Underkofler, Severson, and Goering[38] to effect savings over malt. Le Mense, Sohns, Corman, Blom, Van Lanen, and Langlykke,[15] and recent publications on plant-scale tests[3,40] have estimated a savings of 6 to 10¢ per bushel of grain mashed or 3 to 4¢ per gallon of alcohol produced by utilizing mold-amylase liquor instead of malt.

Fungal amylases, as well as amylases from other sources, have many and varied industrial and practical applications in addition to their present and potential use in the grain-alcohol fermentation industry. Some of the important uses, with literature references, are discussed in more detail in Chapter 4.

The first important market for fungal amylase concentrates was for the pharmaceutical trade, and this continues to be an important outlet.

Fungal amylases are extensively used in the food industries.

Usually highly saccharifying types of fungal-amylase concentrates, which contain α-amylase and amyloglucosidase, are employed in manufacturing various food products. However, they are also used for removal of starch, which may cause cloudiness, from fruit extracts, and in the production of pectin from apple-pomace. For this last use, the enzyme preparation must not contain pectolytic enzymes.

One important industrial application of fungal amylases is in the conversion of acid-modified starches to sweet sirups. Langlois[13a] has recently discussed the application of enzymes in corn-sirup production. Acid hydrolysis of starch is a random action and the composition of the resulting sirup is a function only of the degree of conversion. Enzyme hydrolysis is a patterned action and the composition of the resulting sirup is a function both of the degree of hydrolysis and the kind of enzyme used. The three diastatic enzymes, fungal α-amylase and amyloglucosidase, and malt β-amylase, show distinctly different patterns of hydrolysis. One or more of these enzymes may be used in conjunction with acid in the hydrolysis of starch to produce sirups of very specific composition and having definite physical and chemical characteristics. Sirups differing in sweetness, fermentability, viscosity, osmotic pressure, and humectant character can be produced as desired. Three products have been produced commercially. One is converted to about 63 D.E. (dextrose equivalent) and has high sweetening power with no crystallizing tendency. Another is converted to about 55 D.E. and has moderate sweetness and high fermentability. A third is converted to over 90 D.E. and has a degree of sweetness almost equal to pure glucose. It crystallizes to a solid mass and is palatable without further purification. Corn sugars of this degree of conversion prepared by acid hydrolysis alone are not palatable because of bitter reversion products.

Recently fungal-amylase concentrates have been suggested and extensively tested for fortification of flour. Johnson and Miller[12,13] reviewed the status of fungal enzymes in the baking industry. They reported that new standards of identity for bread, proposed by the Federal Security Agency after extensive favorable hearings, allow the addition of enzyme preparations from *A. oryzae* to bread doughs. They give the following advantages in using fungal enzymes: reduction of "buckiness" in doughs by increasing their mobility during fermentation; control of dough consistency; and improvement of loaf volume and of crumb characteristics, including increase in softness of crumb and elimination of possible sticky crumb due to an excessive amount of supplements. They state that today, fungal enzyme supplements are available with any desired ratio of proteinase to amylase, giving new control over doughs prior to make-up.

More recently Skovholt[26a] has discussed the use of fungal enzymes in bread production, and Reed[19b] the manufacture of industrial fungal enzymes for use in bread making. Pence[17a] in reviewing the current status of problems in the panary fermentation indicated that recent advances in the knowledge of the action of amylases and proteinases on flour components during fermentation and baking have stimulated a rapidly increasing use of fungal enzymes in bread production.

In the textile and paper industries, dextrinizing fungal amylase concentrates may be employed for converting starches in preparing sizing materials, although greater heat stability gives bacterial amylase preference for these uses. Such modified starches are used for sizing cotton and other textile fibers before weaving and in paper coatings. For removing starchy sizing from fabrics after weaving, fungal amylases are frequently employed. The properties of the dextrinizing enzyme make it also suitable for the production of dextrin-type adhesives.

Amylases, mixed with proteolytic enzymes, are employed in dry-cleaning preparations. These agents are used for hand spotting small food stains before cleaning. For badly soiled articles, it is customary to immerse the garment in an enzyme bath which removes the starch or gelatin size and with these the stains and to resize the article after cleaning. Although probably none of the commercial products sold for this purpose contain fungal enzymes, limited tests have shown mixtures containing fungal amylase to have merit.

COMPETITIVE OPERATIONS AND PROCESSES

Fungal amylases, being enzymes which solubilize starch and eventually convert it to sugar, are directly competitive with other amylolytic substances. Acid hydrolysis is, of course, probably the cheapest method of converting starch. Therefore, it has long been employed in preparing starch sirups and sugars, for example, corn sirup and glucose. Much fungal amylase is used in the corn-sirup industry to hydrolyze residual dextrins after acid hydrolysis and thus increase the maltose and glucose content of the sirups.

In the fermentation industry, malt enzymes have been used conventionally in this country for saccharifying starchy fermentation mashes, such as those from grains and potatoes. Fungal enzymes have been shown to be at least equivalent and probably superior to malt for this purpose. Mold bran has been employed in an industrial-alcohol plant and submerged fungal enzyme has been studied on a commercial scale in at least one plant for this purpose as mentioned previously. (See also Chapter 2 of Volume I.)

Malt amylases are also competitive with fungal enzymes for desizing textiles and for preparing sizes for fibers and paper. However, for the latter purpose, malt is at severe disadvantage as compared with fungal enzymes since malt enzymes produce more rapid initial saccharification rather than dextrinization.

Bacterial amylases are also directly competitive with fungal amylases in several applications. Where only dextrinization is required, bacterial amylase is usually preferred to fungal amylase. This is particularly true where relatively high temperatures and a pH near the neutral point are necessary. For use at lower temperatures and in the more acid pH range, fungal amylase is superior. Several of the industrial concerns which produce fungal enzymes also produce bacterial enzymes. From these sources, a range of microbial amylase products is available which fit a wide variety of processing conditions and requirements. From this large group of fungal and bacterial enzymes of different properties, the variety can be selected which is most suitable for any particular purpose.

MINOR CONSIDERATIONS

The availability of fungal carbohydrases has contributed much to our understanding of starch structure and of the mechanisms of the degradation of starch by enzymes. Progress in the knowledge of the amylolytic content of fungal-enzyme preparations has been rapid in recent years. At one time, it was considered that the principal amylolytic enzyme of fungi was α-amylase and it was well known that little or no β-amylase, so characteristic of cereal grains and malt, was present in fungal amylase preparations. Recently the importance of fungal amyloglucosidase in starch splitting has been emphasized. Schwimmer,[23] Corman, and Langlykke[4] and Corman and Tsuchiya[5] have demonstrated the importance of this factor in the production of fermentable sugars from starch by fungal-enzyme preparations. The fungal amyloglucosidase, frequently designated as "maltase" or "glucogenic enzyme,"[4] is not specific for maltose, but also produces hydrolysis of starch and glucose oligosaccharides having 1,4-glucosidic linkages. Recently Phillips and Caldwell[18] have attempted the purification of this enzyme—which they have called "gluc amylase"—from *Rhizopus delemar* and obtained an enzyme solution virtually free from α-amylase. They[19] studied the action of the enzyme on a number of substrates. The gluc amylase produced 100% of the theoretical glucose from maltose, 95 to 96% from the linear and branched fractions of corn starch and of waxy maize starch, 92% from glycogen, and 89% from residual beta dextrins. The enzyme did not attack alpha or beta Schardinger dextrins, isomaltose (brachiose), or a bacterial dextran having predominantly 1,6-α-D-glucosidic linkages.

Another carbohydrase produced by fungi is limit dextrinase, which hydrolyzes limit dextrins resulting from extended action of malt or bacterial amylases on starch. It has been postulated that the main function of this enzyme is the hydrolysis of 1,6-glucosidic linkages. However, there is no experimental evidence that this is the case.

Investigators are attempting the preparation and purification of the fungal carbohydrases and when these will become available in pure form, they should make possible even more rapid strides in the understanding of enzyme function and starch structure. Schwimmer,[23] Roy, and Underkofler,[20] and Phillips and Caldwell[18] have described the purification and concentration of fungal amyloglucosidase. Recently Underkofler and Roy[37] reported the isolation and crystallization of α-amylase and of limit dextrinase from a submerged-culture liquor of *Aspergillus oryzae*. A little later Fischer and de Montmollin[8a] announced the purification and crystallization of α-amylase of *A. oryzae* from commercial takadiastase.

BIBLIOGRAPHY

1. Adams, S. L., B. Balankura, A. A. Andreasen, and W. H. Stark, *Ind. Eng. Chem.*, **39**, 1615 (1947).
2. Anon., *Chem. Eng.*, **58**, No. 3, 244 (1950).
3. Anon., *Chem. Ind. Week*, **68**, No. 3, 19 (1951).
3a. Corman, J., U. S. Patent 2,557,078 (1951).

4. Corman, J., and A. F. Langlykke, *Cereal Chem.,* **25**, 190 (1948).
5. Corman, J., and H. M. Tsuchiya, *Cereal Chem.,* **28**, 280 (1951).
5a. Dingle, J., and G. L. Solomons, *J. Appl. Chem. (London),* **2**, 395 (1952).
6. Dorrell, W. W., and R. M. Page, *J. Bact.,* **53**, 360 (1947).
7. Erb, N. M., and F. M. Hildebrandt, *Ind. Eng. Chem.,* **38**, 792 (1946).
8. Erb, N. M., R. T. Wisthoff, and W. L. Jacobs, *J. Bact.,* **55**, 813 (1948).
8a. Fischer, E. H., and R. de Montmollin, *Helv. Chim. Acta,* **34**, 1987 (1951).
9. Goering, K. J., E. H. Le Mense, and R. L. Ogden, Private Communication (1949).
10. Hao, L. C., E. I. Fulmer, and L. A. Underkofler, *Ind. Eng. Chem.,* **35**, 814 (1943).
11. Jeffreys, G. A., *Food Industries,* **20**, 688 (1948).
12. Johnson, J. A., and B. S. Miller, *Food Ind.,* **23**, No. 3, 80 (1951).
13. Johnson, J. A., and B. S. Miller, *Food Eng.,* **23**, No. 4, 161 (1951).
13a. Langlois, D. P., *Food Technol.,* **7**, 303 (1953).
14. Le Mense, E. H., J. Corman, J. M. Van Lanen, and A. F. Langlykke, *J. Bact.,* **54**, 149 (1947).
15. Le Mense, E. H., V. E. Sohns, J. Corman, R. H. Blom, J. M. Van Lanen, and A. F. Langlykke, *Ind. Eng. Chem.,* **41**, 100 (1949).
15a. Lockwood, L. B., *Trans. N. Y. Acad. Sci., Ser. II,* **15**, 2 (1952).
16. Moyer, A. J., P. A. Wells, J. J. Stubbs, H. T. Herrick, and O. E. May, *Ind. Eng. Chem.,* **29**, 777 (1937).
17. Pan, S. C., A. A. Andreasen, and P. Kolachov, *Ind. Eng. Chem.,* **42**, 1783 (1950).
17a. Pence, J. W., *Agr. Food Chem.,* **1**, 157 (1953).
18. Phillips, L. L., and M. L. Caldwell, *J. Am. Chem. Soc.,* **73**, 3559 (1951).
19. Phillips, L. L., and M. L. Caldwell, *J. Am. Chem. Soc.,* **73**, 3563 (1951).
19a. Pool, E. L., and L. A. Underkofler, *Agr. Food Chem.,* **1**, 87 (1953).
19b. Reed, G., *Trans. Am. Assoc. Cereal Chem.,* **10**, 21 (1952).

20. Roy, D. K., and L. A. Underkofler, *Cereal Chem.*, **28**, 72 (1951).
21. Sandstedt, R. M., E. Kneen, and M. J. Blish, *Cereal Chem.*, **16**, 712 (1939).
22. Savage, G. M., and M. J. Vander Brook, *J. Bact.*, **52**, 385 (1946).
23. Schwimmer, S., *J. Biol. Chem.*, **161**, 219 (1945).
24. Shellenberger, J. A., *Chem. Eng.*, **54**, No. 2, 130 (1947).
24a. Shu, P., *Can. J. Botany*, **30**, 331 (1952).
24b. Shu, P., *Agr. Food Chem.*, **1**, 1119 (1953).
25. Shu, P., and A. C. Blackwood, *Can. J. Botany*, **29**, 113 (1951).
26. Skinner, C. E., C. W. Emmons, and H. M. Tsuchiya, *Henrici's Molds, Yeasts and Actinomycetes*, p. 55, New York, John Wiley, 1947.
26a. Skovholt, O., *Trans. Am. Assoc. Cereal Chem.*, **10**, 11 (1952).
27. Takamine, J., U. S. Patent 525,820 (1894).
28. Takamine, J., U. S. Patent 525,823 (1894).
29. Takamine, J., U. S. Patent 991,560 (1911).
30. Takamine, J., U. S. Patent 991,561 (1911).
31. Takamine, J., U. S. Patent 1,054,324 (1913).
32. Takamine, J., *Ind. Eng. Chem.*, **6**, 824 (1914).
33. Tauber, H., *The Chemistry and Technology of Enzymes*, p. 401, New York, John Wiley, 1949.
34. Tsuchiya, H. M., J. Corman, and H. J. Koepsell, *Cereal Chem.*, **27**, 322 (1950).
35. Tsuchiya, H. M., E. M. Montgomery, and J. Corman, *J. Am. Chem. Soc.*, **71**, 3265 (1949).
36. Underkofler, L. A., E. I. Fulmer, and L. Schoene, *Ind. Eng. Chem.*, **31**, 734 (1939).
37. Underkofler, L. A., and D. K. Roy, *Cereal Chem.*, **28**, 18 (1951).
38. Underkofler, L. A., G. M. Severson, and K. J. Goering, *Ind. Eng. Chem.*, **38**, 980 (1946).
39. Underkofler, L. A., G. M. Severson, K. J. Goering, and L. M. Christensen, *Cereal Chem.*, **24**, 1 (1947).
40. U. S. Dept. Agriculture, "Methods and Costs of Producing Alcohol from Grain by the Fungal Amylase Process on a Commercial Scale," *Tech. Bull. No. 1024* (1950).
41. Van Lanen, J. M., and E. H. Le Mense, *J. Bact.*, **51**, 595 (1946).
42. Woolner, A., Jr., and A. Lassloffy, U. S. Patent 923,232 (1909).

CHAPTER 4

MICROBIAL ENZYMES OTHER THAN FUNGAL AMYLASES

J. C. Hoogerheide

Enzymes may be defined as biocatalysts which bring about, in their natural habitat, specific biochemical reactions forming part of the metabolic processes of the living cell. Enzymes are highly specific in their action on substrates and often many different enzymes are required to bring about, by concerted action, the sequence of metabolic reactions performed by the living cell.

Certain enzymes may be isolated from the living cell and retain their specific activities *in vitro;* others, however, have not yet been separated. Several enzymes have been purified and obtained in crystalline form; they were found to be protein in nature, the composition of which did not differ significantly from that of nonenzymic protein. It is not known at present what makes a protein an enzyme, but it is significant that certain investigators believe that practically all protein present in microorganisms, such as yeast, is enzyme protein.[42]

Only a relatively few enzymes have been studied intensively. Best known are the hydrolases, which bring about various hydrolytic reactions. The enzyme systems which catalyze the syntheses occurring in living cells are still largely unknown, although investigations are now quite active in this field.

The practical application and industrial use of enzymes to accomplish certain biochemical reactions apart from the cell dates back many centuries and was practiced long before the nature or function of enzymes was understood. The use of malted barley for starch conversion in brewing, the addition of saliva to starchy products by primitive tribes in preparing fermented liquors, the use of dung for bating or puering of hides in leather making are examples of ancient uses of enzymes. It was not until the turn of this century that the agents became known which are responsible for bringing about such biochemical reactions. Crude preparations from certain animal tissues (pancreas, mucosa of the stomach) or from plant tissues (papaya fruit) were prepared which found technical applications in the textile, leather, and brewing industries. Once the favorable results with such enzyme preparations were established, a search began for better, less expensive, and more readily available sources of the enzymes. It was found that certain microorganisms excrete enzymes similar to the amylases of malt and pancreas, or to the proteases of the pancreas and papaya fruit. This led to the development of processes for the production of such microbial enzymes on a commercial scale.

Takamine,[86] in 1894, was probably the first to realize the technical possibilities of cultivated enzymes and to introduce them to industry. Whereas this investigator was mainly concerned with enzymes from fungi, Boidin and Effront,[14,15] in 1917, pioneered in the production of commercial enzyme preparations from bacterial sources.

Technological progress in this field during the last decades has been so great that for many uses, cultivated enzymes have replaced the animal or plant enzymes, such as pancreatin, papain, or ficin, in industrial applications. At present only a relatively small number of microbial enzymes have found commercial application, chiefly the amylases and proteases of bacteria and molds, the invertase from yeast, and the pectinases and tannases from mold species. With increasing knowledge of enzymes, this field undoubtedly will be much expanded in the near future.

MICROBIAL AMYLASES

Classification of Amylases

Amylases are conventionally divided into α- and β-amylases.

α-Amylases are capable of breaking down the structure of the starch molecule at random into large fragments (dextrins) consisting of a series of glucose anhydride units. This type of breakdown causes a tremendous drop in the viscosity of a starch solution and leads to a rapid disappearance of the typical blue coloration of starch on addition of iodine. Only when this type of attack has been accomplished are the dextrins further broken down into smaller fragments by this enzyme. The rate of attack on dextrins, however, is much slower than that on starch. α-Amylase is primarily a starch-liquefying and starch-dextrinizing enzyme.

β-Amylases attack the starch molecule from the ends only, cutting off maltose units until their action is blocked by a side chain or phosphate group in the starch molecule. Then the enzyme cannot attack the residue further, except when a new end is created, for example, by the action of an α-amylase. β-Amylase is primarily a saccharifying enzyme; it is found exclusively in plants being especially abundant in some ungerminated grains, malts, and sweet potatoes.

Both bacterial and mold amylases are α-amylases, though by no means identical. Microbial amylases are widely different for each species, or even strain, especially in their attack on dextrins and in the ultimate amount of fermentable sugars formed from starch.

The industrial significance and the manufacture of mold amylases have been discussed in the preceding chapter and, therefore, only the amylases of bacterial origin will be discussed here.

Bacterial Amylases

Bacteria Which Produce Amylase

When certain strains of bacteria are streaked on a nutrient agar medium containing starch, often a colorless halo may be observed around the colonies when dilute iodine is poured over the plate after completion of the growth process. From the many species and strains which show this phenomenon, relatively few have been investigated as to the type of enzyme they excrete. The following species have been studied in recent years: *Bacillus subtilis* and related species, *Bacillus diastaticus, Aerobacillus macerans,*[11,12,60,87,88,89] *Aerobacillus polymyxa,*[79,88] *Clostridium acetobutylicum,*[37,43] and other butyric acid bacteria, *Bacterium cassavanum,*[85] *Phytomonas destructans,*[17] and *Actinomyces microflavus.*[17]

From these species, only *B. subtilis* is known to be used in this country, at present, for the commercial production of bacterial amylase. However, in Russia, *B. diastaticus* seems to have found commercial application for amylase production.[39]

AMYLASE PRODUCTION BY *Bacillus subtilis* IN TRAY FERMENTORS

Source and Selection of Cultures. *B. subtilis* is one of the most widely distributed bacterial species; it may be isolated with little difficulty from soil, dust, air, surface water, or compost from almost any part of the world. An interesting study of a large number of strains of this organism, isolated from different sources, was made by Peltier and Beckord[70] and by Kneen and Beckord.[48] These investigators found that strains of *B. subtilis* can be divided into two distinctly different groups. The great majority of strains isolated from soil or from plant material varied greatly in amylase production from none to relatively high levels. The amylases excreted by these strains were of the saccharifying type and gave rather pronounced conversion of starch to fermentable sugars. However, the yield of the amylase excreted by these *B. subtilis* strains is low and such enzymes have not found industrial application as yet.

However, practically all strains of *B. subtilis* isolated from ropy bread were able to excrete large amounts of amylases. These amylases were of the nonsaccharifying type, with a very high starch-liquefying and dextrinizing activity, but forming little fermentable sugars even after 24 hours of incubation. It is this type of bacterial amylase that is produced commercially and has found wide industrial application. Besides ropy bread, suitable strains for the production of this amylase may be isolated from grains, flours, and other starchy products.

The selection of a suitable strain is of the utmost importance for successful amylase production. Such factors as yield, time required to obtain maximum yield, composition of the medium, reproducibility of results, ease of filtration and purification, properties and stability of the final product are all important in determining the choice of the strain. Once this selection has been made, all possible care must be taken to prevent degeneration, since it is a well known fact that strains of *B. subtilis* are subject to considerable variation beyond the control of the investigator. Lyophilized cultures as well as spore cultures in dry sterile soil (see

Chapter 12) are often used for maintenance and preservation of the stock culture in its original condition.

Medium for Commercial Production of Amylase. It is a rather generally accepted rule in enzymology that production of specific enzymes by microorganisms is stimulated when the culture medium in which the organism grows contains the substrate to be attacked by the enzyme. It is assumed that enzymes are produced by microorganisms for the modification of potential nutrient substrates to bring them into a form which can be assimilated by the organism. Frequently an enzyme may not be produced in any considerable amount by an organism unless the presence of a potential nutrient source requires it. However, amylase production by *B. subtilis* cannot be explained by such a hypothesis. It has been pointed out by Boidin and Effront[14,15] that abundant amylase secretion occurs in the complete absence of starch. In fact, many excellent amylase-producing strains are not able to utilize the breakdown products of their amylases and no real reason can be advanced why such strains excrete an abundance of amylase.

Whereas addition of carbohydrates, including starch, to the medium is not essential for the formation of amylase by *B. subtilis,* a prerequisite for good amylase production is a high percentage of assimilable protein, preferably of plant origin. Defatted soybean flour, peanut or linseed press cake, casein, distillers' slop, asparagus butt juice, dried brewers' yeast, corn-steep liquor, or a mixture of such ingredients, in 5 to 10% concentration, were found to give excellent results and are used most frequently for commercial production of *B. subtilis* amylase.[1,6,14,15,80,81,97,98] In order to make more of the insoluble proteins of the previously mentioned substrates available to the organism, the suspension is first hydrolyzed with boiling dilute acid or by enzymic treatment. Often after neutralization, about 0.1% each of dipotassium phosphate, calcium chloride, and magnesium chloride is added as well as traces of manganese and iron salts. The medium is then filtered, the pH adjusted at 6.5 to 7.0, and sterilized for 30 to 60 minutes at 15 lb steam pressure.

The equipment used for sterilization of the medium is preferably glass-lined or stainless-steel pressure vessel provided with agitator and with a jacket or coils for heating and cooling. Batches up to 1,000 gal are prepared for commercial amylase production, depending on the size of the culture vessels. Extreme care must be taken that no contaminating organisms can enter the tank during cooling and a slight positive pressure is maintained in it at all times.

After cooling to 35°C, the medium is inoculated with a pure culture of *B. subtilis*. The inoculum may be prepared in bottles containing the same nutrient medium as mentioned previously and is grown for 1 to 3 days at 30°C to 35°C. About 1 l of inoculum is sufficient for 1,000 gal of medium. The inoculated culture is then ready for transfer to the special culture vessels.

The Importance of Pellicle Formation during Growth. *B. subtilis* grows on the surface of a nutrient medium forming a more or less heavy, wrinkled pellicle; very little growth occurs in the medium itself underneath this pellicle. On incubation at 30° to 35°C, the pellicle becomes visible often within 24 hours. It increases in thickness and reaches its maximum within 2 to 3 days. On longer incubation, the gray-white, wrinkled pellicle starts to autolyze, becomes very brittle, and abundant spore formation can be observed.

Boidin and Effront,[14,15] in 1917, stressed the importance of a good heavy pellicle for amylase production. When the pellicle is disturbed and becomes submerged, little amylase is formed unless a new pellicle is produced. Gentle circulation of the medium underneath the pellicle promotes amylase excretion, but is difficult to accomplish on a commercial scale without danger of damage to the pellicle.

The thickness of the nutrient layer is of critical importance, 2.5 to 4 cm being most commonly used. A thinner layer usually gives somewhat higher yields, but reduces the over-all yield of a fermentor. A bacterial pellicle does not continue to produce amylase, even when the nutrient medium underneath is replaced by a fresh medium.[6] Usually production ceases within 6 days. This is one of the main reasons why the depth of the nutrient layer cannot be increased above a certain level without merely diluting the total amount of amylase produced.

Culture Vessels. In order to obtain the large surfaces required for good surface growth of *B. subtilis*, special culture vessels had to be designed. The equipment proposed by Boidin and Effront[14,15] was found to be very suitable and is still used, although considerably improved in detail.[1,97,98] It consists of a vertical, circular tank with removable head. The dimensions of the culture tanks range up to 8 ft in height and 6 ft in diameter. A heavy shaft goes through the center and passes through a stuffing box in the tank head. Fastened to this central shaft is a series of shallow circular trays, spaced one above the other and filling the tank from top to bottom. Each tray has an overflow enabling the operator to fill all trays from the sterilizer, starting with the top tray and gradually descending until the bottom tray starts to spill over.

The culture vessel, together with the connecting line to the sterilizer and an adjacent air filter, can be sterilized with live steam under pressure before the charging operation. The air filter provides sterile air during cooling and later during the growth process. When cool, the culture vessel may be charged with the inoculated medium by applying a slight overpressure with sterile air in the sterilizer.

Propagation and Harvesting. Since *B. subtilis* is an obligate aerobic organism with a high growth rate, a large amount of oxygen is consumed and carbon dioxide is produced during the growth period. In order to supply this oxygen demand and to eliminate dissimilation products, it is necessary to circulate fresh, sterile air across the pellicle during active growth. This can be done by means of a pipe system with outlets over each tray. A slight overpressure is maintained in the culture vessel throughout the run to reduce the danger of contamination. The temperature is maintained at 25° to 30°C. During the active growth period, much heat is developed which must be dissipated either by cooling the circulating air or by equipping the tank with a suitable jacket for water cooling. Incubation is continued for 3 to 7 days, at which time, maximum potency is reached. The pH gradually increases from 6.5 to about 8.5. It is not advisable to continue the incubation too long, since a slow but steady decrease in potency occurs at this relatively high pH.

Harvesting is done by rotating the shaft, to which the trays are attached, by means of a crank on the tank head, thus spilling the liquid over the edges of the trays and collecting it via the bottom valve of the tank. In order to remove bacterial debris, the culture is then filtered by means of a plate-and-frame filter-press or by centrifugation.

The Final Product. The clear liquid obtained from filtration of the fermented beer is concentrated from three to ten times, depending on the potency of the culture liquid and the final potency desired. Heating above 50°C for prolonged periods should be avoided during this concentration process; the pH should be maintained close to neutrality.

Usually the enzyme is marketed as a liquid concentrate. In order to prevent putrefaction, 5 to 15% of sodium chloride is added, often in combination with suitable antiseptics, such as pine oil or thymol. If a dry product is desired, the liquid may be further concentrated until a heavy sirup is obtained, from which the active amylase is precipitated quantitatively by addition of 2 to 3 volumes of cold alcohol or acetone. By comparing various concentrations of water-miscible solvents at different temperatures, Gates and Kneen[32] found that isopropyl alcohol, in concentrations of 40 to 60%, gave excellent recoveries of amylase without the necessity of working below 30°C.

Also, ammonium sulfate (about 50% saturation) is frequently added for the precipitation of active amylase. The precipitate may be dried by means of a vacuum tray drier at low temperature. Solid amylase concentrates obtained by precipitation with alcohol or ammonium sulfate, are too potent for most industrial purposes and are usually diluted to a standard potency by blending with inert substances, such as sodium chloride or sugar.

Amylase Production by *Bacillus subtilis* in Submerged Culture

In a recent patent, Smythe, Drake, and Neubeck[83] describe an entirely new process for the production of bacterial amylase, namely, a deep-culture fermentation method, similar to that used for the production of antibiotics. Strains of *B. subtilis* are propagated submerged in the medium, using large fermentors, provided with both aeration devices and mechanical agitation.

In order to obtain high yields of amylase, this process requires a medium considerably different from the classical media recommended by Boidin and Effront for the stationary method of amylase production. In addition to the usual protein sources used in the older process, such as 6% corn-steep liquor, 3% cottonseed meal, or 3% soybean meal, the medium recommended for deep-culture fermentation contains 4% starch plus 10% hominy feed or other finely divided cereal-grain products. The presence of this large amount of starchy substances makes it impossible to sterilize the medium unless the starch is first liquefied. This is accomplished by heating the medium to the gelatinization temperature of the starch, and then adding a starch liquefying enzyme. After liquefaction is complete, the medium is sterilized, cooled, and inoculated with a culture of *B. subtilis;* 1 l of culture per 1,000 gal of medium is sufficient.

The temperature of the fermentor is controlled at 30°C to 40°C during the growth period and aeration is at the rate of more than 1 volume per volume of medium per minute. After 24 to 48 hours of incubation, maximal amylase yields are obtained. Filtration and further treatment are identical with those for the stationary method described previously.

This deep-culture process is of considerable interest, since it had been previously reported that *B. subtilis* forms little or no amylase when cultivated in a submerged medium. The process may also be used for the production of bacterial protease according to the claims of the inventors. It is known that at least three manufacturers are currently producing bacterial enzymes by the submerged-culture process.

Lockwood[52a] has pointed out that culture variability is a formidable problem in the submerged-culture process. If a variant occurs in still cultures, it does not influence other cultures in the culture vessel. In agitated-aerated cultures, the variant will outgrow the original culture throughout the tank and may replace it entirely in a few hours. Such rapid shifts in population are not readily detected in small-scale control laboratory operations. Therefore, use of a culture that has developed a large masked variation during laboratory culturing may mean serious loss in the plant.

Other Methods For Producing *B. subtilis* Amylase

Bacterial amylase may be prepared also by a technique similar to that described for mold-bran production in the preceding chapter. According to Beckord, Kneen, and Lewis,[5] wheat bran is moistened with 1.5 to 2.5 parts of a dilute phosphate buffer, giving the moist bran an initial pH of 6.0. It is spread in thin layers and, after steam sterilization and cooling, is inoculated with a liquid culture of *B. subtilis.* Due to the large surface, rapid growth occurs and after incubation for 48 hours at 37°C, maximum yields are obtained. Certain strains of *B. subtilis* give much higher yields than others; therefore, a strain adapted to this process must be used. The same type of general equipment that is used for the commercial production of mold bran is suitable also for this process. After completion of growth, the bran culture can be dried rapidly by means of warm air without loss in activity. The dried culture may be used as such, or the active enzyme may be extracted with water. A clear enzyme solution can be obtained by addition of 0.5% calcium chloride to the aqueous extract, followed by filtration.

Using a special strain of *B. subtilis* of the high saccharifying type, a bran culture was obtained that, when dried, compared favorably with malt as a saccharifying agent. Especially in the presence of calcium ions, this conversion agent was considerably more heat stable than malt and thus could be used for conversion at higher temperatures. Maltose was the main breakdown product, and the enzyme had its optimum pH at 6.6 to 7.0. To the writer's knowledge, this process has not yet been used commercially, but undoubtedly offers possibilities for special purposes.

In a patent, issued to Christensen,[19] a method is described for increasing the diastatic power of barley malt or soybeans. By the addition of an equal volume of water and subsequent inoculation of the moist product with a culture of *B. subtilis,* bacterial amylase is produced in addition to the original plant amylase. After 24 to 48 hours of incubation, the culture may be inoculated with an amylase-producing mold, such as *Aspergillus oryzae, Rhizopus* or *Mucor* strains, which, after another 24 hours of incubation, adds a mold amylase to the mixture. In order to preserve the original amylolytic activity of the malt, no sterilization is possible and it is doubtful that such a complicated process would give consistently reliable results under nonsterile conditions. The basic principle of combining amylases of different origin has considerable merit, since often a more rapid and more complete starch digestion is obtained by the combined activities of two or more amylases of different origin than by any amylase alone.

Commercial Production of Bacterial Amylase with *B. diastaticus*

In Russia, bacterial amylase is apparently produced commercially by utilizing an obligate aerobic thermophilic spore-forming microorganism, classified as *B. diastaticus.*[39] The strain of the organism used in this process is not available in the United

States at present. Growth occurs in media of strictly neutral pH at temperatures between 37° and 70°C, with an optimum at 60°C.

Following is a brief description of the commercial process as obtained from the Russian literature. The medium consists of a potato decoction. Five kg of cleaned, chopped potatoes and 100 g of calcium carbonate are used per 100 l of water. The suspension is autoclaved for 30 minutes at 22 psi pressure, then cooled to 65°C, and the supernatant liquor is decanted into the culture vessel. This is a coated, glass-lined or tinned steel tank, provided with coils for maintaining a temperature of 60°C. Heavy metals, such as copper and nickel, as well as sulfates and chlorides, are quite inhibitory at the high temperature necessary for growth. A sparger system, consisting of pipes with small openings, covers the bottom of the tank and is used for aeration with purified air. By using a 6 to 8% inoculum, full potency is reached in as little as 4 to 5 hours after inoculation, during which period, the culture is thoroughly aerated. There is only a slight cloudiness evident during growth and no development of any odor. Contamination, if it occurs, sharply reduces final potency and may usually be detected by the presence of large rods with terminal spores among the thin rods of *B. diastaticus.* Very simple equipment is used in the process. The impression is gained that the final fermentation is done under aseptic, but not necessarily sterile, conditions and that the medium from the autoclave is not resterilized in the culture vessel before inoculation. The inoculum, however, is cultivated under sterile conditions in bottles which are provided with porous candle aerators. The inoculum bottles, in turn, are inoculated from static cultures in 5% potato decoction. The organism is maintained on a nutrient agar, consisting of 0.5% peptone, 0.1% calcium carbonate, and 20% potato extract.

Without previous filtration, the culture is concentrated under slight vacuum at 75°C or lower. The concentrate may be dried at 70°C or the activity may be precipitated with ammonium sulfate (60% saturation), yielding dry products of high activity. The enzyme preparation ("superbiolase") obtained is used as a desizing agent and for other industrial purposes. Its properties, as far as can be checked, are quite similar to those of commercial bacterial amylase produced by *B. subtilis.* The enzyme is exceptionally heat stable and desizing temperatures of 80° to 85°C are recommended.[84]

PURIFICATION AND CRYSTALLIZATION OF BACTERIAL AMYLASE

Commercial bacterial amylase, apart from the adjuncts added as preservatives or diluents, is seldom a pure product. Although usually bacterial debris is removed, it still contains the soluble components and breakdown products of the nutrient medium. It also contains enzymes other than amylases; among these are usually proteolytic enzymes.

Low-molecular-weight products can be removed by dialysis in cellophane bags and considerable purification can be achieved by conventional methods of repeated fractional precipitation with alcohol or acetone at low temperatures or by ammonium sulfate precipitation.

By applying these methods Di Carlo and Redfern[23] obtained a highly purified enzyme preparation with nine hundred times the potency of the original culture of *B. subtilis.* The enzyme contained 5.3% ash, 12.1% nitrogen, and 0.05% inositol. It was inactivated by ficin, which indicates that the enzyme is a protein; the nitrogen content also suggests this. Meyer, Fuld, and Bernfeld[59] succeeded in crystallizing bacterial amylase. Although commercially such highly purified enzyme preparations are unnecessary, they are of great importance for a systematic study of their properties.

ACTIVATORS AND INHIBITORS

According to Di Carlo and Redfern,[23] there is no indication for the necessity of the presence of a coenzyme for the action of bacterial amylase. Studies of inactivation by iodine, silver ions, and p-Cl-Hg-C_6H_4-COOH, with subsequent reactivation by cysteine or hydrogen cyanide, indicated the presence of essential -SH groups. Oxidizing agents, such as potassium permanganate, potassium ferricyanide, iodine and quinones, and heavy metal ions, such as those of silver, mercury, lead, copper, zinc, and iron, have a detrimental effect and markedly inhibit the action of bacterial amylase. Other inhibitory substances are maleic acid, mono-iodoacetic acid, hydrazine, phenylhydrazine, semicarbazide, nitrites, oxalates, and fluorides. Other cell and enzyme poisons, such as azide, salicylate, thiosulfate, cyanide, sulfide, and cysteine have no effect. Exposure to a pH of less than 4 or higher than 9 causes rapid inactivation of the enzyme.

Industrial Importance of Bacterial Amylase

Properties of Bacterial Amylase. As pointed out before, commercial bacterial amylase is mainly a starch-liquefying and starch-dextrinizing enzyme. Within a few minutes after addition of the enzyme to a 5% gelatinized starch paste, liquefaction becomes evident and progresses rapidly to completion. Soon after complete liquefaction, the degraded starch no longer gives a blue color with iodine solution. From there on, the breakdown reaction slows down considerably, but does not stop until substantial amounts of reducing sugars are formed, among which glucose, maltose, and an unidentified fermentable trisaccharide are prevalent.[82] The degree of ultimate starch breakdown depends on the enzyme concentration and the temperature; it is maximum at 55°C and a relatively high enzyme concentration. However, even under such optimal conditions, the breakdown is never complete and 15 to 20% of unfermentable dextrins remain behind. Bacterial amylase differs from the amylases of fungi, malt, saliva, and pancreas in that it produces less saccharification for the same liquefying power.

The enzyme has its optimum activity at a pH of 6.5 to 7.0 and (at low temperatures) has a pH range for its activity between 5.0 and 9.0. The optimum temperature for starch digestion is at 55°C. Slow inactivation becomes evident at temperatures higher than 60°C. The enzyme can be employed successfully at such an elevated temperature as 75°C for rapid starch digestion. Even at 100°C, at neutral pH, its short-lived action is sufficient to accomplish complete liquefaction when added to a boiling starch solution.

The presence of calcium ions furnished, for example, by 0.1% calcium chloride solution, markedly protects bacterial amylase against thermal destruction. The presence of starch also has a favorable influence. The outstanding heat stability of bacterial amylase, which is far better than that of fungal or pancreatic amylase, has drawn attention to this enzyme for an increasing number of industrial applications involving starch breakdown.[30]

Industrial Applications of Bacterial Amylase

One of the major applications of bacterial amylase is in the textile industry for the removal of starch sizing from the woven cloth and for the preparation of sizing agents. Fiber thread, especially the warp, is subjected to considerable linear strain and abrasion during the weaving process, causing constant danger of breaking. In order to reduce this danger, the thread is usually reinforced first by treating it with a mixture of adhesive and lubricant which, on drying, strengthens the thread considerably, yet keeps it pliable. Although there are many sizing agents used for this purpose, starch has always been one of the most widely used and cheapest one, especially on cotton, rayon, and numerous mixed yarns.

To give good adhesion and penetration, the starch must be modified by a process involving a slight breakdown of the starch molecules, which causes a sharp reduction in viscosity and formation of dextrins. Modified starches may be prepared by a mild treatment with acids by the starch manufacturer. Often, however, the textile mill does its own modification by an enzyme treatment of the cooked starch.[30] A more uniform, better-penetrating product is obtained by enzymic modification and no chemicals have to be added which might deteriorate the yarn. Bacterial amylase is the preferred enzyme for such a modification because of the desired limited breakdown it produces and its high-temperature activity, which makes it possible to modify the starch without appreciable cooling immediately after gelatinization by heating.

After the weaving process is completed, the starch present in the cloth must be removed, since it would seriously interfere with subsequent processing. This is done almost exclusively by enzymic treatment, the purpose of which is to bring the starch into a soluble form which can easily be washed out with water. Bacterial amylase, in liquid or solid form, is at present most widely used. One of the simplest processes to accomplish desizing is to pass the cloth through a dilute enzyme solution maintained at a temperature of 50° to 70°C, during which the cloth picks up about its own weight of enzyme solution. The wet cloth is then stored for several hours in bins, during which time the starch is converted to soluble dextrins or water-soluble lower carbohydrates which may be removed easily by subsequent washing. Due to its fast action at high temperatures, a slightly higher enzyme concentration reduces the necessary holding time to a few minutes. This has opened the way to the introduction of a continuous desizing process in the textile industry.

Other important industrial applications for bacterial amylase are in the preparation of sizing pastes from starch for paper coatings.[18,21] Reproducible, uniform results and the absence of low molecular carbohydrates are prerequisites for such sizings. The enzyme has also found extensive application for the liquefaction of heavy starch pastes which form inevitably when starchy products in high concentration must be heated, as is the case in corn-sirup and chocolate-sirup manufacture,[4,100] in microbiological solvent production from cooked starches or grains, and in many other processes. Because of its extreme heat stability, the enzyme may be added during heating of the starch suspension, just before gelatinization occurs. Liquefaction occurs as soon as the starch becomes gelatinized, thus preventing serious thickening of the mash.

In the brewing industry, especially in European countries, bacterial amylase has found use as an aid to the often low diastatic power of malt and for starch conversion of adjuncts, such as corn or rice. It is also used, with proteases, for the manufacture of a highly dextrinous, low-fermentable adjunct, prepared from wheat and sometimes employed in the brewing industry.[91,92]

MICROBIAL PROTEASES

Proteolytic Enzymes

Proteolytic enzymes are excreted by a wide variety of microorganisms. This can be demonstrated by the simple technique of streaking pure cultures of such organisms on a suitable solidified nutrient gelatin medium. A zone of liquefaction, observed around the colonies formed, indicates marked proteolytic activity. A few of the many bacterial species which give pronounced liquefaction zones are: *Pseudomonas* species, such as *Ps. putida* and *Ps. aeruginosa; Bacillus* species, such as *B. subtilis;* many sporulating anaerobes, such as *Cl. sporogenes, Cl. histolyticum,* and *Cl. perfringens;* and divergent bacterial species, such as *Serratia marcescens, Proteus vulgaris, B. prodigiosus, B. proteus,* and *B. pruni.* Among the many mold species giving liquefaction are *Aspergillus niger, A. oryzae, A. flavus,* and *Penicillium roquefortii.*

Microbial protease is seldom one single proteolytic enzyme but consists of a system of proteolytic enzymes (proteases) which include proteinases or enzymes capable of breaking down the intact protein molecule into large fragments (proteoses and peptones), and peptidases (polypeptidases and dipeptidases) or enzymes which attack only the more or less degraded protein molecules and break them down to amino acids. There are two kinds of polypeptidases, one which attacks the peptide linkage at a point nearest the amino end of the molecule (amino-polypeptidase) and another which attacks at a point nearest the carboxyl end (carboxy-polypeptidase). By the combined action of these enzyme systems, the protein molecule is broken down completely into its amino acid components. Microbial proteinases are usually excreted directly into the medium, whereas peptidases are often endocellular enzymes and enter the medium only after autolysis of the cell has taken place.

Our knowledge of the complicated mixture of proteolytic enzymes present in microbial protease liquors is still very limited. No successful separation of such enzymes has been obtained. In general, they form a link between the plant proteinases, such as papain and ficin, and the tryptic enzymes of animal origin. With few exceptions, industrial interest in proteases is aimed at bringing about the initial stages of protein breakdown; for this reason, their proteinase activity is more important than their peptidase activity.

Bacterial Protease

As far as is known to the author, commercial production of bacterial protease is done exclusively with organisms of the *B. subtilis* group, under conditions very similar to those described for the production of bacterial amylase.

In general, *B. subtilis* produces both amylolytic and proteolytic enzymes. However, the ratio in which these enzyme systems are produced may vary considerably with the strain used, the composition of the medium, the pH, the temperature, and the degree of aeration. For this reason, the strain of *B. subtilis* used for commercial production of protease is selected only after a thorough study of protease yields from a large number of strains tested and is not the same strain as used for amylase production. The medium used for protease production is also quite different. Contrary to expectation, it was found that addition to the medium of carbohydrate, such as sugar or starch, markedly increased protease yields and often suppressed amylase excretion. In a recent study of the best media for protease production by *B. subtilis*, Kline, MacDonnell, and Lineweaver[47] came to the conclusion that a maximum yield is obtained when the medium contains 0.75 to 1.0% protein from such sources as soybean meal, peanut meal, or asparagus butt juice and, in addition, 6% carbohydrate, such as glucose, starch, or molasses, and salts of calcium and phosphorus equivalent to 0.025% calcium and 0.05% phosphorus pentoxide. Part of the organic nitrogen can be replaced by inorganic nitrogen without sacrifice in yield. The favorable influence of calcium ions on the production of bacterial protease is also stressed by other investigators.[33,34,58]

Other typical media recommended for protease production are those of Ramon, Richou, and Ramon[76] consisting of 8% wheat bran, 0.4% dried brewers' yeast and 0.6% dried malt extract; of Wallerstein[95] using 12° Balling malt extract obtained by mashing barley malt with addition of autolyzed yeast to bring the protein content up to 1.2 to 1.5%; or of Boidin and Effront[16] which consists of soybean flour, peanut or linseed cake to give a protein concentration of 0.6 to 2.5%, plus 2 to 5% carbohydrate in the form of sugars or starch, and 0.5 to 1% calcium or magnesium sulfate.

The equipment and technique for commercial production of bacterial protease is identical with that used for bacterial amylase production. The initial pH of the sterilized medium is 6.5 to 7.0 and incubation of the culture is at 37°C in shallow layers, with sterile air traveling over the trays for better aeration. Optimal activity is usually reached in 3 to 5 days. A considerable drop in potency occurs when incubation is continued too long, especially when the pH of the culture is high.

In general, bacterial protease is less stable than bacterial amylase and considerable care has to be exercised in the preparation of a commercial protease. The culture filtrate, obtained from the harvested culture by the same technique as described for amylase production, is adjusted to neutral pH and concentrated under reduced pressures at temperatures lower than 40°C. By addition of 2 to 3 volumes of alcohol or acetone, or 1 volume of saturated ammonium sulfate, all the active enzyme can be precipitated. The precipitate is separated from the mother liquor by filtration or centrifugation and is air dried at low temperature in vacuum shelf dryers. For special purposes, the concentrate may be absorbed on certain inert materials, such as sawdust, and dried at low temperature. Once bacterial protease is in dry form, it is reasonably stable.

As pointed out before, commercial bacterial protease preparations can be expected to contain considerable amylase activity and vice versa. Although it is possible, by selective adsorption methods, to separate the two enzyme systems to some degree, this step is generally not practiced. Usually the presence of these additional enzyme systems does not interfere with the activity of the main enzyme and often their additional action is desirable.

The final form in which proteases are marketed depends entirely on their intended use. One of the main applications is in the leather industry for "bating" hides. The aim of this bating process is to modify slightly certain constituents of the hide by enzymic digestion, for giving the resulting leather a finer grain and texture, more pliability, and a far better general quality than is obtained when the bating process is omitted. The opinions as to the chemical changes which occur during bating differ widely. However, a marked physical change takes place; this is apparent in a considerable decrease in the swelling of the wet skins. For centuries, and long before the enzymic nature of the process was understood, this digestion was done by soaking the hides for various lengths of time in suspensions of dog or fowl manures.

Proteolytic enzymes, at first pancreatic preparations, but at present also bacterial or mold proteases, have taken over the functions of the dungs. By using standardized enzyme preparations, the danger of overbating, inherent in the old process and often resulting in serious damage to the leather stock, is almost entirely eliminated. Commercial bates contain 1 to 20% enzyme preparation, depending on the potency, with the remainder being inert matter, usually sawdust and ammonium salts. Since the bating process follows the dehairing of the hides by soaking in a lime solution, it is obvious that the lime-treated skins, even after repeated washings with water or dilute acid, are still very alkaline. This high pH might cause a complete destruction of the protease, before its action is completed, but the ammonium salts tend to offset this high alkalinity by acting as buffer and deliming agent. The bating bath is usually in the pH range of 8.0 to 8.6 and the temperature may vary between 25° and 35°C. Bacterial protease, because of its marked activity at relatively high pH (approximate pH range of activity is 6.0 to 9.0), is very suitable as a bating enzyme and excels fungal protease and plant proteases whose activities do not extend far above a pH of 8.0. However, the optimum pH for bacterial protease is 7.0 to 7.5, depending, to some extent, on the type of substrate; its optimum temperature is 37°C.

Another application of microbial protease is in the textile

industry for the removal of proteinaceous sizings, such as gelatin or casein, which are often used as fortifying agents in the weaving of silk or rayon. It is also used in liberating the silk fiber from the naturally occurring proteinaceous material (sericin) in which the fiber is embedded. Other interesting industrial uses for such enzymes are for the recovery of silver from used photographic films, for deproteinizing rubber, for the digestion of fish livers, thus liberating the fish oil,[50] as a spot remover in the dry cleaning industry, and for tenderizing meat. For more detailed information on technical application of enzymes the papers by Gale[30,31] and patents of Triggs[90] and Wallerstein[97] may be consulted. According to Lockwood,[52a] the monetary value of bacterial protease marketed exceeds $1,000,000 annually.

Fungal Protease

As has been pointed out previously, proteolytic enzymes are produced by a wide variety of fungi, including, among the best known, strains of *Aspergillus niger, A. flavus-oryzae, A. parasiticus, A. tamarii, A. ochraceus, A. fumigatus, A. wentii, A. effusus, Penicillium roquefortii,* and the mushroom fungi *Psalliota campestris* and *Cantharellus cibarius.* Bernhauer and Knobloch[9] give detailed data on the many types of proteolytic enzymes produced by fungi.

As a rule, there is a remarkable difference in proteolytic activity between strains of the same species of fungi, even among strains which seem morphologically alike. Therefore, a considerable amount of screening must precede the choice of a suitable strain for commercial production of fungal protease. According to Oshima and Church,[65] and also Ayers and Tobie,[2] the screening may be done by cultivating the strains on such media as skim milk-agar or soy-milk-agar. The strains which rapidly hydrolyze these protein substrates and produce a large clear halo around the colonies are also usually very active protease producers when cultivated under the proper technical conditions.

In the genus *Aspergillus,* there is a large proportion of very active protease producers, especially in the intermediate strains of the *A. flavus-oryzae* group. It may be recalled that this group also contains the most active amylase-producing strains. According to Oshima and Church,[65] this group contains strains which produce high protease, but poor amylase yields; other strains are poor protease, but good amylase producers; and again other strains produce both enzymes in high yields. By using different strains, one may, therefore, obtain different types of enzyme mixture. The choice of a suitable strain is also determined, to a certain extent, by the use for which the enzyme is intended. For instance, in the chillproofing of beer, it is desirable that the protease is active in a pH range of 4 to 5; for bating of hides, however, an enzyme active in a pH range of 8 to 9 is required. As early as 1918, Waksman[91] pointed out that most fungal proteases have an exceptionally wide pH range of activity, much wider than that of animal or bacterial proteases. For instance, Berger[7,8] found that protease of *A. parasiticus* was just as active at pH 5.5 as at pH 7.0 and that protease of *A. oryzae,* with an optimum at 7.0, still exhibited 40% of its maximum activity at pH 4.3. This is in good agreement with an earlier report by Oshima,[61] who found this enzyme active in the pH range of 3.5 to 9.0. Whereas most fungal proteases have a pH optimum around neutrality, some exhibit maximum activity at pH 4 to 5.[95]

Commercial Production of Fungal Protease

For commercial production of fungal protease, a variety of selected strains of *A. flavus, A. oryzae, A. wentii, Mucor delemar, Amylomyces rouxii,* and others are used.[95] The medium on which these fungi are cultivated is usually wheat bran,[50,65,92] soybean cake,[92] alfalfa meal,[92] broken grain, middlings, brewers' grain,[95] yeast, or mixtures of these compounds. Methods of preparing such media, the technique of cultivation, and the equipment used are identical with those for the production of fungal amylase by the mold-bran process (see Chapter 3) and, therefore, need no further discussion. Layers of 1 to 2 in. are used if the culture remains undisturbed and 2 or 3 ft if agitated.[95] Cultivation is continued for 3 to 5 days in a moist atmosphere at 30° to 40°C.

As soon as abundant spore formation occurs, all proteolytic enzymes are no longer intracellular but can be recovered quantitatively from the medium.[64,65] Harvesting must then take place since a considerable decrease in enzyme potency is observed on further incubation. The mold bran is dried at low temperatures or the active protease may be recovered by extraction with water or buffer at pH 6.5. The extract is concentrated under vacuum at temperatures below 40°C and the active enzyme may be precipitated by addition of 3 volumes of alcohol or acetone, or with ammonium sulfate, by a method similar to that described for bacterial protease.

Recent Investigations of Fungal Proteinase Production

In 1950, Maxwell[53a] reported a laboratory investigation made for the purpose of selecting suitable strains of *A. oryzae* and satisfactory conditions for large-scale production of proteinase for fellmongering wool from sheepskins. She reported that wheat bran containing 60% moisture was the best medium. The best conditions were sterilization by steaming for 3 hours and, after inoculation, incubation in 2 cm layers at 28°C for 89 hours. Best enzyme production was obtained with sparsely-growing, heavily-sporulating cultures obtained by drying near the termination of the growth period.

Considerable interest has developed recently in submerged culturing of molds for proteinase production. Dworschack, Koepsell and Lagoda[24a] surveyed four hundred ninety-one strains of the *A. flavus-oryzae* group and one hundred thirty-one strains of other *Aspergilli* in shake-flask cultures on 2% corn meal–1% soybean meal–0.5% calcium carbonate medium. Eighty strains of the *A. flavus-oryzae* group produced sufficient proteinase to warrant further study. Certain strains were found to be far superior and the enzyme from individual mold strains varied in ability to attack the various proteinaceous substrates tested, including gelatin, casein, and milk. Ability to hydrolyze gelatin appeared frequently; milk-clotting activity was rare. High amylase activity was found very infrequently in the cultures tested and was never associated with high proteinase activity. Culture filtrates from high-yielding strains were concentrated by vacuum evaporation and then dried to powder by lyophilization. The resulting solid enzyme preparations had higher casein-digesting ability than twelve commercial proteinase preparations tested for comparison. It was concluded that proteinase production by submerged cultivation of certain strains of the *A. flavus-oryzae* group is feasible. By selection of particular mold strains and culture conditions, it may be possible to favor some types of proteolytic activity over others.

Uses of Fungal Proteases

Fungal protease has optimal activity at 50°C,[64] and is used for the same purposes as bacterial protease. When digestion at a low pH is required, it has advantages over bacterial protease. In Oriental countries, for many centuries, fungal proteolytic enzymes have played the main role in the manufacture of such food products as soy sauce, tamari sauce, miso and tofu. The last two products are used as breakfast foods, the former being a soy-cereal paste and the latter a cheeselike product. These are prepared from cooked or steamed soybeans, often mixed with roasted wheat or rice, by mixing with *koji* (or starter) prepared on similar media inoculated with selected strains of *A. flavus-oryzae* or *A. tamarii.* During incubation in thin layers on trays, under controlled humidity and temperature, the beans become covered with mold mycelium. At the time of maximum enzyme production, brine is added and the enzymic digestion is allowed to proceed for a considerable length of time, often several months. Detailed descriptions of such processes have appeared in the literature.[20,45,60,77]

Mold proteases are also important components of the so-called "chillproofing agents" used in the brewing industry for clarification and maturing of beer and ale and for correcting hazes which often develop in beer during storage. Such turbidities, thought to be due to protein-tannin complexes, often develop under adverse conditions of temperature, especially under repeated agitation and cooling. Addition of proteolytic enzymes, active at relatively low pH, causes a slight digestion of these proteins and thus prevents the formation of insoluble complexes with tannin. Introduced in 1910 by Wallerstein,[93] modern chillproofing agents are usually mixtures of animal, plant, and microbial proteases, containing pepsin, papain, bacterial and fungal protease.[95] According to Wallerstein,[99] a tannase of fungal origin, added to such a proteolytic enzyme mixture, gives an even greater degree of stability by causing digestion of both components of the complex.

MICROBIAL PECTOLYTIC ENZYMES

Pectin-hydrolyzing enzymes have found considerable industrial application for the pretreatment of fruit juices and related liquids as an aid in their clarification. Commercial pectolytic enzymes are usually known by their trade names, "Pectinols" or "Filtragols." Little technical information is available on the manufacture of such enzyme preparations, but it is likely that they are obtained from mold strains of *Penicillium* or *Aspergillus* species. The Pectinols are seldom pure enzyme preparations and often contain more than 90% inert matter, such as sugar, apparently as a diluent.[29] The active part contains at least two types of pectin-decomposing enzymes. More detailed discussion of the pectic enzymes, with regard to nomenclature, occurrence, methods of determination, purification, specificity, and mode of action, may be found in the reviews by Phaff and Joslyn,[73] Kertesz,[46] and Lineweaver and Jansen.[51]

The nomenclature of pectolytic enzymes has been somewhat confused and the two most important ones are known by several names.

The enzyme which hydrolyzes the methoxyl groups in the soluble pectin molecule, with formation of methanol and polygalacturonic acid, has been variously designated as pectinesterase, pectin-methylesterase, pectase, pectin demethoxylase, pectin methoxylase, and pectolipase.[51,54] The first two names seem most suitable for the enzyme that deesterifies pectin, since they indicate the nature of the enzyme action, while pectase is the oldest name. This enzyme will be referred to in the following discussion as pectinesterase.

The enzyme which hydrolyzes soluble polygalacturonic acid to monogalacturonic acid has been called polygalacturonase, pectin-polygalacturonase, pectinase, and pectolase.[46,51] Although the older name, pectinase, is frequently used to designate this enzyme, it has also been used to designate pectolytic enzyme mixtures. As the first two names are more descriptive of the enzyme action, the term polygalacturonase will be employed here.

A third pectolytic enzyme, protopectinase, which hydrolyzes the insoluble protopectin of the cell wall of the plant into soluble pectin, is also present.[22] However, the existence of a separate enzyme for this reaction is not proved beyond doubt and protopectinase activity of commercial pectinols may be exerted by their polygalacturonase.[26]

Pectinesterase and polygalacturonase can be separated by fractional adsorption on cationic-exchange resins at pH 6.3, or by destruction of the pectinesterase activity at pH 0.6 and 25°C for 20 minutes.[41] Besides pectolytic enzymes, commercial pectinols contain also diastatic enzymes, invertase and maltase.[29]

Pectolytic enzymes are produced by several bacterial species, including *B. carotovorus,*[44] *B. subtilis,*[28] *B. mesentericus* var. *fuscus,*[27] *Cl. felsineum, Cl. acetoethylicum,*[3] *Ps. marginalis*[68] when cultivated under proper conditions. Very little is known about the properties of such bacterial enzymes. They may play an important role in the retting of flax.

Of much more industrial interest are the pectolytic enzymes of fungi. Many *Aspergillus* species, such as certain strains of *A. niger,*[3,22,101,102] *A. flavus,*[101,102] *A. oryzae,*[75,91,92] *A. fumigatus,*[101,102] *A. parasiticus,*[101,102] and *A. wentii*[101,102] are good producers of such enzymes. *Penicillium* species, such as *P. glaucum,*[101,102] *P. ehrlichii,*[26] and *P. chrysogenum,*[72] *Rhizopus* species, such as *R. tritici,*[22,35,101,102] and *R. nigricans,*[75,101,102] *Monila fructigena,*[57] *Botrytis cinerea,*[3,22,28,57,75] *Fusarium chromiophthoron,*[57] *Fusarium fructigenum,*[57] and *Mucor* species[75] have all been reported to excrete active pectolytic enzymes.

It has been found that production of pectolytic enzymes by molds is usually greatly stimulated by the presence of pectin or pectin-containing compounds in the nutrient medium.[3,25,35] Pectic acid, D-galacturonic acid, mucic acid, L-galacturonic acid, and galacturonic-acid-containing gums, such as gum tragacanth, also stimulate production of pectolytic enzymes with certain mold strains.[72] It seems reasonable, therefore, to assume that pectin- or galacturonic-acid-containing compounds are probably one of the main carbon sources in the industrial production of such enzymes. According to Lockwood,[52a] the commercial pectin-hydrolyzing enzymes are produced by *Penicillia* grown on pectin-containing substrates, such as sugar-beet cossets.

As a nitrogen source ammonium sulfate,[72] yeast or malt extract,[28] potato decoction, asparagin, peptone, ammonium tartrate,[28] and gelatin or casein[13] have been recommended. According to Fernando,[28] the best nitrogen sources (for *Botrytis* strains) are those which cause a pH shift to the alkaline side. As a result of the action of the excreted enzymes on pectin, the pH drops considerably due to the liberation of free carboxyl groups.[75]

The ratio of the two pectolytic enzymes in the culture filtrates varies greatly with the fungus used.[73] Often the rate of pectin breakdown by commercial preparations is limited by their low pectinesterase content. Jansen and MacDonnell[41] were able to increase the activity seven times by the addition of extra plant pectinesterase to such preparations. Phaff and Joslyn[73] suggest, therefore, the combination, in proper proportion, of the enzymes of two fungi, one rich in pectinesterase, the other rich in polygalacturonase.

There is also a great variation in the excretion of pectolytic enzymes by different strains of the same mold species,[71] even when cultivated on the same medium under identical conditions. The choice of a high-yielding strain is, therefore, of prime importance.

The pectolytic-enzyme system is partially excreted into the medium and partially remains in the cell.[28] Certain mold species, such as *R. chinensis* and *R. microsporus,* excrete almost all their enzymes into the culture medium.[35] In order to obtain the enzyme from the mold mycelium, this must be dried, ground, and then extracted with water. The enzyme system may be obtained in dry form from such a mycelium extract, or from the culture medium, by precipitation with 3 to 4 volumes of alcohol or acetone.[25]

The optimum pH range for fungal pectinesterase activity is 3.0 to 4.5; that of fungal polygalacturonase activity is between 3.7 and 4.5. Optimum temperature for both enzymes varies between 37° and 55°C.[29,71] At neutral pH, or at temperature of 75°C, the enzymes are rapidly destroyed. Bacterial polygalacturonase has a much higher pH optimum, close to neutrality,[28,63] making it less suitable for clarification of acid fruit juices.

The first commercial application of pectolytic enzymes was for the pretreatment of fruit juices, wines, vinegars, sirups and jellies, as a clarification aid for such products.

Most fruit juices, obtained by expressing fresh fruits, such as apples, apricots, oranges, and grapes, are quite turbid due to the presence of finely dispersed colloidal matter. The market value of certain of these, such as apple cider, is much higher when the juice is sparkling clear.[53] However, filtration to a clear cider is almost impossible, unless the colloidal matter responsible for the turbidity is first eliminated or coagulated. Since the food value as well as the original flavor of fresh fruit must be retained, the use of processing aids is rather limited and experience has shown that a proper enzyme treatment gives the best results. Cloudiness in fruit juices is greatly increased by the presence of pectins and pectinlike substances which act as protective colloids for suspended particles. Moreover, the presence of pectins gives the juice a high viscosity which also tends to stabilize the colloidal suspension. If these pectins are broken down by enzymes, the colloidal particles lose their stability, and a flocculent precipitate is formed which can be easily removed by filtration, yielding a clear fruit juice.

Another recent application of pectolytic enzymes is the removal of pectins from fruit juices which are to be concentrated. For example, an apple juice which is concentrated without removal of the pectins will jell. Removal of pectins is necessary in such concentrated fruit juices to keep them liquid.

In several cases, it has been found advantageous to add pectolytic enzymes to crushed fruits before expressing the juice. Breakdown of the pectin at this stage often increases the yield of juice considerably and facilitates subsequent expressing.[36] Addition of pectolytic enzymes to crushed grapes results in greater ease of pressing and greater yield of juice.[36,106] Treatment of certain grape juices with pectolytic enzymes prior to fermentation has been described; the resultant wines clarify much sooner than those made from untreated grape juice.[10,36] (See Chapter 7 of Volume I.)

Another possible application for pectolytic enzymes is in the production of D-galacturonic acid.[78] Galacturonic acid is a useful compound for a number of organic syntheses, including that of vitamin C. Less valuable pectins, such as those from sugar beets or citrus waste products, might be used for such a process.[25,38,40,78]

The application of tomato pectinesterase for the preparation of polygalacturonic acid for use as low-solids gels or pectinate coatings has been patented.[104] Some glucosidic degradation by polygalacturonase in the tomato-enzyme preparation can be obviated by using citrus-peel pectinesterase[66] which is free of polygalacturonase. Such low-methoxyl pectins are too sensitive to precipitation by calcium to be used in the practical preparation of low-solids gels, but they form very viscous solutions and may have useful applications, e.g., as paper coatings and agar substitutes.[67]

The use of commercial pectolytic enzymes to determine "pectic enzyme soluble substances" of foods and food products has been recommended.[40] However, since the commercial pectolytic-enzyme preparations contain several other polysaccharide-hydrolyzing enzymes, such as amylase and inulase,[29,52] the use of purified polygalacturonase in combination with a deesterifying agent could be more desirable for this purpose, if available.[52]

INVERTASE

Invertase (sucrase or saccharase) is the enzyme responsible for the biochemical hydrolysis of sucrose into glucose and fructose. The enzyme also hydrolyzes certain tri- and tetra-saccharides by removing either the fructose or the glucose end of the molecule. Invertase occurs widely in many animal and plant tissues, as well as in a great variety of microorganisms, such as numerous yeast species, filamentous fungi (*A. oryzae, A. niger, P. glaucum*), and bacteria (*E. coli, Thermobact. mobile*).

As a rule, fungi contain less invertase than yeasts and fungal invertase differs fundamentally from yeast invertase. Yeast invertase is a fructosidase (β-*h*-fructosidase, fructofuranosidase), attacking the fructose end of the molecule, whereas fungal invertase is a glucosidase (α-*n*-glucosidoinvertase, α-glucopyranosidase), attacking the glucose end of the molecule. This explains why both types attack sucrose, but differ in their behavior toward certain tri- and tetrasaccharides. For instance, the trisaccharide raffinose (fructosylglucosylgalactoside) is split into fructose and melibiose by yeast invertase, but is not attacked by fungal invertase, because the glucose is blocked by the glucosidic linkage with galactose.[49]

Invertase is an endoenzyme and can be isolated only after mechanical disruption of the cell wall or after autolysis. Commercially, it is prepared from bakers' or brewers' yeast.

The invertase content of yeasts varies widely with the yeast strain and with the conditions of growth. The invertase yield of yeasts may be increased five to ten times by cultivation in the presence of sucrose.[105] One of the procedures for accomplishing this is as follows: To a 5% suspension of beer-yeast cake, 0.2% each of diammonium and dipotassium phosphate and 0.05% of magnesium sulfate are added. After adjustment of the pH to 4.5, the yeast suspension is vigorously aerated for about 8 hours at 30°C. During this period, the equivalent of 3% sucrose is added gradually, maintaining the pH at 4.5. Other manufacturers of yeast invertase, instead of using treated commercial bakers' or brewers' yeast, cultivate a special strain of *Saccharomyces cerevisiae* of exceptionally high invertase content at controlled pH and continuous aeration, using sucrose as a nutrient.[97,98]

After this pretreatment, resulting in a considerable increase in invertase yield, the yeast is filtered and the yeast cake is subjected to a plasmolyzing process, followed by autolysis. Chloroform, ether, ethyl acetate, or toluene is usually used as a plasmolyzing agent. The following is an example: To 100 lb of treated beer yeast cake 10 lb of toluene is added and the mixture is allowed to liquefy for 3 hours. Then 12 gal water is added and the pH adjusted to 5.8. Autolysis is allowed to proceed for 24 hours at 30°C. After completion of the autolysis, the cell debris may be separated from the autolysate by centrifugation, after adjusting the pH to 4.7. The invertase, which remains in the autolysate, may be precipitated by addition of 1.25 to 1.5 volumes of alcohol at pH 4.7 and 0°C. Better precipitating agents, which cause less potency loss, are isopropyl and isobutyl alcohols[94] or water-soluble ethers of the dioxane type.[62] The precipitate is either dried at low temperature or dissolved in 60% glycerol, ethylene glycol, mannitol, or sucrose.

Invertase has its optimum activity in the pH range 3.5 to 5.5 and is also most stable in this range. The more the enzyme is purified the more unstable it becomes. This is why most commercial invertase preparations are not highly purified.

Invertase is used in the manufacture of artificial honey and invert sugar, the second being much more soluble than sucrose. This is an advantage in confectionery manufacturing and in the preparation of liqueurs and ice creams where crystallization of high sugar concentrations must be avoided. A carefully standardized invertase preparation is usually incorporated in the formula for making chocolate-coated, soft-cream-center bonbons. The molding and coating of the bonbon is done while the content is still firm and hard; enzyme conversion and liquefaction of the sucrose to a smooth and stable cream takes place later.[69] Enzymically produced invert sugar is also used in the paper industry as a plasticizing agent.[24] Neuberg and Roberts[63] have published a detailed literature review on invertase.

MISCELLANEOUS MICROBIAL ENZYMES

Within the last few years several other microbial enzymes have appeared on the market. Few details are available as to the industrial methods used in producing these enzymes, but they are nevertheless worthy of mention.

Catalase, an enzyme which decomposes hydrogen peroxide to

oxygen and water is marketed in liquid preparations. It is used in bleaching and dyeing furs, in bleaching human hair to remove residual hydrogen peroxide, and in the manufacture of surgical gut. At least one commercial product is obtained by growing a mold in aerated-agitated culture.[52a]

Penicillinase is manufactured by aerated-agitated culturing of *B. cereus* or *B. subtilis*.[52a] It is precipitated from the clarified culture liquor with an organic precipitant, and is marketed as a dry enzyme concentrate. It finds greatest use in penicillin-production control where the Ford[29a] rapid assay method is used.

A new enzyme preparation was placed on the market in 1951 which contains a mixture of two enzymes, streptokinase and streptodornase.[1a] The preparation is made by hemolytic streptococci grown in aerated culture. It is used clinically in dressing wounds and burns as it disintegrates and clears away blood clots, pus, and dead tissue without attack on healthy tissues. Several production problems have been mentioned, including culture selection, purification of the enzymes, and effect of medium composition on enzyme yields.[1a,9a,26a]

The most recent commercial enzyme is glucose-oxidase. Is is marketed by at least two companies.[1b,1c] It is obtained in association with catalase by submerged culture of *A. niger*[52a] and is marketed in the form of a soluble powder. A principal use of the enzyme preparation is to remove glucose from egg whites and whole eggs before drying to prevent browning and deterioration. Oxidation of the glucose by the glucose-oxidase forms gluconic acid and hydrogen peroxide, the hydrogen peroxide then being decomposed by the catalase.

BIBLIOGRAPHY

1a. Anon., *Chem. Week,* **69**, No. 14, 22 (1951).
1b. Anon., *Chem. Week,* **72**, No. 20, 88 (1953).
1c. Anon., *Agr. Food Chem.,* **1**, 193 (1953).
1. Avery, J., and W. Burger, *Brit. Intell. Obj. Sub-Committee Rept., No. 1156* (1945).
2. Ayres, G. B., and W. C. Tobie, *J. Bact.,* **45**, 18 (1943).
3. Barinova, S. A., *Microbiology (U.S.S.R.),* **15**, 313 (1946).
4. Beaver, A. E. A., British Patent 561,706 (1944).
5. Beckord, L. D., E. Kneen, and K. H. Lewis, *Ind. Eng. Chem.,* **37**, 692 (1945).

6. Beckord, L. D., G. L. Peltier, and E. Kneen, *Ind. Eng. Chem.*, **38**, 232 (1946).
7. Berger, J., M. J. Johnson, and W. H. Peterson, *Enzymologia*, **4**, 31 (1937).
8. Berger, J., M. J. Johnson, and W. H. Peterson, *J. Biol. Chem.*, **117**, 429 (1937).
9. Bernhauer, K., and H. Knobloch, *Methoden der Fermentforschung*, **2**, 1324 (1941).
9a. Bernheimer, A. W., *Trans. N. Y. Acad. Sci.*, **14**, 137 (1952).
10. Besone, J., and W. V. Cruess, *Fruit Products J.*, **20**, 365 (1941).
11. Blinc, M., *Arch. Mikrobiol.*, **12**, 183 (1941).
12. Blinc, M., *Kolloid-Z.*, **101**, 126 (1942).
13. Bohne, A., German Patent 652,490 (1937).
14. Boidin, A., and J. Effront, U. S. Patent 1,227,374 (1917).
15. Boidin, A., and J. Effront, U. S. Patent 1,227,525 (1917).
16. Boidin, A., and J. Effront, U. S. Patent 1,882,112 (1932).
17. Bois, E., and J. Savary, *Can. J. Research*, **23B**, 208 (1945).
18. Casey, J. P., and E. P. Gillan, *Tappi, Monograph 3*, p. 1 (1947).
19. Christensen, L. M., U. S. Patent 2,359,356 (1944).
20. Church, M. B., *U. S. Dept. Agr. Bull. 1152* (1923).
21. Craig, W. L., U. S. Patent 2,360,828 (1944).
22. Davison, R. R., and J. J. Willaman, *Botan. Gaz.*, **83**, 329 (1927).
23. DiCarlo, F. J., and S. Redfern, *Arch. Biochem.*, **15**, 333, 343 (1947).
24. Diehm, R. A., *Paper Trade J.*, **106**, No. 24, 36 (1938).
24a. Dworschack, R. G., H. J. Koepsell, and A. A. Lagoda, *Arch. Biochem. Biophys.*, **41**, 48 (1953).
25. Ehrlich, F., *Handbuch der biologischen Arbeitsmethoden*, Sec. IV, Vol. 2, p 2405, Berlin, Urban und Schwarzenberg, 1936.
26. Ehrlich, F., *Enzymologia*, **3**, 185 (1937).
26a. Elliott, S. D., *Trans. N. Y. Acad. Sci.*, **14**, 137 (1952).
27. Fabian, F. W., and E. A. Johnson, *Michigan Agr. Exp. Sta. Bull. 157* (1938).
28. Fernando, M., *Ann. Botany* (N.S.) **1**, 727 (1937).
29. Fish, V. B., and R. B. Dustman, *J. Am. Chem. Soc.*, **67**, 1155 (1945).
29a. Ford, J. H., *Anal. Chem.*, **19**, 1004 (1947).
30. Gale, R. A., *Wallerstein Labs. Commun.*, **4**, 112 (1941).

31. Gale, R. A., *Nat. Cleaner Dyer,* **32**, 18 (1941).
32. Gates, R. L., and E. Kneen, *Cereal Chem.,* **25**, 1 (1948).
33. Haines, R. B., *Biochem. J.,* **25**, 1851 (1931).
34. Haines, R. B., *Biochem. J.,* **27**, 466 (1933).
35. Harter, L. L., and J. L. Weiner, *J. Agr. Research,* **21**, 609 (1921); **22**, 371 (1921); **24**, 861 (1923); **25**, 155 (1923).
36. Hickinbotham, A. R., and J. L. Williams, *J. Dept. Agr. South Australia,* **43**, 491, 596 (1940).
37. Hockenhull, D. J. D., and D. Herbert, *Biochem. J.,* **39**, 102 (1945).
38. Hollander, C. S., U. S. Patent 2,370,961 (1945).
39. Imshenetskii, A. A., and L. I. Solntseva, *Microbiology (U.S.S.R.),* **13**, 54 (1944).
40. Isbell, H. S., and H. L. Frush, *J. Research Nat. Bur. Standards,* **33**, 389, 401 (1944).
41. Jansen, E. F., and L. R. MacDonnell, *Arch. Biochem.,* **8**, 97 (1945).
42. Johnson, M., *Proc. Yeast Symposium,* Milwaukee, 1948.
43. Johnston, W. W., and A. M. Wynne, *J. Bact.,* **30**, 491 (1935).
44. Jones, L. R., *N. Y. State Agr. Exp. Sta. Bull.* **11** (1909).
45. Kellner, O., *Coll. Agr. Tokyo Imp. Univ.,* **5**, 9 (1888).
46. Kertesz, Z. I., *The Enzymes,* edited by J. B. Sumner and K. Myrbäck, Vol. I, Part 2, New York, Academic Press, 1951.
47. Kline, L., L. R. MacDonnell, and H. Lineweaver, *Ind. Eng. Chem.,* **36**, 1152 (1944).
48. Kneen, E., and L. D. Beckord, *Arch. Biochem.,* **10**, 41 (1946).
49. Kuhn, R., *Nature,* **11**, 732 (1923).
50. Lennox, F. G., and M. E. Maxwell, Australian Patent 118,850 (1944).
51. Lineweaver, H., and E. F. Jansen, *Advances in Enzymology,* Vol. XI, New York, Interscience, 1951.
52. Lineweaver, H., R. Jang, and E. F. Jansen, *Arch. Biochem.,* **20**, 137 (1949).
52a. Lockwood, L. B., *Trans. N. Y. Acad. Sci.,* **15**, 2 (1952).
53. Marshall, R. E., *Mich. Agr. Exp. Sta. Bull. 14,* 208 (1938).
53a. Maxwell, M. E., *Australian J. Appl. Sci.,* **1**, 348 (1950).
54. McColloch, R. J., and Z. I. Kertesz, *J. Biol. Chem.,* **160**, 149 (1945).
55. Mehlitz, A., and H. Maass, *Biochem. Z.,* **276**, 86 (1935).
56. Mehlitz, A., and M. Scheuer, *Biochem. Z.,* **268**, 345 (1934).
57. Menon, K. P. V., *Ann. Botany,* **48**, 187 (1934).

58. Merrill, A. T., and W. M. Clark, *J. Bact.,* **15**, 267 (1928).
59. Meyer, K. H., M. Fuld, and P. Bernfeld, *Experientia,* **3**, 411 (1947).
60. Morikawa, K., Unpublished thesis, Massachusetts Institute of Technology, Cambridge, Mass., 1926.
61. Myrbäck, K., and L. G. Gjorling, *Arkiv. Kemi. Mineral., Geol.,* **20**, No. 5, 13 (1945).
62. Neuberg, C., and I. S. Roberts, U. S. Patent 2,406,624 (1946).
63. Neuberg, C., and I. S. Roberts, *Invertase Monograph,* New York, Sugar Research Foundation, 1946.
64. Oshima, K., *Coll. Agr. Hokhaido Imp. Univ.,* **19**, 125 (1928).
65. Oshima, K., and M. B. Church, *Ind. Eng. Chem.,* **15**, 67 (1923).
66. Owens, H. S., R. M. McCready, and W. D. Maclay, *Ind. Eng. Chem.,* **36**, 936 (1944).
67. Owens, H. S., R. M. McCready, and W. D. Maclay, *Food Technol.,* **3**, 77 (1949).
68. Oxford, A. E., *Nature,* **154**, 271 (1944).
69. Paine, M., and J. Hamilton, U. S. Patent 1,437,816 (1922).
70. Peltier, G. L., and L. D. Beckord, *J. Bact.,* **50**, 711 (1945).
71. Phaff, H. J., Unpublished thesis, University of California, Berkeley, California, 1943.
72. Phaff, H. J., *Arch. Biochem.,* **13**, 67 (1947).
73. Phaff, H. J., and M. A. Joslyn, *Wallerstein Labs. Commun.,* **10**, 133 (1947).
74. Poore, H. D., *Fruit Products J.,* **14**, 170, 201 (1935).
75. Proskuryakov, N. I., and F. M. Osipov, *Biokhimiya,* **4**, 50 (1939).
76. Ramon, G., R. Richou, and P. Ramon, *Compt. rend.,* **220**, 341 (1945).
77. Ramsbottom, J., *Brit. Assoc. Adv. Sci., Ann. Rept.,* 1936.
78. Reitz, E., and W. D. Maclay, *J. Am. Chem. Soc.,* **65**, 1242 (1943).
79. Rose, D., *Arch. Biochem.,* **16**, 349 (1948).
80. Schultz, A. S., and L. Atkin, U. S. Patent 2,159,678 (1931).
81. Shellenberger, J. A., *Chem. Engr.,* **54**, No. 2, 130 (1947).
82. Smits van Waesberghe, F. A. M. J., Thesis, University of Delft, Netherlands, 1941.
83. Smythe, C. V., B. B. Drake, and C. E. Neubeck, U. S. Patent 2,530,210 (1950).
84. Surovaya, A. V., *Tekstil. Prom.,* **4**, 14 (1944).

85. Takokora, T., *J. Chem. Soc. Japan,* **62**, 1255 (1941); **63**, 751 (1942).
86. Takamine, J., U. S. Patent 525,823 (1894).
87. Tilden, E. B., and C. S. Hudson, *J. Am. Chem. Soc.,* **61**, 2900 (1939).
88. Tilden, E. B., and C. S. Hudson, *J. Bact.,* **43**, 527 (1942).
89. Tilden, E. B., M. Adams, and C. S. Hudson, *J. Am. Chem. Soc.,* **64**, 1432 (1942).
90. Triggs, W. W., British Patent 559,588 (1944).
91. Waksman, S. A., *J. Bact.,* **3**, 509 (1918).
92. Waksman, S. A., U. S. Patent 1,611,700 (1927).
93. Wallerstein, L., U. S. Patents 995,824 and 995,825 (1911).
94. Wallerstein, L., U. S. Patent 1,919,675 (1933).
95. Wallerstein, L., U. S. Patents 2,077,447; 2,077,448; 2,077,449 (1937).
96. Wallerstein, L., U. S. Patent 2,116,089 (1938).
97. Wallerstein, L., *Ind. Eng. Chem.,* **31**, 1218 (1939).
98. Wallerstein, L., *Wallerstein Labs. Commun.,* **7**, 5 (1939).
99. Wallerstein, L., U. S. Patent 2,223,753 (1940).
100. Wallerstein, L., U. S. Patents 1,854,353; 1,854,354; 1,854,355 (1932).
101. Willaman, J. J., and Z. I. Kertesz, *N. Y. State Agr. Exp. Sta. Bull. 589,* (1930).
102. Willaman, J. J., and Z. I. Kertesz, *N. Y. State Agr. Exp. Sta. Bull. 178* (1931).
103. Willaman, J. J., and Z. I. Kertesz, U. S. Patent 1,932,833 (1933).
104. Willaman, J. J., H. H. Mottern, C. H. Hills, and G. L. Baker, U. S. Patent 2,358,430 (1944).
105. Willstätter, R., and C. D. Lowry, *Z. physiol. Chem.,* **146**, 158 (1925); **150**, 287 (1925).
106. Yang, H. Y., G. E. Thomas, and E. H. Wiegand, *Wines & Vines,* **31**, No. 4, 77 (1950).

Part IV. THE PRODUCTION OF VITAMINS

CHAPTER 5

PRODUCTION OF RIBOFLAVIN BY FERMENTATION

R. J. Hickey

INTRODUCTION

Microbiologically produced riboflavin, or lactoflavin as it is also known, has long been available in yeast and in related preparations in association with many other vitamins, particularly those of the B-complex (see Chapter 6). The general history of riboflavin has been widely covered in reviews, such as those of Rosenberg,[67] Schopfer,[69] and Van Lanen and Tanner.[92]

Riboflavin is a unique vitamin in that is can be totally synthesized to very high concentrations rather rapidly by certain microorganisms. Reports on authentic precursors or adjuvants for biosynthetic riboflavin have not appeared, but the mechanism of formation has had recent study by Plaut.[47a] Fermentations have been reported in which the riboflavin content of the fermented beer amounted to 1,500 to 2,000 μg per ml and higher. The organisms involved in such processes are the *Ascomycetes, Eremothecium ashbyii,*[40] and *Ashbya gossypii.*[54] Fermentation riboflavin has a good position in competition with synthetic riboflavin.

In 1948, the United States Tariff Commission reported the production of 130,400 lb riboflavin for human, animal, and poultry consumption. Of this, 113,000 lb were sold, valued at $7,033,200 or about $62 per pound. These figures were not broken down into production by fermentation and chemical synthesis, but it is known that a very considerable proportion of this quantity was fermentation riboflavin. The price of riboflavin in 1953 was in the range of $45 to $60 per pound.[7a]

Riboflavin is synthesized, to some degree, by many various types of microorganisms, including bacteria, yeasts, and molds.[92] Other organisms require riboflavin for growth and may, therefore, be used for assays.[79] The organisms which have been found to produce sufficient riboflavin to be, or to have been, of interest from a commercial standpoint are relatively few. Those of prime importance for riboflavin formation rather than for the B-complex are noted in Table 13. In this table, the relation of iron concentration to riboflavin formation by the organism is also noted, since in some processes, the iron content of the medium is a factor controlling the riboflavin yield. Other bacterial genera have also been reported[91,92] which produce quite low levels of riboflavin.

TABLE 13. MICROORGANISMS PRODUCING CONSIDERABLE RIBOFLAVIN AND THE EFFECT OF IRON ON THE BIOSYNTHESIS

Organism	Riboflavin in culture fluid, μg per ml	Optimum iron concentration, μg per ml	Reference
Mycobacterium smegmatis	57.5	Not critical	32
Clostridium acetobutylicum	97.0	1 to 3	34
Mycocandida riboflavina	200.0	Not critical	29a
Candida flareri	567.0	0.04 to 0.06	26
Ashbya gossypii	1760.0	Not critical	54, 56
Eremothecium ashbyii	2480.0	Not critical	40

Commercial fermentation processes for production of riboflavin or riboflavin concentrates are relatively recent, having been developed since about 1937. Aside from food yeasts, the first organism employed primarily for riboflavin production was *Clostridium acetobutylicum,* which was used both in the United States and in Japan. Following the butyl alcohol-acetone fermentation of grain or molasses media, the solvents were removed by distillation and the still residues were dried to yield products containing about 20 to 65 μg of riboflavin per gram of residue. Such products were of considerable value not only because they could be used profitably as a feed supplement, but also because drying the slops provided a very good solution to a stream-pollution problem of some manufacturers of solvents.

The *Cl. acetobutylicum* fermentations of grain or lacteal products, e.g., whey, were improved during the succeeding years after it was found that by controlling the iron concentration in the media, dried residues containing riboflavin to 4,000 to 5,000 μg per g and higher could be obtained.

About 1940, it was discovered in the United States that an *Ascomycete, Eremothecium ashbyii,* was able to convert nutrient media and, in particular, cheap waste products, such as residues from the alcoholic fermentation of grains, into riboflavin solutions of 200 to 400 μg per ml or higher. The fermentation could be conducted in submerged culture in about 4 to 6 days and the iron content of the medium was not particularly critical to the riboflavin yield. Ordinary steel equipment was satisfactory and the resulting "beers" were of sufficient riboflavin potency to make recovery of crystalline riboflavin practicable.

Within the last few years, improved *E. ashbyii* fermentation processes have been described in the patent literature. Yields of 1,000 to 2,000 μg per ml or higher have been reported. Since 1946, another *Ascomycete, Ashbya gossypii,* has also been reported as being capable of producing high-riboflavin beers from cheap raw materials. Yields higher than 1,000 to 1,500 μg per ml have been reported. The *Ascomycete* processes are now assuming a fairly important position in the riboflavin-manufacturing field.

COMMERCIAL FERMENTATION PROCESSES FOR RIBOFLAVIN PRODUCTION

Clostridium Acetobutylicum Processes

The *Clostridium acetobutylicum* processes for the formation of riboflavin are anaerobic, in contrast to the *Candida, Eremothecium ashbyii,* and *Ashbya gossypii* processes which are aerobic. As for the *Candida,* appreciable riboflavin formation by *Cl. acetobutylicum* is dependent on a low iron content of the medium. The *Ascomycete* processes (*E. ashbyii* and *A. gossypii*), however, are relatively unaffected by reasonable concentrations of iron, such as might be introduced by handling in iron or steel equipment.

There are primarily three methods which have been or are being used for riboflavin production by *Cl. acetobutylicum.* The classification is based largely on the medium, or substrate, and are: (a) the molasses process, (b) the grain process, and (c) the milk-product processes. Relative yields of some of these processes are indicated in Table 14.

TABLE 14. BUTANOL-ACETONE PROCESSES FOR RIBOFLAVIN PRODUCTION

Medium	Riboflavin yield μg per ml	Reference
Molasses	1 to 2	35
Grain, corn	25 to 30	2
Grain, corn and rice	40 to 50	93
Whey or skim milk	20 to 50	33
Whey and xylose	70 to 97	34

Since contamination of the media with iron is a major factor which may limit riboflavin production by *Cl. acetobutylicum,* various means of controlling the iron content have been studied and employed. The best method, of course, is to prevent entry of iron into the medium in the first place. In the grain processes,[2,93] excess iron is removed by mechanically cleaning the grain. In the milk products processes, the lacteal raw material (e.g., whey) is produced under conditions[33] which limit the iron content of the medium. Molasses generally contains considerable amounts of inorganic matter, including iron, and fermentations usually yield a low-riboflavin "beer" containing about 1 to 2 μg of riboflavin per ml. Control of iron concentration in the medium by the use of 2,2′-bipyridine (α,α′-dipyridyl)[17,19] to bind it as biologically unavailable iron was unsuccessful in producing higher riboflavin yields in molasses media, although this method worked satisfactorily in grain media. The reason for the difference is not clear.

Other methods for obtaining improved riboflavin yields have been reported. Yamasaki[101] added calcium carbonate or acetate to grain media during the first 7 hours and obtained higher yields. Van Lanen and Tanner[92] consider this to be a practical application of the methods of Steinberg[80,81,82] who removed trace elements, in this case iron particularly, by adsorption on calcium carbonate. Legg and Beesch[24] described a process which employed 0.005 to

0.05% sodium sulfite or other sulfites in grain mashes for improved yields.

Recently, Leviton,[27,28] using synthetic-type media, found that the addition of certain chemicals to a medium containing an excessive amount of iron would allow fermentation to proceed with a good yield of riboflavin. Among these agents were catalase, potassium iodide, sodium hydrosulfite, and yeast dialysate. Leviton suggested that hydrogen peroxide may be involved in riboflavin destruction when appreciable iron is present. No peroxide was found in the beer, however, and a transient existence was postulated. The fact that the presence of catalase is effective in promoting higher riboflavin yields certainly suggests the presence of a peroxide. Leviton showed that iron catalyzes the decomposition of riboflavin in the presence of hydrogen peroxide. The aerobic, high-riboflavin-producing *Ascomycetes,* such as *Eremothecium ashbyii,* undoubtedly produce catalase. These organisms also perform well in the presence of sufficient iron to cause low riboflavin levels in *Cl. acetobutylicum* processes. They yield riboflavin in excess of 1,000 μg per ml. The *Candida* processes are aerobic and are very sensitive to iron. Catalase production would also be expected here.

The Molasses Process

The fact that riboflavin was produced during the butyl alcohol-acetone fermentation of molasses media was described by Miner[35] in a patent in 1940. He pointed out that the dried residue of the molasses fermentation was a good source of riboflavin and other members of the B-complex. The riboflavin content of the "beer" is 1 to 2 μg per ml.

When a high-test molasses medium is employed, following fermentation for about 72 hours at 32°C, the solvents are stripped in a still and the residue, or "slop," is concentrated and dried. Spray or drum driers can be used. The dried product may contain about 50 to 60 μg of riboflavin per g, along with other B-complex vitamins. The product was found to be quite useful and profitable as a feed supplement.

Several organisms have been described in the patent literature for use in the molasses process. An example is *Cl. saccharo-butyl-acetonicum-liquefaciens-delta.*[35] Such organisms are considered by Bergey[3] to be types of *Cl. acetobutylicum.*

As was previously noted, control of iron content of the media at low levels did not result in improved riboflavin yields,[17] in contrast to the findings in the grain and lacteal product fermentations. The fermentations are generally carried out in the customary steel equipment.

The Grain Processes

The fact that acetone-butyl alcohol fermentations of grain yielded a yellow product under some conditions had been known for many years.

About the same period that Miner[35] noted that riboflavin was formed in the butyl fermentation of molasses, Yamasaki[97,98,99,100,101] observed that cereal grain fermentations could yield a yellow pigment which was identified as riboflavin. He obtained a United States Patent[101] on the process in 1942. Yamasaki's method[101] employs *Cl. acetobutylicum* for the fermentation of 4 to 8% cereal-grain mashes, to which, between 0 and 7 hours after inoculation, a small amount of calcium carbonate or calcium acetate is added. The use of calcium salts resulted in considerably increased riboflavin yields. In addition to grain mashes, Yamasaki also employed whey and other carbohydrate media with good results.

In 1943, Arzberger[2] pointed out that one of the basic factors controlling riboflavin formation in the *Cl. acetobutylicum* fermentation of cereal-grain mashes is the concentration of certain metallic ions in the medium. In particular, Arzberger cited iron, cobalt, nickel, copper, zinc, and lead salts as having a deleterious effect on riboflavin formation if present in sufficient concentration. His data show that, of the salts mentioned, those of iron and cobalt suppress riboflavin formation at the lowest levels cited, i.e., 3.2 mg of the salt per l of medium. Increasing concentrations of salts of the other metals, except nickel, did not strongly suppress riboflavin formation until the fermentation was suppressed as indicated by low yields of solvents. Nickel acetate, at about 320 mg per l, depressed the riboflavin yield without appreciably affecting the solvents, but one-tenth and one-hundredth of this level did not greatly affect riboflavin synthesis. It is possible that the riboflavin suppressive effect of the nickel acetate at about 320 mg per l was due largely to the presence of iron and cobalt impurities in the nickel salt. Some of Arzberger's data are given in Table 15.

Table 15. EFFECT OF SOME HEAVY-METAL SALTS ON RIBOFLAVIN FORMATION IN 5% CORN MASH BY *Cl. acetobutylicum*

Salt concentration,	Riboflavin and solvents yields[a] at the salt concentrations noted									
	0		3.2		32.0		320.5		3205.0	
mg per salt added	Riboflavin	Solvent	Riboflavin	Solvent	Riboflavin	Solvent	Riboflavin	Solvent	Riboflavin	Solvent
Ferrous sulfate ($7H_2O$)	2,720	26.4	810	26.6	197	27.8	69	28.0	145	24.3
Cobaltous acetate ($4H_2O$)	3,150	29.0	620	27.7	242	27.9	64	12.8	—	—
Nickel acetate ($4H_2O$)	3,150	29.0	2,850	28.7	2,240	28.9	680	28.9	38	6.6
Cupric acetate ($1H_2O$)	2,890	28.8	3,180	31.0	2,030	31.0	41	6.7	—	—
Zinc acetate ($2H_2O$)	2,890	28.8	3,430	27.6	2,945	29.4	2,440	29.2	55	4.0
Lead acetate ($2H_2O$)	2,890	28.8	2,500	29.6	2,295	29.8	2,110	28.4	39	5.3

[a] Riboflavin yields are expressed as micrograms per gram of dried filtrate. Solvents yields are expressed as percentage of original corn, dry basis.

Source: Arzberger.[2]

Riboflavin is formed in the corn-fermentation process, to the extent of 1 to 2 μg per ml, when operated in iron or steel equipment. In glass or aluminum equipment, mashes prepared from cleaned grain, low in iron, yield about 25 μg per ml or more riboflavin after fermentation.

Walton[93] described a modification of Arzberger's method in which mixtures of corn and rice were used in the mash instead of corn alone, as cited by Arzberger. Riboflavin yields of 40 to 45 μg per ml were obtained. Legg and Beesch[24a] used wheat flour with other flour and obtained about 26 μg riboflavin per ml. In the processes of both Arzberger and Walton, the grain was subjected to special handling, such as blowing with air and magnetic treatment for removal of iron-containing substances. In both processes, *n*-butyl alcohol, acetone, and ethyl alcohol were by-products of economic value. Walton cited riboflavin contents of up to 4,670 μg per g in the dried filtrates.

The optimum iron concentration for maximum riboflavin formation by *Cl. acetobutylicum* is in the range of 1 to 3 μg per ml,[92] which is fifty to one hundred times as high as the optimum level for the *Candida* processes.

The general fermentation procedures are typical of those of the regular butyl alcohol-acetone fermentation of grain (see Chapter 11 of Volume I), except that excess iron contact must be avoided. Aluminum equipment has been recommended.[2]

The Milk-Product Processes

In 1938, Yamasaki[101] found that if a whey-type medium is subjected to the butyl alcohol-acetone fermentation and a calcium salt, such as the carbonate, is added during the first 7 hours of fermentation, considerable improvement in the riboflavin yield will result.

In 1945 and later, Meade, Pollard, and Rodgers[53,64,65,66] obtained several patents[33,34,49,50,51,52] describing methods for riboflavin formation by the *Cl. acetobutylicum* fermentation of lacteal materials. The basis of the method was primarily control of the inorganic nutrition of the organism during fermentation, though the supplementary use of xylose[34] also resulted in substantial increases in riboflavin yield. Methods of handling were also important.

Lacteal products, such as whey and skim milk, contain members of the B-vitamin complex, including riboflavin, along with other nutrients, such as lactose and casein.[33] The riboflavin content of whey is approximately 1.7 μg per ml. If whey is prepared in a standard commercial manner, i.e., in iron or steel vessels, the iron content may be as high as 10 or 12 μg per ml (or ppm). Whey prepared in the laboratory by rennet treatment of skim milk was found to contain about 0.37 μg per ml iron. Meade, Pollard, and Rodgers[33] found that the commercial high-iron whey underwent the *Cl. acetobutylicum* fermentation quite vigorously with good lactose utilization and good solvents yield, but the riboflavin yield was poor, being around 1.8 μg per ml. The low-iron whey (0.37 μg per ml) was found to ferment sluggishly and incompletely, leaving considerable lactose unfermented after 48 hours at 38°C. The riboflavin yield was about 6 μg per ml. If the total iron was adjusted to 1.5 to 2.0 μg per ml before sterilization, the fermentation was quite vigorous and complete and the riboflavin yield was 20 to 30 μg per ml.

It was mentioned[33] that it was of advantage to include trace amounts of salts of strontium, tin, manganese, lithium, and zinc in the medium. The recommended concentration was 0.00003 molar. The medium was adjusted to pH 6 to 7 prior to fermentation. The media were sterilized for 15 to 20 minutes at 121°C. The process yielded up to 50 μg per ml of riboflavin. Some of the data are summarized in Table 16.

TABLE 16. EFFECT OF IRON CONTENT OF WHEY MEDIUM ON RIBOFLAVIN PRODUCTION BY *Cl. acetobutylicum*

	Concentration of Agent in Medium			
Medium	Sugar (as lactose), %	Protein, %	Iron, μg per ml	Riboflavin μg per ml
Whey, unfermented	4.70	0.90	variable	1.7
Whey, fermented, low iron	3.20	0.80	0.37	6.0
Whey, fermented, high iron	0.53	0.74	5.00	1.8
Whey, fermented, optimum iron	0.71	0.74	1.5 to 2.0	20 to 30

Source: Meade, Pollard, and Rodgers.[33]

In an improved process, in 1947, Meade, Rodgers, and Pollard[34] showed that the inclusion of xylose in the lacteal medium results in improved riboflavin yields from the *Cl. acetobutylicum* process. Xylose, in a concentration of 0.5 to 1% may be added, though somewhat higher or lower levels are still useful. The media were also supplemented with 0.15% by weight $CaCO_3$, 0.15% $Ca_3(PO_4)_2$, 4.8 μg per ml $ZnSO_4 \cdot 7H_2O$, 4.2 μg per ml $MnSO_4 \cdot H_2O$, and 0.01 μg per ml *p*-aminobenzoic acid. Iron is cited at a concentration of about 1 to 2 μg per ml, though the iron range is claimed between 0.5 and 4.5 μg per ml.

Without the supplementary xylose, riboflavin yields of about 60 to 64 μg per ml were noted, while with about 0.5 to 1.0% xylose added, riboflavin yields of 92 to 97 μg per ml were obtained. The authors felt that xylose did not act simply as another energy source. Xylose can be supplied by the use of sulfite liquor, wood hydrolysates, etc.

Several related patents were issued to Pollard, Rodgers, and Meade. These correlate variable concentrations of iron in the lacteal media to the zinc content as it affects riboflavin formation,[52] to the magnesium content,[51] to the manganese content,[53] and to the ammonium content.[50] The relationship between iron content and pH is detailed in other patents,[49,65] while inoculum development is described in still another.[66]

Fungal Processes

Synthesis by Yeasts

Food Yeasts. Commercial food or bakers' yeasts contain 40 to 85 μg riboflavin per g.[92] Investigations by Massock[31] and by Van Lanen[92] of methods for improving riboflavin formation by *Saccharomyces cerevisiae* were unsuccessful. Others[43,77] have been able to bring about minor alterations, but no methods have been found which would produce riboflavin specifically in high concentration by the use of *Saccharomyces.* This is also true for *Torulopsis utilis.*[92] Detailed information on yeasts and their food value may be found in Chapters 9 and 10 of Volume I, and 6 of this volume.

Synthesis by Candida Yeasts. Burkholder,[5,6,7] in 1943 and later, observed that *Candida guilliermondia* was able to produce significant amounts of riboflavin. In his media, which were largely synthetic, he obtained riboflavin concentrations of up to 75 μg per ml. This fact was likewise observed in 1944 by Schopfer[72] who noted in addition[71] that *C. tropicalis* var Rhaggi was a producer. Some irregularities seemed to occur in the riboflavin biosynthetic abilities of *C. guilliermondia*[86] which were shortly explained by Tanner and associates[86,87,92] on the basis of iron control of the biosynthesis. This effect is similar to that of iron on the *Cl. aceto-*

butylicum synthesis[2] of riboflavin. The optimum concentration of iron for the *Cl. acetobutylicum* synthesis is 1 to 3 μg per ml, while for *C. guilliermondia* it was found to be 0.005 to 0.01 μg per ml, though 0.04 to 0.06 μg per ml was also used successfully.[26] Thus the *Candida* process permits only 1 to 2% of the iron permissible for *Cl. acetobutylicum* processes.

Tanner and coworkers[86,87] found that, with proper control of the iron level of media, two strains of *C. guilliermondia* and one of *C. flareri* were able to produce more than 100 μg per ml riboflavin in 5 to 7 days, *C. flareri* having produced 216 μg per ml. They described a medium containing salts, urea, asparagin, biotin, and glucose. Heavy metals were removed by the 8-hydroxy-quinoline-chloroform method of Waring and Werkman.[94] Iron was then added to the desired level, preferably 0.005 to 0.01 μg per ml and the media were sterilized. The fermentation was conducted with aeration and agitation.

A more practical procedure was also described,[22] in which a 4% glucose solution was passed through a deionizing resin (for example, Nalcite X) for iron removal. Salts, asparagin, and biotin were added, the pH was adjusted to about 5.0, and, after inoculation with *C. guilliermondia,* the medium was incubated for 7 days at 30°C. A riboflavin yield of about 103 μg per ml was obtained in the example cited. The control, which was not resin-treated, yielded 17.5 μg per ml.

In 1949, a modified procedure was described by Levine, Oyaas, Wasserman, Hoogerheide, and Stern[26] who were able to conduct their process without sterilization in plastic vessels. Using *Candida flareri,* they obtained riboflavin yields as high as 567 μg per ml in glucose-containing media, with the average yield being 339 μg per ml. Similar results were obtained when sucrose was used in place of glucose. These authors employed largely synthetic media, such as the one given in Table 17. Satisfactory sterility was generally maintained, since the pH of the medium, which was initially 5 to 6, dropped rapidly to 3.0 to 3.5. This low pH level is antagonistic to most bacterial contamination.

The riboflavin yield paralleled, to a large degree, the yield of yeast. The process was run with aeration and agitation for 4 or 5 days. Dried products were obtained with riboflavin potencies up to 28,000 μg per g for the whole product and up to 97,000 μg per g for the soluble solids.

TABLE 17. MEDIUM FOR RIBOFLAVIN PRODUCTION BY CANDIDA YEASTS

Agent	Quantity
Distilled water, or low-iron tap water	1,000 ml
K_2HPO_4	0.5 g
$MgSO_4 \cdot 7H_2O$	0.2 g
Urea	1.84 g
Sucrose or glucose	40.0 g
Biotin[a]	1.0 μg
Zn	140.0 μg
B, Mn, Cu, and Mo	20.0 μg each
Fe, from all ingredients above	40-60 μg

[a] Cane molasses, 0.5 g per l, may be employed as source of biotin if the iron is removed.

Source: Levine, Oyaas, Wasserman, Hoogerheide, and Stern.[26] [Reprinted by permission from *Ind. Eng. Chem.*, **41**, 1665 (1949).]

The extent of commercial interest in *Candida* processes has not been made public. The critical handling requirement with respect to iron is an obvious disadvantage of these processes not inherent in the *Ascomycete* processes.

In 1951, McClary[29a] reported the use of an organism named *Mycocandida riboflavina* which produced riboflavin in glucose-corn steep-ammonium sulfate media under acid conditions. Yields of 60 to 200 μg per ml were claimed in an agitated process. The organism is evidently not critical with regard to the iron content of the medium, since corn steep liquor and cane molasses have been employed as raw materials.

SYNTHESIS BY ASCOMYCETES

Outstanding among the riboflavin-producing microorganisms are the two *Ascomycetes, Eremothecium ashbyii* Guilliermond and *Ashbya gossypii* (Ashby and Nowell) Guilliermond *(Nematospora gossypii)*. Both organisms are capable of synthesizing very large amounts of riboflavin under appropriate conditions.

In 1947, Moss and Klein[40] reported biosynthesis of riboflavin as high as 2,480 μg per ml by *E. ashbyii* in media containing ground lentils. In 1949, Pridham and Raper[54,56] noted that up to 1,760 μg per ml could be produced by *A. gossypii* under rather rigid fermentation conditions.

Assuming a beer to contain 2,000 μg of riboflavin per ml, calculation shows that 10,000 gal of such a beer would contain about 167 lb of riboflavin. At about $60 per pound,[7a] this quantity would be worth about $10,000, if it were all recovered. With cheap raw materials, such a process is obviously profitable.

In addition to the importance of high-assay beers, it is of great commercial significance that neither organism is particularly adversely affected by moderate levels of iron in the medium, i.e., levels which are excessive for *Candida* and *Clostridium acetobutylicum* processes. For this reason, iron or steel equipment can be used for handling the beers.

SYNTHESIS BY *Eremothecium ashbyii*

Eremothecium ashbyii and *Ashbya gossypii* are plant parasites.[12,14] *E. ashbyii* was first found as a cotton parasite at Berber, Sudan, by R. E. Massey. Confirmation was made by S. F. Ashby. Guilliermond[12,14] described both organisms in 1928 and 1936. *E. ashbyii* is very much like the description of *Eremothecium cymbalariae* Borzi[1,4,13] which was found on an ivy in Italy in 1888, but which has not been found and identified since.

Many studies, leading to the identification of *E. ashbyii* as a riboflavin producer of some importance, were made by Guilliermond, Raffy, and others,[14,15,36,37,38,57,58,59] starting about 1935. Guilliermond, Fontaine, and Raffy[15] observed in 1935 that yellowish crystals were found at times in the vacuoles of cells of *E. ashbyii*. This yellow substance was identified by Raffy[57] in 1937 as riboflavin. Recently Grob[16] showed spectroscopically that the riboflavin in the culture media exists chiefly in the free form.

Culture. *E. ashbyii* is a heterothallic *Ascomycete,* but only one sexual form has been described as yet. This is in contrast with *A. gossypii* which is homothallic.

Cultural variations of *E. ashbyii* were discussed in some detail by Schopfer and Guilloud[75] in 1945. These consisted mainly of yellow, flavin-producing variants, a poorly sporulating, cream-colored form, which produced relatively little flavin, and white forms, which produced only traces. The white form was described in 1944 by Ritter.[62] The culture may be carried on wort and other nutrient agar media. Schopfer and Guilloud[75] found that the yellow form constantly produced white variants, but the white form was stable and did not yield yellow variants. Desieve[9] observed that the organism gradually lost its flavogenetic ability.

Nutrition. Studies on nutritional requirements of *E. ashbyii* have been made by a number of investigators and not all of the published information is in agreement. Intensive studies were undertaken by Schopfer[70,73] in 1944 and later. Under his conditions, biotin was reported to be essential for proliferation while *i*-inositol and thiamin were found to be supplementary factors. These observations were disputed by Dulaney and Grutter[9a] who found an absolute requirement for *i*-inositol, but not for biotin or thiamin. These latter findings were supported by Yaw.[102] Other, unidentified factors for growth or riboflavin formation by *E. ashbyii* were noted in Norite-treated peptone,[70,73] wheat germ,[70,73] potato extracts,[70,73] yeast,[23a] rice embryo extracts,[23a] milk products,[9,19a,19b,45,46] lipids,[46] gliadin,[19b] and liver extract.[19b]

In their search for satisfactory carbohydrate nutrients for *E. ashbyii,* Schopfer and Guilloud[75] found glucose, sucrose, levulose, and mannose to be useful. Tabenkin,[85] however, employed, in addition to those mentioned, glycerol, mannitol, sorbitol, maltose, invert sugar, and certain others. In 1952 Larsen[23b] showed that in stillage media, the use of a glucose plus a maltose supplement resulted in higer riboflavin yields than were obtained with either sugar supplement used separately. For example, in a medium employing wheat-flour stillage from the ethyl alcohol fermentation, a 7-day, aerobic *E. ashbyii* fermentation resulted in a riboflavin yield of 418 μg/ml in the control with no sugars added, up to 813 μg/ml with glucose supplements of up to 5 g/100 ml, and 1,350 μg/ml with 0.5 g/100 ml of glucose and 2.5 g/100 ml of maltose. Lower yields were obtained with maltose alone. The maltose-glucose mixture was employed effectively also in casein-containing and certain other media.[19a,19b]

Studying nitrogen metabolism, Schopfer and Guilloud[75] found that arginine, asparagine, ammonium tartrate, ammonium citrate, and ammonium succinate were among the useful simple nitrogen sources. The same authors[74] were unable to grow *E. ashbyii* on a medium containing glycine, glucose, mineral salts, biotin, thiamin, and *i*-inositol. The addition of L-leucine or L-arginine or both supported growth to some extent. About 25 μg riboflavin was formed per ml. Studies were also made, in 1944 and 1945, by Renaud and Lachaux[61] who obtained 159 μg/ml riboflavin in 24 days, using nonsynthetic media.

The cultures of Dulaney and Grutter[9a] did not utilize glycine as a nitrogenous substrate either, under the conditions described, but L-(—)-proline, L-(+)-arginine, and DL-glutamic acid allowed growth and low-level riboflavin formation if *i*-inositol was present. Yaw[102] found histidine to be advantageous in combination with methionine. She also studied D-ribose, alloxan, and other potential precursors and found none to be effective.

In 1950, the Japanese investigators Minoura and Kozima published a series of papers[23a,35a,35b,35c] on nutrition of *E. ashbyii* and on the relationship of growth to riboflavin production. Minoura[35a] found that both *i*-inositol and phytin, which occur in rice bran, act as growth stimulants. Phytin, however, exerted its effect only in media containing suitable nitrogen sources. It was presumed that phytase formation by *E. ashbyii* is dependent on the utility of the nitrogen source in the medium, which, therefore, controls the conversion of phytin to inositol. Minoura[35b] also sought an active nutritional principle in peptone. He observed that since deaminized peptone hydrolyzate affects growth only slightly, but does not affect riboflavin formation, the principle is probably an amino compound. Asparagine and arginine were found to affect growth, but not riboflavin production. Kozima,[23a] studying extracts of a number of natural products, found that yeast and rice-embryo extracts were particularly effective growth promoters. Vitamin requirements were also studied,[35c] with vitamins B_1, B_6 and pantothenic acid showing some enhancing action beyond that observed for inositol and biotin. It was concluded that riboflavin formation did not parallel the extent of mycelial growth.

Perhaps some of the apparently contradictory findings by different investigators might be related to the use of different or modified strains of *E. ashbyii* which may have different requirements. In some instances, conclusions were drawn on the basis of studies in which relatively low levels of riboflavin were produced under the optimum conditions cited. Multiple deficiencies may have been involved, complicating the problem considerably. The importance of lipids[46] in nutrition has not always been recognized.

Very recently, MacLaren[28a] studied the effects of pyrimidines and purines on riboflavin formation by *E. ashbyii* and suggested precursor action. Xanthine, adenine, guanine, uric acid, and perhaps hypoxanthine promoted riboflavin formation, with xanthine being the most effective. In contrast, uracil was observed to significantly suppress riboflavin formation. Competitive inhibition was suggested. Yaw[102] found xanthine to be ineffective.

The existence of a heat-labile factor for *E. ashbyii* was noted by Desieve.[9] This was found in milk products, such as casein, and was destroyed on extended sterilization. (See also under *A. gos-*

sypii.) In 1949, Phelps[45] found that casein had a unique nutritional advantage over other milk ingredients in promoting riboflavin formation by *E. ashbyii.* Lactalbumin was not as effective. Phelps also showed[46] that certain lipids, when added to several proteinaceous media, produced very marked improvements in riboflavin yield, *i.e.*, from about 500 μg per ml without the lipid to about 1,000 or more μg per ml when a lipid, such as corn oil or butterfat, was included in the medium to the extent of 0.6 to 1.2%. Rudert[68] had previously observed that lipids could replace carbohydrates as energy sources in media, but Phelps'[46] work shows a growth-factor type of response.

A "synthetic" medium for the cultivation of *E. ashbyii,* which allowed formation of over 500 μg riboflavin per ml was described by Tabenkin[85] in 1950. (See under Commercial Methods.) This medium contained sucrose, ammonium salts as nitrogen sources, and other salts. Nitrogen, in the form of nitrate or nitrite, was nutritionally ineffective. The sterilized medium was inoculated with 4% by volume of a 2-day liquid culture (not detailed). This was then aerated and agitated, yielding 571 μg riboflavin per ml in 5 days. The inclusion of *i*-inositol in the medium raised the yield to 714 μg per ml in 6 days. Essential elements in assimilable form, in addition to carbon, phosphorous, sulfur, and ammonium nitrogen, were indicated as zinc, potassium, copper, iron, and magnesium. Calcium carbonate or certain other agents were employed as buffers, the preferred pH range being 5.5 to 6.0. It is of interest that Tabenkin added no biotin or thiamin as such to his media. These factors may have been introduced into the medium by way of the carbohydrates or in the inoculum.

In 1953, studies were described[19b] which resulted in a nearly "synthetic" or reproducible medium which was capable of supporting a riboflavin yield of up to about 1,000 μg/ml. The "synthetic" basal medium involved is given in Table 18. This consisted of sugars, vitamins, salts, an amino acid, and an oleate and resulted in riboflavin yields of up to 300 μg/ml in *E. ashbyii* fermentations. Replacement of 1 g of the 20 g/l glutamate with certain complex nitrogenous agents, e.g., casein and gliadin, resulted in increases in riboflavin titer to about 800 μg/ml. Certain other proteinaceous agents were not effective, e.g., peptone, gelatin, and zein; the last two agents even suppressed riboflavin formation to a level of 80 μg/ml. Not all casein samples, i.e., from different sources, were equally effective; one of the best studied was a "vitamin-free" casein. Strepogenin did not appear to be involved. Replacement of all the glutamate of the medium of Table 18 with casein resulted in riboflavin yields of up to 1,500 μg/ml. At this casein level, 20 g/l, the type of casein had relatively little effect on the riboflavin yield in comparison with differences noted where 1.0 g casein plus 19 g monosodium glutamate per liter was used.

TABLE 18. GLUTAMATE BASAL MEDIUM FOR RIBOFLAVIN PRODUCTION BY *Eremothecium ashbyii*

Agent	Quantity per liter
Monosodium D-glutamate	20 g
Maltose	10 g
Glucose	15 g
K_2HPO_4, KH_2PO_4, and NaCl[a]	1 g each
$MgSO_4 \cdot 7H_2O$	0.7 g
$CaCl_2$[a]	0.5 g
$MnSO_4 \cdot 4H_2O$[a] and H_3BO_3[a]	20 mg each
Sodium oleate	0.6 g
i-Inositol	30 mg
Thiamin	1.5 mg
Biotin	0.7 μg
Distilled water	to 1 liter
pH = 6.2	

[a] Essentiality not established.
Source: Hickey.[19b]

It was also observed[19b] that while oleate appeared to act effectively as a growth factor or metabolite for *E. ashbyii* in an otherwise reasonably adequate medium, yeast extract was more effective than sodium oleate on a weight basis in stimulating riboflavin production. It was concluded that there were probably at least two as yet unidentified nutritional factors for *E. ashbyii,* one being supplied by certain complex nitrogenous agents and the other by yeast extract, although oleate in appreciable quantity can substitute for lesser amounts of yeast extract. Further studies may result in an adequate, chemically completely defined medium which will support maximum riboflavin formation by *E. ashbyii.*

Commercial Methods. Descriptions of commercial methods for riboflavin production are primarily limited to patent literature. It would not be surprising to find that there are secret methods or secret modifications of described methods.

Some of the methods to be described here may be in operation today. Perhaps some of them were operated and have been changed or modified because of changes in prices of raw materials or for other reasons. The methods described can be considered as commercial, potentially commercial, or the bases of operating commercial processes.

Methods for cultivation of *E. ashbyii* may be surface methods exposed to air, or submerged methods in which air is dispersed throughout the mash, with or without supplementary agitation.

In 1943, Okabe, Hatsuda, and Ueda[42] described a method employing a rice-germ medium for surface culture in Japan. In 1949, in the United States, Martin[29] patented a method for surface culture of *E. ashbyii* on a porous, wet, solid medium, such as a wheat-bran or oat-hull medium containing additional nutrients. Whole dried residues were obtained containing riboflavin to a concentration of 3,000 to 4,000 μg per g, along with other B-complex agents. Other related procedures have been reported in which soybean-oil cake, other seed cakes, rice and barley germ, and different brans were used as supporting media.[21b,90a]

Most patented methods employ submerged culture of *E. ashbyii* in aerated liquid media. A good liquid or submerged-culture process has the obvious advantages over a solid- or semisolid-medium process of efficient use of space and investment and ease of transmission of mash or fermented medium ("beer") from one site to another. The following methods will emphasize the fact that *E. ashbyii* is a truly remarkable and most valuable microorganism.

In 1946, Piersma[47] claimed riboflavin yields up to 500 μg per ml in liquid media composed of animal proteins, carbohydrates, and malt extract. The proteinaceous substances were 2 to 4% glandular materials, tankage, fish meal, etc. Sugar-containing substances, such as molasses and corn sirup were employed at about 1%, and malt extract, at 1.5 to 2%. The initial pH was 5.5 to 6.5 and incubation was at 27°C with aeration.

In 1947, Foster,[11] in a British patent, described a process illustrated by the following example. The medium was composed of 1% yeast extract, 6% crude dextrose, 0.2% potassium phosphate, and 0.05% magnesium sulfate at an initial pH of 4.6 to 5.6. This was sterilized for 15 minutes at 15 lb steam pressure, cooled, and inoculated with 1% by volume of an aerated culture of *E. ashbyii*. The medium was then aerated for 5 days at 27° to 30°C. At the end of this period, centrifugation yielded a supernatant medium containing 412 μg of riboflavin per ml. Foster noted that other B-vitamins, particularly pantothenic and nicotinic acids, were also obtained following fermentation.

In 1949, Stiles[84] patented a process for the production of riboflavin by the fermentation of several nutrient media with *E. ashbyii.* The patent discloses the use of stillage ("slops") from the ethyl alcohol and butyl alcohol-acetone fermentations in aerated, submerged culture with or without nutritional adjuncts. Also covered, in addition to the fermentation residues, are fish meal, dried yeast, skim milk, cottonseed meal, soybean meal, peas, tankage and other agents. When these were employed, the enzyme, "Clarase," was cited as being used to treat the raw materials prior to fermentation. Sterilized media were inoculated with, for example, 0.7% by volume of a culture of *E. ashbyii* and the fermentation was allowed to proceed in submerged culture with aeration for about 90 hours at about 29°C. Some examples of the Stiles process are given in Table 19. Stiles also mentioned the use of lipids with protein substances (see Rudert[68]). The highest riboflavin yield shown by Stiles was 436 μg per ml which was produced in a supplemented, thin, grain-stillage medium. Moore and de Becze[39] also noted the use of stillages in 1947; a patent[8a] was issued in 1951 describing the use of fortified thin stillage to yield

TABLE 19. SOME RAW MATERIALS EMPLOYED IN EXAMPLES OF STILES' PROCESS FOR RIBOFLAVIN PRODUCTION BY *E. ashbyii*

Medium	pH		Riboflavin yield,
	Initial	Final	μg per ml
Thin grain stillage (TGS), alcoholic	5.3	6.9	290
TGS + 0.3% high-test molasses	5.4	7.4	352
TGS + 0.3% corn-steep liquor + 0.2% glucose added at 42 and 73 hours	5.8	7.5	436
2.25% meat scraps + 0.75% glucose	6.5	7.6	138
3.0% hard red wheat	6.5	7.6	158
2.25% tankage + 0.75% glucose	6.5	7.6	162
3.0% ground field beans	5.5	8.1	170
3.0% corn-gluten meal	6.5	7.1	140
3.0% peas	6.5	7.8	180
3.0% soybean meal	6.5	7.8	130
3.0% cottonseed meal	6.5	7.6	105
2.25% fish meal + 0.75% glucose	6.5	7.9	225
3.0% wheat middlings	6.5	7.6	214
3.0% linseed-oil meal	6.5	8.0	118
3.0% dried yeast	6.5	7.6	207
Skim milk (hand skimmed)	7.0	7.5	406
Commercial skim milk (liquid)	7.0	7.7	250

Source: Stiles.[84]

a product containing 2,000 to 10,000 μg of riboflavin per gram of solids. A process employing whole peanut flour, supplemented with a carbohydrate, was described recently by Beesch and Frazer.[2b] It is of interest that stillages were also employed as substrate for B-vitamin synthesis by *Aerobacter* species.[41]

Another process involving the use of cereal grain was patented in 1950 by James.[21a] In this procedure, a highly malted wheat mash was employed a a substrate for the aerobic cultivation of *E. ashbyii.* Riboflavin yields in excess of 1,000 μg per ml were obtained from media containing about 12% initial grain solids.

In 1948, a British patent issued to Chas. Pfizer & Co., Inc.,[44] described one of the higher yields reported for *E. ashbyii.* A medium cited contained 5% brown sugar, 3% neo-peptone (Difco), 1% wheat germ, 0.3% beef extract (Difco), 0.3% KH_2PO_4 and 0.25% NaCl made up in distilled water. The initial pH was 6, the inoculum 10%, the temperature 28°C, the fermentation period 7 days, and the yield 1,800 μg per ml. Agitation was employed in addition to aeration.

More recently, Larsen[23b] demonstrated that in the fermentation of stillages from the alcoholic fermentation of cereal grains, the use of sugar mixtures as more advantageous than that of single sugars. The preferred mixtures contained maltose supplemented with glucose, sucrose, or levulose. In an example, wheat-flour stillage, supplemented with 0.5 g glucose and 4 g maltose per 100 ml of medium, yielded 1,610 μg per ml riboflavin after 10 days of submerged fermentation with *E. ashbyii.*

In 1949, Phelps obtained two patents[45,46] which are of considerable interest. One of these[45] describes media containing casein or other milk products, commercial malt extract, and glucose, which, on fermentation with *E. ashbyii* in submerged, aerated culture, yielded in some cases over 600 μg riboflavin per ml. In one example, a basal medium containing 1.75% commercial malt extract and 0.5% cerelose (glucose hydrate) was supplemented with several milk products which contained casein. After 88 hours at 30° to 32°C, the riboflavin yields were 510 to 690 μg per ml.

Phelps[45] demonstrated that casein was the primary constituent of milk which was responsible for high riboflavin yields. He showed that the malt extract-cerelose medium (see before), when supplemented with skim milk, casein, whey plus casein, casein plus lactalbumin, or whey plus casein plus lactalbumin, after fermentation, yielded riboflavin in the range of 532 to 612 μg per ml. When the supplements were whey or lactalbumin, the yields were 192 to 232 μg per ml.

In his other patent, Phelps[46] supplemented his skim milk, malt extract, glucose type of medium with butterfat and showed that when about 1% butter fat was added to the medium, the riboflavin yield amounted to 1,120 μg per ml in 88 hours, compared with a control containing no butterfat which yielded 455 μg per ml. Phelps also studied the effect of other lipids. These were added to a basal medium which contained 1.75% (dry basis) malt extract, 0.5% glucose, and 1% casein solubilized by treatment with ammonia. (A 10% aqueous suspension of casein was heated for 1 hour at 45° to 50°C in the presence of 0.7 ml of 26% ammonia per g of casein.) The lipids studied included corn oil, soybean oil, cocoanut oil, lard oil, oleomargarine, lecithin, menhaden oil, and others. They were added to the medium at a concentration of 0.6 to 1.25%. Riboflavin yields ranged from 810 to 1,200 μg per ml, as compared with a control having no lipid supplement which yielded 420 μg riboflavin per ml.

In 1952, Hickey[19a] described a patented process which employed sodium oleate, along with other nutrients, in casein media and resulted in riboflavin yields of up to 1,370 μg per ml after 5 days of submerged, aerated cultivation of *E. ashbyii*. Both maltose and glucose were employed as carbohydrates.

Phelps[46] also described the effect of lipids on proteinaceous media other than casein or lacteal media. Table 20 shows the effect on the riboflavin yield of supplementing several types of proteinaceous media with several lipids. It is evident from the data of Table 20 that the addition of lipids to various media will stimulate riboflavin synthesis by *E. ashbyii* to the extent of a 100% increase or more in some instances.

In 1950, Tabenkin[85] described a "synthetic" medium for riboflavin production by *E. ashbyii*. His basal medium contains, for example, per liter of medium: 50 g sucrose, 3 g $(NH_4)2HPO_4$, 5 g $CaCO_3$, 2 g K_2HPO_4, 0.5 g $MgSO_4 \cdot 7H_2O$, 0.2 g $FeSO_4 \cdot 7H_2O$, 0.04 g $ZnSO_4 \cdot 7H_2O$, 0.005 g $CuSO_4 \cdot 5H_2O$. A pH range of 5.5 to to 6.0 was preferred. This medium, after sterilizing, was inoculated with a 4% by volume of a 2-day liquid inoculum of *E. ashbyii*. The medium, on vigorous aeration and agitation at 28° to 30°C for 5 days, yielded 571 μg of riboflavin per ml. The same operation, but with a supplement of 0.05 g of inositol per 1 yielded 714 μg per ml in 6 days. It is of interest that Tabenkin used essentially no complex nitrogen in the medium.

A process for riboflavin synthesis by *E. ashbyii* in excess of 2,000 μg per ml was covered by a British patent in 1947. This was the procedure of Moss and Klein[40] of Roche Products Ltd., who found that good yields of riboflavin could be produced by *E. ashbyii* in submerged culture and media containing lentils *(Lens esculenta)* and a fermentable carbohydrate. Control of pH at about 7 by intermittent addition of cane molasses, and the use of ammonium salts of certain organic acids favorably influenced results. Incubation was at 28° to 30°.

TABLE 20. ENHANCEMENT OF RIBOFLAVIN YIELD BY *Eremothecium ashbyii* BY INCLUSION OF LIPIDS

Medium	88-hour riboflavin yield, μg per ml: No lipid added	Cream added	Corn oil added	Peanut oil added
Basal[a] + 3% fish meal	815	1,295	1,120	1,095
Basal[a] + 2% peptone	290	375	440	430
Basal[a] + 5% distillers' solubles	560	985	1,040	985
Basal[a] + 3% liver cake[b]	410	830	815	1,065
Basal[a] + 1% solubilized casein[c]	530	960	1,015	930

[a] Basal contained (dry basis) 1.75% malt extract and 0.5% glucose; all media adjusted at pH 6.5 to 7.0 before sterilizing.
[b] Digested at pH 9.5 with a proteolytic enzyme.
[c] Solubilized by ammonium hydroxide.
Source: Phelps.[46]

Examples of the method of Moss and Klein are presented in Table 21. Inoculum was 10% by volume of a heavy mycelial suspension. Octadecanol in paraffin oil was employed as an antifoam.

It is possible that riboflavin yields by *E. ashbyii* may exceed 2,500 μg per ml. The high yields reported are quite amazing. They are also unique in the field of microbiological vitamin synthesis.

SYNTHESIS BY *Ashbya gossypii*

Ashbya gossypii (Ashby and Nowell) Guilliermond, also called *Nematospora gossypii,* is a parasite of some importance to cotton, coffee, and certain other plants.[12,14,54,56,83] In 1928, Guilliermond[12] had observed the similarities of *E. ashbyii* and *A. gossypii.* In 1935, Guilliermond, Fontaine, and Raffy[15] had found that riboflavin was produced rather weakly by *A. gossypii.* This was also

TABLE 21. RIBOFLAVIN PRODUCTION BY *Eremothecium ashbyii*

Medium	Fermentation, days	Temperature, °C	Riboflavin found, μg per ml
1. 4% ground lentils, 1% cane molasses, 0.5% NaCl[a]	8	28	1,400
2. 4% ground lentils, 5% cane molasses, 0.5% NaCl[b]	7	28	1,100
3. 4% ground lentils, 0.75% D.B.[c] sucrose, 0.25% inverted D.B. sucrose, 0.3% ammonium succinate, and 0.5% NaCl	8	28	1,700
4. 4% ground lentils, 0.75% D.B. sucrose, 0.25% inverted D.B. sucrose, 0.27% ammonium succinate, 0.23% sodium succinate, and 0.5% NaCl	7	28	1,810
5. 2% ground lentils, 1.125% D.B. sucrose, 0.375% inverted D.B.[d] sucrose, 0.5% ammonium succinate, and 0.5% NaCl	5.7	28	2,480

[a] Medium was autoclaved for 30 minutes at 120°C and 2 days later again.
[b] Lentils had been steamed for 3 hours prior to use.
[c] Dark brown.
[d] Sterile molasses or brown sugar solutions were added from time to time as means of maintaining the pH between about 6.8 and 7.5. Without this supplementation, the pH rose above 7.5.
Source: Moss and Klein.[40]

the finding of Schopfer and Guilloud.[76] In 1946, Wickerham, Flickinger, and Johnston[95] reported that a strain of *Ashbya gossypii* (NRRL Y-1056), which they obtained in 1943 from W. J. Robbins, yielded a variant whose colonies on agar produced a bright, orange-yellow color after a few days at room temperature. In 1950, Pridham and Raper[56] published a review on *A. gossypii* which covered occurrence, pathogenicity, taxonomy, cultural, morphological, and physiological characteristics, along with brief information on riboflavin production.

Wickerham, Flickinger, and Johnston[95] pointed out that the orange-yellow variant[56] produced riboflavin to a significant extent in the presence of moderate amounts of iron in contrast with *Candida* yeasts and *Clostridia* which required very low iron levels to form important amounts of riboflavin. They showed that in a medium composed of 0.3% powdered-yeast extract, 0.5% peptone, and 2% Cerelose (glucose) at pH 6.8 to 7.0 in aerated laboratory fermentations, yields up to about 380 μg per ml were obtained in 8 days. The addition of 0.3% distillers' solubles and

0.1% $CaCO_3$ improved results somewhat. They also reported that a different lot of yeast extract gave poor riboflavin yields, perhaps because of deficiency in certain factors.

In 1947, Tanner and Van Lanen[88] noted briefly preferable conditions for riboflavin formation by *A. gossypii*. The general procedure was patented in 1948 by Tanner, Wickerham, and Van Lanen.[89] Tanner, Vojnovich and Van Lanen[90] detailed their findings more completely in 1949, with laboratory and pilot-plant-scale studies made at the Northern Regional Research Laboratories.

The optimum temperature range for riboflavin production was found to be between 26° and 28°C. Although the organism grew well over a fairly broad temperature range, it was found that at higher temperatures, the yield was reduced, while at lower temperatures, excessive fermentation periods were required.

An initial pH of 6.0 to 7.0 was found to be preferable. A starting ph of 4.5 to 5.5 allowed a good growth, but the riboflavin yield was less satisfactory. At an initial pH of 4.0, both growth and riboflavin formation were quite poor.

A. gossypii was found to be rather discriminating regarding satisfactory carbohydrate substrates. Glucose was quite satisfactory. Fairly pure forms of sucrose and maltose were also employed satisfactorily, while crude agents, such as cane or beet molasses or hydrol, were unsatisfactory because of poor riboflavin yields, although mycelial growth was quite heavy. Pentoses did not support growth well, nor did starch or modified starches. During the fermentation, glucose is largely consumed before riboflavin starts to appear in quantity.[90] Other carbohydrates have also been studied.[56] Carbohydrate metabolism of *A. gossypii* was studied in 1950 by Mickelson.[34a]

Regarding media, Farries and Bell[10] reported nutritional information in 1930. In the following years, studies were made by a number of other investigators. These were reviewed in 1950 by Pridham and Raper.[56] Essential for the growth of *A. gossypii* are *i*-inositol, thiamin, and biotin. Robbins and Schmidt[63] found that *A. gossypii* grew rapidly in a synthetic-type medium composed of glucose, asparagin, biotin, thiamin, *i*-inositol, and several salts. A significant amount of riboflavin did not appear to form. Nitrate is evidently not assimilated, and the ammonium ion appears to have little nutritional value.[56] More complex media,[95] containing peptone, yeast extract, and sugar, resulted in much better riboflavin formation. Lipids may be of importance for high riboflavin production by *A. gossypii.* The use of lipids was indicated,[89] but details were not supplied. Lipids were shown by Phelps[46] to be of considerable importance in an *Eremothecium ashbyii* process. *E. ashbyii* was also reported[85] to give good riboflavin yields in media in which the nitrogen was supplied by ammonium salts.

Tanner, Vojnovich and Van Lanen[90] of the Northern Regional Research Laboratory showed, in 1949, that corn-steep liquor was a satisfactory and economic agent which could replace yeast extract in the media. Distillers' solubles could also be used. As substitutes for peptone, vegetable proteinaceous agents, such as soybean meal, glutens, linseed, and cottonseed meals, were investigated and were found to give results which were erratic and inferior to those obtained in a corn steep-peptone type medium. Proteinaceous agents from animal sources were found to operate quite effectively in corn steep-glucose media. Among the useful agents were tankage, animal stick liquor and meat scraps. Acid- or enzyme-hydrolyzed proteinaceous agents gave considerably better yields than the unhydrolyzed product.[89,90] Fish stick liquor was unsatisfactory because of very low yields of riboflavin. As a control, a medium containing 4% glucose, 0.5% corn-steep liquor, and 0.5% peptone yielded 395 μg of riboflavin per ml in 8 days. When beef tankage was substituted for peptone, a 7-day yield of 356 μg per ml was reported. It was mentioned that by the generally prescribed procedure, with some modifications which were not specified, yields in excess of 1,000 μg per ml were obtained. The highest assay reported by these investigators was 1,420 μg per ml which was obtained in a 30-l fermentation. Very recently, Beesch and Firman[2a] reported a process using unhydrolyzed plant proteins which yielded about 1,000 μg/ml.

Along with the importance of nutrition of *A. gossypii* for obtaining good riboflavin yields, Tanner, Vojnovich, and Van Lanen[90] pointed out in 1949 that excessive sterilization of media reduced greatly the eventual riboflavin yield. Data showing this are presented in Table 22. Flash-sterilization or continuous-cooking techniques would, no doubt, be important here; sterilization of ingredients separately was one means of avoiding the effect. Brief sterilization on a large scale is probably best handled by some means other than batch cooking.

The same authors also found in inoculum-development studies

TABLE 22. EFFECT OF DIFFERENT STERILIZATION PERIODS ON RIBOFLAVIN YIELD BY *Ashbya gossypii*

Autoclave period at 121°C, minutes	Fermentation age	
	8 days, μg per ml	10 days, μg per ml
0 (Seitz filtration)	678	676
15	648	700
30	680	700
45	494	586
60	308	346
75	288	360
90	248	324

Source: Tanner, Vojnovich and Van Lanen.[90]

that the age and proportion of inoculum influenced the riboflavin yield considerably. A 24-hour inoculum was found to be much preferable to 72-, 96-, or 120-hour inocula. The preferred quantity of inoculum was 0.5 to 1.0% by volume. Higher levels were less satisfactory and a 10% by volume inoculum resulted in a much lower yield. Efficient aeration was essential.

Pridham and Raper[54,55,56] and Pridham[53a] pointed out, in 1949, 1950, and 1951 that with very critical control of methods and with a higher nutrient level, including corn-steep liquor, yields of 1,200 μg per ml were obtained from *A. gossypii* quite consistently. With glucose feeding, a peak of 1,760 μg per ml was obtained. An inoculum of 0.25% by volume was used. They found no superior strains through strain selection.

In 1952 Pridham and Raper[56a] described studies on variation and mutation of *A. gossypii*. They noted a general tendency of the organism to form degenerate types which are less effective as riboflavin producers. The inclusion of 0.1% sodium dithionate in stock-culture media enhanced pigment formation.

Pfeifer, Tanner, Vojnovich, and Traufler[48,48a] described the *A. gossypii* process on a pilot-plant basis in 1949 and 1950. Details were also presented in a 1950 pamphlet of the Northern Regional Research Laboratory by Pridham, Hall, and Pfeifer.[53b] Fermentations were operated satisfactorily in glass, aluminum, copper, and stainless-steel equipment. A medium was used which contained 2% glucose, about 2% corn-steep liquor, and 1% animal stick liquor at pH 4.5. As an antifoam, 0.1 to 0.2% soybean oil was included. This was sterilized for 5 minutes at 275°F by the continuous procedure. The medium was inoculated with about

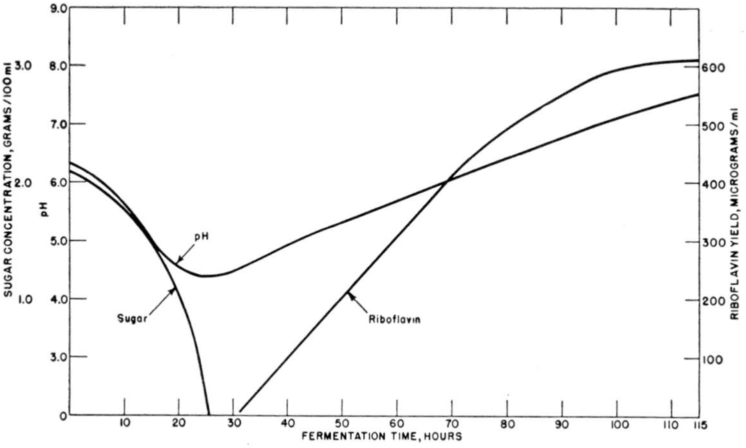

FIGURE 6. *Changes in Medium during Fermentation* [Reprinted by permission from a paper by Pfeifer, Tanner, Vojnovich, and Traufler, *Ind. Eng. Chem.*, **42**, 1776 (1950)]

0.5 to 1% by volume of seed culture. The medium was aerated and mildly agitated for 4 or 5 days at 28° to 29°C. The riboflavin yield was 500 to 600 μg per ml. The resulting product was concentrated and dried on a drum drier, yielding a 2.5% riboflavin product estimated to cost 3.75¢ per gram of riboflavin. Dried fer-

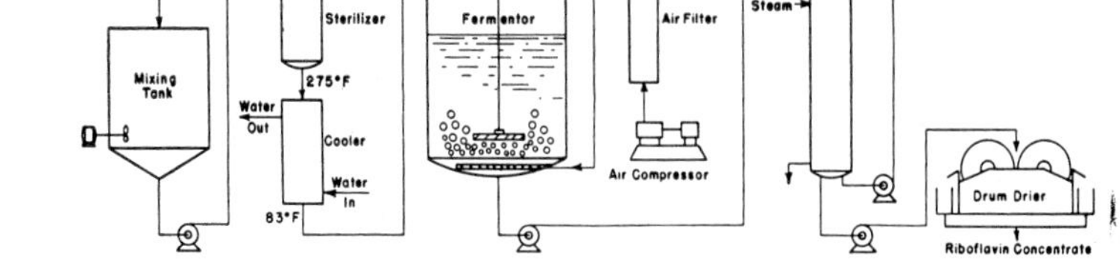

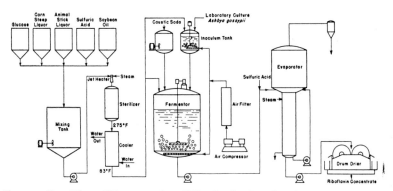

FIGURE 7. *Flow Diagram for Riboflavin Production by Fermentation* [Reprinted by permissison from a paper by Pfeifer, Tanner, Vojnovich, and Traufler, *Ind. Eng. Chem.*, **42**, 1776 (1950)]

mentation products having a riboflavin potency of 25,000 to 30,000 μg per g are reported.[56] Crystalline riboflavin can be obtained. Figure 6 shows characteristic changes in the medium during fermentation and Figure 7 is a flow sheet of the general process.[48a]

Recently, Smiley and coworkers[78] found that *A. gossypii* produced vitamin B_{12} or a B_{12}-like agent; also the *Lactobacillus bulgaricus* factor[78,96] is formed in quite appreciable amounts.

Although *Eremothecium ashbyii* and *A. gossypii* are morphologically similar[14] and both evidently require inositol, biotin, and thiamin,[23,63,70,92] the general types of media used for highest riboflavin yields from the two organisms are considerably different.

RECOVERY OF RIBOFLAVIN FROM FERMENTED PRODUCTS

Fermentation liquors containing riboflavin may be recovered for use as animal- or human-food supplements simply by drying and powdering. This is true of butyl alcohol-acetone fermentation residues, among others. Drying is generally accomplished where practicable by drum- or spray-drying procedures. These preparations usually contain other members of the B-complex,[35] along with minerals and proteinaceous substances.

Some liquors, or "beers," may contain sufficiently high concentrations of riboflavin to make isolation of pure, crystalline riboflavin profitable. This is particularly true of liquors from the *E. ashbyii* and *A. gossypii* processes which may exceed 1,000 μg per ml in potency.

Two methods were described in 1945 by Hines,[20,21] who found first that by bacteriological reduction, up to 90% of the riboflavin in *E. ashbyii* fermentation liquors could be precipitated as a reddish-brown, amorphous product of 80 to 90% purity.[18] *Streptococcus faecalis, S. liquefaciens,* and certain other bacteria were used. The method was described further in 1946 by Hickey,[18] who discussed the nature of the precipitate. Hines then found[20] that a similar result could be obtained more rapidly, almost immediately in fact, by the use of chemical reducing agents, such as sodium dithionite ($Na_2S_2O_4$), stannous and chromous chlorides, and certain other reducing compounds. A modification of the chemical precipitation-recovery system was reported by McMillan,[30] who precipitated the riboflavin in the presence of a filter aid, such as Super-Cel. This allowed a very easy and rapid recovery of the precipitate by press filtration and also assisted settling where this was desirable. Of the chemical reducing agents, sodium dithionite (hydrosulfite) has economic and low toxicity advantages over some of the other operative reducing agents, such as stannous or chromous chlorides.

The crude precipitate which is obtained by such means can be readily converted to crystalline riboflavin, for example, by the method of Dale.[8] In this process, the precipitate is dissolved in a hot polar solvent, such as 50% aqueous isopropyl alcohol. The mixture is filtered and yields a greenish filtrate, containing riboflavin in a reduced state. The reduced riboflavin is oxidized, by air, for example, and the oxidized riboflavin precipitates as yellow crystals.

There have been many methods for recovering riboflavin from various sources. Some of these have been reviewed by Rosenberg.[67] Many of the methods were applied to raw materials having a very low potency compared with preparations from *E. ashbyii* or *A. gossypii* which may contain 1,000 to 2,000 μg or more per ml.

The chemical precipitation methods of Hines[20] and McMillan[30] operate primarily on solutions containing much riboflavin; at least 20 μg per ml. *E. ashbyii* beer filtrates operate well, but other riboflavin solutions can also be precipitated.[18]

A recent method for the recovery of riboflavin from low-potency preparations has been described by Keresztesy and Rickes,[22] who employed butyl alcohol for extraction of riboflavin directly from the fermentation substance. Repeated extractions were necessary.

In 1949, Legg and Cranmer[25] described a method for application to high- or low-concentration (i.e., less than 20 μg per ml) preparations. The effective operation of this method depends on the fact that reduced or leuco riboflavin is much easier to extract from aqueous solution by certain alcohols, e.g., *n*-butyl alcohol, than is the oxidized form. Alcohols which can be used are only partially miscible with water. Isobutyl, secondary butyl, isoamyl, benzyl, and cyclohexyl alcohols are among those cited, but *n*-butyl alcohol is preferred. An example of the method is as follows: A stillage containing about 10.6 μg of riboflavin per ml and about 0.91% of other solids was reduced with about 0.02%

sodium dithionite (hydrosulfite), and was mixed with an equal volume of *n*-butyl alcohol in an atmosphere of carbon dioxide. After shaking and standing, the phases separated. Assay following reoxidation showed about 76% of the riboflavin in the alcohol phase. The potency on a dry basis was 10,700 μg per g compared with 1,164 μg per g in the original substance. When the operation was run without reduction, only 37% of the riboflavin appeared in the *n*-butyl alcohol phase. The solids in this extract assayed 5,500 μg per g, or about half the potency of the dry product prepared by the reduction-extraction procedure. Reducing agents other than dithionite may also be used.

BIBLIOGRAPHY

1. Arnaud, G., *Bull. soc. mycol., France,* **33**, 572 (1917).
2. Arzberger, F. C., U. S. Patent 2,326,425 (1943).
2a. Beesch, S. C., and M. C. Firman, U. S. Patent 2,631,120 (1953).
2b. Beesch, S. C., and B. W. Frazer, U. S. Patent 2,647,074 (1953).
3. Bergey, D. H., R. S. Breed, E. G. D. Murray, A. P. Hitchens, and others, *Manual of Determinative Bacteriology,* 6th ed., Baltimore, Williams and Wilkins, 1948.
4. Borzi, A., *Nuovo gior. bot. Ital.,* **20**, 452 (1888).
5. Burkholder, P. R., *Arch. Biochem.,* **3**, 121 (1943).
6. Burkholder, P. R., *Proc. Nat. Acad. Sci. U. S.,* **29**, 166 (1943).
7. Burkholder, P. R., U. S. Patent 2,363,277 (1944).
7a. *Chemical & Engineering News,* **31**, 4934 (1953).
8. Dale, J. K., U. S. Patent 2,421,142 (1947).
8a. DeBecze, G. J., H. N. Moore, and E. Schraffenberger, U. S. Patent 2,543,897 (1951).
9. Desieve, E., *Milchwissenschaft,* **2**, 141 (1947).
9a. Dulaney, E. L., and F. H. Grutter, *Mycologia,* **42**, 717 (1950).
10. Farries, E. H. M., and A. F. Bell, *Ann. Botany,* **44**, 423 (1930).
11. Foster, J. W., British Patent 593,027 (1947).
12. Guilliermond, A., *Rev. gén. botan.,* **40**, 606, 690 (1928).
13. Guilliermond, A., *Compt. rend.,* **200**, 1556 (1935).
14. Guilliermond, A., *Rev. Mycol.* (N.S.), **1**, 115 (1936).
15. Guilliermond, A., M. Fontaine, and A. Raffy, *Compt. rend.,* **201**, 1077 (1935).
16. Grob, E. C., *Intern. Z. Vitamin-Forsch.,* **20**, 178 (1948).

17. Hickey, R. J., *Arch. Biochem.*, **8**, 439 (1945).
18. Hickey, R. J., *Arch. Biochem.*, **11**, 259 (1946).
19. Hickey, R. J., U. S. Patent 2,425,280 (1947).
19a. Hickey, R. J., U. S. Patent 2,605,210 (1952).
19b. Hickey, R. J., *J. Bact.*, **66**, 22 (1953).
20. Hines, G. E., Jr., U. S. Patent 2,367,644 (1945).
21. Hines, G. E., Jr., U. S. Patent 2,387,023 (1945).
21a. James, R. M., U. S. Patent 2,498,549 (1950).
21b. Kawano, Y., and S. Honda, Japanese Patent 4100 (1951).
22. Keresztesy, J. C., and E. Rickes, U. S. Patent 2,355,220 (1944).
23. Kögl, F., and N. Fries, *Z. physiol. Chem.*, **249**, 93 (1937).
23a. Kozima, T., *J. Fermentation Tech.* (Japan), **28**, 129 (1950); *Biol. Abst.*, **26**, 9222 (1952).
23b. Larsen, D. H., U. S. Patent 2,615,829 (1952).
24. Legg, D. A., and S. C. Beesch, U. S. Patent 2,370,177 (1945).
24a. Legg, D. A., and S. C. Beesch, U. S. Patent 2,581,419 (1952).
25. Legg, D. A., and J. T. Cranmer, U. S. Patent 2,464,243 (1949).
26. Levine, H., J. E. Oyaas, L. Wasserman, J. C. Hoogerheide, and R. M. Stern, *Ind. Eng. Chem.*, **41**, 1665 (1949).
27. Leviton, A., *J. Am. Chem. Soc.*, **68**, 835 (1946).
28. Leviton, A., U. S. Patent 2,477,812 (1949).
28a. MacLaren, J. A., *J. Bact.*, **63**, 233 (1952).
29. Martin, J., U. S. Patent 2,491,927 (1949).
29a. McClary, J. E., U. S. Patent 2,537,148 (1951).
30. McMillan, G. W., U. S. Patent 2,367,646 (1946).
31. Massock, H. E., M. S. Thesis, University of Wisconsin, 1943.
32. Mayer, R. L., and R. Rodbart, *Arch. Biochem.*, **11**, 49 (1946).
33. Meade, R. E., H. L. Pollard, and N. E. Rodgers, U. S. Patent 2,369,680 (1945).
34. Meade, R. E., N. E. Rodgers, and H. L. Pollard, U. S. Patent 2,433,232 (1947).
35. Miner, C. S., U. S. Patent 2,202,161 (1940).
35a. Minoura, K., *J. Fermentation Tech. (Japan)*, **28**, 125 (1950); *Biol. Abst.*, **26**, 9227 (1952).
35b. Minoura, K., *J. Fermentation Tech. (Japan)*, **28**, 186 (1950); *Biol. Abst.*, **26**, 9228 (1952).
35c. Minoura, K., *J. Fermentation Tech. (Japan)*, **28**, 308 (1950); *Biol. Abst.*, **26**, 9229 (1952).
36. Mirimanoff, A., and A. Raffy, *Compt. rend.*, **206**, 1507 (1938).
37. Mirimanoff, A., and A. Raffy, *Bull. soc. chim. biol.*, **20**, 1166 (1938).

38. Mirimanoff, A., and A. Raffy, *Helv. Chim. Acta,* **21**, 1004 (1938).
39. Moore, H. N., and G. de Becze, *J. Bact.,* **54**, 40 (1947).
40. Moss, A. R., and R. Klein, (Roche Products Ltd.), British Patent 615,847 (1947).
41. Novak, A. F., U. S. Patent 2,447,814 (1948).
42. Okabe, M., Y. Hatsuda, and Z. Ueda, *J. Agr. Chem. Soc. Japan,* **19**, 571 (1943).
43. Pett, L. B., *Biochem. J.,* **29**, 937 (1935).
44. Pfizer, Chas., and Co., Inc., British Patent 593,953 (1948).
45. Phelps, A. S., U. S. Patent 2,473,817 (1949).
46. Phelps, A. S., U. S. Patent 2,473,818 (1949).
47. Piersma, H. O., U. S. Patent 2,400,710 (1946).
47a. Plaut, G. W. E., *Federation Proc.,* **12**, 254 (1953).
48. Pfeifer, V. F., F. W. Tanner, C. Vojnovich, and D. J. Traufler, Abst. of Papers, 116 Meeting, A.C.S., p. 22A, Atlantic City, 1949.
48a. Pfeifer, V. F., F. W. Tanner, C. Vojnovich, and D. J. Traufler, *Ind. Eng. Chem.,* **42**, 1776 (1950).
49. Pollard, H. L., N. E. Rodgers, and R. E. Meade, U. S. Patent 2,433,063 (1947).
50. Pollard, H. L., N. E. Rodgers, and R. E. Meade, U. S. Patent 2,449,140 (1948).
51. Pollard, H. L., N. E. Rodgers, and R. E. Meade, U. S. Patent 2,449,141 (1948).
52. Pollard, H. L., N. E. Rodgers, and R. E. Meade, U. S. Patent 2,449,142 (1948).
53. Pollard, H. L., N. E. Rodgers, and R. E. Meade, U. S. Patent 2,449,143 (1948).
53a. Pridham, T. G., U. S. Patent 2,578,738 (1951).
53b. Pridham, T. G., H. H. Hall, and V. F. Pfeifer, *Production of Riboflavin (Vitamin B_2) with Ashbya gossypii,* Peoria, Ill., Northern Regional Research Laboratory, 1950.
54. Pridham, T. G., and K. B. Raper, *Am. J. Botany,* **36**, 814 (1949).
55. Pridham, T. G., and K. B. Raper, General Program, 6th N. Y. Meeting of AAAS, p. 200, December 1949.
56. Pridham, T. G., and K. B. Raper, *Mycologia,* **42**, 603 (1950).
56a. Pridham, T. G., and K. B. Raper, *Mycologia,* **44**, 452 (1952).
57. Raffy, A., *Compt. rend. soc. biol.,* **126**, 875 (1937).
58. Raffy, A., *Compt. rend. soc. biol.,* **128**, 392 (1938).
59. Raffy, A., *Compt. rend.,* **209**, 900 (1939).

60. Raffy, A., and A. Mirimanoff, *Bull. soc. chim. biol.*, **20**, 1166 (1938).
61. Renaud, J., and M. Lachaux, *Compt. rend.*, **221**, 187 (1945); see also *Ibid.*, **219**, 498 (1944).
62. Ritter, W., *Schweiz. Z. Path. u. Bakt.*, **7**, 370 (1944).
63. Robbins, W. J., and M. B. Schmidt, *Bull. Torrey Botan. Club*, **66**, 139 (1939).
64. Rodgers, N. E., Ph.D., Thesis, University of Wisconsin, 1942.
65. Rodgers, N. E., H. L. Pollard, and R. E. Meade, U. S. Patent 2,433,064 (1947).
66. Rodgers, N. E., H. L. Pollard, and R. E. Meade, U. S. Patent 2,449,144 (1948).
67. Rosenberg, H. R., *Chemistry and Physiology of Vitamins*, New York, Interscience, p. 153, 1945.
68. Rudert, F. J., U. S. Patent 2,374,503 (1945).
69. Schopfer, W. H., *Plants and Vitamins*, Chronica Botanica Co., Waltham, Mass., 1943.
70. Schopfer, W. H., *Helv. Chim. Acta.*, **27**, 1917 (1944).
71. Schopfer, W. H., *Compt. rend. soc. phys. hist. nat. Genève*, **61**, 233 (1944).
72. Schopfer, W. H., *Compt. rend. soc. phys. hist. nat. Genève*, **61**, 147 (1944).
73. Schopfer, W. H., *Antonie van Leeuwenhoek J. Microbiol. Serol.*, **12**, 133 (1947).
74. Schopfer, W. H., and M. Guilloud, *Experientia*, **1**, 22 (1945).
75. Schopfer, W. H., and M. Guilloud, *Schweiz. Z. Path. u. Bakt.*, **8**, 521 (1945).
76. Schopfer, W. H., and M. Guilloud, *Experientia*, **1**, 332 (1945).
77. Singh, K., G. N. Agarwal, and W. H. Peterson, Abstracts of papers, 111th Meeting, A.C.S., p. 9-10A, April 1947.
78. Smiley, K. L., M. Sobolov, F. L. Austin, R. A. Rasmussen, M. B. Smith, J. M. Van Lanen, L. Stone, and C. S. Boruff, *Ind. Eng. Chem.*, **43**, 1380 (1951).
79. Snell, E. E., *Physiol. Revs.*, **28**, 255 (1948).
80. Steinberg, R. A., *J. Agr. Research*, **51**, 413 (1935).
81. Steinberg, R. A., *J. Agr. Research*, **57**, 261 (1938).
82. Steinberg, R. A., *Botan. Rev.*, **5**, 327 (1939).
83. Steyaert, R. L., *Sci. Monthly*, **63**, 268 (1946).
84. Stiles, H. R., U. S. Patent 2,483,855 (1949).
85. Tabenkin, B., U. S. Patent 2,493,274 (1950).
86. Tanner, F. W., C. Vojnovich, and J .M. Van Lanen, *Science*, **101**, 180 (1945).

87. Tanner, F. W., and J. M. Van Lanen, U. S. Patent 2,424,003 (1947).
88. Tanner, F. W., and J. M. Van Lanen, *J. Bact.*, **54**, 38 (1947).
89. Tanner, F. W., L. J. Wickerham, and J. M. Van Lanen, U. S. Patent 2,445,128 (1948).
90. Tanner, F. W., C. Vojnovich, and J. M. Van Lanen, *J. Bact.*, **58**, 737 (1949).
90a. Takata, R., and T. Nagata, *J. Fermentation Tech. (Japan)*, **27**, 279, 281, 285, 287 (1949).
91. Tittsler, R. P., and F. O. Whittier, *J. Bact.*, **42**, 151 (1941).
92. Van Lanen, J. M., and F. W. Tanner, *Vitamins and Hormones*, Vol. VI, p. 163, New York, Academic Press, 1948.
93. Walton, M. T., U. S. Patent 2,368,074 (1945).
94. Waring, W. S., and C. H. Werkman, *Arch. Biochem.*, **1**, 303 (1942).
95. Wickerham, L. J., M. H. Flickinger, and R. M. Johnson, *Arch. Biochem.*, **9**, 95 (1946).
96. Williams, W. L., E. Hoff-Jørgensen, and E. E. Snell, *J. Biol. Chem.*, **177**, 933 (1949).
97. Yamasaki, I., and W. Yoshitome, *Biochem. Z.*, **279**, 398 (1938).
98. Yamasaki, I., *Biochem. Z.*, **300**, 160 (1939).
99. Yamasaki, I., *J. Agr. Chem. Soc. Japan*, **16**, 169 (1940).
100. Yamasaki, I., *Biochem. Z.*, **307**, 431 (1941).
101. Yamasaki, I., U. S. Patent 2,297,671 (1942).
102. Yaw, K. E., *Mycologia*, **44**, 307 (1952).

CHAPTER 6

PRODUCTION OF VITAMINS OTHER THAN RIBOFLAVIN

J. M. Van Lanen

INTRODUCTION

In recent years, remarkable advances have been made in our knowledge of the vitamins—their distribution in nature, chemistry, biological functions, and biosynthesis. Microorganisms, as tools of study, have come to play a prominent part in gathering this knowledge and applying it to practical nutrition problems. They have been especially useful in the discovery and elucidation of various members of the B-complex family, since these vitamins are not only common to all living cells, but they perform essentially the same biochemical roles in each, i.e., they serve as functional constitutents of enzyme systems which provide the cell with energy and the required structural units. The selection and utility of microorganisms in pursuing nutritional problems is based on this fundamental similarity in cellular metabolism, coupled with the ease and rapidity with which they can be cultivated on substrates of known chemical composition.

For the determination of vitamins and the exploration of specific biochemical reactions, necessarily widespread use has been made of fastidious organisms which are highly dependent on the external environment for their growth requirements. This has served to focus attention on the vitamin deficiencies of fungi. However, within the past decade, it has been demonstrated, both by surveying numerous species for their vitamin requirements and by the actual assay of microbial products, that the great majority of microorganisms is capable of appreciable vitamin synthesis. Those organisms with exceptional biosynthetic ability are receiving ever greater attention both in industry and in the research laboratory.

Vitamins formed during the growth of microorganisms are retained within the cells, degraded during metabolism, or excreted into the surrounding medium in the same manner as other metabolites. As a consequence, in certain environments, both natural and artificial, which support abundant microbial development, vitamins often are demonstrable in substantial amounts. In nature, such loci are waters of high organic content, sewage, soils, especially in the proximity of higher plant life, and the intestinal tracts of animals. The extent to which vitamins so formed contribute to the nutrition of animals and conceivably to the early development of some higher plants is not fully known; however, it is true that all the B-vitamin needs of ruminants can be satisfied once the rumen flora becomes established.

On artificial substrates, i.e., pure-culture propagation, vitamin synthesis occurs together with growth or fermentative activity. Where large volumes of media are processed, as in industrial fermentations, this synthesis substantially enhances the nutritional value of either the fermentation products or the by-products. In recent years, increasing attention has been given to the full recovery and proper utilization of these values. In this chapter, those microbial processes will be discussed in which the vitamins synthesized have commercial or potential commercial importance. For consideration here, these have been grouped as follows:

(1) Biosynthetic processes which result in the vitamin enrichment of the fermentation product, such as yeasts grown for food, pharmaceutical, or feed purposes.

(2) Biosynthetic processes accompanying other fermentations in which the vitamins occur in and add value to the fermentation residues. Several industrial fermentations fall into this

group, the vitamins resulting in vitamin-rich by-products whose recovery has often come to bear importantly on the over-all economics of the process involved. An outgrowth of the recognition of vitamin synthesis in certain fermentations, for example, the acetone-butanol process, has led to a more intensive study of the biosynthesis of vitamins and to improved methods for the production of riboflavin (see Chapter 5).

(3) The use of microorganisms to demonstrate the existence of previously unidentified vitamins and to carry out their synthesis in primary fermentation processes, i.e., those in which the vitamin is the principal product of the fermentation.

HISTORICAL

Microorganisms have been associated with the nutritional well-being of animals dating back to a period which coincides with the origin of the science of nutrition. Yeast was fed to livestock with favorable results during the latter part of the last century and in 1900, reports began to appear concerning its nutritional merits.[47] The work of Hopkins,[26] in 1906, supplied partial confirmation for these claims, since he showed that milk and yeast contained "accessory factors" for rats placed on purified rations. The first specific nutritional entity to be found a component of any microorganism was the antineuritic principle which, in 1910, Schaumann[54] demonstrated to be present in yeast. A few years later, while searching for the antineuritic principle, Funk[16] isolated nicotinic acid from yeast. Although nicotinic acid was inactive against polyneuritis, its isolation undoubtedly attracted further attention to yeast as a source of vitamins. From these early observations until the present time, yeast has continued to rank among the best natural carriers of B vitamins and in several cases (folic acid, *p*-aminobenzoic acid, coenzyme A, etc.), it has been used as a source material for isolation of specific factors. Several reviews have dealt with the vitamin content and general nutritional value of various types of yeast.[45,47,76]

The conviction that other microorganisms synthesize vitamins likewise had an early origin. In 1911, Osborne and Mendel[41] noted that a diet inadequate for rats permitted growth when access was had to the feces of animals maintained on a complete diet. They attributed this action to the synthetic activities of microbial flora engendered by the complete diet. In 1914, Cooper[12] showed that alcoholic extracts of hen and rabbit excreta possessed antineuritic activity which led him to postulate synthesis by intestinal flora. Further evidence along this line, and the first with ruminants, was the report of Theiler, Green, and Viljoen[72] in 1915 that the rumen contents of cattle contained antineuritic principle although the diet lacked this substance. They likewise implicated rumen microflora in the acquisition of the vitamin.

Reports, in 1927, by Fridericia, Freudenthal, Gudjonnsson, Johansen, and Schoubye[15] and by Roscoe[50] demonstrated that on diets rich in starch, some rats became completely independent of a dietary B-vitamin supply. This relationship between diet and microbial activity in the intestinal tract has been amply confirmed in recent years.[36] The complete independence of ruminants of dietary B vitamins was shown by Bechdel, Honeywell, Dutcher, and Knutsen[4] who found that calves could be reared to maturity and were capable of normal reproduction on diets which were deficient in B vitamins for laboratory animals. The importance of both intestinal and rumen microflora in satisfying the nutritional requirements of animals has been discussed from biological and biochemical viewpoints in recent articles.[24,28,31a,36,66a]

That specific bacteria are capable of synthesizing vitamins was first suggested by the finding of Pacini and Russel,[42] in 1918, that culture media, in which typhoid organisms had been cultivated, stimulated the growth of rats. In 1923, Damon[13] found that vitamin B was synthesized by several mycobacteria. The production of vitamin B by *Bacillus vulgatus* also, was reported by Scheunert and Schieblich[55] in 1927. Beginning about 1930, an increasing number of microorganisms was shown to be capable of producing definite members of the vitamin B complex. During the past two decades, attention has been directed both to the quantitative aspect of synthesis by organisms in industrial use and to the mechanisms and intermediates of vitamin formation by fungi, in general. These studies which have been summarized by Knight,[31] Peterson and Peterson,[45] and Van Lanen and Tanner,[75] provide much evidence to support our present concepts regarding the wide distribution and importance of B vitamins in metabolism.

SYNTHESIS OF B VITAMINS DURING INDUSTRIAL FERMENTATIONS

Yeast Fermentations

Yeasts are the principal fermentation product in bakers', fodder, and food yeast production and a by-product of brewing and certain distillery operations. The details of their cultivation in these fermentations are described in Chapters 6, 9 and 10 of Volume I. Whether fermentation main-product or by-product, their main use is in the formulation of foods, pharmaceuticals, and feedstuffs and thus vitamin content and uniformity are important factors in yeast propagation.

The vitamin composition and the range of potencies of the commonly grown yeasts is shown in Table 23. It is obvious from this table that yeasts are a comparatively rich source of B vitamins, except vitamin B_{12}, but that there are appreciable variations both within and between species. While these differences are as yet imperfectly understood, they can be accounted for largely by one or more of the following variables: (a) medium composition, (b) propagating conditions, and (c) species or strain used. The manner and extent to which each of these influences vitamin potencies is outlined briefly as follows.

TABLE 23. VITAMIN B COMPLEX CONTENT OF COMMERCIAL YEASTS[a]

	Bakers' compressed	Bakers' active dry	Primary grown	Brewers'	*Torula utilis*
Thiamine	16-40	12-29	120-200	60-180	5-53
Riboflavin	25-80	31-70	40-80	21-80	26-70
Pantothenic acid	150-330	100-150	112-150	42-200	40-180
Niacin	240-700	180-300	250-700	281-1,000	153-600
Pyridoxine	25-65	—	32-40	23-100	35
Folic acid	19-80	—	—	19-30	4-31
Biotin	0.8-0.24	0.4-1.5	2.5	0.8-1.1	0.5-3.6
p-Aminobenzoic acid	24-175	15	10-16	9-102	17-62
Inositol	4,320	—	4,000	2,700-5,000	2,700-3,600
Choline	2,100	—	4,000	2,500-5,000	—

[a] μg per g of dry cells.

MEDIUM COMPOSITION

It has been well established by the use of synthetic media and other substrates of known vitamin content that yeasts are capable of the total synthesis of B vitamins. However, higher vitamin potencies are usually obtained when media are supplemented with natural products, such as grain extracts, malt sprouts, molasses, etc. Such supplements have both direct and indirect effects on vitamin formation. They supply inorganic elements and growth-promoting substances which indirectly promote vitamin synthesis. Directly and, more important, quantitatively, they provide intact vitamins and vitamin precursors which the cells absorb or convert to vitamins. Compared with other fungi, yeasts are capable of both absorption and synthesis from precursors to a high degree and each of these properties has been utilized industrially to prepare enriched products of predetermined vitamin content.

ABSORPTION

Thiamine, niacin, and biotin are avidly absorbed by yeasts; pyridoxine and inositol are taken up to a lesser extent, while pantothenic acid and some other B vitamins may be absorbed under limited physiological conditions. In unfortified media, absorption is an important contributory factor to high vitamin potencies only under anaerobic conditions where the yeast crop is low in relation to the available vitamins. The brewing process is a good example of fortification under anaerobic conditions and the higher levels of thiamine and niacin in brewers' yeast (see Table 23 and Norris[39]) can be explained on this basis. In fortified media, however, propagation may be either aerobic or anaerobic and enrichment with thiamine can be brought about merely by aerating a cell suspension in the presence of the vitamin.

In one of the patented methods of enriching yeast with thiamine,[20] rice polishings, malt sprouts, or other natural vitamin-containing materials are extracted and the extracts are combined. Carbohydrate is added to yield the yeast crop and vitamin potency desired and the medium is fermented in the usual manner. Another patented process[60] involves cultivation of yeast in media fortified with crystalline thiamine. The thiamine potency can be increased in this way from the normal level up to 600 to 800 μg per g with a high efficiency and cells containing over 6,000 μg per g have been produced with a low efficiency. Absorbed thiamine, especially under anaerobic conditions, is converted largely to cocarboxylase.

Yeasts also absorb niacin[73] and a four- to fivefold enrichment can generally be obtained by adding this vitamin to the medium. Potencies of 3,000 μg per g have been secured by adding a large excess of niacin. The uptake of niacin, in contrast to thiamine, requires fermentable carbohydrate and, under anaerobic conditions, absorbed niacin is rapidly released following carbohydrate utilization.

Biotin absorption is appreciable with some yeasts, but this property varies considerably between species and strains.[77] One *Saccharomyces cerevisiae* culture was increased in biotin content from 0.5 to over 200 μg per g by fortifying the medium. Although biotin, niacin, and thiamine absorption takes place in both synthetic and natural media, the process is more complete in the second due apparently to the fact that they supply factors which favor absorption.

Riboflavin values are fairly uniform for the yeasts of commerce, since any excess of this vitamin is released into the medium rather than retained following synthesis. A similar behavior is exhibited in the case of *p*-aminobenzoic acid which is synthesized and excreted into the medium.[32] Certain yeasts, i.e., *Hansenula anomala* and *Mycotorula lipolytica*,[44] have been found to produce as much as 1 mg of *p*-aminobenzoic acid per gram of cells. Yeasts of commerce are very low in vitamin B_{12} content, ranging from 0.01 to 0.2 mμg per gram.

Synthesis from Precursors

Microorganisms are frequently limited in the synthesis of a vitamin because of inability to fabricate a particular portion of the molecule. Once this limiting portion or precursor is supplied, the synthesis is completed, sometimes in substantial amounts. The first indication of biosynthesis from precursors was noted in studies on thiamine production in 1937.[43] It was found that thiamine, after heat inactivation, was regenerated when added to cultures of growing yeast. Knight,[30] about the same time, reported that pyrimidine and thiazole could be substituted for thiamine as a growth factor for *Staphylococcus aureus*. In 1938, Schultz, Atkin, and Frey[58] found that these compounds replaced thiamine for yeasts which were incapable of synthesizing their growth requirements. These observations culminated in the development, patenting, and commercial use of a number of quite similar processes for enriching yeasts with thiamine.[23,35,59] The synthesis of thiamine from the intermediates involves formation of a methylene bridge between the pyrimidine and thiazole moieties:

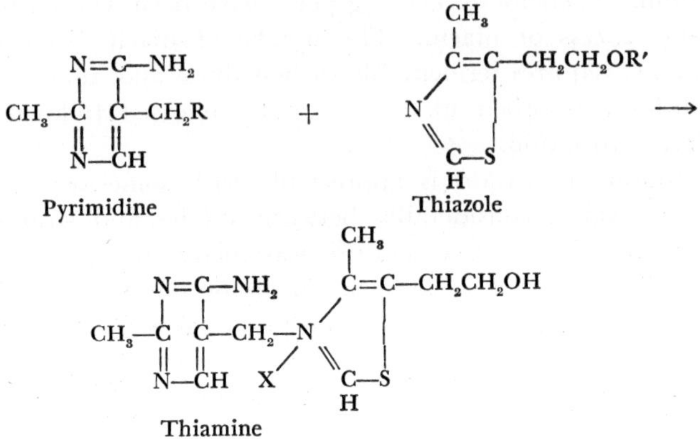

Compared with the filamentous fungi, yeasts are relatively exacting as to the pyrimidines and thiazoles which they can convert to thiamine. Only substitutions in the positions marked R and R′ in the preceding formulas are permissible. R may be an aminomethyl, ethoxymethyl, oxymethyl, hydroxymethyl, aminoethyl, thioformyl, or cyano group.[3,14,23,59] R′ may be hydrogen, an acyl group, or another group hydrolyzable to yield the hydroxyl group.

To carry out the biosynthesis, the pyrimidine and thiazole are added to the wort during propagation of the yeast, preferably during the later stages of growth. By this method, cells can be enriched from the normal thiamine content of 20 to 40 μg per g up to 1,200 μg per g. However, commercial products commonly have been standardized at approximately 650 μg per g. At this level of enrichment, the intermediates are coupled almost quantitatively.

Although various types of yeast couple pyrimidine and thiazole, bakers' types are particularly efficient. Table 24 shows the synthesis of thiamine from pyrimidine and thiazole by one strain of bakers' yeast.[74] It will be noted that some synthesis occurred by adding thiazole alone. However, microorganisms vary widely in their ability to respond to thiamine intermediates when only one is supplied.[31,57]

The synthesis of thiamine from pyrimidine and thiazole is favored by aerobic growth conditions, a temperature of 20° to 30°C, and by the presence of unidentified substances in various natural products.

TABLE 24. SYNTHESIS OF THIAMINE FROM PYRIMIDINE AND THIAZOLE BY BAKERS' YEAST

Intermediate added to the medium			
Pyrimidine,[a] millimoles per liter	Thiazole,[b] millimoles per liter	Yield of dry yeast, grams per liter	Thiamine content, of dry cells, micrograms per gram
0.000	0.000	7.27	65
0.005	0.005	7.60	261
0.010	0.010	7.62	430
0.015	0.015	7.55	667
0.020	0.020	7.60	802
0.050	0.050	7.67	835
0.100	0.100	7.77	795
0.000	0.020	7.80	150
0.020	0.000	7.27	62

[a] 2-Methyl-5-ethoxymethyl-6-aminopyrimidine.
[b] 4-Methyl-5-β-hydroxyethylthiazole.

Early studies with pantothenic-acid-deficient yeasts showed that β-alanine promoted their growth and suggested that this compound was a precursor of pantothenic acid.[78] This was confirmed subsequently by a number of investigators working with several different types of organisms. One process by which yeast is enriched with pantothenic acid by adding β-alanine to the medium was patented.[61]

Industrial applications to date have been confined to precursors for thiamine and pantothenic acid and conjugated forms of pantothenic acid. However, the same general technique has been adopted extensively to determine the pathway and intermediates involved in the synthesis of other vitamins (folic acid, biotin, niacin, etc.).

CONDITIONS OF CULTIVATION

Highly aerobic conditions, such as result from combined air sparging and agitation, favor conversion of carbohydrate to yeast cell substance. Accordingly, per unit of carbohydrate utilized, aerobic conditions result in a greater yield of vitamins. However, optimum conditions for attaining maximum vitamin potencies vary with each factor. For example, Singh, Agarwal, and Peterson[64] demonstrated with three species of yeasts that thiamine and riboflavin potencies decreased as aeration and agitation increased. Niacin, however, increased directly with improved aeration, while folic acid could not be correlated with aeration rates.

SPECIES DIFFERENCES

Species variations become apparent when cultures are grown under similar conditions. This is shown in Table 25 in which results are given for the vitamin content of four species cultivated on identical substrates.[64] It may be seen that species differences in thiamine, niacin, and folic acid potencies are most pronounced. Similar observations in the folic acid content of a number of yeasts have been reported by MacKenzie, Noble, and Peppler.[33] In this study, *Candida krusei* attained an unusually high folic acid content.

TABLE 25. VITAMIN SYNTHESIS BY DIFFERENT SPECIES OF YEASTS CULTIVATED UNDER SIMILAR CONDITIONS

Yeast	Yield based on sugar fermented %	*Vitamin content, micrograms per gram of dry cells* Thiamine	Riboflavin	Niacin	Folic acid
Beet molasses					
S. cerevisiae	41	37	44	296	34
T. utilis	72	26	53	213	10
C. arborea	74	22	58	301	12
O. lactis	74	14	40	196	12
Cane molasses					
S. cerevisiae	38	28	45	402	34
T. utilis	74	21	54	282	15
C. arborea	71	16	60	313	20
O. lactis	69	12	41	195	15

These investigations emphasize the importance of selecting both the proper organism and proper operating conditions to assure the maximum production of any desired vitamin by yeast.

Vitamin Synthesis in Other Industrial Fermentations

For many years, it was common practice to discharge the bulk of fermentation wastes into lakes and streams with only small amounts being fed to livestock, largely in wet form. As the brewing and distilling industries were expanded in the middle thirties, this method of waste disposal caused serious water-pollution problems and eventually it became necessary to impose preventative regulations. As a consequence of these restrictions, various means of utilizing fermentation wastes were explored, including a thorough investigation of their general nutritive value and content of individual vitamins. The second phase of this study was greatly facilitated by the development of microbiological-assay procedures. The high nutritive value of residues from grain distillation and acetone-butanol production, which are now fully recognized, was indicated by these investigations. These residues are currently recovered in dry form and marketed as feed supplements. The chemical composition of various fermentation wastes, methods of recovery, and the development of the distillers' by-product feed industry have been outlined by Boruff.[9]

The vitamin content of fermentation residues is a reflection of (a) the amount of fermentation products and volatile by-products removed during fermentation and recovery and (b) the actual synthesis of vitamins which has occurred during the fermentation process. For example, in the ethanol fermentation of grain, there is approximately a threefold concentration of residual matter through the production of carbon dioxide and the removal of alcohol by distillation. This concentration, coupled with the synthesis of vitamins, of which riboflavin, pantothenic acid, and choline are outstanding, accounts for the high nutritive value of distillers' by-products. The vitamin contents of fermentation residues are shown in Table 26. Distillers' Dried Solubles represent the more soluble fraction separated from the whole residue by screening or screening followed by centrifugation. Distillers' Dried Grains are the coarse fraction, which is termed Distillers' Light Grains, if dried directly, and Distillers' Dark Grains, if mixed with the solubles fraction after evaporation to 30 to 35 percent solids. It will be noted from the table that Distillers' Dried Solubles are higher in vitamin content.

During the acetone-butanol fermentation, riboflavin and pantothenic acid syntheses are appreciable, with the content of the first being increased about tenfold over that initially present. Pyridoxine and pantothenic acid are also formed in this fermentation. These are recovered by evaporating and drying the whole residue.

In the commercial penicillin fermentation in which members of the *Penicillium notatum-chrysogenum* group are employed, there is a marked synthesis of pantothenic acid (about 18 μg per ml) and an increase in biotin and pyridoxine.[67] On a dry basis, the residues from this fermentation are comparable with, or slightly higher in vitamin potency than, Distillers' Dried Solubles. The dried mycelium

TABLE 26. VITAMIN CONTENT OF FERMENTATION RESIDUES[a]

Fermentation process	Material fermented	Type of residue	Thiamine	Riboflavin	Pantothenic acid	Niacin	Pyridoxine	Biotin	Choline
Ethanol	Corn-malt	Distillers' Dried Solubles	6-9	15-20	29-36	104-160	8-10	0.3-07	6,000-7,000
Ethanol	Corn-malt	Distillers' Dark Grains	3-4	7-10	12-15	70-90	4-6	0.2-0.3	4,000-5,000
Ethanol	Molasses	Condensed solubles[b]	—	5-7	5-25	40	25	1.0-1.5	—
Butanol	Molasses	Whole residue	15	60	300	—	30	—	—
Penicillin	Corn steep -lactose	Mycelium	3-7	20-48	88-212	107-180	21-25	0.6-1.5	3,000-4,000
Penicillin	Corn steep -lactose	Mycelium-free liquor	—	7-29	592-805	232-333	82-114	0.6-0.8	—
Butanediol									
A. polymyxa	Wheat	Whole residue	6	—	153	—	—	—	—
A. aerogenes	Wheat,	Whole residue	19	19[c]	121	—	—	—	—
	molasses, and corn		23-37	45-132	—	—	—	—	—

[a] μg per g, dry basis.
[b] A sirup of 40 to 45% solids.
[c] Partial inactivation during product recovery.

has been found nontoxic to poultry[38] and is being recovered by some producers as a feed supplement. The synthesis of relatively high levels of pantothenic acid appears to be characteristic of the penicillia since other species including *P. utricae, P. intricatum,* and *P. stoloniferum* also produce unusual amounts.[68]

Two bound forms of pantothenic acid, i.e., coenzyme A and the *Lactobacillus bulgaricus* factor (LBF), have been found widely distributed in microorganisms. Some commercial fermentation products are excellent sources and have been used to prepare purified preparations for identification and investigation purposes. Coenzyme A has been obtained from *S. fradiae* (neomycin) fermentations[40a] and is currently being recovered from brewers yeast.[4a] LBF, a component of the coenzyme A molecule, whose reduced form has been termed pantetheine [N-(pantothenyl)-2-aminoethanethiol][65b] is produced in high potency by *A. gossypii*[47a,65] and *B. subtilis*[75a] and has been purified from both sources.

Riboflavin, pantothenic acid, and probably other B vitamins are synthesized during the aerobic (*Aerobacter aerogenes*) and anaerobic (*Bacillus polymyxa*) 2,3-butanediol fermentations.[1,11] *A. aerogenes* has been found particularly active in grain mashes in which it produced as much as 9.3 μg of riboflavin per ml. This organism has also been successfully applied to the fermentation of grain stillage in order to increase its vitamin potency.[40]

The formation of vitamins during other fermentations, such as the production of organic acids, has not been extensively studied. However, biotin[56] and rhizopterin, a derivative of folic acid,[83] have been found in media fermented to fumaric acid with *Rhizopus* species. Except for vitamin B_{12}, which will be discussed later, little information is available on vitamin syntheses accompanying recently developed antibiotic processes.

A few bacteria and molds, especially *Bacillus prodigiosus* and *Aspergillus niger,* have been observed to produce ascorbic acid.[75] Long incubation periods are usually required and the yields are markedly affected by the conditions of cultivation. The highest yield reported is approximately 14 mg per 100 ml of medium.

MICROORGANISMS IN THE DISCOVERY AND PRODUCTION OF NEW VITAMINS

The requirement by both animals and microorganisms for

accessory growth substances was established toward the end of the last century. Those factors required for animals were termed "vitamines" by Funk[17] in 1912 and the name "bios" was applied some years earlier by Wildiers[80] to the principle required for the growth of yeast. For many years, investigations of the isolation and characterization of these substances were continued along parallel but completely independent lines. During this period, vitamins A, B_1, B_2, C, D, E, and K were elucidated and certain bios constituents, such as inositol, were isolated and identified. Although Williams[81] had reported, as early as 1919, that yeast growth was stimulated by vitamin B and suggested that this might form the basis for a vitamin assay, the microbiological approach was not pursued until many years later.

Begining about 1930, a number of investigations disclosed that microorganisms required vitamins for growth and, conversely, that animals responded to certain microbial growth factors. Williams and Roehm[82] observed that thiamine was required for the growth of certain yeasts and Knight[30] showed that niacin, the anti-pellagra vitamin, was essential for the development of *Staphylococcus aureus*. Similarly, biotin, a growth-promoting substance for yeast, was found to be identical with Vitamin H, the anti egg white injury principle. Subsequently, pantothenic acid and folic acid, which had been recognized earlier as microbial growth factors, were demonstrated to have vitamin properties. These developments served to emphasize the close relationship between the various fields of nutrition, with the result that a general biological significance is implied for any new factor regardless of the original test species.

Recent studies on the nutritional requirements of various bacteria, yeasts, and molds have revealed the existence of a number of new growth substances. Some of these are finding practical applications in nutrition. Indeed, it appears that required nutrients brought to light through the study of microbial nutrition may constitute the principal source of new vitamins. The sequence of utilizing microorganisms for this purpose has followed a rather well-established pattern which involves the following steps:

(1) A particular microorganism is found to require (either naturally or following a mutation) a previously unidentified substance for growth or a specific biochemical function. This permits the development of an assay procedure and the evaluation of natural products for potency.

(2) Not uncommonly, other microorganisms which do not require the factor are found to synthesize it, and some in quantity, thus making available material for chemical fractionation.

(3) By the use of microbially synthesized substances, broad nutritional studies to establish vitamin activity are conducted with various species and different types of rations.

(4) Final purification and determination of structure are carried out with microbial substances. Then, the vitamin can be prepared commercially by either chemical or biological methods, depending on the relative cost.

This application of microorganisms to problems of animal nutrition is well exemplified in the elucidation of the anti pernicious anemia factor (APA). This procedure was also used to good advantage in the study of folic acid[27] and the conjugated forms of pantothenic acid.

Pernicious anemia is a deficiency disease characterized by defective erythrocyte formation. Liver for many years was the conventional source of the curative factor and by clinical assay procedures, purification had proceeded to the point where as little as 1 mg of concentrate represented a daily curative dose. However, purification and standardization of potency were tedious, since they were dependent on the use of human patients during regression from anemia. With the discovery of Shorb[62] that *Lactobacillus lactis* Dorner responded to liver extracts in proportion to their APA activity, a rapid assay procedure was made possible and purification of the active substance soon followed.[48]

This compound, vitamin B_{12}, is one of a family of complex coordination compounds containing trivalent cobalt which have been termed cobalamins.[65a] Vitamin B_{12} itself contains a cyano group in coordination with cobalt (cyanocobalamin) and has an approximate empirical formula[49] of $C_{61-64}H_{86-92}N_{14}O_{13}PCo$. At least three members of the vitamin B_{12} family have been found in microorganisms. A few micrograms of vitamin B_{12} elicit a positive hematological response in pernicious anemia patients. Nutrition studies with animals have shown that this vitamin is essential for animal growth and must be added to diets lacking in animal protein. Vitamin B_{12}, therefore, is an important constituent of the "animal protein factor" (APF).

Several observations suggest that the microbiological synthesis of vitamin B_{12} is a significant factor in satisfying the requirement of

TABLE 27. QUANTITATIVE SYNTHESIS OF VITAMIN B_{12} BY VARIOUS MICROORGANISMS

Organism	Medium	Highest vitamin B_{12} potency reported $\mu g/ml$	References
Flavobacterium devorans	Glucose, soybean meal, corn steep liquor, cobalt and other inorganic salts	0.6	22[a]
Flavobacterium solare	Yeast extract, malt extract, glucose, salts or maltose, penicillin-mash residue	0.6	45[a], 46
Pseudomonas sp. and Proteus sp.	Soybean meal or other protein materials	0.4	25[a]
Saccharomyces cerevisiae and *Aerobacter cloacae*	Grain-ethanol stillage or penicillin-fermentation residues, inorganic salts	—	2, 13[b]
Bacillus megatherium	Beet molasses, ammonium phosphate, cobalt and other inorganic salts	0.45	18, 18[a]
Propionobacter freudenreichii	Glucose, N-Z-Amine, yeast extract, lactic acid, cobalt	3.0	32[a]
Streptomyces fradiae	Glucose, brewers' yeast, soybean meal, cobalt and other inorganic salts	0.7	37
Streptomyces griseus	Beef extract, N-Z-Amine, cobalt or glucose, soybean meal, distillers' solubles, cobalt	0.3	25, 83[a]
Streptomyces olivaceus	Glucose, distillers' solubles, cobalt or soybean meal, cobalt and other inorganic salts	3.3	20[b], 21
Streptomyces vitaminicus	Glucose, animal stick liquor or other protein materials and cobalt	2.0	17[a]

animals for this vitamin. Rubin and Bird[52] found cow manure to be a good source of the factor probably owing to rumen and intestinal synthesis. Chicken feces, although low in the factor when voided, also acquired activity on incubation at temperatures favoring bacterial growth. Substantiation of microbial synthesis was supplied by Stokstad and associates[66] with the finding that a bacterial isolate from chick feces synthesized the factor in media of known composition.

Within the past few years, microorganisms have been intensively surveyed for their ability to biosynthesize vitamin B_{12}.[9b,20a,22,52a,63,71] The distribution of this vitamin in fungi and plant tissues has been the subject of a recent review.[13a] The surveys have revealed that a high percentage of bacteria and actinomycetes, but only a few yeasts and filamentous fungi produce vitamin B_{12}. Table 27 lists those organisms which synthesize amounts sufficient to receive commercial consideration.

Current requirements for pharmaceutical- and feed-grade vitamin B_{12} are being met by primary fermentation methods and by recovery as a by-product of certain antibiotic fermentations, e.g., streptomycin, aureomycin, and terramycin fermentations. Another potential source is activated sludge from municipal

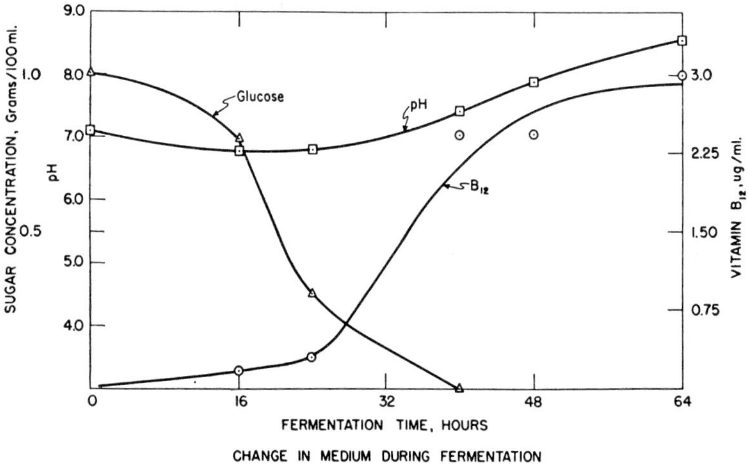

FIGURE 8. *Course of Vitamin B_{12} Fermentation with* Streptomyces olivaceus [Reprinted by permission from *J. Appl. Microbiology,* **1,** 124 (1953)]

sewage-treatment systems.[35a] Of the primary fermentations, *Streptomyces olivaceus* has been used widely, particularly for the manufacture of animal feed supplements. Figure 8 from the report by Hall et al.[21] shows the course of a typical deep-tank fermentation with this organism. Cobalt, a component of the vitamin B_{12} molecule, is present in inadequate amounts in the usual fermentation substrates and must be added for maximum biosynthesis. The fortification of media with both cobalt[83a] and cyanide[34a] has been patented. Vitamin B_{12} is retained within the cells of certain species which produce it[18a,32a] and it is absorbed by other species.[9a] This property offers a method of concentration. The chemistry of vitamin B_{12}, its purification from natural sources, and its role in animal nutrition have been discussed in detail elsewhere.[65a,85]

FAT-SOLUBLE VITAMINS AND PROVITAMINS

The fat-soluble vitamins and provitamins, in contrast to the B complex vitamins, are sporadically distributed among the fungi. Ergosterol (provitamin D) occurs widely in yeasts and molds and in certain bacteria. The carotenoids, of which several are partially or wholly convertible to vitamin A in the animal body, are produced by a number of chromogenic bacteria[19a,75] as well as by representatives of the following genera of yeasts and molds: *Saccharomyces, Rhodotorula, Sporobolomyces, Mucor,* and *Phycomycetes.* Vitamin K (K_2) is limited to certain lactics and coli-aerogenes organisms, while vitamin E is not an established constituent of fungi. It should be pointed out, however, that the fat-soluble factors are not uncommonly produced in relatively large amounts by fungi. For example, ergosterol[6] may comprise as much as 2.7% and carotene[57] as much as 0.25% of the dry-cell substance of certain species. Nevertheless, except for ergosterol, which will be discussed later, no concerted effort has been made to explore the industrial aspects of microbiologically synthesizing fat-soluble vitamins, since these substances are more economically available from other natural and synthetic sources. Several books and reviews contain discussions on the distribution and significance of the fat-soluble vitamins and related compounds in fungi.[7,19a,29,51,57,75]

Ergosterol

Ergosterol, which is the principal sterol of fungi, was first isolated from ergot by Tanret[69] in 1889. In the years immediately following, it was observed to be widely present among various microorganisms. Interest in this substance was greatly stimulated with the discovery, in 1927, that it became antirachitic on irradiation. Assays of various natural products for ergosterol potency revealed that yeasts and molds were exceptionally rich sources. The following values, taken from the literature, show the ranges which have been reported for the three general groups of microorganisms: bacteria 0 to 0.49%, yeasts 0.14 to 2.9%, and molds 0.02 to 2.3%.

It is evident that certain yeasts and molds offer commercial potentialities as ergosterol sources, provided they can be economically propagated or recovered as by-products of other fermentations. Originally, the mycelium produced in citric acid manufacture and brewers' yeast were processed. These have been supplanted for the most part by special methods of yeast propagation which result in products of higher ergosterol content. In addition, some attention has been given recently to the ergosterol content of the mycelium of molds produced in antibiotic fermentations, particularly penicillin, which might be utilized under certain circumstances. The utilization of various sterols as intermediates in cortisone production has renewed interest in the types and amounts of these compounds found in fungi.

Irradiated ergosterol (vitamin D_2, calciferol) is of low activity for the chick, but is highly active for mammals. It is used extensively in foods and pharmaceuticals and in feed supplements for four-footed animals. The production of this vitamin in the United States is shown in Table 28.

TABLE 28. PRODUCTION OF IRRADIATED ERGOSTEROL IN THE UNITED STATES

Year	U.S.P. units $\times 10^9$
1941	728[a]
1942	5,892
1943	14,248
1944	33,906
1945	27,301
1946	33,448
1947	15,610
1950	30,346
1952	26,447

[a] 18.14 billion units per lb.
Source: U. S. Tariff Commission.

The ergosterol content of microorganisms is a function of (a) the particular species or strain employed, (b) the medium composition, and (c) the physicochemical environment. Several methods for increasing ergosterol synthesis, based on one or more of these features, have been developed and patented. Current practices, although never fully disclosed, probably represent combinations of exceptional yeast strains in conjunction with favorable physical and chemical conditions.

Species and Strain Variations

When cultivated under similar conditions, both yeasts and molds may vary widely in ergosterol content. Bills, Massengale, and Prickett[8] found, for example, that in twenty-nine yeasts examined, the ergosterol content ranged from a trace with *Saccharomyces logos* to 2% with *S. carlsbergensis*. Similarly, different isolates from one species varied from 0.2 to 1.4%. Variations of similar magnitude have also been observed with numerous molds. Because of its characteristically high ergosterol content, *S. carlsbergensis* is probably the species most widely adopted commercially.

Physicochemical Conditions

Several physicochemical factors have been explored for their effect on ergosterol synthesis by yeasts. Of these the most critical is the degree of aerobiosis. The marked effect of aeration was first noted by Bills, Massengale, and Prickett,[8] who found that brewers' yeast, with only a trace of ergosterol, contained 1.4% when propagated aerobically. Likewise, Bennett[5] has observed that certain nontoxic oxidizing agents, such as peroxides, persulfates, etc., combined with aeration and a reduced nitrogen supply, further enhanced ergosterol formation. These modifications, in addition to supplementing the medium with ethanol and incubating at a higher temperature during the latter part of the fermentation, are claimed[6] to raise the ergosterol level to as much as 2.7%. Neutral and slightly alkaline conditions also have been reported favorable to ergosterol formation.

Media

In view of the lipid nature of sterols, it has been suggested that conditions which promote the formation of fats might be expected to increase sterol synthesis. However, attempts to relate the production of total lipid to that of sterols have been largely unsuccessful. More commonly, the relationship is an inverse one. For the yeasts, at least, a correlation has been noted between sterol content and the particular type of carbohydrate supplied, di- and trisaccharides yielding more sterol than monosaccharides.[34] Ascorbic acid has also been observed to enhance ergosterol synthesis by yeasts. With *Aspergillus fischerii*,[79] 10% glucose, a glucose-urea ratio of 20:1 and an initial pH of 2 were found most favorable.

As mentioned before, commercial ergosterol originates mainly from yeasts, but might also be derived from the by-product mycelium of mold fermentations. Examples are the mycelium of *Penicillium notatum*,[10,84] *Aspergillus flavus-oryzae* cultured on cereal bran, and, as stated previously, *Aspergillus niger* from the production of citric acid. *P. notatum*, however, varies widely in potency, depending on the particular strain and the conditions of propagation. For example, mycelium produced under submerged conditions contained only 0.22%,[53] while that grown on the surface of unagitated media yielded 2.3% when incubated for several days.[70] Thus, the submerged method of penicillin production results in mycelium of low potency, but the variability between strains offers some possibility of selecting more active organisms.

The conversion of ergosterol to vitamin D is carried out at any one of several stages, depending on the final use of the irradiated product. Yeast cells, after centrifuging and washing, are commonly irradiated directly with ultraviolet light and then pressed and dried if to be used in feeds. For pharmaceuticals and food fortification, the cellular material is first autolyzed or hydrolyzed and then saponified with alkali.[19] Ergosterol is then extracted with ether and further purified. This purified product is irradiated in suitable solvents, such as benzene or xylene. Rosenberg[51] has discussed in some detail the irradiation of ergosterol as well as the chemistry and physiology of the activated compound.

BIBLIOGRAPHY

1. Adams, G. A., and J. D. Leslie, *Can. J. Research*, **24F**, 12 (1946).
2. Ansbacher, S., and H. H. Hill, Absts. 116th Meeting Am. Chem. Soc. 31A, 1949.
3. Ashida, K., *J. Agr. Chem. Soc., Japan*, **18**, 723 (1942).

4. Bechdel, S. I., H. E. Honeywell, R. A. Dutcher, and M. H. Knutsen, *J. Biol. Chem.*, **80**, 231 (1928).
4a. Beinert, H., R. W. Von Korff, D. E. Green, D. A. Buyske, R. E. Handschumacher, H. Higgins, and F. M. Strong, *J. Biol. Chem.*, **200**, 385 (1953).
5. Bennett, W. G., U. S. Patent 2,059,980 (1936).
6. Bennett, W. G., U. S. Patent 2,776,710 (1942).
7. Bills, C. E., *Physiol. Rev.*, **15**, 1 (1935).
8. Bills, C. E., O. N. Massengale, and P. S. Prickett, *J. Biol. Chem.*, **87**, 259 (1930).
9. Boruff, C. S., (In Rudolfs' *Industrial Wastes,* ACS Monograph No. 118), New York, Reinhold, (1953).
9a. Burkholder, P. R., *Arch. Biochem. Biophys.*, **39**, 322 (1952).
9b. Burton, M. O., and A. G. Lochhead, *Can. J. Botany,* **29**, 352 (1951).
10. Cavallito, C. J. *Science,* **100**, 333 (1944).
11. Commercial Solvents Corp., British Patent 598,177 (1948).
12. Cooper, E. A., *J. Hyg.*, **14**, 12 (1914).
13. Damon, S. R., *J. Biol. Chem.*, **56**, 895 (1923).
13a. Darken, M. A., *Botan. Reviews,* **19**, 99 (1953).
13b. de Becze, G. I., U. S. Patent 2,636,823 (1953).
14. Fink, H., and F. Just, *Biochem. Z.*, **313**, 39 (1942).
15. Fridericia, L. S., P. Freudenthal, S. Gudjonnsson, G. Johansen, and N. Schoubye, *J. Hyg.*, **27**, 70 (1927).
16. Funk, C., *J. Physiol.*, **45**, 75 (1912).
17. Funk, C., *J. State Med.*, **20**, 341 (1912).
17a. Garey, J. C., J. F. Downing, and W. H. Stark, Abstracts of papers, 119th ACS Meeting, 22A, 1951.
18. Garibaldi, J. A., K. Ijichi, J. C. Lewis, and J. McGinnis, U. S. Patent 2,576,932 (1951).
18a. Garibaldi, J. A., K. Ijichi, N S. Snell, and J. C. Lewis, *Ind. Eng. Chem.*, **45**, 838 (1953).
19. Goering, K. J., U. S. Patent 2,395,115 (1946).
19a. Goodwin, T. W., *Botan. Reviews,* **18**, 291 (1952).
20. Gorcica, H. J., and H. Levine, U. S. Patent 2,295,036 (1942).
20a. Halbrook, E. R., F. Cords, A R. Winter, and T. S. Sutton, *J. Nutrition,* **41**, 555 (1950).
20b. Hall, H. H., U. S. Patent 2,643,213 (1953).
21. Hall, H. H., R. G. Benedict, C. F. Wiesen, C. E. Smith, and R. W. Jackson, *J. Appl. Microbiol.*, **1**, 124 (1953).
22. Hall, H. H., J. C. Benjamin, H. M. Bricker, R. J. Gill, W. C. Haynes, and H. M. Tsuchiya, Proc. 50th Meeting Soc. Am. Bacteriologists, p. 21, 1950.

22a. Hall, H. H., and H. M. Tsuchiya, U. S. Patent 2,561,364 (1950).
23. Harrison, H. E., U. S. Patent 2,359,521 (1944).
24. Hastings, E. G., *Bact. Rev.,* **8**, 235 (1944).
25. Hendlin, D., and M. L. Ruger, *Science,* **111**, 541 (1950).
25a. Hodge, H. M., C. T. Hanson, and R. J. Allgeier, *Ind. Eng. Chem.,* **44**, 132 (1952).
26. Hopkins, F. G., *Analyst,* **31**, 385 (1906).
See also *J. Physiol.,* **49**, 425 (1912).
27. Hutchings, B. L., E. L. R. Stokstad, N. Bohonos, N. H. Sloane, and Y. SubbaRow, *Ann. New York Acad. Sci.,* **48**, 265 (1946).
28. Johansson, K. R., and W. B. Sarles, *Bact. Rev.,* **13**, 25 (1949).
29. Karrer, P., and E. Jucker, *Carotenoide,* Basel, Verlag Birkhauser, 1948.
30. Knight, B. C. J. G., *Biochem. J.,* **31**, 966 (1937).
31. Knight, B. C. J. G., *Vitamins and Hormones,* **3**, 105 (1945).
31a. Knodt, C. B., *Antibiotics and Chemotherapy,* **3**, 442 (1953).
32. Lampen, J. O., I. L. Baldwin, and W. H. Peterson, *Arch. Biochem.,* **7**, 277 (1945).
32a. Leviton, A., and R. E. Hargrove, *Ind. Eng. Chem.,* **44**, 2651 (1952).
33. MacKenzie, F. B., W. M. Noble, and H. J. Peppler, *Chemurgic Digest,* **8**, 10 (1949).
34. Massengale, O. N., C. E. Bills, and P. S. Prickett, *J. Biol. Chem.,* **94**, 213 (1932).
34a. McDaniel, L. E., and H. B. Woodruff, U. S. Patent 2,650,896 (1953).
35. Mead, M. W., Jr., and J. Lee, U. S. Patent 2,328,025 (1943).
35a. Miner, C. S., and B. Wolnack, U. S. Patent 2,646,386 (1953).
36. Najjar, V. A., and R. Barret, *Vitamins and Hormones,* **3**, 23 (1945).
37. Nelson, H. A., K. M. Calhoun, and D. R. Colingsworth, Abstracts of papers, 118th Meeting Am. Chem. Soc., 16A, 1950.
38. Newell, G. W., W. H. Peterson, and C. A. Elvehjem, *Poultry Sci.,* **26**, 284 (1947).
39. Norris, F. W., *Brewers Digest,* **23**, 45, 96 (1948).
40. Novak, A. F., U. S. Patent 2,447,814 (1948).
40a. Novelli, G. D., *Federation Proceedings,* **12**, 675 (1953).
41. Osborne, T. B., and L. B. Mendel, *Carnegie Inst. Wash. Pub.,* **156**, part 2 (1911).
42. Pacini, A. J. P., and D. W. Russel, *J. Biol. Chem.,* **34**, 43 (1918).

43. Pavcek, P. L., W. H. Peterson, and C. A. Elvehjem, *Ind. Eng. Chem.*, **29**, 536 (1937).
44. Peterson, W. H., *Yeasts in Feeding Symposium*, Milwaukee, Wis., November 8-10, 1948.
45. Peterson, W. H., and M. S. Peterson, *Bact. Rev.*, **9**, 49 (1945).
45a. Petty, M. A., U. S. Patent 2,595,605 (1952).
46. Petty, M. A., and M. Matrishin, *Proc. 49th Meeting Soc. Am. Bacteriologists*, p. 47, 1949.
47. Prouty, C. C., *Wash. Agr. Expt. Sta. Bull., No. 484* (1947).
47a. Rasmussen, R. A., K. L. Smiley, J. G. Anderson, J. M. Van Lanen, W. L. Williams, and E. E. Snell, *Proc. Soc. Exptl. Biol. and Med.*, **73**, 658 (1950).
48. Rickes, E. L., N. G. Brink, F. R. Koniuszy, T. R. Wood, and K. Folkers, *Science*, **107**, 396 (1948).
49. Rickes, E. L., and T. R. Wood, U. S. Patent 2,563,794 (1951).
50. Roscoe, M. H., *J. Hyg.*, **27**, 103 (1927).
51. Rosenberg, H. R., *Chemistry and Physiology of the Vitamins*, New York, Interscience, 1942.
52. Rubin, M., and H. R. Bird, *J. Biol. Chem.*, **163**, 387 (1946).
52a. Saunders, A. P., R. H. Otto, and J. C. Sylvester, *J. Bact.*, **64**, 725 (1952).
53. Savard, K., and G. A. Grant, *Science*, **104**, 459 (1946).
54. Schaumann, H., *Arch. Schliffs-u. Tropen-Hyg.* Beihefte, **14**, (8) 5-397 (1910).
55. Scheunert, A., and M. Schieblich, *Biochem. Z.*, **184**, 58 (1927).
56. Schivek, A. I., U. S. Patent 2,359,443 (1944).
57. Schopfer, W. H., *Plants and Vitamins*, Waltham, Mass., Chronica Botanica Co., 1943.
58. Schultz, A. S., L. Atkin, and C. N. Frey, *J. Am. Chem. Soc.*, **60**, 490 (1938).
59. Schultz, A. S., L. Atkin, and C. N. Frey, U. S. Patent 2,262,735 (1941).
60. Schultz, A. S., L. Atkin, and C. N. Frey, U. S. Patent 2,285,465 (1942).
61. Schultz, A. S., L. Atkin, and C. N. Frey, U. S. Patent 2,377,044 (1945).
62. Shorb, M. S., *Science*, **107**, 397 (1948).
63. Shull, G. M., and J. B. Routien, Abstracts of papers, 119th Meeting Am. Chem. Soc., 22A (1951).
64. Singh, K., P. N. Agarwal, and W. H. Peterson, *Arch. Biochem.*, **18**, 181 (1948).
65. Smiley, K. L., M. Sobolov, F. L. Austin, R. A. Rasmussen,

M. B. Smith, J. M. Van Lanen, L. Stone, and C. S. Boruff, *Ind. Eng. Chem.*, **43**, 1380 (1951).

65a. Smith, E. L., H. H. Fantes, S. Ball, J. G. Waller, W. B. Emery, W. K. Anslow, and A. D. Walker, *Biochem. J.*, **52**, 389 (1952).

65b. Snell, E. E., and G. M. Brown, *Advances in Enzymology*, **14**, 49 (1953).

66. Stokstad, E. L. R., A. Page, Jr., J. Pierce, A. L. Franklin, T. H. Jukes, R. W. Heinkle, M. Epstein, and A. D. Welch, *J. Lab. Clin. Med.*, **33**, 860 (1948).

66a. Stokstad, E. L. R., *Antibiotics and Chemotherapy*, **3**, 434 (1953).

67. Tanner, F. W., Jr., S. E. Pfeiffer, and J. M. Van Lanen, *Arch. Biochem.*, **8**, 29 (1945).

68. Tanner, F. W., Jr., and J. M. Van Lanen, U. S. Patent 2,449,340 (1948).

69. Tanret, C., *Compt. rend.*, **108**, 98 (1889).

70. Tappi, G., *Gazz. chim. ital.*, **78**, 311 (1948).

71. Tarr, H. L. A., B. A. Southcott, and P. W. Ney, *Food Technol.*, **4**, 354 (1950).

72. Theiler, A., H. H. Green, and P. R. Viljoen, *South African Direct Vet. Research Dept.*, **3-4**, 7 (1915).

73. Van Lanen, J. M., *Arch. Biochem.*, **12**, 101 (1946).

74. Van Lanen, J. M., H. P. Broquist, M. J. Johnson, I. L. Baldwin, and W. H. Peterson, *Ind. Eng. Chem.*, **34**, 1244 (1942).

75. Van Lanen, J. M., and F. W. Tanner, Jr., *Vitamins and Hormones*, **6**, 163 (1948).

75a. Vitucci, J. C., N. Bohonos, O. P. Wieland, D. V. Lefemine, and B. L. Hutchings, *Arch. Biochem. Biophys.*, **34**, 409 (1951).

76. Weber, P. J. F., R. V. Siebel, and E. Singruen, *Brewers' Yeast a Victory Food*, Chicago, E. A. Siebel, 1943.

77. Wei-Shen Chang and W. H. Peterson, *J. Bact.*, **58**, 33 (1949).

78. Weinstock, H. H., Jr., H. K. Mitchell, E. F. Pratt, and R. J. Williams, *J. Am. Chem. Soc.*, **61**, 1421 (1939).

79. Wenck, P. R., W. H. Peterson, and E. B. Fred, *Zentr. Bakt. Parasitenk, Abt. II*, **92**, 330 (1935).

80. Wildiers, E., *La Cellule*, **18**, 3 (1901).

81. Williams, R. J., *J. Biol. Chem.*, **38**, 465 (1919).

82. Williams, R. J., and R. R. Roehm, *J. Biol. Chem.*, **87**, 581 (1930).

83. Wolf, D. E., R. C. Anderson, E. A. Kaczka, S. A. Harris, G. E. Arth, P. L. Southwick, R. Mozingo, and K. Folkers, *J. Am. Chem. Soc.*, **69**, 2753 (1947).

83a. Wood, T. R., and D. Hendlin, U. S. Patent 2,595,499 (1952).
84. Zook, H. D., T. S. Oakwood, and F. C. Whitmore, *Science,* **99**, 427 (1944).
85. Zucker, T. F., and L. M. Zucker, *Vitamins and Hormones,* **8**, 1 (1950).

PART V. THE PRODUCTION OF PHARMACEUTICALS

CHAPTER 7

THE PENICILLIN FERMENTATION

John C. Sylvester and Robert D. Coghill

INTRODUCTION

Brief History

Penicillin is the name applied by Fleming[23] to the bacteriostatic principle produced by a mold which was later identified as *Penicillium notatum*. Since Fleming's discovery, in 1929, it has been found that penicillin is produced by a variety of molds belonging to other species and genera and also that there is a series of closely related penicillins, all of which have approximately the same antibiotic characteristics. The word penicillin has, therefore, become a generic term which can be properly applied to a large number of closely related chemical entities.

Following Fleming's original work, Clutterbuck, Lovell, and Raistrick[15] undertook to study the chemistry of penicillin. The results of their work, published in 1932, indicated that penicillin was an organic acid which was extractable into organic solvents from aqueous solutions of low pH, but that it was extremely labile to hydrogen ion, heat, and even disappeared on evaporation of a solution to dryness. In view of this great lability and the very low yields (less than 1% of current practice), work on penicillin was abandoned in favor of more promising projects. After this, except for a paper by Reid[42] in 1935, nothing further appeared on penicillin for a period of 8 years.

In 1940, stimulated by the urgent need for better agents for the treatment of burns and other wounds, Chain and Florey,[11] with their collaborators, undertook the reinvestigation of penicillin. They cultured Fleming's organism in surface culture on what amounted to a small pilot-plant scale. By keeping a low temperature during their extraction procedure, they were able to concentrate the penicillin a thousandfold and produce a dry powder, in the form of a salt of penicillin, which had reasonable stability on storage. This represented the first great advance in penicillin production and furnished the material which was used to demonstrate the phenomenal curative properties of this antibiotic.

Because of the virtual impossibility of quickly producing a significant amount of penicillin under wartime conditions in England, Florey and Heatley came to the United States in July of 1941 in order to enlist the aid of the American government and pharmaceutical industry. The story[16,17] of their success in this direction has been told before and there is no need to repeat it here. Suffice it to say that by the time of the invasion of France in 1944, adequate amounts of penicillin were available on all fronts to effect a tremendous saving of life among the wounded. Penicillin was most certainly one of the few benefits accruing from World War II. Had it not been for the urgent need for such a curative agent, for the wartime ease of securing government money for such research projects, and for the excess profits tax which made it possible for industry to make a terriffic gamble with fifteen-cent dollars, we might not have had penicillin even today.

Therapeutic Usefulness

Penicillin is without doubt the most outstanding drug ever discovered for the treatment of infections. Its high order of activity, combined with its virtual nontoxicity, serves to make it the drug of choice for the treatment of most conditions caused by susceptible organisms. Its only bad feature is the fact that approximately 5% of those receiving the drug may suffer from some degree of allergic response. This condition, however, can frequently be averted by the use of a suitable antihistaminic drug.

Penicillin is active, in general, against Gram-positive microorganisms and inactive, with certain notable exceptions, against Gram-negative forms. Although occasional resistant strains are encountered, penicillin is markedly effective against pathogenic strains of the following genera: *Micrococcus, Streptococcus, Diplococcus, Neisseria, Clostridium, Treponema, Borrelia, Leptospira, Corynebacterium, Bacillus,* and *Actinomyces.* In addition, certain strains of other types of microorganisms, including some of the larger viruses and *Rickettsia* (such as *Miyagawanella psittacii),* are susceptible. In general, penicillin is ineffective against such genera as *Escherichia, Aerobacter, Klebsiella, Proteus, Salmonella, Shigella, Pasteurella, Eberthella, Pseudomonas, Vibrio, Brucella, Hemophilus, Mycobacterium, Rickettsia,* and yeasts, molds, and viruses.

Although the previous lists are incomplete, they give an approximate idea of the range of effectiveness of penicillin and serve to indicate its wide area of applicability—including many fields in which there has before been no completely satisfactory remedy. The necessary dosages of penicillin vary greatly from the 50,000-unit, single-dose treatment of gonorrhea to the several million units per day over a long period of time necessary in the treatment of subacute bacterial endocarditis.

Penicillin may be used topically, orally, or parenterally. When used orally, the dosage must be much larger (at least fivefold) because of destruction in the digestive tract as well as incomplete and variable absorption. When used parenterally, the intramuscular route is the usual method of choice.

Various salts of penicillin are in current use. The potassium, sodium, and calcium salts are all highly water soluble and can be used interchangeably. The calcium salt has not yet been crystallized and is available only in the crude, amorphous form. The crystalline potassium salt of penicillin G will undoubtedly attain the widest distribution among all soluble penicillin salts because of lower manufacturing costs as compared with the crystalline sodium salt or any of the amorphous forms. It is an example of a curious situation that a relatively pure crystalline substance is cheaper to produce than the impure amorphous compound.

More recently, the insoluble procaine salt of penicillin G has achieved popularity in repository dosage forms which permit the injection of penicillin at much less frequent intervals.

The Chemistry of Penicillin

The chemistry of penicillin has been discussed very comprehensively in a recent monograph.[14] In this chapter, only those phases of the subject will be covered which are essential to a proper understanding of the problems arising in penicillin production by fermentation.

Penicillin, as mentioned before, is a generic term which refers to a class of compounds, the known members of which have the structure:

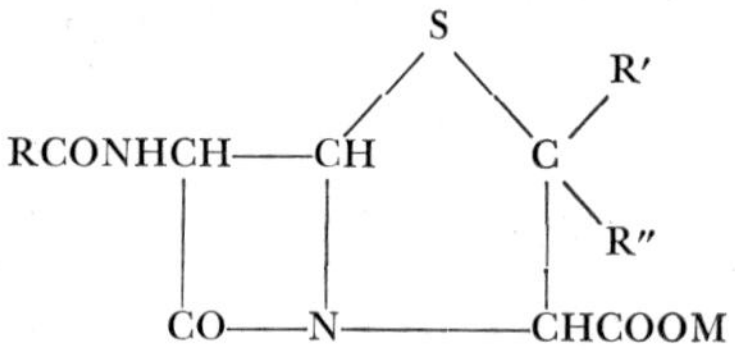

Penicillin

where R may be a simple substituted alkyl or aryl group, R′ and R″ are simple alkyl groups, and M is a hydrogen atom or an equivalent of base (organic or inorganic). In all of the known mold-produced penicillins, R′ and R″ are methyl groups and the M depends on the method of isolation. The chemical nomenclature in current use assumes that R′ and R″ are methyl groups, the distinctive name of a particular penicillin being derived from the nature of the R group. When the R is the benzyl group ($C_6H_5CH_2$—), the specific name becomes benzylpenicillin and the sodium salt is called sodium benzylpenicillinate. There is, however, another system of nomenclature which has a historical rather than a chemical background, but which is the only system used in the penicillin trade. Under this system, benzylpenicillin is termed penicillin G and its sodium salt, sodium penicillin G.

The common penicillins known to be produced by fermentation are listed in Table 29, along with their designations under both systems of nomenclature. As this chapter is designed primarily for those interested in producing penicillin by fermentation, only the trade terminology will be used.

TABLE 29. PENICILLIN TERMINOLOGY

R	Chemical Name	Trade Name
$CH_3CH_2CH{=}CHCH_2$-	2-Pentenylpenicillin	Penicillin F
$CH_3(CH_2)_3CH_2$-	*n*-Pentylpenicillin	Dihydropenicillin F
$CH_3(CH_2)_5CH_2$-	*n*-Heptylpenicillin	Penicillin K
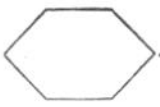—CH_2-	Benzylpenicillin	Penicillin G
HO 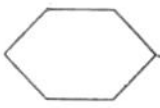—CH_2-	*p*-Hydroxybenzylpenicillin	Penicillin X

Penicillin measurement was originally accomplished by the so-called "cylinder plate method."[14,25] Based on this technique, an arbitrary unit, the Oxford unit, was established. In October of 1944, however, at an international meeting held in London, the International Unit was established. As defined there, an International Unit is that amount of penicillin which is equivalent in activity to 0.6 μg of the International Standard, as measured against either of two specified strains of *Staphylococcus aureus* (one of them being the C-203 strain of the United States Food and Drug Administration). The International Standard consisted of a sample of crystalline sodium penicillin G which was as pure as was possible to prepare at that time, and which, in retrospect, we know to have contained more than 98% sodium penicillin G.

The nature of the R group or, in other words, the type of penicillin produced, can be determined, to a large extent, by incorporating in the medium a precursor which may be a derivative (salt, amide, or ester) of the corresponding acid, RCH_2COOH, or an amine of the formula $RCH_2CH_2NH_2$. Not all compounds having these formulas will serve as precursors of the corresponding penicillins, but literally dozens of them have been so used. In the absence of an effective precursor, the type of penicillin produced seems to be a function of the mold strain and, in general, one obtains a mixture of several penicillins. Fleming's original strain of *P. notatum*, for instance, seems to have produced on his medium largely penicillin F. It is thus highly probable that this was the penicillin which he discovered and which the English workers subsequently isolated in the pure crystalline state. However, the currently used commercial strains, i.e., nonpigment-secreting descendents of *P. chrysogenum* Q176, when cultivated in the absence of precursors, produce a mixture of penicillins which consists largely of penicillin K, along with smaller amounts of dihydropenicillin F.

Penicillin manufacturers, however, strive to produce penicillin G solely, as this is the only penicillin in demand by the medical profession. This demand arises from the fact that penicillin G has better pharmacological properties and stability than penicillins K, F, and dihydro F, and is cheaper to produce than penicillin X. Other penicillins, equally as good as G, could probably be made cheaply and easily by the proper choice of precursors, but as they differ only *quantitatively,* rather than *qualitatively,* in their effectiveness against pathogenic organisms, it is unlikely that anyone will ever try to overcome the tremendous inertia of the literature and medical practice which placed penicillin G on the pedestal it occupies today.

Sodium penicillin G is the salt of a strong hydrophobic acid whose pK is 2.75. The free acid can, therefore, be extracted from water into a wide variety of solvents when the pH of the solution has a value of 2 to 3. The inorganic salts of penicillin are soluble in water, methanol, and ethanol, but are insoluble in anhydrous solvents, such as acetone, ether, chloroform, and esters. Addition of small amounts of water, however, greatly increases the solubility of penicillin salts in these solvents. Sodium penicillin G has a specific rotation ($[\alpha]_D^{20-25}$ in water) of approximately $+300°$.

Penicillin is quite unstable in the presence of hydrogen ions. Under such circumstances, it rapidly rearranges to a biologically inactive isomer, known as penillic acid, which has the following structure:

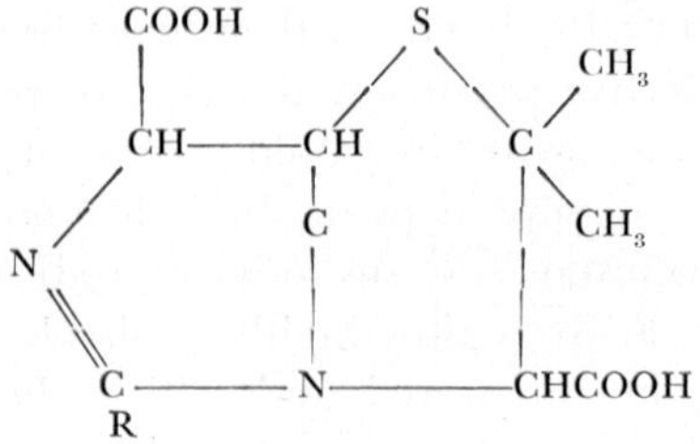

Penillic acid

At pH 2 to 3, this reaction goes almost to completion in 3 to 4 hours at room temperature. It is therefore necessary, in the isolation of penicillin, to minimize the time during which it is exposed to low pH and to maintain as low a temperature as possible during such exposure.

Penicillin is also very unstable at an alkaline reaction. At pH 12 at room temperature, the β-lactam ring opens quickly to yield a salt of what is known as penicilloic acid, having the following formula:

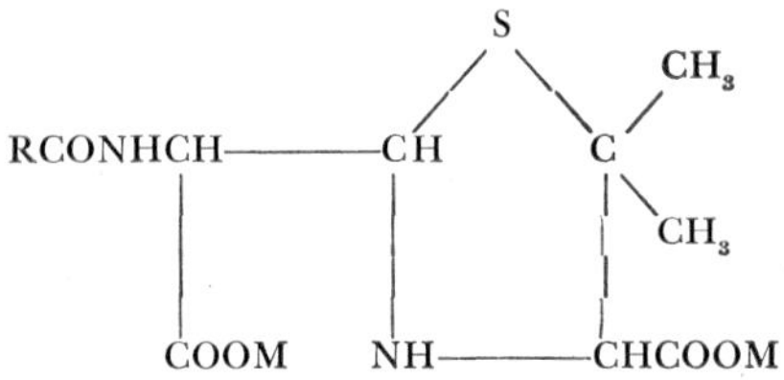

Penicilloic acid

For this reason, it is unwise to expose penicillin to a pH greater than 8. This reaction is catalyzed by many heavy metal ions, in particular Zn and Cu, and one must exercise every precaution to exclude even faint traces of these ions from processing operations.

The same reaction as produced by alkali, in all probability, is also brought about by exposure of penicillin to penicillinase, an enzyme commonly produced by many bacteria which are laboratory and plant contaminants. At room temperature, in the neighborhood of neutrality, small amounts of this enzyme can quickly and completely inactivate tremendous quantities of penicillin. Although the end product of the reaction has not been isolated, the production of a new acidic and a new basic group, having the same pKs as the corresponding groups in penicilloic acid, as well as other chemical evidence, leads strongly to this interpretation of the reaction.

Primary and secondary alcohols also lead to the inactivation of penicillin, methanol and ethanol being the worst in this respect. The reaction is again one of opening the β-lactam ring, the half esters of the penicilloic acids having been isolated from the reactions. In the case of methanol, the inactivation product has the following structure:

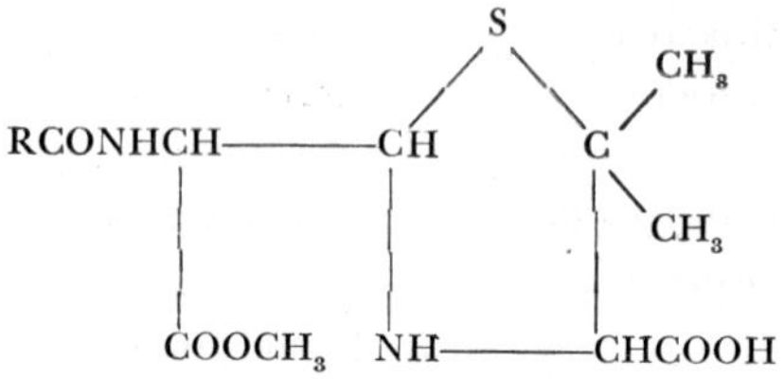

α-Methyl penicilloate

In this reaction, also, heavy metal ions act as powerful catalysts. The higher primary alcohols, as well as secondary and tertiary alcohols, react in this manner at such a low rate that, with proper timing and temperature, it is possible to use them in penicillin-recovery operations.

A similar reaction with primary amines also leads quickly to penicillin inactivation. The end products of this reaction are the corresponding α-amides of the penicilloic acids. Presumably secondary amines will react in the same fashion at a lower rate. It is thus important, in making the salts of penicillin with organic bases, to use tertiary amines. An exception to this, however, has been the procaine salt, where the weak basicity of the NH_2 group presumably slows down the reaction, so that there is little inactivation before the insoluble salt has come out of solution.

Finally, one must consider the inactivation of penicillin by heat. Amorphous penicillin, whether impure or very pure, is very unstable to dry heat, 1 or 2 hours at 100°C leading to virtually complete inactivation. The nature of this reaction has not been clearly demonstrated, but it probably is a reaction with absorbed water. The dry crystalline sodium or potassium salts of the penicillins are quite heat stable; sodium penicillin G, for example, is completely stable at 100°C for as long as 30 days.

Economic Considerations

The economic importance of penicillin far exceeds that of any other product of the pharmaceutical industry and also of the fermentation industry, with the possible exception of the alcohols. The United States Tariff Commission has reported that United States penicillin production for 1946 and 1947 was valued at approximately $90,000,000 per year and for 1948 at $137,652,000. The value for 1951 was about $137,517,000. No other pharmaceutical has ever approached these figures. Of this amount, about $50,000,000 worth is exported annually (as of 1951). Exports in 1952 were lower because of much lower prices and foreign production.

The world's production of penicillin is centered almost entirely in the United States with the exception of England, where production is about 10% of that of the United States. No other country approaches even 1% of our capacity. It is true, however, that modern penicillin plants are being built in many countries and the United States will probably not be able to maintain its current large exports which support production to a considerable extent.

Table 30 gives the United States production[46] of penicillin since its inception in 1943, as compiled by the United States Department of Commerce, except where otherwise noted.

TABLE 30. PRODUCTION OF PENICILLIN IN THE UNITED STATES

Year	Units $\times 10^9$
1943	21
1944	1,633
1945	7,125
1946	25,809
1947	30,640
1948	90,501
1949[a]	138,100
1950[b]	219,903
1951[a]	324,293

[a] From U. S. Tariff Commission.
[b] *Chem. Eng. News,* **29,** 3050 (1951).

Penicillin prices have followed the usual pattern of increasing production. Table 31 gives the prices of vials of 100,000 units of penicillin which have been certified for parenteral injection.

TABLE 31. PRICES OF PENICILLIN

Date	Wholesale price per 100,000 unit vial[a]
July 1943	$20.00
October 1943	15.00
January 1944	7.00
April 1944	3.75
July 1944	2.75
April 1945	0.78
January 1946	0.51
January 1947	0.36
January 1948	0.31
January 1949	0.20
January 1950	0.13
January 1951	0.15
January 1952	0.15
June 1952	0.13

[a] Through April 1945, this represents the price on government contracts, as no penicillin was sold to civilians.

These data depict a truly remarkable happening. In a period of 5½ years, the price of penicillin has dropped to less than 1% of its first quoted price and, during that time, the quality improved from that of a relatively unstable amorphous powder containing 10 to 20% of penicillin to that of a stable crystalline salt of penicillin G of 90 to 99% purity.

Bulk penicillin has had an equally precipitous drop in price and, as of November 1953, the following prices[5] per million units (in large quantities, bulk) prevailed: Crystalline potassium penicillin G, sterile, 19 to 22.5¢, crystalline procaine penicillin G, sterile, 18 to 20¢.

Crystalline potassium penicillin G at 4¢ per 100,000 units represents a price of approximately 64¢ per gram. Assuming all penicillin to have been of this quality and price, the United States production for 1950 had a weight of approximately 300,000 lb and a value of $90,000,000. When one considers that the major portion was sold as higher-priced procaine penicillin and that the price of the many dosage forms (tablets, oil suspensions, inhalation powders, etc.) varies from two to five times the bulk price, it is evident that the penicillin business at the wholesale level amounts to several hundred millions of dollars a year.

THE PENICILLIN-PRODUCING ORGANISMS

The mold described by Fleming,[23] in 1929, and subsequently identified as a strain of *Penicillium notatum* by Dr. Charles Thom of the United States Department of Agriculture, was used in the early developmental stages of the surface-culture process. The yields with this culture were poor and considerable work was carried out in several laboratories in an attempt to obtain a higher-yielding strain. Several strains,[34] including a descendant of Fleming's culture, were found which produced higher yields of penicillin. These new strains, together with other improvements in the process, combined to give yields of penicillin in the range of 200 units per ml by the surface-culture method. None of these cultures, however, performed well in submerged culture.

Many strains of *P. notatum* and *P. chrysogenum* were tested in an effort to find one which would produce good yields in the submerged process. One of the first good cultures found was a strain of *P. notatum* NRRL 832 of the Northern Regional Research Laboratory.[35] This strain was used by most producers during the early stages of the development of the submerged process and was capable of producing in the range of 100 units of penicillin per ml.

Continued work at the Northern Regional Research Laboratory[39] resulted in the isolation of *P. chrysogenum* NRRL 1951, which gave somewhat higher yields than did the strain NRRL 832. By plating NRRL 1951, followed by selective isolation, strain NRRL 1951•B25 was obtained. This strain was significantly better than the parent. Demerec,[20] at the Carnegie Institute, then subjected NRRL 1951•B25 to X-ray irradiation and isolated the survivors. His strains were screened at the University of Minnesota and further tested at the University of Wisconsin and one of them, X1612, was found to be significantly superior to the parent NRRL 1951•B25. This culture was subjected to ultraviolet irradiation by Backus and Stauffer[6] at the University of Wisconsin and a strain designated Q176, which proved superior to X1612, was isolated. This strain was adopted by most of the penicillin producers and monospore isolates from it are in general use.

Raper and Alexander[39] have reviewed very completely the history of the various strains of *Penicillia* that have been of significance in the development of penicillin. Their paper also includes detailed descriptions of the strains and a discussion of their extreme variability and methods of maintaining stock cultures.

The penicillin-producing molds are characterized by unusual variability; in general, the greater the productivity of the strain the more unstable it is. The extreme variability of these organisms presents the microbiologist with a difficult problem in maintaining active stock cultures of consistent penicillin-producing capacity. This problem is of the utmost importance in commercial production where the cost of the finished product varies almost inversely with the yield in the fermentors.

Stock cultures can be maintained as agar slants, in dry soil, or in lyophilized form (see Chapter 12). Stocks carried on agar slants are perhaps most liable to variation since successive transfer of the organism provides greater opportunity for variation to occur. Frequent transfers tend to propagate selectively those portions of the culture population that sporulate most readily and profusely. If agar slants are used, transfers should be relatively infrequent. A medium that provides minimum vegetative growth and maximum sporulation should be employed and large numbers of conidia should be used as inoculum. One company has found the preservation of stocks by freezing agar slant cultures, after good sporulation is obtained on the slants, to be very satisfactory.

Soil stocks are prepared by adding a heavy aqueous suspension of spores to tubes of sterile soil, allowing the soil to dry at room temperature, and then storing at 2° to 4°C. Stocks carefully prepared in this fashion will keep indefinitely and it is very little work to prepare enough tubes at one time to last for many years. This procedure has proved satisfactory for the penicillin-producing molds in most cases. However, one instance is known in commercial production in which stocks of *P. chrysogenum* X1612 preserved on soil gradually deteriorated over a period of about 1 year to the point where yields of less than 10 units per ml were obtained from stocks which originally had produced several hundred units per ml. A similar experience in a research laboratory is reported by Gailey, Stefaniak, Olson, and Johnson.[24]

It is believed by many that lyophilized stocks are best for preserving penicillin-producing cultures. To prepare stocks, a mass of conidia is suspended in a small amount of whole blood serum, milk, or peptone solution. The spore suspension is then dispensed into small tubes, dried from the frozen state under high vacuum, and the tubes sealed. An excellent description of this technic is given by Raper and Alexander.[40] As in the case of soil stocks, sufficient tubes can be prepared at one time to last for many years.

The microbiologist charged with the maintenance of stock cultures of penicillin-producing molds should constantly keep in mind the instability of these organisms. He should be continually alert to morphological variation and deterioration in penicillin-producing capacity. It may be necessary, when such changes occur, to isolate from the stock culture, by plating, a number of strains which are then tested. It is possible, in many instances, to recover from a deteriorated stock strains which are equivalent to the original in penicillin-producing ability. Churchill and Roegner,[12] working under the direction of Backus and Stauffer, have studied the population patterns and variability of *P. chrysogenum* Q176 and related strains. They describe five types of colonies and point out that all five types show instability. A wide variation in yield of single-spore isolates of *P. chrysogenum* Wis. Q176 is described by Reese, Sanderson, Woodward, and Eisenberg.[41]

CHEMICAL CHANGES DURING THE FERMENTATION

While the production of penicillin may be considered a fermentation process, the product, penicillin, is not a true fermentation product in the same sense as are gluconic acid, lactic acid, alcohol, etc. True fermentation products are usually the result of the energy metabolism of the cell and, as such, the mechanism of their formation is comparatively clearly defined. Penicillin and most other antibiotics, however, are apparently not directly related to the energy metabolism. In the lactic acid fermentation, 90% of the sugar fermented may appear as lactic acid; in contrast, in the penicillin fermentation, based on the carbohydrate metabolized, a yield of only 3 to 5% of penicillin is obtained. Also, it is possible to obtain, with an otherwise normal fermentation, little or no penicillin. It seems likely that penicillin results from the metabolism of the amino acids and other complex organic compounds when a natural medium is used. However, the mold can synthesize penicillin when grown in synthetic media containing only an energy source and inorganic constituents. The general opinion is that penicillin is a minor by-product of the normal metabolism of the organism and plays no essential role in mold metabolism. This comment is applicable to most antibiotics as well as to the excessive production of vitamins by certain organisms. The question of why and how penicillin is produced cannot be satisfactorily answered on the basis of our present knowledge. It is hoped that the intensive studies being carried out on this fermentation will soon provide some explanation of the process.

The chemical changes which occur during the penicillin fermentation vary widely, depending on the composition of the medium. While these variations are of the greatest interest from a research standpoint and lead toward an elucidation of the mechanism of penicillin production, for the purpose of this review, it will suffice to consider only those changes which take place in the medium generally in use for commercial production. While the specific media employed by the various penicillin producers are carefully guarded secrets, it is believed that all manufacturers employ variations of a similar formula composed primarily of the following substances:

	%
Corn-steep liquor solids	1-5
Lactose	1-3
Glucose	0-1
Sodium nitrate	0-0.5
Calcium carbonate	0-1
Phenylacetic acid or a derivative	0.01-0.1
Water	

While the exact composition of corn-steep liquor is unknown, the proximate analysis,[8] on a dry basis, is as follows:

	%
Total solids	40-60
Total nitrogen	7.4-7.8
Amino nitrogen	2.6-3.3
Lactic acid	12-27
Reducing sugars (as glucose)	1.5-14
Ash	18-20

In addition to these ingredients, small amounts of phenylalanine, tyrosine, β-phenylethylamine, tyramine, and similar compounds of significance in the penicillin fermentation are present. The concentrations of lactic acid and sugar will vary in different lots of corn-steep liquor, depending on the amount of fermentation that has taken place. In general, however, the amount of sugar plus the amount of lactic acid is approximately constant. While the composition of different batches of corn-steep liquor will vary, the components which are important for the fermentation are present in adequate amounts in most lots. These substances are lactic acid, organic nitrogen, precursors for penicillin G and other penicillins, and mineral salts. The chemical changes of a corn-steep liquor medium during a typical fermentation by *P. chrysogenum* Q176, as described by Brown and Peterson,[9] are shown in Figure 9. The chemical changes and the metabolism of the penicillin-producing molds are reviewed very excellently in papers by Peterson[37] and Johnson.[28]

Carbohydrate Utilization

A wide variety of carbohydrates can be utilized by *P. chrysogenum* for growth. The group of compounds which can provide

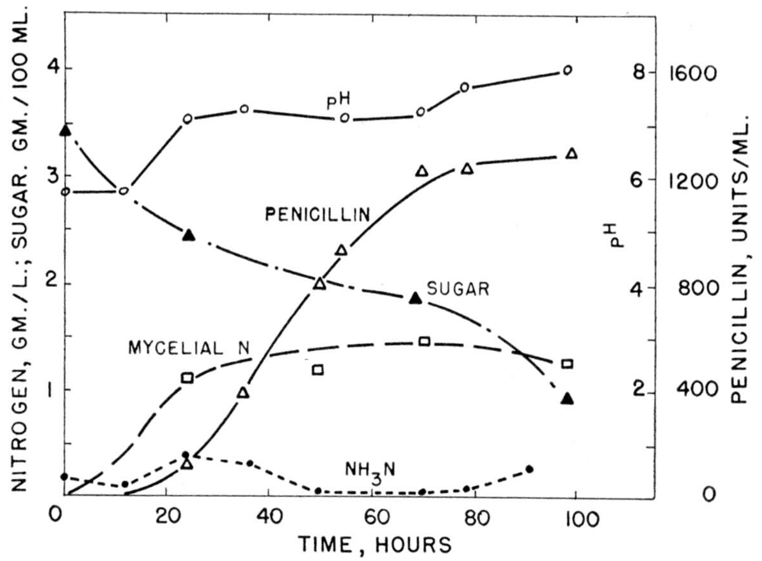

FIGURE 9. *Chemical Changes in a Typical Fermentation with Added Phenylacetic Acid*[9]

energy and carbon includes various organic acids (such as lactic, acetic, and formic), amino acids, starch, dextrins, glucose, sucrose, brown sugar, lactose, and many other sugars. Lactose gives the highest yields of penicillin and is generally used commercially. Brown sugar, while somewhat inferior to lactose, is suitable for commercial production. It is possible, under special conditions, as, for example, very high aeration in glucose media, to obtain good yields on carbohydrates other than lactose. However, such special conditions are not suitable for industrial application. Recently Hosler and Johnson,[27a] employing a basal synthetic medium, obtained high yields of penicillin from glucose by feeding the sugar in solution continuously at the rate of 0.03% per hour. The pH was controlled in the optimum range 7.0 to 7.4 by automatic addition of gaseous ammonia.

The superiority of lactose is presumably due to its slow availability. No acids accumulate during its utilization, probably because any acids formed are more readily available than lactose and are utilized as rapidly as they appear. When carbon sources other than lactose are used, the concentration of the various medium constituents must be such that the pH changes during the fermentation favor good growth and penicillin production. In a corn-steep liquor medium, much of the carbon metabolized is supplied by the corn-steep rather than by the added carbohydrate.

In a typical medium, containing as sources of carbon, glucose, corn-steep liquor compounds, lactic acid, and lactose, the mold uses first the glucose and simultaneously some of the corn-steep carbon. When the glucose is completely utilized, the lactic acid is metabolized, together with more of the corn-steep compounds. Lactose utilization begins when other carbon sources are exhausted. The period of lactose utilization is the main phase of the penicillin-producing cycle. The utilization of the carbohydrate sources in the described manner tends to produce a characteristic pH curve which will be discussed later.

When the lactose is exhausted, no additional available nutrients remain. If the fermentation is allowed to continue, autolysis begins and the penicillin concentration decreases. In commercial operations, the fermentations are harvested before this point because of the difficulties encountered in filtering autolyzed cultures and in order to avoid losses of penicillin.

Nitrogen Utilization

Organic nitrogen, in the form of amino acids, peptones, and proteins and inorganic nitrogen, as ammonia and nitrate, can be readily utilized for growth and penicillin production by *P. chrysogenum*. Ammonia is most readily assimilated, as is indicated by the initial drop in the ammonia-nitrogen level of a fermenting culture. A significant portion of the nitrogen assimilated by the mold is in the form of ammonia nitrogen, formed during the fermentation by the deamination of nitrogenous corn-steep liquor compounds. Usually, an initial increase in ammonia-nitrogen occurs in the medium because it is produced at a faster rate than it is utilized. This is followed by a very rapid decrease, which continues until the end of the fermentation. Occasionally, a second rise in ammonia nitrogen occurs, as a result of autolysis of the mold at the end of the fermentation cycle. Nitrate nitrogen appears to be utilized only when organic nitrogen is no longer available.[30] In a high-corn-steep, low-nitrate medium, the organic nitrogen serves as the main source of nitrogen. If the corn-steep concentration is reduced and the nitrate increased, the nitrate supplies most of the organism's requirements. The nitrogen incorporated in the cells rises rapidly during the early stages of the fermentation and then levels off. A late decrease in mycelial nitrogen may occur if autolysis takes place.

pH Changes

The pH changes of a typical penicillin fermentation follow a more or less characteristic pattern. At the beginning of the fermentation, the pH is between 5 and 6, depending on the amount and kind of corn-steep liquor used in preparing the medium and the neutralizing agents used. During the first several hours, no change in pH occurs. Glucose and corn-steep compounds are being utilized to build mycelium and any ammonia produced is immediately used. Following this period, the lactic acid is assimilated and ammonia begins to accumulate. This results in a marked rise in pH up to 7 to 7.5, primarily because of the utilization of the lactic acid with the resultant freeing of inorganic cations. As the mold begins utilizing the lactose, the accompanying ammonia assimilation may cause a slight pH drop. Often, however, a stable plateau or a slow rise is observed. Toward the end of the fermentation, a secondary rise in pH to 8 or above may occur, when the carbohydrate of the medium is exhausted and autolysis begins. The pH changes during the fermentation will vary considerably with different media. For example, in a glucose fermentation, as described by Johnson,[28] the pH does not rise until the glucose is exhausted. The pH then rises sharply, followed by autolysis because of lack of nutrients. There is no pH plateau similar to that occurring in a lactose medium. In general, pH values below 7 or much above 8, during the penicillin-synthesis phase of the fermentation, are not associated with good penicillin yields.

Mycelial Development

The development of the mycelium occurs primarily during the first 20 to 30 hours of the fermentation, coinciding with the period of rapid glucose and corn-steep carbon utilization. This growth can be followed by mycelial-nitrogen determinations. In a typical fermentation, the growth is quite appreciable and considerable thickening of the culture takes place. In some cases, a 24-hour culture is so heavy that a sample placed in a 25-mm diameter test tube will remain in the tube when it is inverted. The growth may be in the form of indefinite clumps of mycelium or definite pellets, ranging from 0.5 to 2 mm in diameter. During the main penicillin-producing phase of the fermentation, little or no mycelial growth takes place. When the nutrients have become exhausted, autolysis of the mycelium begins and its weight decreases.

Pigment Formation

The older strains of the penicillin-producing molds usually synthesize, together with penicillin, a number of yellow pigments which impart a bright-yellow color to the fermenting culture. While these pigments are usually associated with good yields of penicillin, their presence in the culture is not always indicative of this condition. Research by Churchill and Roegner,[12] working in the laboratory of Backus and Stauffer on irradiation of *P. chrysogenum* by ultraviolet rays, has resulted in strains which are capable of good penicillin production, but which produce no appreciable amount of pigment. Recently workers from the University of Wisconsin have reported on the genealogy and biochemical performance of the races of pigmentless mutants obtained from *P. chrysogenum* Q176 by treatment with ultraviolet light and nitrogen mustard.[3a] Non-pigment-secreting strains are in general use in industrial operations.[48]

Penicillin Formation

The production of penicillin begins as soon as a good amount of mycelial growth has taken place. However, during the first 20 to 30 hours, when growth is rapid, the rate of antibiotic production is comparatively low. The maximum rate of penicillin synthesis occurs during the main phase of the fermentation, while the lactose is being utilized. Throughout this period, the rate of production is rather uniform and the yield curve rises sharply. When the medium nutrients are exhausted, penicillin production levels off and, if the fermentation is allowed to continue, the penicillin concentration in the medium will begin to decrease. The yield of penicillin produced in commercial fermentations is unknown, as these figures are considered confidential by the various manufacturers. However, yields of 1,500 units per ml have been reported in pilot-plant fermentations[37] and it is generally believed that industrial-fermentation yields exceed this figure.

Oxygen Utilization

Since the penicillin-producing molds are highly aerobic organisms, a large amount of oxygen must be supplied during the fermentation cycle. This is accomplished by blowing a continuous stream of sterile air into the fermenting medium. Koffler, Emerson, Perlman, and Burris[30] reported uptakes of 109 to 134 ml of oxygen per hour per liter of culture with strain NRRL 832 on a lactose medium. On a glucose medium, the oxygen uptake was considerably higher. The oxygen consumed by the mold is utilized in the aerobic oxidation of carbon compounds and results in the development of large quantities of carbon dioxide during the fermentation. The rate of carbon dioxide development varies proportionately with the oxygen supplied. Johnson[28] reported the production of 2.7 volumes of carbon dioxide per minute per 1,000 volumes of medium at an aeration rate of 0.15 volumes per minute per volume of medium and 10.0 volumes of carbon dioxide at an aeration level of 1.5 volumes. The rate of sugar fermentation obviously depends on the available oxygen and, as is indicated by the carbon dioxide development, varies with the oxygen supply.

Precursors

The types of penicillin produced in the fermentation can be influenced considerably by the presence of certain precursor compounds in the medium. When phenylacetic acid compounds are present,[36] the mold utilizes the phenylacetic residue directly,[7] resulting in a greatly increased production of penicillin G and a decreased production of other types. Other compounds of the proper structure will act as precursors for penicillin X, penicillin F, penicillin K, and dihydropenicillin F. It has also been found possible to produce "unnatural" penicillins, such as *p*-nitrobenzylpenicillin,[21] *p*-iodobenzylpenicillin, *p*-tolylmercaptomethylpenicillin, phenylmercaptomethylpenicillin,[2] *p*-bromophenylmercaptomethylpenicillin, and many others, by supplying the mold with the proper precursor. The effect of precursors on the total yield of penicillin G and on the relative amounts of the respective penicillins is described by Singh and Johnson.[44] A summary of the work on precursors for penicillin is presented in the monograph, *The Chemistry of Penicillin.*[14]

It should be pointed out that very little information is available on the course of the fermentations in industrial practice and it is very probable that the characteristics of the fermentations of various producers differ considerably. Sufficient data on small-scale operations are available to indicate that minor changes in such factors as medium, aeration rate, type of inoculum, etc., will result in significant alteration of the chemical changes during the fermentation. The discussion presented here can only be based on data that are available and, at best, can indicate the fundamental characteristics of the fermentation. However, it is felt that the published work does cover the basic mechanism of the penicillin fermentation and that, although it is certain that various producers have improved processes, these improvements are the result of variations which do not involve changes in the fundamental aspects of the fermentation.

COMMERCIAL PRODUCTION OF PENICILLIN

Different procedures for the cultivation of the penicillin-producing molds have been studied in considerable detail and several technics developed which involve the surface growth of the mold on a solid menstruum. The use of wood shavings loosely packed in a column was one of the procedures that received some attention. In this technic, the nutrient medium was trickled slowly through a column of wood shavings on which the mold was growing, in a manner similar to that used for the production of acetic acid. Another method developed was the propagation of the mold on bran moistened with nutrient solution and contained in a slowly revolving drum. This technic was developed by one company to a full-scale commercial production unit. However, apart from the surface-culture procedure on a liquid menstruum and the submerged-culture process, no other method of production ever achieved any economic significance and this discussion will be limited to the two methods which have produced large quantities of penicillin and were of real economic importance.

The Surface-Culture Fermentation

Although the surface-culture method for the commercial production of penicillin is no longer used in the United States or England, a brief description of this process will be given, since it was responsible for the first commercial production of penicillin and rather large production units were operated successfully in this manner. Despite the fact that the surface-culture process was expensive, laborious, and commercially and economically impractical, it served to bridge the period between the initial interest in penicillin and the development of the submerged-culture process and made penicillin commercially available when it was desperately needed.

It was natural that the first attempt to produce penicillin commercially should have been by the surface process, since this method is based on the natural tendency of molds to grow as a thin layer on the surface of liquid media. Many variations of the process, with respect to the type of container in which the mold was grown, were developed by different laboratories. The fundamental requirement for the cultivation of the mold under surface conditions is that the depth of the medium should be approximately 2 cm. Any container which will provide a shallow layer of medium is satisfactory. Production units were built from various types of bottles, flasks, and metal or glass trays and pans. Many novel and unusual methods were used for sterilization of the containers and inoculation of the media. Basically, however, all the various production procedures were similar. Mass spore cultures of a suitable strain of *Penicillium* (NRRL 1249-B21, NRRL 1951-B25 and others) were produced on large agar slants in bottles, or on moistened bran or other grain products, such as oat hulls, cracked corn, etc. If desired, an aqueous suspension of the spores was prepared by adding sterile water to spore culture and vigorously agitating to break up the clumps of spores and suspend them in the water. The sterile medium in a suitable container was inoculated by spraying the aqueous suspension of spores into the container or by simply adding a few milliliters of the suspension to the container from a pipette. When dry spores on a grain carrier were employed for inoculum, the material was blown into the container or a few grams merely poured into it. The dry spores gave somewhat more satisfactory results since many spores floated on the surface of the medium, which resulted in more rapid development of the mat. After inoculation, the medium was incubated at 24° to 28°C for 6 or 7 days. Usually, the maximum production of penicillin was obtained in this period and the cultures were then harvested. The containers were emptied and the broth collected, after separation of the mold. In some plants, the medium was withdrawn from the container, without greatly disturbing the mold mat, and replaced with fresh medium. Maximum yields were obtained in the second lot of medium several days earlier than in the original fermentation. This procedure could be repeated several times. The fermented medium was then worked up by procedures similar to those that will be described later in the discussion on the recovery of penicillin.

The basic principles of the surface-culture process are well illustrated in the following description of a laboratory experiment.

Sixty-five milliliters of a medium containing, per liter, 50 g corn-steep solids, 30 g technical lactose, 3 g $NaNO_3$, 0.5 g KH_2PO_4, 0.25 g $MgSO_4$, 0.2 g $C_6H_5CH_2COOH$, 0.044 g $ZnSO_4$, and NaOH to pH 5.5, were placed in a 300-ml Erlenmeyer flask. The flask was stoppered with a cotton plug and sterilized 30 minutes in an autoclave at 15 psi steam pressure. After cooling, the flask was inoculated with one loopful of dry spores from a 6-day agar-slant culture of *P. notatum* NRRL 1249-B21. The flask was incubated at 24°C and samples were removed for assay after 5, 6, 7, and 8 days. The following results were obtained:

	5 days	*6 days*	*7 days*	*8 days*
Penicillin, units per ml	65	142	193	136
pH	7.2	7.6	8.1	8.2

The difficulties encountered in the operation of industrial production units, using the surface-culture process, were tremendous. Any significant volume of production required a great many containers, huge incubators, a large labor force, and a relatively large investment in buildings and supporting equipment. Some plants handled 50,000 or more bottles daily. Even plants using trays or pans had to process thousands of them a day to achieve any significant production and, despite the enormous amount of material, space, and labor that were used in the surface-culture plant, the daily production of fermented medium amounted to only a few thousand gallons. The sterility problems involved in the surface culture alone were great enough to make this method unfeasible for economical large-scale production. Because of the great difficulties encountered and the inherent disadvantage of the process, as soon as the submerged-culture process was developed to the point of large-scale operation, the surface-culture method was abandoned by all penicillin producers.

The Submerged-Culture Fermentation

Although the submerged-culture process for the commercial production of penicillin is fundamentally simple, the actual production unit is a very intricate and complex mechanical system. The modern penicillin plant is a very highly engineered unit, which depends for successful operation on the perfect integration and control of a large amount of machinery and is essentially a mechanical process requiring human effort only to coordinate the functioning of the machinery and to keep it running smoothly. The first impression one gets on visiting a penicillin plant is that there seem to be very few people required for the operation of a relatively great amount of mechanical equipment. From an engineering standpoint, the development of the process required a tremendous amount of engineering research, including the design of special unique equipment as well as the adaptation of much standard machinery to rather unusual uses. A relatively modest submerged-culture plant requires an investment of several million dollars in building and equipment and the services of highly skilled personnel to supervise its construction and operation.

The development of the submerged-culture process, made possible by the discovery of strains of *Penicillium* suitable to this type of operation, was the greatest single factor responsible for the truly large-scale production of penicillin, which has made possible the manufacture of tremendous quantities of this remarkable chemotherapeutic drug at an extremely low cost. Although this development took place within a relatively short period of time, in terms of research effort, tremendous investment was required. More than twenty laboratories, both governmental and private, were concerned with this work and contributed to its successful completion. Even yet, much continuous study is being devoted to the submerged-culture process and improvements in yields are being achieved constantly.

Fundamentally, the submerged-culture process is as simple as the surface method, consisting essentially of growing a suitable strain of *Penicillium* homogenously dispersed throughout a volume of liquid medium. The process is illustrated by the following description of a shake-flask experiment: One hundred and fifty milliliters of medium, containing, per liter, 12.5 g corn-steep solids, 20.0 g lactose, 5.0 g Cerelose, 3.0 g $NaNO_3$, 2.0 g KH_2PO_4, 0.04 g $ZnSO_4$, 1.0 g $CaCO_3$, and 0.5 g phenylacetic acid derivative, contained in a 1 qt Mason fruit jar, Kerr type, was sterilized 30 minutes in an autoclave at 15 psi steam pressure. The flask was then inoculated with 6 ml of a 48-hour vegetative culture of *P. chrysogenum* Q176. (The inoculum culture was prepared by inoculating a flask of the described medium with 1 ml of an aqueous spore suspension made by suspending the spores from a 6-day agar-slant test-tube culture in 5 ml of sterile water; the flask was then incubated 48 hours at 24°C in the shaking machine.) The inoculated flask was incubated at 24°C in a rotary shaker having a 2.25 in. diameter movement and rotating at 180 rpm. Samples were taken at 3, 4, 5, 6, and 7 days for penicillin assay. The results were as follows:

	4 days	*5 days*	*6 days*	*7 days*
Penicillin, units per ml	492	602	545	560
pH	7.4	8.1	8.6	8.3

Although each penicillin manufacturer developed his own method and the details of the various producers' processes are not available for comparison, it is generally thought that these processes are essentially similar and the differences that do undoubtedly exist are ones of minor technical nature. In general, the submerged-culture fermentation process consists of three steps: (1) Preparation of inoculum, (2) production fermentation, and (3) harvesting the fermented medium.

In this review, it will be impossible to describe in detail all the various types of production units in operation because of the limited information available. Each manufacturer's plant differs in some degree from all others, because each was developed more or less independently. However, since the various plants are fundamentally the same, it will suffice for our purpose to describe what may be considered a typical production unit, one which will illustrate the basic design of a submerged-culture penicillin-production plant. Figure 10 is a diagrammatic flow sheet of this typical unit and shows the various steps in the process. Recently, another flow sheet and brief description of the process has been published.[4]

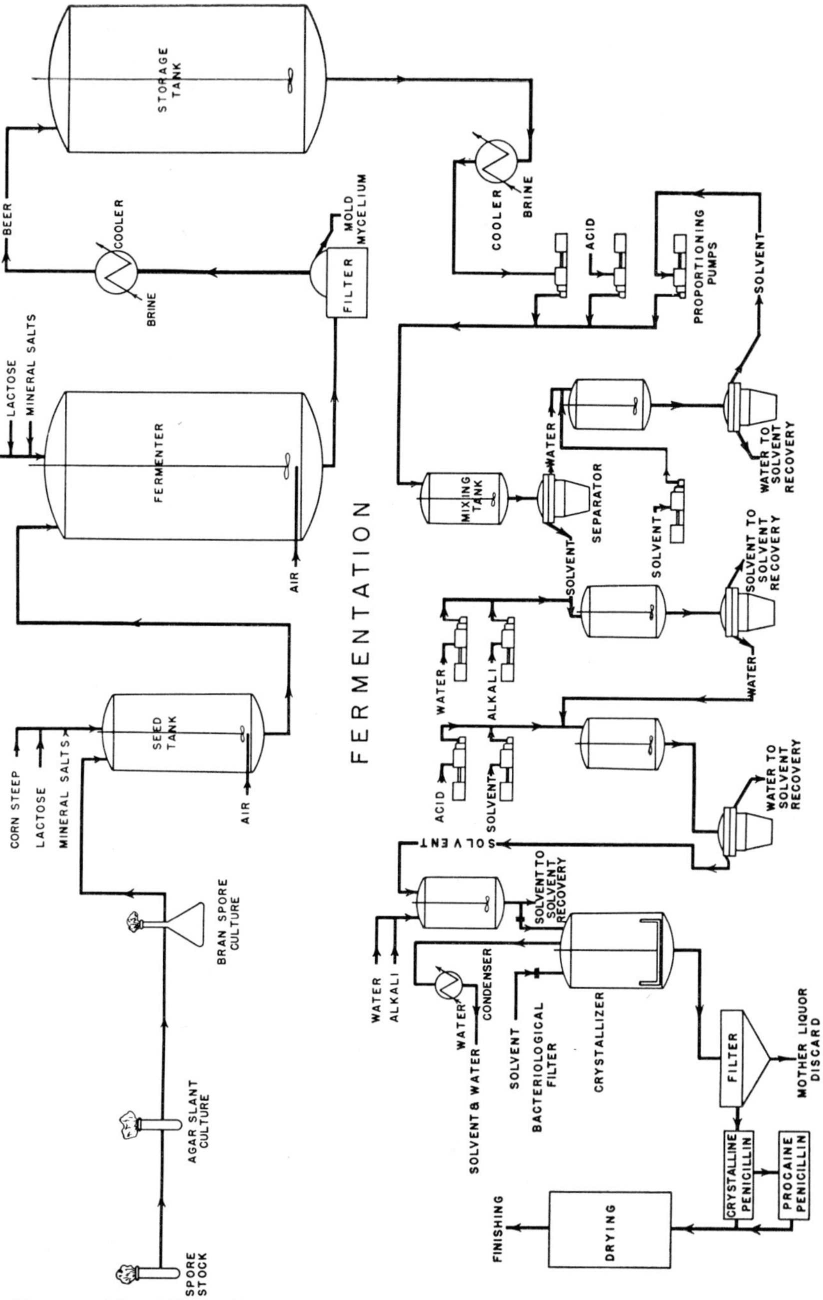

FIGURE 10. *Flow Sheet of Penicillin Submerged-Culture Process*

The various phases of the process will be considered in the order that they function in the commercial production unit.

PLANT EQUIPMENT

The equipment for growing the penicillin-producing molds consists essentially of various-sized fermentation tanks which are used for two purposes: to grow mold for inoculum and to grow mold for penicillin production. The tanks used to grow inoculum are referred to as seed tanks, while those in which the penicillin is produced are called fermentors. Usually, the seed tanks and fermentors are of identical design, with the exception of certain minor modifications required for their particular function. The tanks may be constructed of carbon steel, stainless steel, Inconel-clad steel, or plastic-coated steel. The last three materials have certain advantages over plain carbon steel from the standpoint of cleaning and corrosion and may be slightly superior from the standpoint of yield. The tanks are equipped with mechanical agitators, a sparging system for aeration, and usually with a coil connected to high-pressure steam for sterilization and to water for cooling. Where the medium is sterilized before being put into the tank, the cooling system may consist of a jacket around the tank, the coil being eliminated.

Many different types of agitators are in use. They are usually of a turbine type to impart a definite rotary motion to the liquid. Baffling may or may not be used, depending on the type of agitator employed. High-speed agitation which will finely disperse the air in the liquid is required, necessitating relatively large motors and heavy drives. The tanks are cylindrical and their height is usually two to three times the diameter. The seed tanks commonly have a capacity of about 10% of the fermentor. Table 32 indicates, in a general way, some of the characteristics of the fermentation tanks.

TABLE 32. FERMENTOR CHARACTERISTICS

Type of tank	Total capacity gal	Working capacity gal	Agitator rpm	Motor hp
Seed	500	350	240	5
Fermentor	5,000	4,000	120	50

As aeration causes a great deal of foaming during the growth of the mold, the tanks are equipped with automatic devices which add defoamer whenever the foam reaches a certain level.

Various means are used to distribute the air that is introduced into the tanks. Usually, the air outlets are immediately below the agitators, as this arrangement provides better dispersion of the air. The air outlet may consist of a pipe with small holes along the sides and arranged in the form of a square or circle, of jets, shower heads or, even of a single pipe open at the end. Carborundum blocks or other air-dispersing materials of extremely fine porosity are not suitable since the mold would grow in the small openings, causing plugging during the run. Figure 11 shows the interior of a 5,000-gal fermentor.

FIGURE 11. *Interior of a 5,000-gal Fermentation Tank* (Courtesy —Cutter Laboratories)

The seed tanks and fermentors are connected by piping in such a way that they form an entirely closed system, adequately protected from the atmosphere and capable of operation without exposing the medium to contamination. The inoculum is transferred from the seed tank to the fermentor by air pressure. During operation, a 5 to 10 psi air pressure is maintained on the tanks. When not in use, all piping connecting the seed tank and fermentor, and all other connections to the tanks, are maintained under 15 to 30 psi steam pressure. It is absolutely essential in the operation of the fermentation tanks that the system is kept free of contamination.

Large quantities of compressed air are required for the growth of the mold and, since this air must pass through a system of considerable resistance, supply pressures in the range of 20 to 50 psi are required. Air compressors capable of delivering several thousand cubic feet of air per minute at these high pressures are employed. Centrifugal blowers can be used if a multi-stage system, capable of building up the desired pressure, is employed.

Several different methods are used to sterilize the air. One of the most common is by filtration through carbon or glass wool. A filter using these materials consists of a large pipe which is packed with the filtering agent and provided with a steam inlet for sterilization. The filter is sterilized under 15 to 30 psi steam pressure directly on the filtration agent. A typical filter capable of sterilizing up to 500 cfm of air is constructed of 18 in. steel pipe, 10 ft in length with flanged ends. It contains a column of carbon (10×30 mesh) supported on a fine wire mesh and will operate satisfactorily for many months without changing the carbon. This filter is sterilized each time a run is made in the tank. Other methods of air sterilization employ scrubbing towers, containing germicidal solutions, and incineration. One plant apparently obtains satisfactory operation by placing the air compressors close to the tanks and depending on the heat generated during the compression of the air for sterilization.

Preparation of Inoculum

Since the *Penicillium* strains used for the manufacture of penicillin are subject to extreme variation, the handling of the culture and the preparation of inoculum for the production fermentors must be carefully controlled. A master stock of the organism *P. chrysogenum,* in the form of soil stocks, lyophilized stocks, or frozen slants serves as the starting point for each fermentor run. Inoculation of a production fermentor from another production fermentor is not used except as an emergency measure if the inoculum prepared for that particular run cannot be used. The first step in the preparation of inoculum, in a typical process, consists of transferring a few spores from the master stock to a test-tube agar slant of a medium that will provide optimum sporulation. The sporulation medium, developed by Moyer at the Northern Regional Research Laboratory, gives excellent sporulation. This medium contains, per liter, 12.5 g molasses, 7.5 g glycerol, 5.0 g peptone, 4.0 g NaCl, 0.25 g $CaCO_3$, 0.05 g $MgSO_4$, 0.06 g KH_2PO_4, 0.003 g $FeSO_4$, 0.001 g $CuSO_4$, and 15.0 g agar. The agar-slant culture is incubated at 24° to 26°C until good sporulation occurs, usually 5 to 7 days. The spores from the tube are then suspended in sterile water and used to inoculate a bottle or flask containing wheat bran moistened with a nutrient solution. The formula for the bran medium contains 667 g wheat bran moistened with 333 ml nutrient solution which has the same composition as the sporulation medium except that it contains no agar and the concentration of all ingredients is increased fourfold.

The bran cultures are incubated at 24°C until heavy spore formation has taken place (5 to 7 days). These spores are used to inoculate the seed tank containing liquid medium. Various media may be used in the seed tanks and the exact composition must be determined experimentally. However, satisfactory results may be obtained employing media similar to that used in the production fermentor. The medium is sterilized prior to inoculation and the transfer of the spores to the tank is made with careful aseptic precautions. After inoculation, the tanks are held at a constant temperature and continuously agitated and aerated. The temperature control is achieved through the use of automatic control devices. This factor is critical. The number of spores used for inoculation, the medium, and the incubation temperature vary, depending on the time cycle desired and the characteristics of the seed for the particular plant. These factors have to be determined experimentally on the production equipment. Usually, the time cycle is relatively short, ranging from 24 to 48 hours. Very heavy mycelial growth is obtained in this period under the proper conditions. Seed tanks for growing inoculum are illustrated in Figure 12.

An alternate method of preparing inoculum consists of developing sufficient vegetative mycelium in small containers for inoculating the seed tanks. This can be done, for example, by inoculating shake flasks, containing 100 to 150 ml of liquid medium, with spores from an agar slant and incubating the flasks for 48 to 72 hours on a shaking machine at 22° to 26°C. The vegetative growths from several flasks are then combined and used to inoculate the seed tank. Under good conditions, about 1,000 ml of vegetative mycelium will suffice for 200 to 500 gal of seed medium.

FIGURE 12. *Seed Tanks for Growing Inoculum* (Courtesy—Abbott Laboratories)

If the production fermentors are large, in the range of 20,000 gal, it may be necessary to build up the inoculum through two steps instead of one. In this case, the first seed tank is used to inoculate the second larger tank, which then is used to inoculate the production fermentor.

During the development of inoculum in the seed tanks, samples are withdrawn periodically and examined for growth and contamination. Both microscopic examination and subcultures to broth medium are made to check on possible contamination with bacteria or yeasts. Only when it is certain that the seed culture has grown satisfactorily and is free of contamination is it used.

PRODUCTION FERMENTATION

The production fermentation is carried out in tanks of 2,500 to 20,000 gal capacity. The tanks are filled to about 75% of capacity with the medium, which may be sterilized in the fermentor or continuously as it is fed into a presterilized tank. A tank containing 4,000 gal of medium requires approximately a 4-hour sterilization cycle when 100 psi steam pressure is used in the coil. Approximately 90 minutes are required to bring the temperature of the medium to 120°C, where it is held for 30 minutes. An additional 120 minutes are needed to cool the medium sufficiently to permit inoculation. Sterilization with batch cookers, followed by cooling with tubular heat exchangers, is also employed. The exact composition of the medium used may vary, depending on a number of

FIGURE 13. *Production Fermentation Tanks* (Courtesy–Abbott Laboratories)

factors, and each producer has his own formula. Fundamentally, however, most production media are similar. A typical medium contains, per liter, 30.0 g corn-steep solids, 30.0 g lactose, 5.0 g glucose, 3.0 g $NaNO_3$, 0.25 g $MgSO_4$, 0.044 g $ZnSO_4$, 0.05 g phenylacetamide, and 3.0 g $CaCO_3$.

The various ingredients are mixed in a small tank with sufficient water to make an easily pumpable solution. The charge is then pumped into the fermentor and water added to the desired volume. The use of a medium-mixing tank has many advantages over mixing in the fermentation tank and is often used. If a continuous sterilization process is used, the medium is prepared in its final form in a reservoir tank and pumped through a sterilizer into the fermentor. Typical production fermentors are shown in Figure 13.

When the sterilized medium is ready for inoculation, the contents of the seed tank are blown into the fermentor. The inoculated medium is then continuously agitated and aerated for a period of several days. The temperature is maintained constant in the range of 22° to 28°C by the use of automatic controls. The length of the fermentation cycle varies from 40 to 96 hours in different plants, depending on the manner in which the plant is operated. The various factors affecting the fermentation are so controlled that the maximum yield will be obtained by the time specified for harvest. Most plants are operated on a definite schedule which does not permit much variation in the over-all fermentor cycle. Samples are

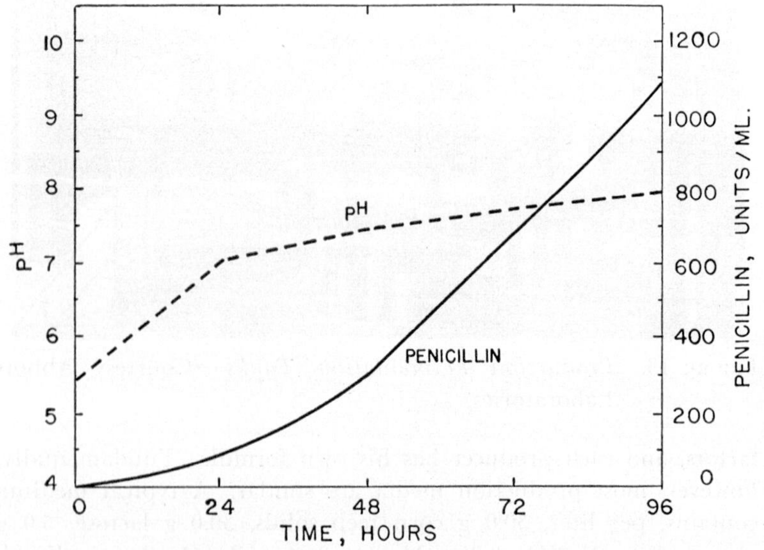

FIGURE 14. *Fermentation Cycle in a 5,000-gal Fermentor*

removed from the fermentor at periodic intervals for penicillin assay and examination for contamination. A typical fermentation cycle in a 5,000 gal fermentor is shown in Figure 14.

Harvesting the Fermented Medium

At the time of harvest, the fermented culture is light tan and of about the consistency of light sludge. The culture is blown or pumped into a reservoir tank which feeds a continuous, rotary vacuum filter. The mycelium is separated from the liquor, washed on the filter, and discharged as a thick blanket, resembling paper pulp. The penicillin-containing beer is pumped from the filter, through a heat exchanger which cools it at 2° to 4°C, into a storage tank. As the filtration step is not a sterile operation, it is desirable to chill the beer as soon as possible after it leaves the filter. Between runs, the filter is usually sterilized with steam or germicidal solution. It is essential that growth of bacteria is not allowed to build up in

Figure 15. *Rotary Vacuum Filter for Separating the Mycelium from the Broth* (Courtesy–Abbott Laboratories)

the filtration system, because this can result in severe destruction of penicillin by the action of bacterial penicillinases. The capacity of the rotary filter varies with the size of the drum, ranging upward from 2,000 to 3,000 gph for a drum 4 ft in diameter and 28 in. wide. A typical rotary vacuum filter is shown in Figure 15.

To produce 1,000 gal of fermented culture (5 to 6 pounds of penicillin) by the submerged-culture process requires approximately 500 lb of nutrients, 7,500 lb of steam, 10,000 gal of water, 1,000 kwh of electricity, and 250,000 cu ft of air.

Recovery of Penicillin

Two processes for the recovery of penicillin have been used in commercial production. One process is based on the adsorption of penicillin from the beer on carbon, while the other is based on the extraction of penicillin from the beer at an acid pH with water-immiscible organic solvents. The carbon process was widely used during the early period of penicillin production, but has largely been supplanted by the solvent process. The primary advantage of the solvent process over the carbon process is that for the second method, increased amounts of carbon are required as the concentration of penicillin in the beer increases. In contrast, increased volume of solvents is not generally necessary with higher-potency beers. Also, there are several handling and operating disadvantages inherent to the use of carbon on a large scale in the production of penicillin.

The Carbon Recovery Process

The filtered beer is pumped into an agitated tank, the desired amount of activated carbon added, and the slurry mixed for 15 to 30 minutes. The exact amount of carbon used depends on the concentration of penicillin in the beer and will range from 2 to 5% by weight. One gram of carbon will adsorb approximately 20,000 units of penicillin. The type of carbon used is determined on the basis of its relative bulk, wetting ability, penicillin adsorption and elution, and filtration characteristics. Such carbons as Nuchar C-190-N and Darco G-60 give very satisfactory results. No pH adjustment is made in this step, because the normal pH of the filtered beer (pH 7 to 8.5) is satisfactory for good adsorption. When adsorption is complete, the carbon is collected by filtration[10] in a plate-and-frame press or by centrifugation in a basket centrifuge. The plate-and-frame press has many advantages over centrifugation. Continuous, rotary filters may also be employed for this filtration. Approximately 90% of the penicillin is adsorbed under good conditions, leaving about 10% unadsorbed. After washing with water, the carbon adsorbate is eluted. The elution may be accomplished in several different ways, the most preferable being with 80% acetone. This eluting solution has a short contact time, high filtration rate, good volume reduction, and high elution efficiency, as well as other advantages over other eluting agents. The volume of eluting solution used varies from 0.05 to 0.5% on the volume of the beer. Either stirred elutions or static elutions may be used. In the first method, the carbon adsorbate is placed in an agitated tank, containing the eluting solution, and stirred for 30 to 60 minutes. The carbon is then filtered off and the eluate collected. In the static elution, the carbon adsorbate is left in the press and the eluting solution pumped through the press. Recycling of the eluting solution may or may not be used, depending on the particular equipment that is employed. In either method, a second elution can be used which is then employed as the first elution of the next batch. While the recovery of penicillin in the eluate will vary, depending on a number of factors, in both methods, 60 to 80% of the penicillin will be eluted in the first elution. The second elution removes an additional 10 to 20% giving an over-all elution of 70 to 90%. The total recovery of penicillin from the beer to this point is usually about 60 to 70%.

The 80% acetone eluate is next concentrated, either by repeated extractions with a water-immiscible solvent[32] or by distillation or evaporation under vacuum[45] at 60° to 90°F. After the acetone is removed, the remaining aqueous solution of penicillin is cooled at 2° to 3°C, acidified at pH 2 to 2.5, and extracted with amyl acetate or a similar solvent in the same manner as will be described in the discussion of the solvent process.

The Solvent Recovery Process

The solvent-extraction process is based on the fact that penicillin is preferentially soluble in water or organic solvents depending on whether it is in the form of the free acid or a salt. An idealized distribution curve of penicillin between water and a solvent is shown in Figure 16.

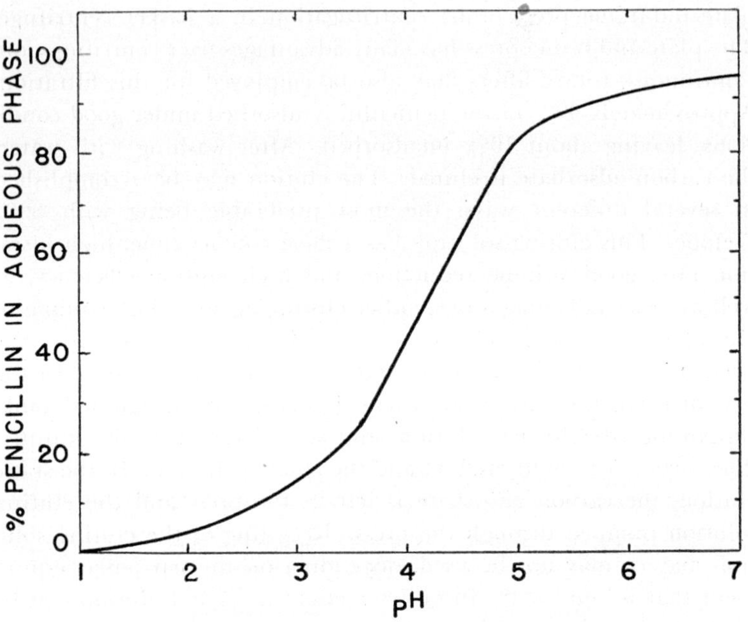

FIGURE 16. *Distribution of Penicillin between Water and Amyl Acetate*

A very good discussion of the theoretical aspects and principles of the solvent extraction of penicillin has been given by Rowley, Steiner, and Zimkin.[43] Whitmore[47] described a laboratory, continuous, counter-current extraction process.

Several solvents have been used for the extraction of penicillin on a large scale, but the most widely used solvents are amyl acetate, methyl isobutyl ketone, and butyl acetate. The choice of a suitable solvent depends on a number of factors, such as cost, solubility in water, flash point, toxicity for humans, ease of handling, etc. Although many solvents, as for example, methyl isobutyl ketone, methylcyclohexanone, and others have greater solvent power for penicillin than the commonly used solvents; these latter appear to be more suitable from an all-around standpoint. Certain specific conditions of the process, such as pH, type of deemulsifier, etc., are dependent on the solvent employed. Basically, however, the process is the same, regardless of the solvent used.

The filtered beer, chilled at 2° to 3°C, is first clarified by filtration with 1 to 1.5% Hyflo through a plate-and-frame filter press or a rotary filter. The removal of certain dissolved solids can be effected by adding to the beer, before filtration, aluminum sulfate,[31] tannic acid,[31] or similar substances that precipitate the dissolved solids, which are then removed with the suspended solids in the filtration. This clarification step is not essential, but it reduces the emulsion difficulties encountered in the extraction steps and also lowers the amount of insolubles which collect in the centrifuges. Where the Podbielniak extractor (made by Podbielniak, Inc., 341 E. Ohio Street, Chicago, Ill.) is used, it is very important to clarify the beer. The amount of solids removed by the clarification step varies widely in different beers and while, in certain instances, it is possible to process a beer without clarification, many are encountered which cannot be handled without polishing. Most plants, as a routine, clarify their beers before processing and, although a few plants apparently operate satisfactorily without this step, it is felt that clarification should be adopted as standard practice.

Proportioned flows of broth, acid, and organic solvents are required in the penicillin-recovery process. This is accomplished by the use of proportioning pumps of the positive-displacement piston-type, which accurately deliver any desired liquid flow. The clarified beer, at 2° to 3°C, is fed continuously through the proportioning pumps into the solvent-extraction system. The organic solvent and acid (phosphoric or sulfuric) are similarly proportioned into the system. Usually, the volume of solvent used is approximately 0.2 to 0.4 of the volume of the beer. The acid flow is such that a pH of 2.0 to 2.5 is maintained. A constant-recording pH meter is standard equipment, because the pH must be very carefully controlled. The proportioned flows are fed continuously to a mixing tank where they are intimately mixed and discharged as a fine emulsion. The mixing tank is relatively small (15 to 20 gal) and is equipped with a high-speed agitator. A small mixing tank is necessary because a hold-up time of 60 to 90 seconds in the tank is desired.

The water-organic solvent emulsion is separated into two phases by centrifugation in either disc-type or bowl centrifuges. The through-put capacity is somewhat higher with the disc-type machines. The aqueous phase is discharged from one spout, the solvent phase from the other. The standard-size centrifuges have a capacity in the range of 1,000 gph of emulsion. It is essential to have a small amount of a good deemulsifying agent (Ultrawet 30E or Santomerse S) present in the solutions being processed, to facilitate the splitting of the emulsion. The deemulsifying agent can be added to the beer before it enters the system, or it can be proportioned into the system before the mixing tank. The aqueous phase discharged from the centrifuge is fed into a second mixing tank, along with fresh solvent, and then passed to a second centrifuge. The aqueous discharge from this machine goes to a solvent-stripping column for solvent recovery and is then neutralized with caustic soda and discharged to the sewer. The solvent from the second centrifuge passes to a pump feeding the first stage of the system and is mixed with the incoming fresh beer. Thus a two-stage extraction is accomplished. If desired, a three- or four-stage system can be used. The solvent discharge from the first centrifuge is pumped to a storage tank preparatory to further processing.

A battery of centrifugal separators used in the solvent-recovery process is shown in Figure 17.

The Podbielniak centrifugal, counter-current solvent extractor[29] is used in some plants in place of the mixing tank and centrifuge. This machine is essentially a counter-current extractor with a multiple-stage effect. The main component of the machine is a stainless-steel rotor which revolves at approximately 2,000 rpm. The rotor speed used varies with the solvent. Actually, the principle of the Podbielniak extractor is the same as that of a counter-current extraction column or a multiple-stage, counter-current system using mixing tanks and centrifuges. In a system employing Podbielniak extractors, the chilled, clarified beer, containing a deemulsifier and acidified to the proper pH, is pumped under pressure into the rotor of the extractor. The solvent under pressure is also pumped into the rotor, but at an opposing point to the beer. The two liquids pass rapidly through the rotor in opposite directions and are discharged separately. Since the Podbielniak extractor is equivalent to several extraction stages, one passage is sufficient for complete extraction. The aqueous phase passes to the solvent-recovery unit and the solvent phase to a storage tank.

A considerable amount of purification is effected in the first

FIGURE 17. *Centrifugal Separators Used in the Solvent-Recovery Process* (Courtesy—Chas. Pfizer and Co.)

extraction of penicillin. In the clarified beer, the solids have a penicillin potency of about 30 units (18 μg) per mg. The solids in the solvent solution have a potency of 200 to 400 units per mg. Under satisfactory operating conditions, 90 to 95% of the penicillin is extracted.

If it is desired, the solvent may be treated with a small amount of carbon (0.25 to 0.5%) to remove certain pigments and other impurities. The importance of this step depends largely on the beer being processed and may not be necessary.

The next step in the process is to extract the penicillin from the organic solvent into water. The volume of water used is generally 0.02 to 0.1 of the volume of solvent. The solvent, water, and sufficient alkali (sodium or potassium hydroxide), or a buffer solution to maintain a pH of 6.5 to 7.0, are proportioned into a system identical to that used in the first extraction step, except that only a single stage is employed. A Podbielniak extractor may also be used for this step. No deemulsifier is required. After intimate mixing, the two phases are separated; the solvent passes to the recovery unit and the aqueous phase to the next processing step. A 90 to 95% recovery of penicillin is obtained in this extraction and the potency of the penicillin is increased to approximately 600 to 800 units per mg. The volume of solution at this point is roughly 0.01 to 0.004 of the original beer volume.

The chilled aqueous solution is next acidified and extracted with an organic solvent. This may be done continuously in the same type of equipment as already described or it may be done batchwise in agitated tanks. No emulsion problems are encountered in this step and the two phases separate readily without centrifugation. Usually a different solvent is employed in this step, because a greater degree of purification is achieved if a solvent other than the one used in the first extraction step is employed. Chloroform, ether, and other solvents will give satisfactory results.

The penicillin is again extracted into water at pH 6.5 to 7.2 by titration with a solution of base. The base used here depends on which salt of penicillin is desired as a final product. Commercially, the sodium, potassium, and small amounts of calcium salts are produced. If the amorphous salts of sodium, potassium, and calcium are desired, the solution obtained at this point is freeze-dried and finished without further processing. Most of the penicillin, however, is processed to the crystalline sodium or potassium salt.

Several widely different procedures[1,14,18,19,22,32,38] are described for producing crystalline salts of penicillin and it is doubtful that any two commercial producers employ identical methods. One method, which apparently gives very satisfactory results, consists of filtering an aqueous solution of sodium or potassium penicillin (100,000 to 200,000 units per ml) through a bacteriological filter and placing it in a crystallizer. A water-immiscible solvent, such as butanol, also filtered, is added to the crystallizer and an azeotropic distillation under vacuum is carried out. As the last of the water is removed, the penicillin salt crystallizes and is then collected aseptically on a filter. The crystallizing operation must be carried out under aseptic conditions to avoid contamination of the product with microorganisms. The crystalline salt is then removed from the filter and dried *in vacuo* at 200 to 400 microns pressure. The dried salt is screened to size and then finished into the various pharmaceutical forms (Figure 18).

FIGURE 18. *Stainless-Steel Turntable Used in Finishing Operations* (Courtesy—Chas. Pfizer and Co.)

Procaine penicillin may be made from either the potassium or sodium salt by dissolving the crystals in water and adding procaine hydrochloride. Procaine penicillin crystallizes from the aqueous solution and is filtered off and dried *in vacuo.* An alternate procedure that may be used consists of preparing a solution of penicillin (free acid) in an organic solvent and adding a solution of the free procaine base. The procaine penicillin salt crystallizes readily from the solvent solution.

Penicillin Standards

After manufacture, all penicillin salts must be carefully tested for quality and conformity to the standards set up by the various governments. In the United States, sodium or potassium penicillin G conforms to the following standards, all tests being performed by the methods stipulated by the USP: (1) It is sterile; (2) it is nontoxic; (3) it is nonpyrogenic; (4) it contains not less than 85% by weight of the sodium or potassium salt of penicillin G; (5) its moisture content does not exceed 1.5%; (6) its pH in aqueous solution, at a concentration of 30 mg per ml, is not less than 5.0, nor more than 7.5; (7) the aqueous solution at a concentration of 30 mg per ml is substantially free of any turbidity or undissolved substances.

Disposal of Residues

In the production of penicillin, large quantities of residual waste, consisting of mycelium, extracted broth, and wash water from equipment, are obtained. These residues have relatively high b.o.d. values and present a serious disposal problem for many plants. Heukelekian[26,27] reported that the stripped penicillin beer from one plant had an average b.o.d. of 4,380 ppm. A significant portion of this b.o.d. was contributed by the residual amyl acetate in the broth. The b.o.d of the wash water from the same plant averaged 3,420 ppm. The mycelium also has a high b.o.d.

The methods employed for disposing of the penicillin waste varies from plant to plant. Those which are located in or near large metropolitan areas are able to dispose of the residues through the sanitary sewage system. However, there are a number of plants in areas where the capacity for sewage disposal is insufficient to handle the penicillin wastes. These plants have to dispose of the residues by other means, some of which involve considerable expense. Usually the mycelium is handled separately from the extracted broth and wash water. A few plants dry the mycelium by drum or spray drying and sell the dried mycelium as a feed supplement. This material has found a ready market at a price which at least pays for the cost of drying. Its approximate composition is as follows.

Composition of Dried Penicillium Mycelium

Protein (6.25 × TN)	32.0%
Moisture	8.0
Fat	7.0
Ash	20.0

Crude fiber	7.0
Nitrogen-free extract	26.0
Choline	3,700.0 μg per g
Thiamine	6.0
Riboflavin	37.0
Pantothenic acid	64.0
Nicotinic acid	140.0
Folic acid	7.0
Biotin	5.0
Pyridoxine	13.0
Vitamin B_{12}	0.02

At least one plant combines the mycelium and extracted broth and drys the mixture. Several plants haul the mold to some isolated area and bury it in the ground. The liquid wastes are disposed of by evaporation and burial or by treatment in a sewage-disposal plant installed for this purpose. It is probable that in the next few years, improved methods of handling penicillin wastes will be developed and the recovery of as much material as possible in an economically valuable form will be the general practice in the penicillin industry.

BIBLIOGRAPHY

1. Abbott Laboratories, British Patent 649,644 (1951).
2. Abbott Laboratories, unpublished data.
3. Abraham, E. P., E. Chain, C. M. Fletcher, H. W. Florey, A. D. Gardner, N. G. Heatley, and M. A. Jennings, *Lancet,* **2**, 177 (1941).

3a. Anderson, R. F., L. M. Whitmore, Jr., W. E. Brown, W. H. Peterson, B. W. Churchill, F. R. Roegner, T. H. Campbell, M. P. Backus, and J. F. Stauffer, *Ind. Eng. Chem.,* **45**, 768 (1953).

4. Anon., *Chem. Eng.,* **58**, No. 4, 174 (1951).
5. Anon., *Chem. Eng. News,* **31**, 383 (1953).
6. Backus, M. P., J. F. Stauffer, and M. J. Johnson, *J. Am. Chem. Soc.,* **68**, 152 (1946).
7. Behrens, O. K., J. Course, R. G. Jones, E. C. Kleiderer, Q. F. Soper, F. R. Van Abeele, L. M. Larson, J. C. Sylvester, W. J. Haines, and H. E. Carter, *J. Biol. Chem.,* **175**, 765 (1948).
8. Bowden, J. P., and W. H. Peterson, *Arch. Biochem.,* **9**, 387 (1946).

9. Brown, W. E., and W. H. Peterson, *Ind. Eng. Chem.,* **42**, 1769 (1950).
10. Callahan, J. R., *Chem. & Met. Eng.,* **51**, 94 (1944).
11. Chain, E., H. W. Florey, A. D. Gardner, N. G. Heatley, M. A. Jennings, J. Orr-Ewing, and A. B. Sanders, *Lancet,* **2**, 226 (1940).
12. Churchill, B. W., and F. R. Roegner, unpublished data, Dept. of Botany, University of Wisconsin, Madison, Wisc.
13. Churchill, B. W., and F. R. Roegner, unpublished data, Dept. of Botany, University of Wisconsin, Madison, Wisc.
14. Clarke, H. T., J. R. Johnson, and R. Robinson, *The Chemistry of Penicillin,* Princeton, Princeton University Press, 1949.
15. Clutterbuck, P. W., R. Lovell, and H. Raistrick, *Biochem. J.,* **26**, 1907 (1932).
16. Coghill, R. D., *Chem. Eng. News,* **22**, 588 (1944).
17. Coghill, R. D., and R. S. Koch, *Chem Eng. News,* **23**, 2310 (1945).
18. Commercial Solvents Corp., British patent 642,370 (1950).
19. Commercial Solvents Corp., British patent 642,371 (1950).
20. Demerec, M., U. S. Patent 2,445,748 (1948).
21. Editorial Board of Monograph on the Chemistry of Penicillin, *Science,* **106**, 503 (1947).
22. Eli Lilly and Company, British patent 657,814 (1951).
23. Fleming, A., *Brit. J. Exp. Path.,* **10**, 226 (1929).
24. Gailey, F. B., J. J. Stefaniak, B. H. Olson, and M. J. Johnson, *J. Bact.,* **52**, 129 (1946).
25. Heatley, N. G., *Biochem. J.,* **38**, 61 (1944).
26. Heukelekian, H., *Ind. Eng. Chem.,* **41**, 1412 (1949).
27. Heukelekian, H., *Ind. Eng. Chem.,* **41**, 1535 (1949).
27a. Hosler, P., and M. J. Johnson, *Ind. Eng. Chem.,* **45**, 871 (1953).
28. Johnson, M. J., *Ann. N. Y. Acad. Sci.,* **48**, 57 (1946).
29. Keko, W. L., and J. L. Martin, U. S. Patent 2,530,883 (1950).
30. Koffler, H., R. L. Emerson, D. Perlman, and R. H. Burris, *J. Bact.,* **50**, 517 (1945).
31. Lovens Kemeske Fabrik, Danish Patent 294 (1949).
32. McKeen, J. E., *Trans. Am. Inst. Chem. Eng.,* **40**, 747 (1944).
33. Merck & Co., British patent 685,285 (1952).
34. Moyer, A. J., and R. D. Coghill, *J. Bact.,* **51**, 57 (1946).
35. Moyer, A. J., and R. D. Coghill, *J. Bact.,* **51**, 79 (1946).
36. Moyer, A. J., and R. D. Coghill, *J. Bact.,* **53**, 329 (1947).
37. Peterson, W. H., *Harvey Lectures, Ser.,* **42**, 276 (1947).
38. Pasternak, R., and P. P. Regna, U. S. Patent 2,430,946 (1947).

39. Raper, K. B., and D. F. Alexander, *J. Elisha Mitchell Sci. Soc.*, **61**, 74 (1945).
40. Raper, K. B., and D. F. Alexander, *Mycologia*, **37**, 499 (1945).
41. Reese, E., K. Sanderson, R. Woodward, and G. M. Eisenberg, *J. Bact.*, **57**, 15 (1949).
42. Reid, R. D., *J. Bact.*, **29**, 215 (1935).
43. Rowley, D., H. Steiner, and E. Zimkin, *J. Soc. Chem. Ind.*, **65**, 237 (1946).
44. Singh, K., and M. J. Johnson, *J. Bact.*, **56**, 339 (1948).
45. Taylor, T. H. M., *Chem. Eng. Progress*, **43**, 155 (1947).
46. U. S. Department of Commerce, *World Trade in Commodities*, Vol. 6, Part 3, No. 17, April 1948.
47. Whitmore, F. C., *Ind. Eng. Chem.*, **38**, 942 (1946).
48. Woodruff, H. B., and A. H. Larsen, U. S. Patent 2,532,980 (1950).

CHAPTER 8

STREPTOMYCIN

H. B. Woodruff and L. E. McDaniel

Streptomycin is a therapeutic agent which was discovered in the course of a planned survey of microorganisms for antibiotic properties. Early in the survey, several antibiotics having little value as chemotherapeutic agents were obtained at the Department of Microbiology of the New Jersey Agricultural Experiment Station, Rutgers University.[39] Those which have been purified and studied in some detail include actinomycin, clavacin, fumigacin, chaetomin, micromonosporin and streptothricin. Toxicity, inactivity *in vivo,* production difficulties, or other factors limited usefulness of these compounds, but the experience gained in their production and isolation, particularly in the case of streptothricin, contributed greatly to the speed with which streptomycin was developed from a laboratory agent to a commercially available chemotherapeutic compound.

Two microorganisms, one (No. 18-16) from a heavily manured field soil and the other (No. D-1) from the throat of a chicken, produced an antibiotic different, in the range of microorganisms inhibited, from all previously described antibiotics. High dilutions of fermented broths of the cultures inhibited growth of many Gram-positive bacteria, Gram-negative bacteria, particularly the coliforms and intestinal parasites, and the acid-fast bacteria. The antibiotic agent was named *streptomycin.* The antibiotic activity of the partially purified product, isolated from such broths, is now known to be due to a mixture of antibiotics, called the streptomycin complex. A bibliography of the literature on streptomycin through 1952 has appeared.[38b]

The microorganisms which produced streptomycin were similar in cultural characteristics and in morphology to *Actinomyces griseus,* isolated from soil and described 28 years previously. According to modern schemes of classification, the two microorganisms are identified as *Streptomyces griseus.*

The production of streptomycin is a strain-specific characteristic of *S. griseus.* Numerous attempts to isolate streptomycin-producing strains from nature have been made, but descriptions have been published of only a few strains which produce streptomycin in appreciable quantity. Many strains produce a small quantity of inhibitory substance which has not been positively identified as streptomycin. A filter-paper strip-chromatogram method has been used to demonstrate chemical similarity of the inhibitory substance from some of these strains to streptomycin.[16]

Streptomycin is associated with a second antibiotic, possibly streptothricin, in the fermented broth of one strain of *S. griseus.*[38] A substance similar to streptomycin is produced by a culture closely resembling *S. griseolus,* which was given the name *S. bikiniensis* on the basis of minor cultural distinctions.[18]

Many of the cultures which are used at present in industrial organizations for the production of streptomycin originated from one of the original isolates of *S. griseus,* No. 18-16, although new isolates from natural sources are used in some plants.

LABORATORY FERMENTATION METHODS

The original description of streptomycin production indicated that both a shallow-layer stationary method and an aerated submerged procedure were satisfactory for growth of *S. griseus* and for streptomycin production.[34] Various organic complexes are satisfactory as components of media. (See the section on raw materials.) Working details of a method for producing streptomycin by surface culture in pint milk bottles were described in a report from the Wellcome Physiological Research Laboratories.[1] Streptomycin yields in the crude-culture filtrates were about 250 μg per ml after 10 to 14 days of incubation.

Working Details of the Surface-Culture Method

Since the procedures for laboratory and pilot-plant production of streptomycin are similar, the second is described here.

An important ingredient used in the media is papain digest of beef which is prepared as follows: 120 lb of beef muscle is placed in 90 l of warm water, brought to 60°C, and 190 g of powdered papain added. The mixture is incubated at 60°C for 4 hours and the digest is then boiled 5 minutes, cooled, and filtered. After storage at 5°C, the fat on the surface is removed by skimming. The spore-production medium contains 6% by volume papain digest of beef, and by weight, 1.5% molasses, 0.58% glycerol, 0.5% NaCl, 0.0015% $FeSO_4 \cdot 7H_2O$, 0.009% $CuSO_4 \cdot 5H_2O$, 0.0013% $(NH_4)_6Mo_7O_{24} \cdot 4H_2O$, 0.00028% $MnSO_4 \cdot 4H_2O$, 0.006% KH_2PO_4, and 0.005% $MgSO_4 \cdot 7H_2O$, in water. After adjusting to pH 7.0 ± 0.2 with sodium hydroxide, the unfiltered medium is distributed in 300 ml quantites into 2.5 l Roux bottles and autoclaved.

The stock culture of *S. griseus* (NCTC 7187) is stored by sealing the cotton plugs of agar-slant cultures with liquefied paraffin and holding at 4°C in the dark. Spores from a single stock culture of *S. griseus* are transferred to a single flask of spore-production medium. After 7 days incubation at 28.5°C, the growth pellicle and spores are thoroughly suspended by shaking with 30 ml of 1% "calsolene" solution and used to inoculate an additional lot of fifty bottles.

The production medium contains by volume, 5% papain digest of beef and by weight, 0.4% $NaNO_3$, 0.1% KH_2PO_4, 0.05% KCl, 0.2% sodium citrate, 0.2% sodium acetate, 0.004% $MnSO_4 \cdot 4H_2O$, 0.00175% $CuSO_4 \cdot 5H_2O$, 4.0% glucose monohydrate, and 0.25% extract of beef (Wilsons). The medium is filtered with Celite, adjusted to pH 7.0, and refiltered.

Milk-bottling machines are used for filling 200 ml of medium into pint milk bottles. The bottles are not plugged with cotton, but are covered by loose-fitting aluminum caps and autoclaved at 15 psi steam pressure for 15 minutes. Extension of autoclaving time to 1 hour makes the medium useless for streptomycin production.

Inoculation is accomplished by means of spray guns which

distribute about 3 ml of spore inoculum as a fine spray into each pint bottle.

Cultures are incubated 10 to 14 days at 28.5° ± 1°C in a slanted position in large, movable dollies.

Harvesting is effected by a mechanical harvester in which bottles are inserted and drained of their contents. Empty bottles are transferred to the bottle-washing and medium-filling machine.

Solid Substrate

Moist solid substrates, such as rice bran and wheat bran, are satisfactory for development of heavily sporulated inoculum of *S. griseus.* For best results, *S. griseus* requires adaptation to growth on rice bran. Repeated transfer of spores from rice-bran cultures to sterile, moist rice bran eventually results in rapid growth. Sporulated rice-bran cultures may be used as an inoculum for production of streptomycin by the stationary-culture method described before.

Streptomycin is produced as a result of growth on a rice-bran medium, but is absorbed on the substrate. The antibiotic may be released by leaching the cultures with dilute acids dissolved in water or alcohol. A 0.1 *N* solution of formic acid is a satisfactory leaching agent.

Supplementation of the moist rice-bran medium with 1% sodium chloride improves the yield of streptomycin. The addition of water equivalent to three fifths of the weight of the dry bran yields a satisfactory consistency of substrate for growth of *S. griseus* in shallow layers in laboratory glassware. In larger volumes, aeration must be maintained at an adequate level, by the use of rotary-drum fermentors or similar methods.

Laboratory yields on a rice-bran substrate averaged 1,400 μg of streptomycin per gram of dry bran in an incubation period of 14 to 21 days.[41] The best yield obtained with the same strain of *S. griseus* in comparative experiments made with stationary liquid cultures was 400 to 500 μg of streptomycin per ml of broth, in 9 to 14 days.

Submerged Cultures

Laboratory studies of the submerged-culture technique are limited mostly to shake-flask cultures.

Details of a successful procedure are as follows. Medium is dispensed in 40 to 80 ml amounts in 250 ml Erlenmeyer flasks. The flasks are plugged with cotton and sterilized at 15 psi steam pressure for 15 to 20 minutes. The medium is inoculated with spores of *S. griseus,* removed from an agar culture, or with a portion of a submerged culture, preferably in the logarithmic phase of growth. The flasks are placed on a shaking-machine platform which rotates in a horizontal plane through a 360° arc of 1.25 in. radius at 220 rpm. The temperature is maintained at 28° to 30°C.

With strains of *S. griseus* similar to those used in examples described before and liquid media containing enzymic digests of animal proteins or ground plant products, such as soybean meal, yields of 200 to 500 μg of streptomycin per ml of crude culture filtrate have been reported, following a 5-day incubation period, using spore inoculum. The incubation period required is reduced by 1 to 2 days if 5 to 10% by volume of a vegetative inoculum is used.

RAW MATERIALS

Recently factors affecting streptomycin yields in submerged culture have been investigated.[14a] Four physiologically different strains of *S. griseus* were tested on a variety of media. Streptomycin yields were quite different from those reported in the literature for the various media. It was concluded that the medium must be specifically designed for the strain if the streptomycin-producing capacity of the organism is to be fully realized. Streptomycin production from media containing peptone, meat extract, glucose, and sodium chloride was studied systematically. The antibiotic yield was markedly affected by the concentration of the medium ingredients and by use of tap water. Over 1,200 units of streptomycin per ml were obtained when the selected highest-yielding strain was grown on a medium containing 1% peptone, 0.25% meat extract, 1% glucose, and 0.5% sodium chloride in tap water.

A streptomycin-producing "precursor," has been postulated.[40] Beef extract and corn-steep liquor were reported to contain this factor, which has been called the "activity factor," in contradistinction to the term "growth factor." These results have not been reconciled with the equivalent production of streptomycin in synthetic medium (see later). In some media, inorganic salts substitute for beef extract. (Swift's meat extract has been found to contain 35% ash.)

Minor modifications in the protein hydrolyzate included in the medium have considerable influence on streptomycin yields in stationary culture. In addition to Difco peptone and papain digest of beef, described before, Bacto tryptose, Armour beef granules, and yeast extract are satisfactory nitrogen sources, in the presence of beef extract. Other nitrogen sources, including Sheffield N-Z amine, Armour peptonum siccum, and acid-hydrolyzed casein, were found to be unsatisfactory for surface-culture growth of *S. griseus,* because submerged, wet pellicles developed. However, these nitrogen sources were all satisfactory for submerged production of streptomycin.[41]

Supplementing the nitrogenous components, carbohydrates may be supplied in the form of glucose or starch. Glycerol also is satisfactory. Sucrose is not utilized by *S. griseus.* Various glycerides, including lard oil and soybean oil, support normal streptomycin production, when used as a complete replacement for carbohydrates.[27b] Those salts which are essential nutrients must be supplied. Best yields of streptomycin have been obtained in the presence of limited phosphate-ion concentration. Soluble phosphate disappears from the medium during mycelial growth and returns following lysis. If soluble phosphate is present throughout the fermentation, mycelial growth is heavy, but streptomycin production is poor.[12]

Inexpensive media have been devised. Corn-steep liquor is suitable for streptomycin production, but complicates the extraction procedure. Soybean meal, dried fermentation solubles, dried distillers' solubles, groundnut-cake hydrolyzate, or combinations of these satisfactorily serve as the sole source of nitrogen. Acid-hydrolyzed stillage from wheat mash and asparagus-butt juice are satisfactory in a medium supplemented with corn-steep liquor ash. Certain fermentation ingredients which have been found valuable in commercial production have been claimed in patents.[7,10,23,24,25,37] Two examples of media which have proved satisfactory for streptomycin production in submerged culture are listed as follows.

Medium A: 1% glucose, 0.5% peptone, 0.5% meat extract, 0.5% sodium chloride.

Medium B: 1% soybean meal, 1% glucose, 0.5% sodium chloride.[30]

S. griseus grows well in certain media containing inorganic sources of nitrogen, provided the media are balanced to maintain the pH in a range suited for growth. Usually, streptomycin production is poor in such media. *S. griseus* does not utilize nitrate nitrogen in the absence of a supplemental nitrogen supply. Ammonium lactate or diammonium phosphate serve as satisfactory nitrogen sources, since the complete compounds are assimilated. Yields of 150 μg per ml have been obtained from growth of *S. griseus* by the submerged method on a medium containing 10 g glucose, 2 g NaCl, 2 g K_2HPO_4, 1.0 g $MgSO_4 \cdot 7H_2O$, 0.4 g $CaCl_2$, 20 mg $FeSO_4 \cdot 7H_2O$, 4 g $(NH_4)_2HPO_4$, and 1 l distilled water.

A number of amino acids may be utilized for growth of *S. griseus.* Only six, DL-α-alanine, β-alanine, L-histidine, glycine, L-proline, and L-arginine supported yields in excess of 100 μg per ml. Striking results have been obtained with L-proline as a sole nitrogen source. Growth of the culture was heavy and yields of 800 μg per ml were obtained with the same strain which gave the yields in organic media described in the section on laboratory procedures.[11] There is a long lag before initiation of growth of *S. griseus* on a proline medium, but the lag may be reduced somewhat without impairing streptomycin production if a trace of ammonium nitrogen is present.

In a synthetic medium containing glucose, sodium lactate, ammonium, and phosphate, elimination of potassium, magnesium, or zinc limits growth. Iron, magnesium, and copper have no effect on growth when each is separately eliminated from the medium. All these elements, except copper, are required for maximum streptomycin production. The optimum molal levels of salts of each element for streptomycin production in the synthetic medium were found to be: 0.01 disodium phosphate, 0.06 potassium chloride, 0.01 magnesium sulfate, 0.00005 zinc sulfate, 0.00005 ferrous sulfate, and 0.0005 manganous chloride.

Other trace elements, when added to the before-mentioned synthetic medium, either singly or in combination, did not affect production or growth. However, certain elements, when added to a synthetic medium which contained no Zn, Fe, Mn, and Cu, influenced growth and streptomycin production. The elements studied were Co, Mo, As, Hg, Pb, W, Bi, U, Be, Ti, V, Sr, Th, Ba, B, Cr, I and Sb.[36]

A mixture of known vitamins did not affect streptomycin production. It seems probable that the "activity factor," necessary to promote streptomycin formation, is inorganic in nature. A proper concentration of corn-steep ash or potassium-bearing salts, such as potassium phosphate or chloride, has been reported to replace beef extract in media containing protein hydrolyzates.[5] In a medium containing soybean meal, either sodium chloride or sodium sulfate are essential ash constituents and may be supplied by beef extract.

FERMENTATION MECHANISM

Complex Media

The knowledge of metabolic activities of the actinomycetes in submerged or surface culture is quite limited. The chemical changes which take place in submerged fermentation of glucose-meat extract-peptone medium have been measured[12] and are given in Table 33. The fermentation falls into two fairly distinct phases. During the first stage, the mycelium is formed. There is a decrease in soluble nitrogen and carbon in the medium and utilization of glucose and inorganic phosphate is practically complete. The lactic acid content of the medium at first increases and then is fermented. The formation of streptomycin begins.

TABLE 33. CHANGES OCCURRING DURING FERMENTATION OF A GLUCOSE-MEAT EXTRACT-PEPTONE MEDIUM[a]

	Duration of fermentation, days								
	0	1	2	3	4	5	6	7	8
pH	7.35	7.30	7.55	7.50	7.75	8.25	8.55	8.65	8.90
Mycelium, mg per ml	—	0.4	5.1	5.8	5.7	4.8	4.6	4.2	3.8
Streptomycin, μg per ml	—	0	37	194	198	231	270	186	267
Glucose, mg per ml	9.0	8.8	8.0	2.4	1.2	0.6	—	—	—
Soluble carbon, mg per ml	10.2	8.6	7.0	5.1	5.0	4.4	4.6	4.5	4.6
Lactic acid, μg per ml	292	328	114	13	10	16	12	6	15
Oxygen demand, Q_{O2} per ml	—	19	81	82	53	25	—5	—	—
Soluble nitrogen, mg per ml	1.48	1.30	1.10	0.67	0.70	0.73	0.90	0.88	1.14
Mycelial nitrogen, mg per ml	—	0.04	0.44	0.62	0.57	0.49	0.40	0.38	0.29
Inorganic phosphorus, mg per ml	118	108	34	1	5	2	19	24	34
Ammonia nitrogen, μg per ml	66	70	75	63	103	115	170	232	265

[a] Medium: 10 g glucose, 5 g meat extract, 5 g peptone, 5 g sodium chloride per liter.

During the second phase, streptomycin accumulates. There is a rise in pH, accompanied by lysis of the culture. Soluble nutrients increase as a result of liberation during autolysis of cells.

Continued incubation of the culture after lysis has occurred leads to a decrease in streptomycin titer in the broth. The cause of the decrease has not been determined. Although streptomycin is exceptionally resistant to enzymic decomposition by soil microorganisms and probably is not destroyed by enzymic action of lytic products, it has been inactivated by incubation with certain microorganisms isolated from streptomycin-enriched soils.[29]

C-14-labeled glucose and starch have been used for the preparation of C-14-labeled streptomycin by laboratory-scale fermentation.[18a] The labeled product is useful in studying the mechanism of action of streptomycin and in the development of isotopic dilution assays for the antibiotic. Studies with a fermentation medium containing glucose and C-14-labeled glycine and acetate showed that the streptomycin carbons were derived almost entirely from glucose. The small amount of labeled carbon found in the streptomycin was located in the guanidine carbons.[25a]

Oxygen Tension

The oxygen demand of submerged cultures of *S. griseus* is high, not only during the logarithmic phase of mycelial growth, but also during the rather extensive period in which mycelium weight remains constant, previous to lysis. In glucose-meat extract-peptone media of varying composition, the Q_{O_2} per ml (microliters of oxygen taken up per milliliter of culture per hour) reached a plateau at 80 to 120.[12] This demand is reflected in a fall in oxidation-reduction potential of the medium to a constant value, as a result of the growth of *S. griseus*.

Aeration of uninoculated medium at rates of 0.5 to 6 volumes per minute stabilizes the oxidation-reduction potential of the medium at 310 to 375 mv, regardless of the rate. Oxidation-reduction measurements of laboratory cultures of *S. griseus* varied directly in relation to the before-mentioned aeration rates. The different rates had a marked effect on the fermentation process. The pH was higher at high air rates, possibly indicating that the fermentation had proceeded further. Also, about 50% more streptomycin was produced at the higher air rates.[20]

A quantitative study of oxygen transfer and agitation in 5-l laboratory-scale fermentors provided data and techniques which were used with success in translation among laboratory-scale, pilot-plant, and factory fermentors.[3] The data fit a diffusion theory, which was expressed mathematically. Values for the rate of oxygen uptake by the organisms calculated from the equations agreed well with the experimentally determined values.

The rate of oxygen absorption by *S. griseus* cells is independent of the oxygen concentration at the cell-liquid interface if the concentration exceeds about 10^{-8} gram mole of oxygen per ml.

Streptomycin production was found to be related to power input and air-flow rate. It increased with increase in power to a broad maximum, ultimately decreasing at high agitator speed. Productivity with sintered spargers was greater than with constricted pipe spargers. Air-flow rate was also important, but the critical level was dependent on agitator speed. An optimum rate of oxygen uptake at the cell surface is obtained with constant rates of cell growth and antibiotic production above threshold values of k_d (oxygen-transfer coefficient) and of power absorption. The combinations of air-flow rates and mechanical-power input to achieve or exceed these threshold levels are the basis for economic fermentor design.

CULTURE SELECTION AND MAINTENANCE

S. griseus is a very unstable microorganism. Its variants differ in their ability to yield streptomycin. Morphological differences between single-colony isolates are also apparent. Substrains differ in degree of sporulation, soluble-pigment production, and types of growth. Some variants differ so much that they do not fit in the generic classification *Streptomyces,* but more closely resemble *Nocardia.* Old, giant colonies show distinct sectoring. While variation presents a problem in maintenance of stock cultures, use has been made of it to isolate strains of greater productivity. Strains of improved productivity have been selected from natural variants. Induced mutation with ultraviolet light, followed by single-colony selection, has resulted in isolation of strains which yield over 900 μg per ml of streptomycin on ordinary plant-production media.[35] Recently Pittenger and McCoy[27c] studied variants of *S. griseus* induced by ultraviolet irradiation. Five of 98 strains selected for high activity from more than 4,000 colonies screened gave higher submerged-culture yields of streptomycin than the parent culture. One of these five strains excelled the parent in all four media used. Variation in streptomycin production on the different media by an individual strain points up the fact that the choice of a single medium for screening may restrict the detection of high-yielding strains.

Cultures may be stabilized by lyophilization procedures, as described in Chapter 12. Spores of active cultures, suspended in sterile skimmed milk, form an ideal pellet on lyophilization. Tests for activity, growth, spore production, and physical appearance showed no difference between the parent and lyophilized cultures.[6] An alternate method, described in some detail in Chapter 12, is the preparation of soil cultures. Spores of *S. griseus* are mixed with sterile soil and allowed to dry out at room temperature. The stock soil cultures are satisfactory for development of inoculum for experimental purposes.

DEVELOPMENT OF CULTURE

S. griseus is developed for initiation of production batches by aseptic transfer of lyophile pellets or grains of soil cultures to a nutritive medium. A satisfactory sporulation medium is described in the section on procedures for the surface-culture method. A medium which is adequate for initiating submerged growth contains 1% yeast extract, 1% glucose, and 0.5% $Na_2HPO_4 \cdot KH_2PO_4$ buffer of pH 7.0.

PLANT FERMENTATION

Description of the submerged, aerated production process employed at Merck & Co., including a flow sheet, has been published.[28] Campbell[5a] has also presented a flow sheet for the manufacture of streptomycin. Details of the Merck production process are summarized here, and the flow sheet is given in Figures 19a to 19d.

An agar slant, inoculated with a large loop ol lyophilized soil culture of *S. griseus*, is incubated until heavy sporulation occurs. Spores are suspended in water and transferred to initiate growth of three 300-ml shake-flask cultures by incubation as described in the section on submerged-culture laboratory procedures. The de-

FIGURES 19a-19d. *Flow Diagram of Streptomycin Production and Recovery Process* [Reprinted by permission from *Chem. Eng.*, **53**, No. 10, 142 (1946)]

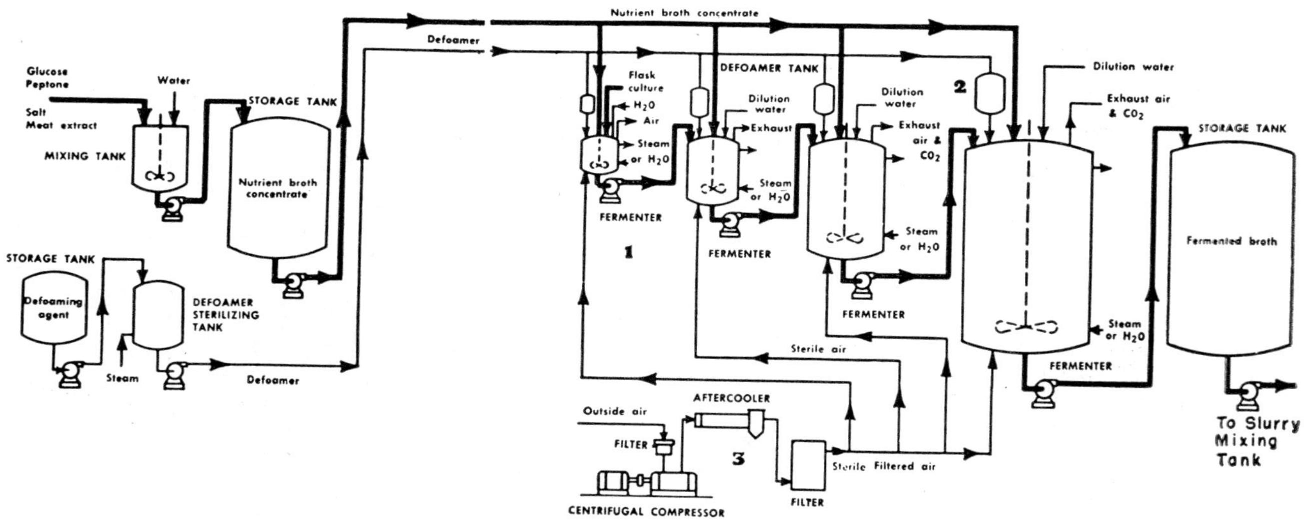

FIGURE 19a. *Medium Preparation and Fermentation*

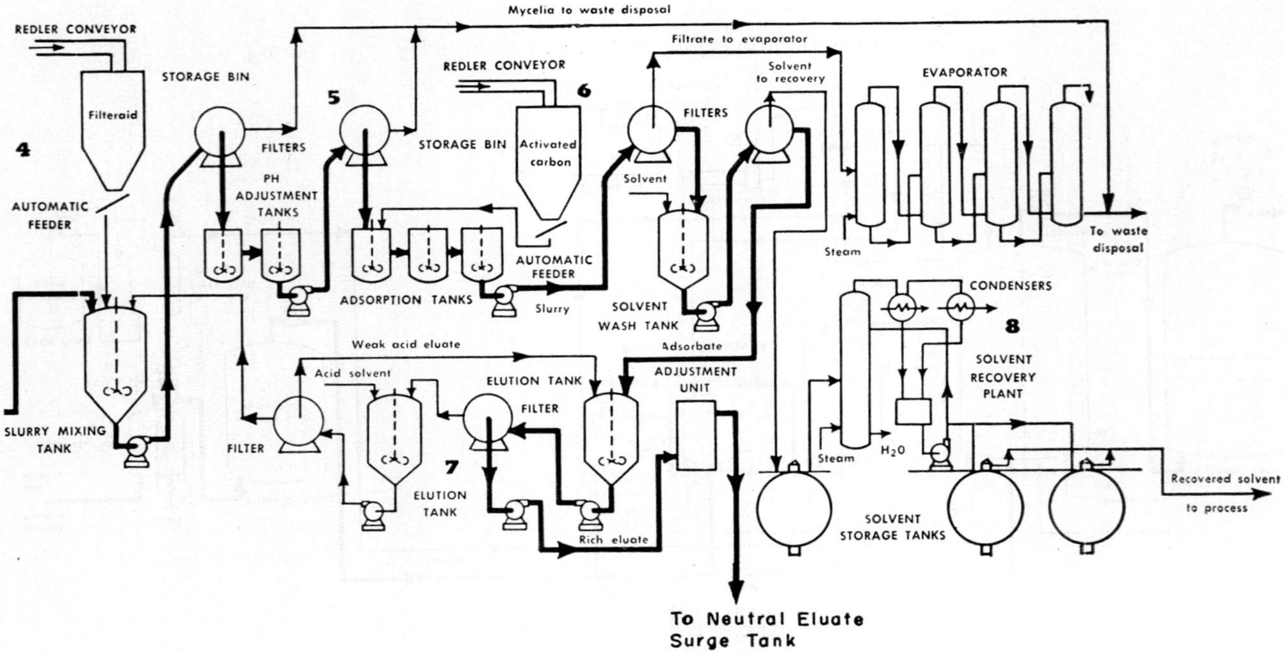

FIGURE 19b. *Adsorption and Elution*

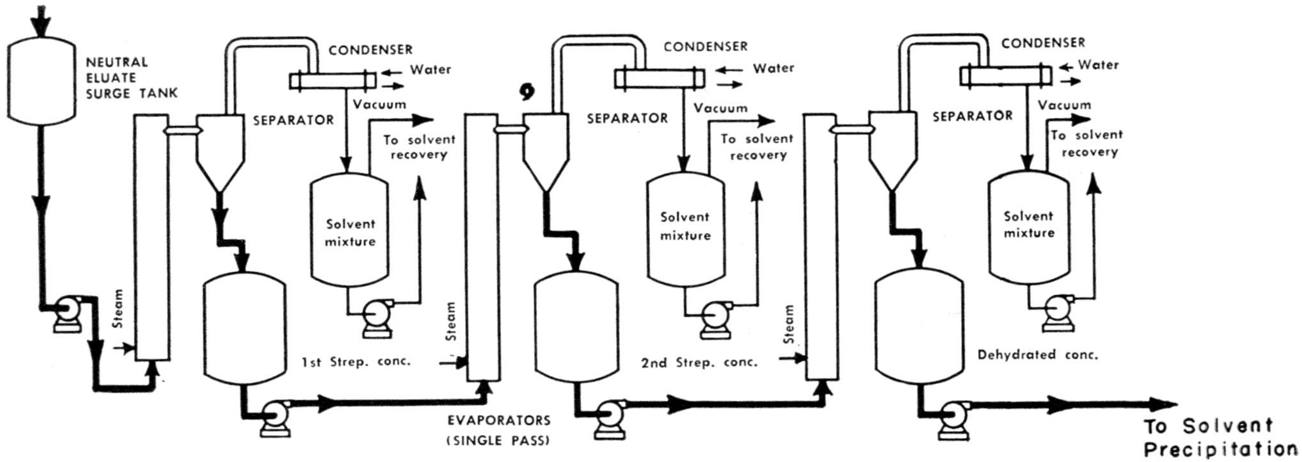

FIGURE 19c. *Concentration and Dehydration*

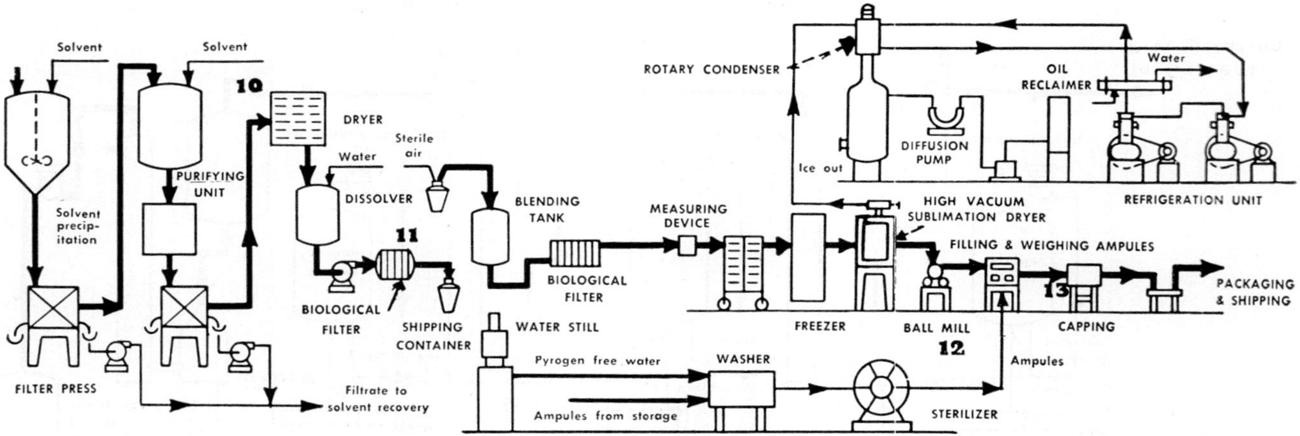

FIGURE 19d. *Precipitation, Purification, and Finishing*

veloped flask culture becomes the starting point for mass production in the plant.

Raw materials of the nutrient medium are mixed with water in a concentration about ten times higher than finally required. The mash is pumped either through a continuous sterilizer or directly to fermentors where, properly diluted, it is sterilized and cooled.

Four sizes of fermentors, constructed of carbon steel, are employed for fermentation. Each vessel in the series was designed to provide culture for efficient inoculation of the next larger fermentor in the series. The laboratory culture serves as inoculant for the smallest tank. During fermentation, the temperature is carefully controlled at 25° to 30°C by cooling with water. Total fermentation time for all four vessels in series amounts to several days, although each stage follows its own assigned schedule. The number of inoculant tanks is such that maximum fermentor capacity is always in operation.

Aeration is accomplished by air from sterilizing filters, supplied to the tanks at approximately 0.03 cu ft per minute per gal of fermentor capacity. Each vessel is equipped with an agitator to promote aeration.

As a result of aeration, considerable foaming occurs. This is controlled by the addition of sterile antifoam agents, which must be selected for lack of toxicity.

The final fermented culture volume is about 15,000 gal. The streptomycin content of broths produced commercially has not been published. Laboratory and pilot-plant fermentor yields ranging from 150 to 1,200 μg per ml in production-type media have been reported.

Fermentation operations must, at all times, be conducted under aseptic conditions to prevent contamination, which might inhibit or reduce streptomycin production or add to extraction difficulties. All joints and connections to the fermentor are protected by steam seals. Valves are steam sealed and all pipes leading to the fermentor are kept under steam pressure except when in use. Streptomycin is not subject to destruction by the usual type of bacterial contamination, so that streptomycin produced before the appearance of contaminants will not be lost.

In addition to precautions against bacterial contamination, it is necessary to guard against phage infestations. Lysis by actinophage is particularly serious when inoculum vessels are invaded, since the inoculum becomes useless for initiating fermentation in the production fermentors. Actinophage infection, initiated in the production tank, leads to early lysis of the *S. griseus* and reduced yields. Carvajal[6a] recently described an extensive study of phage problems in the streptomycin fermentation. He reported that in one manufacturing plant during a 3-week period, 74 of 111 fermentors were infected with a multivalent phage. Effective measures for control and eradication depended on strict sanitary precautions and mainly on development of cultures immune to action of the phage. Actinophage-resistant strains of *S. griseus* have been developed which retain the ability to produce streptomycin and are used in factory processes.[6a,33,42]

Isolation details, as well as the production process, are diagrammed in the flow sheet of Figures 19a to 19d. A series of five operations is used: (1) removal of mycelium by filtration, (2) adsorption of streptomycin on activated carbon, (3) elution of streptomycin from the carbon adsorbate, (4) concentration of the eluate by evaporation and dehydration, (5) double solvent precipitation, followed by filtration and drying. Further purification is accomplished by undisclosed procedures to yield a product of over 90% purity.

Activated carbon is a relatively nonspecific adsorbent and a great deal of other substances becomes adsorbed and eluted with the streptomycin. The procedure may be made more specific by treating streptomycin-containing material with carbon at pH 4.0. Impurities are adsorbed, but most of the streptomycin is not and is recovered in a purified state. Adsorbed streptomycin is eluted and reworked.[22]

Streptomycin has been purified by alumina chromatography and crystallized with a number of reagents. Crystalline streptomycin reineckate and streptomycin helianthate have been described and procedures given for conversion to pure but noncrystalline streptomycin hydrochloride. A crystalline streptomycin trihydrochloride-calcium chloride double salt has been made. Because of its purity, this salt has advantages for clinical application and was marketed in lyophilized form as the final product from the process described previously.

An example of recovery and product purity obtained by various purification steps is shown in Table 34. These data have

TABLE 34. REPRESENTATIVE LABORATORY PURIFICATION PROCEDURES

Experiment	Steps	Weight Recovery, g Taken	Weight Recovery, g Recovered	Recovery of activity, %	Activity, μg per mg
No. 1[a]	*a*. Broth	10.5 (liters)	—	—	82
	b. Acetone ppt from eluate of pH 8 adsorption	—	—	—	30-80
	c. Acetone ppt from eluate of pH 4 adsorption of *b*	—	1.010	11	60-85 (reworked to step *d*)
	d. Unadsorbed portion of step *c* adjusted to pH 6. Acetone ppt of eluate of charcoal adsorption at pH 6	—	0.475	22	350-450
No. 2[b]	*a*. Streptomycin hydrochloride absorbed on 160 g alumina	6.0	—	—	225
	b. First 250 ml methanol elute	—	0.698	23	500
	c. Next 150 ml elute	—	1.031	34	490
	d. Next 250 ml elute	—	0.729	14	290
	e. Next 350 ml elute	—	0.413	4	160
No. 3	*a*. 500 μg/ml streptomycin in methanol solution crystallized with 20.8 g methyl orange	26.3	32.000	87	—
No. 4[c]	*a*. 7 g $CaCl_2$ in methanol added to methanol solution of streptomycin helianthate. Filtrate concentrated and crystallized	30.0	11.500	77	750

[a] LePage and Campbell.[22] [b] Kuehl, Peck, Hoffhine, Graber, and Folkers.[21] [c] Peck, Brink, Kuehl, Flynn, Walti, and Folkers.[26]

been selected from various published laboratory experiments and probably are representative of yields which may be expected from similar processes on a production scale.

A more recent development is the application of ion-exchange methods to the purification of certain basic antibiotics. This development is of great practical value in the commercial preparation of streptomycin. Ion-exchange resins had been suggested on several occasions as adsorbents for streptomycin, but had not been altogether satisfactory because the resins adsorbed not only the antibiotic but various other complex organic compounds invariably found in fermentation broths. Thus, as with activated carbon, the adsorption step served primarily as a concentration step after which numerous complicated purification steps were required for final purification.

Ion-exchange resins having carboxylic polar groups have found widespread application for the adsorption of streptomycin. The product obtained directly from the broth by use of carboxylic acid resins is of such purity as to yield directly a crystalline calcium chloride complex salt from methanol solution. Nonbasic substances are not adsorbed because of the ion-exchange nature of the adsorption. The resins are also highly selective for streptomycin among the various organic ions contained in the fermentation broth. The adsorbed streptomycin is eluted from the resin by practically stoichiometric amounts of acid. It is possible to recover the streptomycin in almost quantitative amounts from the fermentation broth. The resin may be converted to the basic cycle and reutilized indefinitely.

Characteristics of the adsorption with various adsorbents are reported in Table 35.

Various technical modifications have been proposed for efficient use of ion-exchange resins. They are usually packed in columns arranged in series to provide a continuous-flow process.

An example of a laboratory scheme, which is adaptable to plant operation, is quoted from the patent of Howe and Putter[17] as follows:

> Two columns containing 50 g of a granular copolymer of methacrylic acid and divinyl benzene containing 5% divinyl benzene were converted to the sodium cycle by passing through each column 1,000 cc of 10% aqueous sodium hydroxide. The columns were washed with 2,000 cc of water each. The

Table 35. COMPARISON OF VARIOUS ION-EXCHANGE RESINS AND ADSORBENTS IN THE RECOVERY OF STREPTOMYCIN

Type of adsorbent	Capacity of adsorbent mg per g	Eluting agent used	Recovery on elution %	Potency of product obtained units per mg[a]
Sulfonated phenol-formaldehyde resin	10	Aqueous NaCl Aqueous H_2SO_4	30-50 40	ca 100
Silicate-exchange resin	40	Aqueous NaCl	100	ca 50
Sulfonated coal resin	12	Aqueous NaCl Aqueous H_2SO_4	40-60 40	ca 100
Carbon	20	Acidic menthanol-water	60-75	200-350
Fuller's earth	50	Aqueous pyridine hydrochloride	30-40	75
Copolymer of methacrylic acid and divinyl benzene, containing 5% divinyl benzene	600-1,500	Aqueous HCl	100	600
Copolymer of methacrylic acid and divinyl benzene, containing 10% divinyl benzene	300-800	Aqueous HCl	100	600
Carboxylic resin believed to be of the phenol-formaldehyde type known as Permutit XHIC	500-900	Aqueous HCl	100	650

[a] Potency for pure streptomycin calcium chloride complex = 779 units per mg.
Source: Howe and Putter.[17]

columns were connected with rubber tubing to operate in series flow.

Sixty-four and eight-tenths liters of streptomycin broth was acidified to pH 2 with phosphoric acid, filtered, neutralized to pH 7 with aqueous sodium hydroxide, and refiltered. The treated broth assayed 360 units per cc.

The treated broth was then allowed to flow under about 1 lb pressure through the two resin columns in series at a rate of 150 cc/min. The spent broth at the end of the run assayed 4.3 u./cc. or 1.2%. The columns were washed with 1,000 cc of water and then eluted by passing through each column 1,000 cc of 1.0 N aqueous hydrochloric acid. The rich eluate from the first column amounted to 670 cc and assayed 25,000 u./cc. The rich eluate from the second column amounted to 235 cc and assayed 28,000 u./cc. The recovery of streptomycin calculated 100.5%.

The eluate from the first column was concentrated in vacuo at 30°C almost to dryness, treated with 100 cc methanol, and filtered. The solid removed by filtration weighed 2.75 g. The methanol solution was precipitated in 500 cc of acetone; the solid filtered, washed with 50 cc acetone, and dried in vacuo. Weight: 26.3 g, assay: 650 u./mg, ash (sulfated): 15%. The recovery at this point calculated 101.5%.

The acetone precipitated solid was dissolved in 81 cc of methanol and to the solution was added 23.6 g of calcium chloride, anhydrous. The solution was stirred for 40 hours and then filtered. The precipitate was washed with a 10% solution of calcium chloride in methanol, then with ethanol, and dried. The product weighed 17.5 g and assayed 780 u./mg (theoretical for pure streptomycin calcium chloride complex=779 u./mg). The over-all recovery from broth (through the first column) was 80%.

CHEMICAL STRUCTURE

The chemical structure of streptomycin is known, although the steric arrangement of the molecule has not been completely established. Synthesis by chemical means has not been accomplished. Streptomycin hydrochloride has the empirical formula $C_{21}H_{39}N_7O_{12}\cdot 3HCl$. It is a glycoside of the inositol derivative, streptidine, with streptobiosamine, a disaccharide composed of the sugars streptose and *N*-methyl-L-glucosamine.

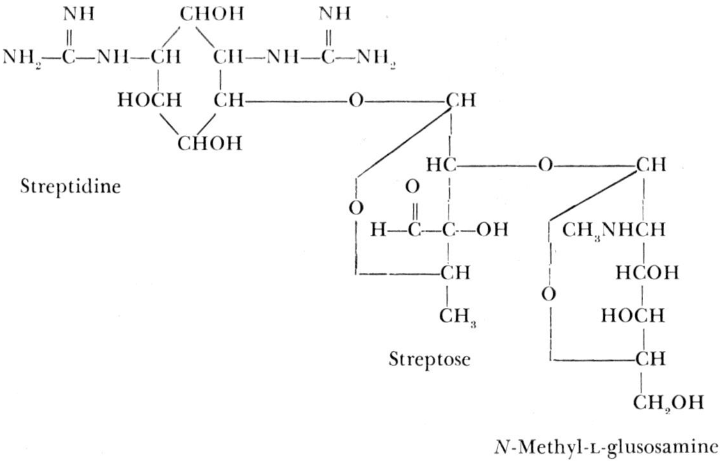

Structure of streptomycin

RELATED THERAPEUTIC COMPOUNDS

By reductive hydrogenation, streptomycin is converted to dihydrostreptomycin.[27] This compound has properties similar to streptomycin, but has a somewhat modified inhibition spectrum against microorganisms and has a modified quantitative toxicity response. Most chemicals which form addition compounds on the carbonyl group destroy the antibiotic properties of streptomycin.

S. griseus produces a derivative of streptomycin, mannosidostreptomycin, in some types of media. This derivative is a glycoside of streptidine with a trisaccharide, composed of streptose, *N*-methyl-L-glucosamine, and mannose. Against many bacteria, it has reduced antibacterial effect.[31]

An enzyme is produced late in the fermentation which converts mannosidostreptomycin to streptomycin. Cultures which are rich in enzyme can be used to hydrolyze the mannose from the mannosidostreptomycin contained in poor fermentations, or that isolated as a by-product in the streptomycin isolation procedure. Streptomycin is formed as the end product of the enzymic reaction and an increased fermentation yield is obtained. The reaction may be carried out by whole cultures of *S. griseus* or by cell-free extracts of the enzyme.[21a]

Another related compound, hydroxystreptomycin, is produced by *Streptomyces griseo-carneus*. The chemical structure of hydroxystreptomycin is similar to that of streptomycin but there is one additional hydroxyl group on the terminal carbon of the streptose portion of the molecule.[4]

PRODUCTION VOLUMES

From the time that streptomycin first became available in appreciable quantity, in August 1946, the production rapidly increased. Through December 1948, the production curve followed a constant slope increase, when plotted as a logarithmic function against time. This is characteristic of an expanding industry which has not been subjected to limitations of sale, raw materials, or construction restrictions. The basic level of production for the United States was at about 8,000 kg per month, for the next 2 years, but domestic consumption has continued to increase. Production of streptomycin for 1951 and 1952 was 159,500 and 175,100 kg, respectively.[27a]

Production data compiled by the U.S. Department of Commerce[9] are shown in Figure 20.

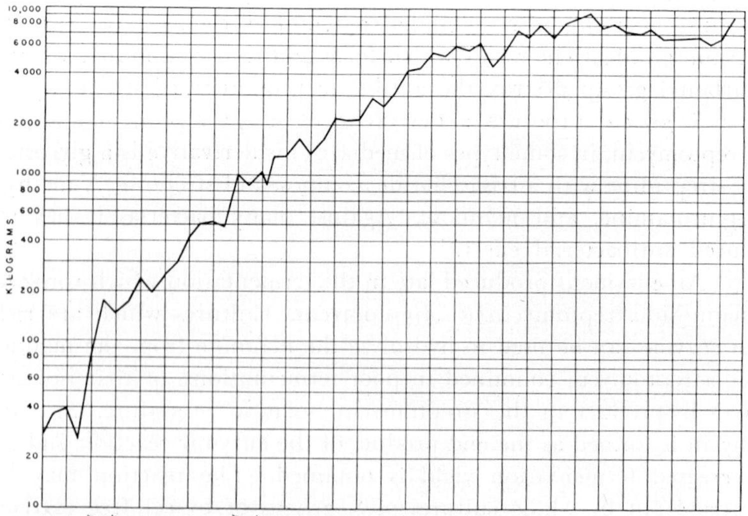

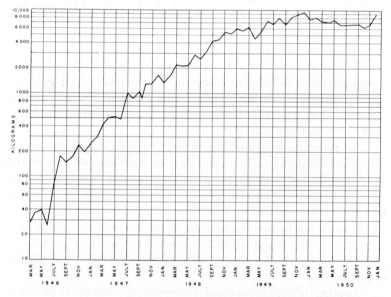

FIGURE 20. *Streptomycin Production by Months*

RESIDUES AND WASTES

The chief residues from the streptomycin fermentation fall into two groups. The first to be obtained is the mycelium of *S. griseus* and the insoluble residues from the medium, which are removed during the filtration stage, before the adsorption of streptomycin. These residues have potential value as animal feeds. *S. griseus* mycelium is rich in proteins, being equivalent to the bacteria in this respect. Although *S. griseus* does not require an exogenous source of B vitamins for growth, it does synthesize appreciable quantities of these substances, in this manner adding to the value of the mycelial residues as an animal feed.

The degree of contamination of the mycelium with insoluble residues is dependent on the fermentation medium. *S. griseus* has strong proteolytic activity and, even in media prepared from ground soybean meal, very little insoluble matter remains at the completion of fermentation. Because of the bacterial nature of the cell growth, much filter aid must be used to obtain a good filtration rate. The filter aid contaminates the mycelial wastes and must be removed to make possible utilization of the residue.

Following adsorption of streptomycin, the spent broth contains approximately 2% soluble organic matter, which can be recovered by drying, particularly if disposal of high b.o.d. wastes present a problem. It is useless as a major component of animal feeds because of its high salt content. With the media which have been proposed for large-scale production of streptomycin, the salt content of the total soluble solids is greater than 25%. It can, however, serve as a source material for isolation of the B vitamins, especially vitamin B_{12}.

Commercially, the most important vitamin by-product of *S. griseus* fermentations is vitamin B_{12}.[14,32] Some modifications of the culture medium, which have no bearing on streptomycin production, are often necessary to obtain optimum by-product yields. Vitamin B_{12} contains cobalt and, since the natural proteinaceous media used for streptomycin production are often deficient in cobalt, the medium is supplemented with a soluble cobalt salt at a concentration just under that toxic for streptomycin production.[15] Vitamin B_{12} may be isolated as a pure compound for its therapeutic properties, or the dried fermentation residues or partially purified concentrates may be marketed as an animal-feed supplement for their APF (animal-protein factor) activity. Some feed supplements are prepared from combinations of a B_{12}-containing fermentation residue and an antibiotic to obtain increased growth response and a degree of protection against infection. (See also Chapter 6.)

SPECIFICATIONS FOR COMMERCIAL STREPTOMYCIN

The specifications for commercial streptomycin have been outlined and are controlled by the Food and Drug Administration[13] of the United States Government. Tests and assays are performed on all lots for potency, pyrogens, toxicity, moisture, pH, histamine-like substances and clarity of solution. Standard methods have been developed for all tests. Potency is measured by biological test for the antibiotic activity of streptomycin against a standard strain of *Bacillus subtilis* by a cup-diffusion method or *Klebsiella pneumoniae* by a turbidimetric method of assay. The potency of streptomycin preparations is designated as the number of "micrograms" (potency unit) contained in a milligram or milliliter of sample. A microgram is defined as the streptomycin activity (potency) contained in 1.38 μg of a streptomycin-calcium chloride complex master standard established by the Food and Drug Administration. Chemical procedures based on maltol formation as a degradation product of streptomycin or polarographic analysis are widely used as supplementary analytical procedures. Carboxylic acid ion-exchange resins may be used for quantitative purification of the streptomycin before application of the analytical procedures.

CLINICAL USES OF STREPTOMYCIN

Streptomycin is indicated in the treatment of a variety of infections due to susceptible Gram-negative or acid-fast organisms. Although it is not a substitute for penicillin in the treatment of infections due to Gram-positive, penicillin-sensitive microorganisms, streptomycin does inhibit effectively the growth of many Gram-positive bacteria. It may be employed successfully in the treatment of certain cases of infection due to penicillin-resistant strains of such pathogens. It should never be used in the treatment of conditions caused by organisms found to be insusceptible to therapeutic concentrations of streptomycin.

Among the infections in which streptomycin has been found of therapeutic value are certain urinary-tract infections, tularemia, influenzal meningitis, bacteremia, and wound infections caused by streptomycin-sensitive organisms. It may often be used advantageously pre- or postoperatively for the prevention or control of infections caused by susceptible pathogens. Intermittent intramuscular injection is the preferred method of parenteral administration, although solutions of the drug may be also injected subcutaneously in selected cases. When indicated, streptomycin may be given intraspinally, locally, or by inhalation. Since negligible amounts of streptomycin are absorbed from the gastro-intestinal tract, oral administration is ineffective for the treatment of systemic infections.

Dosage varies greatly with patient, infection, and route of administration. The rapidity with which many microorganisms develop resistance to streptomycin is a factor which has indicated selection of rather large initial doses. Dosage usually ranges from 0.5 to 4 g per day parenterally, additional amounts being given by such other routes as may be indicated in the individual case.

In addition to its application in treatment of human diseases, streptomycin is now widely used in veterinary medicine. With large-volume production, the cost of the antibiotic is no longer a significant factor in limiting its use.

It is now recognized that streptomycin is a valuable and, in some cases, an essential therapeutic adjunct in certain types and stages of tuberculosis, although it is not a substitute for such well-established methods of treatment as bed rest, special diet, and adequate surgical procedures. The proper combination of therapeutic measures to be employed in the individual case must be the choice of the physician in charge.

Streptomycin therapy is considered mandatory in acute hematogenous miliary tuberculosis and tuberculous meningitis, and its use is strongly indicated in tuberculous pneumonia. Streptomycin is effective also in the treatment of selected cases of tuberculosis of the respiratory, genito-urinary, and gastro-intestinal tracts. Certain types of pulmonary tuberculosis are greatly improved by the proper employment of streptomycin in conjunction with traditional therapeutic methods. In many instances, symptomatic improvement is striking and x-ray findings of favorable import are noted. However, it is by no means advisable to employ streptomycin in all cases of pulmonary tuberculosis; in each instance, the physician must exercise critical clinical judgment to determine the thereapeutic regimen, with or without antibiotic therapy, best suited to the individual case.[2,8,19]

Streptomycin is capable of producing side reactions of varying severity. Most of those occurring early in the course of treatment are transient and not of a serious nature. However, severe toxic reactions may occur in individuals who have become hypersensitive to streptomycin. Although antihistamine drugs or "desensitization" procedures may be effective in such cases, in certain instances, further administration of streptomycin or contact with it may have to be avoided.

When administered in comparatively high dosage over the relatively long periods required for treatment of various types of tuberculosis, certain neurotoxic reactions frequently become an important consideration. Interference with vestibular function (impairment of the sense of balance) is most common and has appeared in 96% of patients in a large series of cases. With a dosage of 1.8 to 2.0 g per day, it usually appeared in the fourth week of treatment. Complete or partial functional recovery by various compensation mechanisms occurs in most cases, even with continued treatment. Impairment of hearing may also occur, but this is relatively rare. Only 1% of a large series of cases showed sufficient toxicity of this type to make cessation of treatment advisable. If streptomycin therapy is continued after impairment of hearing becomes evident, the disability may be expected to progress and become permanent. However, if administration of the drug is discontinued as soon as signs or symptoms of such injury are noted, complete recovery may be expected in most cases.

Early clinical studies indicate that dihydrostreptomycin is probably as effective as the parent drug, streptomycin, in the treatment of certain types of tuberculosis and that it exhibits significantly less neurotoxicity.

Hydroxystreptomycin has been tested for ototoxic effects in cats. It had a neurotoxic effect greater than that of either dihydrostreptomycin or streptomycin and caused a severe hearing loss after very few doses.

Streptomycin is of great effectiveness in clinical medicine as a therapeutic agent. Therefore, it is widely used despite its shortcomings. The two most serious drawbacks to its effective use are the ease which microorganisms develop resistance to streptomycin and the inherent neurotoxicity of the molecule to the animal body on prolonged administration. With the development of dihydrostreptomycin and the adoption of streptomycin dosage regimens involving relatively low daily doses for limited periods of therapy, the problem of neurotoxicity has been greatly reduced. The development of bacterial resistance to streptomycin and dihydrostreptomycin therapy is a serious problem clinically. However, it was found in recent years that streptomycin or dihydrostreptomycin in conjunction with *p*-aminosalicylic acid or isoniazid (isonicotinic acid hydrazide) was generally more effective in tuberculosis chemotherapy than streptomycin used alone.[38a] There is clear evidence that *p*-aminosalicylic acid inhibits the development of resistance of *M. tuberculosis* to streptomycin.

Streptomycin has also found application in the treatment of bacterial diseases of plants.[21b] Streptomycin is absorbed and stored by the plant leaf. Since it is not removed by rainfall, the antibiotic within the leaf acts as a reservoir of protective agent effective against infection. It is especially valuable in bacterial diseases of valuable plants, such as fire blight of pears and walnut blight. It is also reported effective against bacterial wilt of chrysanthemum and halo blight of beans. Much research is needed before the final place of streptomycin in plant-disease control is established. At high concentrations, streptomycin causes chlorosis and, at somewhat lower concentrations, inhibits root development. However, the last-mentioned toxic effect can be overcome by manganese salts.[32a]

The manufacture of streptomycin is one of the important fermentation operations in the United States and abroad. It is useful in clinical medicine and new and broader applications are being found continuously. Further improvements in production, increased knowledge of its activity, and wider applications may be anticipated from the extensive research which is still in progress on streptomycin.

BIBLIOGRAPHY

1. Ainsworth, G. C., A. M. Brown, P. S. S. F. Marsden, P. A. Smith, and J. F. Spilsbury, *J. Gen. Microbiology,* **1**, 335 (1947).
2. Barnwell, J. B., P. A. Bunn, and M. Walker, *Am. Rev. Tuberc.,* **56**, 485 (1947).
3. Bartholomew, W. H., E. O. Karow, M. R. Sfat, and R. H. Wilhelm, *Ind. Eng. Chem.,* **42**, 1801, 1810 (1950).
4. Benedict, R. G., F. H. Stodola, O. L. Shotwell, A. M. Borud, and L. A. Lindenfelser, *Science,* **112**, 77 (1950).

5. Bennett, R. E., *J. Bact.,* **53**, 254 (1947).
5a. Campbell, A. H., *Research,* **6**, 42 (1953).
6. Carvajal, F., *Mycologia,* **38**, 596 (1946).
6a. Carvajal, F., *Mycologia,* **45**, 209 (1953).
7. Colingsworth, D. R., U. S. Patent 2,504,067 (1950).
8. Council on Pharmacy and Chemistry, Annual Rept. of the Committee on Therapy and the Subcommittee on Streptomycin; revised version, *J. Am. Med. Assoc.,* **135**, 641 (1947).
9. Department of Commerce, U. S. A., Bureau of Foreign and Domestic Commerce.
10. Donovick, R., W. L. Koerber, and G. W. Rake, U. S. Patent 2,516,682 (1950).
11. Dulaney, E. L., *J. Bact.,* **56**, 305 (1948).
12. Dulaney, E. L., and D. Perlman, *Bull. Torrey Bot. Club,* **74**, 504 (1947).
13. Food and Drug Administration, compilation of regulations for tests and methods of assay and certification of antibiotic drugs, *Vol. I, Tests and Methods of Assay,* Washington, D. C.
14. Fricke, H. H., B. Lanius, A. F. DeRose, M. Lapidus, and D. V. Frost, *Federation Proc.,* **9**, 173 (1950).
14a. Garner, H. R., H. Koffler, P. A. Tetrault, M. Fahmy, M. V. Mallett, R. A. Faust, R. L. Phillips, and N. Bohonos, *Am. J. Botany,* **40**, 289 (1953).
15. Hendlin, D., and M. L. Ruger, *Science,* **111**, 541 (1950).
16. Horne, R. E., Jr., and A. L. Pollard, *J. Bact.,* **55**, 231 (1948).
17. Howe, E. E., and I. Putter, U. S. Patent 2.541,420 (1951).
18. Johnstone, D. B., and S. A. Waksman, *Proc. Soc. Exptl. Biol. Med.,* **65**, 294 (1947).
18a. Karow, E. D., R. L. Peck, C. Rosenblum, and D. T. Woodbury, *J. Am. Chem. Soc.,* **74**, 3056 (1952).
19. Keefer, C. S., F. G. Blake, J. S. Lockwood, P. H. Long, E. K. Marshall, Jr., and W. B. Wood, *J. Am. Med. Assoc.,* **132**, 70 (1946).
20. Kempf, J. E., and P. Sayles, *J. Bact.,* **51**, 596 (1946).
21. Kuehl, F. A., R. L. Peck, C. E. Hoffhine, Jr., R. P. Graber, and K. Folkers, *J. Am. Chem. Soc.,* **68**, 1460 (1946).
21a. Langlykke, A. F., and D. Perlman, U. S. Patent 2,493,489 (1950).
21b. Leben, C., and G. W. Keitt, *Agri. and Food Chem.,* **2**, 234 (1954).
22. LePage, G. A., and E. Campbell, *J. Biol. Chem.,* **162**, 163 (1946).

23. McDaniel, L. E., U. S. Patent 2,515,461 (1950).
24. McDaniel, L. E., U. S. Patent 2,538,942 (1951).
25. McDaniel, L. E., and D. Hendlin, U. S. Patent 2,538,943 (1951).
25a. Numerof, P., M. Gordon, A. Virgona, and E. O'Brien, *J. Am. Chem. Soc.,* **76**, 1341 (1954).
26. Peck, R. L., N. G. Brink, F. A. Kuehl, Jr., E. H. Flynn, A. Walti, and K. Folkers, *J. Am. Chem. Soc.,* **67**, 1866 (1945).
27. Peck, R. L., C. E. Hoffhine, and K. Folkers, *J. Am. Chem. Soc.,* **68**, 1390 (1946).
27a. Perlman, D., A. E. Tempel, and W. E. Brown, *Ind. Eng. Chem.,* **45**, 1944 (1953).
27b. Perlman D., and G. H. Wagman, *J. Bact.,* **63**, 253 (1952).
27c. Pittenger R. C., and E. McCoy, *J. Bact.,* **65**, 56 (1953).
28. Porter, R. W., *Chem. Eng.,* **53**, No. 10, 94 (1946).
29. Pramer, D., and R. L. Starkey, *Science,* **113**, 127 (1951).
30. Rake, G., and R. Donovick, *J. Bact.,* **51**, 596 (1946).
31. Rake, G., C. M. McKee, F. E. Pansy, and R. Donovick, *Proc. Soc. Exptl. Biol. Med.,* **65**, 107 (1947).
32. Rickes, E. L., N. G. Brink, F. R. Koniuszy, T. R. Wood, and K. Folkers, *Science,* **108**, 634 (1948).
32a. Rosen, W. G., *Proc. Soc. Exptl. Biol. Med.,* **85**, 385 (1954).
33. Saudek, E. C., and D. R. Colingsworth, *J. Bact.,* **54**, 41 (1947).
34. Schatz, A., E. Bugie, and S. A. Waksman, *Proc. Soc. Exptl. Biol. Med.,* **55**, 66 (1944).
35. Stanley, A. R., *J. Bact.,* **53**, 254 (1947).
36. Thornberry, H. H., and H. W. Anderson, *Arch. Biochem.,* **16**, 389 (1948).
37. Trussell, P. C., U. S. Patent 2,541,726 (1951).
38. Trussell, P. C., C. O. Fulton, and G. A. Grant, *J. Bact.,* **53**, 769 (1947).
38a. Veterans Administration; Army-Navy, *Transactions of the Twelfth Conference on the Chemotherapy of Tuberculosis,* Atlanta, Ga., and Washington, D. C., Veterans Administration, 1953.
38b. Waksman, S. A., *Literature on Streptomycin 1944-52,* New Brunswick, N. J., Rutgers University Press, 1952.
39. Waksman, S. A., and A. Schatz, *J. Am. Pharm. Assoc., Sci. Ed.,* **34**, 273 (1945).
40. Waksman, S. A., A. Schatz, and H. C. Reilly, *J. Bact.,* **51**, 753 (1946).
41. Woodruff, H. B., *J. Bact.,* **54**, 42 (1947).
42. Woodruff, H. B., U. S. Patent 2,585,713 (1952).

CHAPTER 9

THE BROAD-SPECTRUM, POLYPEPTIDE, AND OTHER ANTIBIOTICS

William H. Peterson and Mary S. Peterson

INTRODUCTION

Under this title are included antibiotics of microbiological origin other than penicillin, streptomycin and dihydrostreptomycin, since these products are dealt with by other authors in separate chapters. Actidione (cycloheximide), Aureomycin (chlorotetracycline), bacitracin, Chloromycetin (chloramphenicol), Erythrocin (erythromycin), fumagillin, Ilotycin (erythromycin), Magnamycin (carbomycin), neomycin, polymyxin, Terramycin (oxytetracycline), tyrothricin, tetracycline, and viomycin make up the list of antibiotics that are now (February, 1953) market products. Many of these are better known by their trademark names (first letters in capitals) than by their scientific names. The system of double names is confusing and troublesome. It is especially so when trade names that have become well-established in scientific literature, such as Aureomycin and Terramycin, are withdrawn from such use and replaced by a chemical designation.

The bulk of the antibiotics industry is made up of penicillin, streptomycin and the so-called "broad spectrum" antibiotics, Aureomycin, Chloromycetin, and Terramycin. The latter, though scarcely five years old, have established themselves as leaders from an industrial, economic, and clinical point of view. In June, 1952, Welch[287a] reported that the production of "broad spectrum" antibiotics amounted to 24 tons per month as compared to 30 tons for penicillin and 24 tons for streptomycin. In tonnage, the two older antibiotics exceed the "broad spectrum" antibiotics, but in market value the latter appear to equal or outrank the former. The market price of the "broad spectrum" antibiotics in capsule form ranges from $2.13 per gram for Aureomycin to $1.20 per gram for Chloromycetin. These prices are subject to trade discounts, but even allowing for these and for the unequal production of the three antibiotics,[206a] it seems probable that the returns to the producers of the "broad spectrum" antibiotics is well over 200 million dollars annually. The returns from penicillin and streptomycin, when compounded into tablets, injectable solutions, suspensions, and many other dosage forms, is probably of the same order of magnitude. If the income from the minor antibiotics and from the sale of antibiotics as supplements to animal feeds is added to these sums, a total of about 500 million dollars yearly seems a reasonable figure for the size of this huge and expanding industry.

Besides the commercial antibiotics, there are several others that either have not been tested sufficiently to establish their usefulness or for which there still appear some commercial possibilities under certain circumstances even though they are too toxic for widespread use. Some of these are new types of compounds and others contain new examples of old types, e.g., amino acids. Although they may be too toxic for human therapy, some may prove useful in the treatment of animal diseases or in combating insect pests and plant diseases. Many manufacturers have one or more antibiotics in various stages of development but consideration of these will have to wait until information about them has been released. It is probably a reasonable estimate that 100,000 new isolates are tested yearly by the producing companies. Many new, interesting and, hopefully, useful compounds should come out of this vast number of cultures.

Many antibiotics, uncovered in this search but only partly purified, have been abandoned by producers because toxicity or other undesirable features make them too unpromising for further work. If these could be studied and their chemical nature determined (for example, by university laboratories) it would

FIGURE 20a. *Seed and Fermentation Tanks Used in the Production of Antibiotics* (Courtesy—Lederle Laboratories Division, American Cyanamid Company)

provide a body of knowledge that might be very valuable in arriving at some rational explanation for antibiotic activity.

In addition to the antibiotics discussed in some detail in this chapter, there is a second group that is listed in a table giving the most important facts about them. There are probably close to a hundred others for which the data are too scanty to warrant inclusion in this table. For a more complete list of antibiotics, including those of plant origin, reference is made to recent books,[19a,105,151,211] and reviews.[18,26,57,130,153,198,204a,299,305a]

In the discussion that follows, emphasis has been placed on the production, isolation, and chemical nature of the antibiotic rather than on its activity and toxicity. However, some mention of the therapeutic value could not be omitted because clinical use ultimately determines the commercial production of an antibiotic. The antibiotics are discussed in alphabetical order rather than in the order of their industrial importance. The latter arrangement would not be possible, since no figures are available where there is only one producer and the possible importance of the noncommercial products is still uncertain.

In writing this report, the authors are regretfully aware that much of the information will be outdated before these pages appear in print. The printed page is always out of date, but is conspicuously so in the rapidly developing field of antibiotics.

COMMERCIAL ANTIBIOTICS

Actidione (Cycloheximide)

During the commercial production of streptomycin by The Upjohn Company, three of their research workers[295] discovered that the beers used for that production contained also another antibiotic which they named Actidione. This is the trade name and cycloheximide is the chemical name.

PRODUCTION

Whether *Streptomyces griseus* produces greater quantities of streptomycin or of Actidione seems to depend on the medium used for its growth. In a glucose (Cerelose)-salts medium supplemented by dried brewers yeast, high yields of streptomycin were obtained, but little Actidione. In another medium supplemented by soybean meal and Curbay BG (the dried solids in the spent beers from the butanol fermentation of molasses; a U. S. Industrial Chemicals, Inc., product) instead of yeast, the yield of Actidione was higher than that of streptomycin. Tests showed that the Curbay BG was the constituent of the medium that stimulated production of Actidione. Since Curbay BG itself is a very complex substance, it would be interesting to know which of its ingredients promotes the formation of Actidione.

Shaken flasks were used for the fermentation at 24°C and assay for Actidione was made by the paper-disk method. Mutations (x-ray) that destroy the capacity of the organism to produce one of these antibiotics usually destroy the ability to produce the other also, but this is not invariably the case.[294]

Yields of Actidione average 80 to 100 mg per l, but in a few cases 200 to 250 mg per liter has been obtained. A crude product, 30 to 60% pure, was obtained from the beer by extraction with chloroform, adsorption on activated carbon, elution with acetone, and removal of acetone by distillation. Chromatography proved most convenient for isloation of the crystalline compound.[108]

Chemical Properties

An empirical formula, $C_{27}H_{42}N_2O_7$, was reported for Actidione by Leach, Ford, and Whiffen,[163] but further study[108] showed the formula to be $C_{15}H_{23}NO_4$. More detailed investigation[158,159] of its chemical nature, by chemical degradation and transformations and physical studies, established the structural formula:

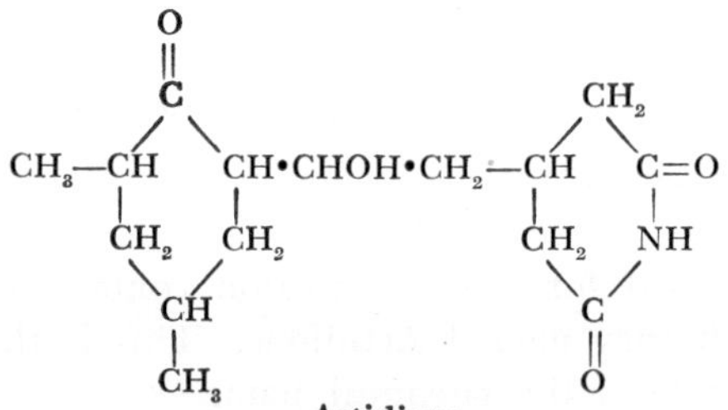

Actidione

From this formula, it is apparent that there is only one ketonic group instead of two, so that the name Actidione is actually a misnomer.

Crystalline Actidione has a melting point of 115° to 117°C. Its specific rotation is—3.0° in methanol and +6.8° in water. It is very soluble in the common organic solvents, except the saturated hydrocarbons. Its solubility in water is 2.1 g per 100 ml at 2°C.

Activity and Toxicity

Actidione has little activity against bacteria, but is very potent against a number of yeasts and some other fungi. It inhibits a fungal pathogen, *Cryptococcus neoformans,* and may prove of value in the treatment of the serious human infections caused by this organism.[127]

Its most promising use seems to be in combating fungal plant diseases, such as those infesting golf greens and other turfs.[10a,267] It is also useful in the laboratory as an aid to the detection and isolation of bacteria from samples contaminated with molds and yeasts. Addition of Actidione to the agar medium inhibits the growth of the fungi, but allows the bacteria to develop on the plates.[118a,122b] The toxicity of Actidione for animals varies greatly with the species, e.g., from $LD_{50} = 2.5$ mg per kg for rats intravenously to 150 mg per kg for mice by the same route. Its toxicity to plants likewise varies with the species.

Aureomycin (Chlorotetracycline)

Source

Aureomycin is produced by the Lederle Laboratories, Inc., a subdivision of American Cyanamid Co. Aureomycin is the trademark name and chlorotetracycline is the generic name. Tetracycline is the basic-structure name agreed on by the producers[247] of Aureomycin and Terramycin (oxytetracycline).

This antibiotic is produced by a streptomyces which was isolated from a soil sample collected from a timothy field in Missouri. It is described[49,81,82] as a new species and named *Streptomyces aureofaciens* because on agar "at a certain stage in the growth of the colony . . . there is typically the production of a golden yellow pigment in the moist hygrophorus substrate mycelium." Several different strains of the organism have been obtained and one of these, A-377, is described in a patent[82] covering the production of the antibiotic. A transfer of *S. aureofaciens,* strain A-377, has been deposited in the collection of the Northen Regional Research Laboratory, Peoria, Illinois, under the number NRRL-2209. The antibiotic is "faintly golden yellow" in color, which feature has given rise to the name Aureomycin.

Production

The original patent[82] states that *S. aureofaciens* can utilize many different kinds of carbohydrate and nitrogenous substances for its growth. As an example of flask fermentations the following is given: 100 ml of medium containing 1 to 3% sucrose, 1 to 2% corn-steep liquor, 0.5% $(NH_4)_2HPO_4$, 1.5% KH_2PO_4, 0.2% $MgSO_4$

$\cdot 7H_2O$, and traces of Mn, Cu, Zn, adjusted to pH 6.2, are placed in a 500-ml Erlenmeyer flask, sterilized, inoculated, and then shaken continuously at 28°C for 24 to 36 hours. From 500 to 2,000 Aureomycin units (usual range 1,000 to 1,500) are produced. A unit of Aureomycin is defined as the quantity of the antibiotic required to inhibit the growth of *Staphylococcus aureus* in 10 ml of a nutrient broth. Judging from later reports on the weight of pure Aureomycin required to inhibit assay strains of *S. aureus,* a unit is probably in the neighborhood of 0.5 μg per ml. The yields of Aureomycin given in the patent, if calculated to a weight basis, are in the order of 500 to 750 mg per liter. Judging from the improvement of penicillin cultures, the current yields could be much greater than the 500 to 750 mg figures. Van Dyck and De Somer[265a] reported yields of 1,300 mg per liter in shaken flasks for an improved culture and tank fermentations should be higher.

In deep-tank fermentations, the medium is somewhat more dilute and also contains 0.1% $CaCO_3$. During fermentation, the medium is kept at 26° to 28°C and aerated at the rate of 0.5 to 1.5 l of air per liter of medium for 24 to 48 hours. The pH of the medium may decrease to 4.5 during fermentation, but in a later patent,[197c] the drop in pH is said to be undesirable and is avoided by using up to 1% $CaCO_3$ to maintain the pH between 6.4 and 7. This patent emphasizes further the importance of calcium carbonate as an aid to increased Aureomycin production, because the calcium ion reacts with Aureomycin, as it is formed, to give a precipitate and thus promote the production of more Aureomycin. The yield reported, 396 μg per ml, seems distinctly low. The patents give no additional information as to the sequence of chemical changes in the medium and the relation of these changes to Aureomycin formation.

A recent biochemical study of the Aureomycin fermentation has been made by Van Dyck and De Somer.[265a] In this study, cultures selected from spores that had been exposed to ultraviolet light were employed. The spores were plated, colonies were examined for morphological characteristics (color, sporulation, and colony form) and those selected were then tested for Aureomycin production. From 1,457 colonies, 9 strains were obtained that gave around 1,300 μg Aureomycin per ml as compared with about 800 for the parent culture. The fermentations were done with 50 ml medium in Erlenmeyer flasks on a rotary shaker at 250 rpm, for about 80 hours at 24° to 30°C. The medium was essentially that given in the Aureomycin patent, but with added peanut meal and molasses. The chemical changes in the medium were as follows: The sucrose content fell rapidly from 3% at the start to about 0.1% after 50 hours. Formation of Aureomycin became marked after 15 hours and continued uniformly for about 35 hours and then leveled off. During this time, about 900 μg per ml of Aureomycin was formed or about 25 μg per ml per hr. This rate compares favorably with that of other antibiotic fermentations. With the exhaustion of the sucrose supply, there was little additional formation of Aureomycin. The pH of the medium remained fairly constant at 6.4 to 6.6 during the first 50 hours and then rose rapidly to 7.1 at 70 hours. The rise in pH and cessation of antibiotic production with exhaustion of sugar follow the familiar pattern observed in penicillin fermentations.

Pidacks and Starbird[209a] describe the recovery of Aureomycin from the broth by chromatographing on a silicate column with an acidulated solvent, collecting the fraction that is yellow under ultraviolet light, concentrating and extracting this with butanol, and finally precipitating the Aureomycin as the hydrochloride with ether.

A later patent[4d] gives a recovery procedure that seems better adapted to large volumes of fermentation broth. The Aureomycin precipitated with lime at pH 8.8 is then extracted from the filter cake with sulfuric acid at pH 1.35. The filtrate is evaporated in vacuum at 30°C to about one fourth its volume, a small quantity of butanol is added, and the solution nearly saturated with sodium chloride. On standing overnight at 2° to 5°C, the Aureomycin crystallizes as the hydrochloride of about 97% purity.

CHEMICAL NATURE AND PROPERTIES[49,82,282]

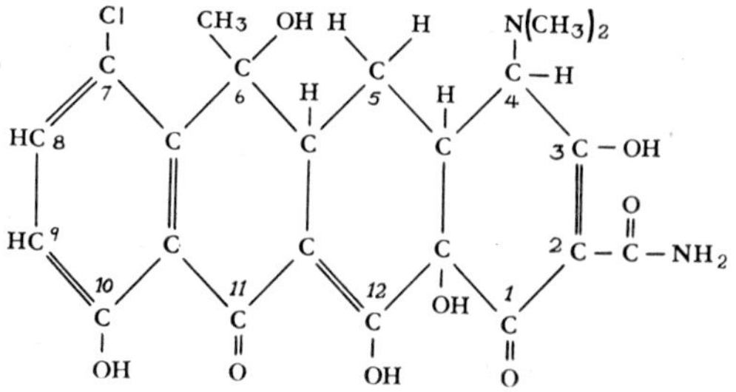

Unraveling the structures of Aureomycin and Terramycin is one of the recent brilliant achievements in organic chemistry. This successful work resulted from the efforts of two large groups of chemists working separately on the respective antibiotics. To date seven papers have appeared on Aureomycin and eight on Terramycin. Only the last paper in each series will be cited, as reference to one or more of the preceding communications is made in the last paper of each series.

The structures of Aureomycin and Terramycin are very similar. Both contain a naphthacene nucleus and all but two of the modifying groups are alike in both antibiotics. Compared with Aureomycin, Terramycin has a hydroxy group at position 5 instead of a hydrogen and a hydrogen at position 7 instead of a chlorine atom. In effect, it has a hydroxyl instead of a chlorine but, oddly enough, the two occupy different places in the naphthacene skeleton. The other hydroxyl, amide, carbonyl, and methyl groups are the same in number and position in both antibiotics. The basic structure common to both Aureomycin and Terramycin is called tetracycline and thus Aureomycin is named chlorotetracycline and Terramycin is designated oxytetracycline.

Aureomycin is a weak base that contains both nitrogen and nonionic chlorine. The free base has a melting point of 168° to 169°C, a levo rotation $[\alpha]_D^{23°} = -275°$ (in methanol), and a solubility in water of 0.5 to 0.6 mg per ml. It is much more soluble in dilute acid or alkali. Its solubility in organic solvents is reported as follows: very soluble in pyridine, Cellosolves, and dioxane; slightly soluble in methanol, butanol, and ethyl acetate; insoluble in ether and petroleum ether. Aureomycin forms a crystalline hydrochloride containing one equivalent of hydrochloric acid per mole of base. In 0.1 N hydrochloric acid, three absorption maxima occur at 230, 262.5, and 367.5 mμ. In sodium hydroxide, the peaks are all shifted to longer wave lengths.

Systematic study has been reported of the stability of Aureomycin in connection with its assay and tests against bacteria.[79,212] These tests show that it is a rather unstable compound which is surprising considering its effectiveness. Even in distilled water, about 50% of the activity was lost in 24 hours at 37°C. In phosphate buffer pH 7.0 or in the presence of peptone, blood serum, and many other substances, deterioration was increased. Acidity (pH 2) and low temperature (4°C) increased the stability.

ACTIVITY

Aureomycin is an unusually versatile antibiotic; it is active *in vitro* against numerous Gram-negative and Gram-positive bacteria and appears particularly potent against the spore-forming aerobes.[37c,202,212] Because it acts on many Gram-positive and Gram-negative bacteria as well as rickettsiae and certain viruses, it is called a "broad spectrum" antibiotic. Chloromycetin and Terramycin also belong in this category. Although Aureomycin is not as effective as penicillin against infections caused by Gram-positive bacteria, its use in such infections has given good results especially in cases that have become resistant to penicillin. Aureomycin appears to be very effective against rickettsial diseases, such as Rocky Mountain spotted fever, scrub typhus fever and typhus fever[98,303] and less effective against Q fever.[166] Notable results have been obtained in the treatment of psittacosis, lymphogranuloma venereum, granuloma inguinale, and virus pneumonia.[303,306] Somewhat promising results have been reported against brucellosis,[75,123a,242,303] amebiasis,[182b] and whooping cough.[24,128e] Aureomycin is apparently ineffective against tuberculosis and also against infections caused by proteus- and pseudomonas-type bacteria.

Microorganisms that develop resistance to Aureomycin are also resistant to Terramycin and vice versa.[203b] This result is to be expected since both antibiotics have the same basic chemical structure. Some of these resistant organisms are also resistant to Chloromycetin.[111b] This effect has been explained on the assumption that all three antibiotics act on microorganisms in the same way. Prolonged ingestion of Aureomycin, Chloromycetin, and Terramycin may lead to marked changes in the fecal flora and to the development of secondary fungal infections in the mouth and gastro-intestinal tract.[201a,203b] This infection is called moniliasis as the organisms causing it are classed as monilia (old name for candida). An example is *Candida albicans.* Such organisms are not inhibited by the broad-spectrum antibiotics. Therefore, when the normal inhabitants of the intestinal tract (the colon-aerogenes bacteria) are suppressed, the monilia have a good opportunity to develop.

TOXICITY

The intravenous LD_{50} for mice is 134 mg per kg.[128] In clinical work, it is usually given by mouth and in some cases nausea and diarrhea have been experienced, but no consistently serious effects have been observed.[98,303]

Bacitracin

Source

Bacitracin is produced by a strain of *Bacillus licheniformis,*[184] formerly classified as *Bacillus subtilis,* strain Tracy. It was isolated from infected tissue removed at the time of surgical operation from the fractured tibia of a 7-year-old child, named Margaret Tracy. The first announcement of the isolation of the organism and naming of the antibiotic was made in 1945 by Johnson, Anker, and Meleney.[147] Since that time, many papers have been published regarding bacitracin and in 1948, it was produced semicommercially and put on the market by Commercial Solvents Corporation.[10] The new bacitracin plant of the company began operating late in 1951.[143a]

A unit of bacitracin is defined as the amount of the antibiotic which when dissolved in 1024 ml of beef infusion broth completely inhibits the growth of a group A hemolytic streptococcus under specified conditions of inoculum and incubation.

Bacitracin, as commercially produced, is a light-gray, hygroscopic powder of bitter taste. Its purity is approximately 45 units per mg.[143a] The purest preparation of bacitracin obtained to date had a potency of 66 units per mg.[70] One unit is, therefore, equivalent to about 15 μg. This weight per unit is very large in comparison with penicillin (one unit equals 0.6 μg) and streptomycin (one unit equals 1.0 μg). Comparison of antibiotics on the basis of units is meaningless and even misleading. An antibiotic similar to bacitracin was obtained by a group of English workers[197] from a different strain of *B. licheniformis.* Originally this antibiotic was called ayfivin, but after careful purification and comparison with bacitracin, it was decided that several of the constituent polypeptides in bacitracin and ayfivin were identical and the name ayfivin has, therefore, been abandoned.

Production

Various media have been used in flask and pilot-plant fermentations.[8,175] One of the most important constituents for high yields is a plant protein, such as soybean or peanut meal. Yields up to

100 units per ml, or about 1.5 mg per ml on a weight basis, have been obtained on such media. This yield is in about the same range as that for penicillin and Aureomycin.

Production on a plant scale has recently been reported[143a] in a new plant built and operated by the Commercial Solvents Corporation. The layout of the plant and a description of its operation from stock culture to finished product are given. A summary of the process will be listed here, but for details, the original report should be consulted. A flow sheet is shown in Figure 21.

The medium used in plant production differs somewhat from that used in the laboratory. Because of the greater aeration in tanks, the bacteria grow more rapidly and bring about more extensive deamination of proteins with accompanying rise in the pH of the medium. To offset this effect, the medium is balanced with carbohydrate. When metabolized, this reduces deamination and also yields acids which counteract the alkaline products resulting from the metabolism of proteins. Dextrose, sucrose, and starch are suitable carbohydrates, but maltose gives low bacitracin titers. From a suitable medium, Lee[165a] reports titers of more than 100 units per ml in 36 to 48 hours. This is a shorter period than that required for mold and streptomyces fermentations.

The fermentation process starts with spores of *B. licheniformis* (*B. subtilis*) kept in sterile soil. These are inoculated into a 4-l flask of peptone or tryptone broth. The flask is incubated at 37°C and shaken continuously for 18 to 24 hours. The contents of the flask are then used to inoculate media in a 150-gal stainless-steel tank. The culture is aerated for 6 hours and is then blown by sterile air into a 1,200-gal seed tank containing previously cooked mash cooled to 37°C. The seed tank is aerated and when proper growth has been attained, the seed is used to inoculate a 24,000-gal fermentor. During fermentation, sterile air is forced through the mash, but apparently, simultaneous agitation (used in most antibiotic fermentations) is not employed. As the fermentation proceeds, the pH of the medium rises gradually. Formation of bacitracin begins about midway in the period, presumably at about 18 hours, and increases rapidly to the end of the fermentation. Unfortunately, little information is given on the chemical changes that take place during the fermentation. In justice to the writers, it should be said that this criticism applies not only to the bacitracin fermentation, but also to those for Aureomycin, Terramycin,

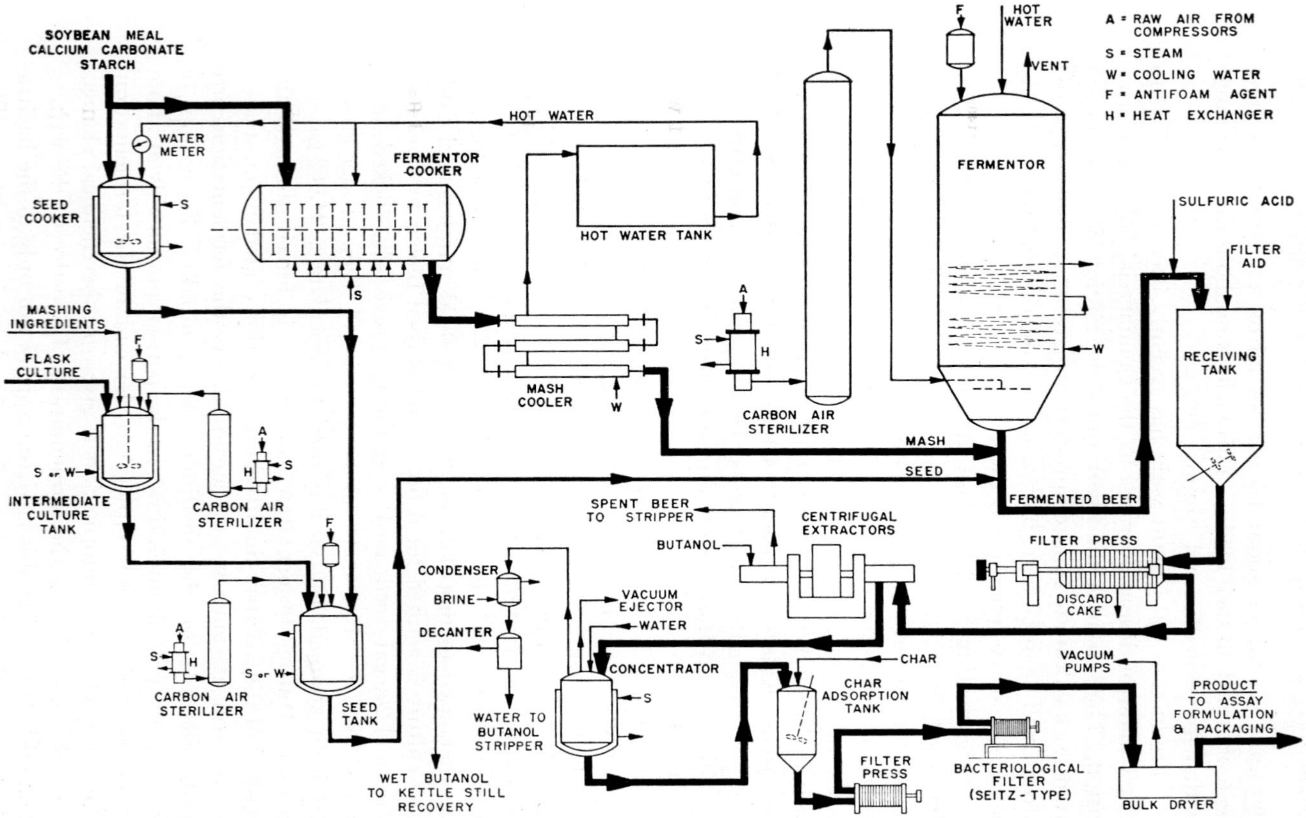

FIGURE 21. *Flow Diagram for Production of Bacitracin* [Reprinted by permission from a paper by Inskeep, Bennett, Dudley, and Shepard, *Ind. Eng. Chem.*, **43**, 1488 (1951)]

and other commercial antibiotics. Such information, the reviewers believe, could be given without disclosing the exact composition of the medium and would add greatly to our knowledge of the metabolism of these important industrial microorganisms.

Bacitracin is recovered from the fermented medium by acidifying the broth, filtering it to remove cells and other solids, and extracting the filtrate with butanol. The solvent is removed in a vacuum below 28°C, with simultaneous addition of water to keep the bacitracin in solution. The aqueous concentrate is further purified with activated charcoal to remove color and other impurities. It is then passed through a Seitz filter to remove any chance bacteria, and finally it is freeze-dried. The light-gray, hygroscopic powder thus obtained is the product of commerce and contains about 45 units of bacitracin per mg. If 60 units per mg is taken as the potency of pure bacitracin, the product has a purity of about 75%.

CHEMICAL NATURE AND PROPERTIES[8,20,70,70a,70b,143a,175,197,210a]

Bacitracin appears to be a mixture of polypeptides, contains about 14 to 15% nitrogen, and gives positive biuret, Hopkins-Cole, and xanthoproteic tests, but negative Millon and Sakaguchi reactions. It is precipitated from solution by many of the common protein reagents, such as heavy metals, trichloroacetic acid, phosphotungstic acid, and even by saturation with sodium chloride.

The commercial product appears to consist of several antibacterial polypeptides designated bacitracin A, B, C, D, E, and F. The A component makes up the main part of the product with varying amounts of minor components. Three of these proved identical with components found in another antibiotic, called ayfivin, but this name is now abandoned in favor of bacitracin.

Commercial bacitracin has been highly purified by countercurrent distribution and the A component obtained free from other components. It has been hydrolyzed with acid and the amino acids determined by ion-exchange-column chromatography. One mole each of L-cysteine, L-histidine, L-leucine, L-lysine, D-phenylalanine, D-glutamic acid and D-ornithine and 2 moles each of DL-aspartic acid and L-isoleucine were found. However, only 90.7% of the total nitrogen and 87.4% of the weight of peptide taken were recovered in the amino acids. The same amino acid composition was reported for bacitracin purified by carrier-displacement chromatography. The molecular weight of bacitracin A obtained by direct determination, 1411 to 1450, is in good agreement with that calculated from the amino acid content.

Bacitracin is soluble in water, methyl and ethyl alcohols, glacial acetic acid, and formamide; slightly soluble in *n*-propyl and other alcohols; and insoluble in ether, chloroform, and several other organic solvents.

Bacitracin is adsorbed from aqueous solution; for example, by charcoal, Super Filtrol, but not by activated alumina. The ultraviolet absorption curve shows a small peak at 253 mμ.

Bacitracin is a very stable antibiotic. In water at 100°C, 85% of the activity remains after 30 minutes and at refrigerator temperatures, full activity is retained for about 3 weeks. It is most stable at pH 4 to 5. At pH 1.6 at 37°C, about 60% of the activity is lost in 24 hours. It is not inactivated by pepsin or trypsin, but seems somewhat labile to papain. It is inactivated by ultraviolet light.

Activity and Toxicity[10,17,76,89,90,145b,147,175,183,235]

Bacitracin resembles penicillin in its bacterial spectrum; it is most active against Gram-positive bacteria. However, many organisms in this group that are resistant to penicillin are sensitive to bacitracin. When used together, the two antibiotics appear to be synergistic, but in other combinations an antagonistic effect has been noted.

The LD_{50} for mice was 240 mg per kg for preparations having an activity of 35 units per mg where this was administered intraperitoneally. It brings about serious kidney damage in mice and most patients given bacitracin show more or less albumin in the urine. For this reason it should not be injected and its use is generally restricted to topical application or to local infiltration into wounds and pus-forming lesions. It is reported to have given good results in many surgical cases and in the treatment of various skin diseases.

Chloromycetin (Chloramphenicol)

Source

Chloromycetin is the trade name and chloramphenicol is the generic name given to this antibiotic by the producer, Parke, Davis and Co.

The actinomycete which produces Chloromycetin is a new

species[91,92] named *Streptomyces venezuelae* and since the compound contains chlorine, it is quite evident how the word Chloromycetin was devised. Three cultures from widely separated geographical regions have been isolated and found to produce Chloromycetin. One of these[91] came from a soil sample collected near Caracas, Venezuela, another[58,122] from a compost soil on the farm of the Illinois Agricultural Experiment Station at Urbana, and a third from Japan.[264] According to Umezawa, Tazaki, and Fukuyama[263] Chloromycetin-producing streptomyces are widely distributed in Japanese soils and cultures can be readily isolated by adding 0.025 to 0.05 mg of Chloromycetin per ml to the agar plate before inoculating with soil samples. The cultures from Venezuela and Illinois have been carefully compared and appear to be identical as judged by morphology, nutrition, antibiotic production, and other tests.

Production

In preliminary trials,[92] the Venezuela culture gave slightly higher yields than the Illinois culture and was the organism used in industrial production.[226] In laboratory fermentations,[240] the best yield (169 μg per ml) was obtained on a medium containing 1.0% glycerol, 0.5% B-Y fermentation solubles (Commercial Solvents Corp.), 0.5% tryptone (Difco), and 0.5% NaCl. Glycerol seems to be superior to sugars as a carbohydrate source and animal products, *e.g.*, hog stomach residue, and hydrolyzed soybean protein are excellent sources of nitrogen. In 2,000-gal vat fermentations, 89 μg per ml has been reported.[93] The Japanese investigators, Umezawa and associates,[260,261,307] have published a number of papers dealing with the production of high-yielding mutants by treatment of the parent culture with ultraviolet light and improvement of the medium by addition of various nutrients. Their best culture, A33-115, which is also nearly free of pigment, gave yields of 440 μg per ml in flask fermentations and 300 μg per ml in 150-l batches. The medium contained 2% starch, 0.5% beef extract, 0.5% peptone, and 1% NaCl. A study of the effect of amino acids on Chloromycetin production showed that phenylalanine, tyrosine, and methionine used singly greatly increased the yield. The collective effect of these amino acids was even greater. For example, 480 μg per ml was obtained with all three amino acids when 0.01% of each was added to the previously described natural medium. Several possible precursors of Chloromycetin were tested, but only dichloroacetic acid gave any response and the results with this compound were irregular.

FIGURE 21a. *Fermentors Used for Production of Antibiotics* (Courtesy—Parke, Davis & Company)

Data on the chemical changes brought about in the medium have been reported for flask fermentations.[201] Production of mycelium and Chloromycetin run parallel and reach a maximum of 5 mg and 74 μg per ml respectively in about 3 days. The parallelism between mycelium and Chloromycetin production is quite different from the relationships found in penicillin and streptomycin fermentations, in which antibiotic production is not marked until after mycelial growth is well developed. In the Chloromycetin fermentation, glycerol was used up and ammonia nitrogen was at its lowest level after 3 days. Autolysis then set in as was evidenced by decrease in weight of mycelium and increase in ammonia nitrogen. There were no marked changes in the pH of the medium. In a fermentation started at 7.4, the pH of the medium decreased to 7.0 and then rose slowly to 8.2 at the end. In an earlier paper,[240] the pH of the medium was given as about 6 or below throughout the fermentation.

Apparently, Chloromycetin can be formed both in acid and slightly alkaline media. An interesting feature of the fermentation was the appearance of an unknown reducing substance simultaneously with the rapid utilization of glycerol.[201] About 70 mg of reducing substance per 100 ml, calculated as glucose, accumulated during the second day of fermentation and disappeared by the end of the third day. During the second day, more than 500 mg of glycerol was used per 100 ml.

A somewhat different sequence of events is described in a later paper by Legator and Gottlieb.[165b] They found that maximum growth was reached about 48 hours before the peak yield of Chloromycetin. An interesting feature of their work was the effect on biosynthesis of adding Chloromycetin to the medium. The same level of Chloromycetin, about 100 μg per ml, was reached with or without additions. In other words, added Chloromycetin merely prevented the formation of equivalent amounts of Chloromycetin. Confirmation of these unexpected results under more varied conditions and with higher yielding strains would seem desirable and worth while. For comparison, the authors made similar experiments with the Terramycin producer, *S. rimosus,* but additions did not prevent the normal synthesis of Terramycin.

Chloromycetin can be obtained readily from the broth[21,22] by filtering off the mycelium and solids of the medium with the aid of Super Filtrol and Hyflo Super-Cel and extracting the antibiotic from the filtrate at pH 8.5 to 9.0 with ethyl acetate. The extract is concentrated until it contains 20 mg of Chloromycetin per ml, a quarter volume of kerosene is added to keep impurities in solution, and, after being washed with dilute acid and alkali, the extract is dried over sodium sulfate. The solvent is partially removed by vacuum distillation and, on standing at 5°C, Chloromycetin crystallizes out. The crystals are purified by a second crystallization from ethylene dichloride and then melt at 149.2° to 150.2°C. Recovery, from broth to final crystals, is about 70%. Second crops of crystals have raised the recovery to over 80%.

Chloromycetin has been synthesized in several ways. An outline of one of these syntheses is as follows: Condensation of benzaldehyde and nitroethanol is followed by reduction of the condensation product to the amino compound, $C_6H_5 \cdot CHOH \cdot CHNH_2 \cdot CH_2OH$; acetylation of the amino and primary alcohol groups gives $C_6H_5CHOH \cdot CH(NHAc)CH_2OAc$; separation of the

DL-erythro form by crystallization, acetylation of the secondary alcohol group, and nitration to $O_2N{\cdot}C_6H_4{\cdot}CH(O{\cdot}Ac){\cdot}CH(NHAc)CH_2OAc$ are followed by hydrolysis, resolution of the base via D-camphorsulfonic acid to $O_2N{\cdot}C_6H_4{\cdot}CHOH{\cdot}CHNH_2{\cdot}CH_2OH$, and condensation with methyl dichloroacetate to give Chloromycetin, $O_2N{\cdot}C_6H_4{\cdot}CHOH{\cdot}CH(NH{\cdot}CO{\cdot}CHCl_2)CH_2OH$. For details of the many steps involved in the syntheses, the reader is referred to the original papers[66,169] and patents.[72]

FIGURE 21b. *Vacuum Evaporators Used for the Concentration of Crude Chloromycetin Extract* (Courtesy—Parke, Davis & Company)

Chloromycetin has the unique distinction among antibiotics of being produced both by fermentation and by chemical synthesis.

At present, it is believed that only 20% or less is produced by fermentation.[165b] It is possible that synthesis will ultimately displace the fermentation process.

Some of the plant equipment used in the fermentative production of Chloromycetin is shown in Figures 21a and 21b.

Formula and Properties[22,66,72,169,222]

The molecular formula for Chloromycetin is $C_{11}H_{12}O_5N_2Cl_2$, and the structural formula:

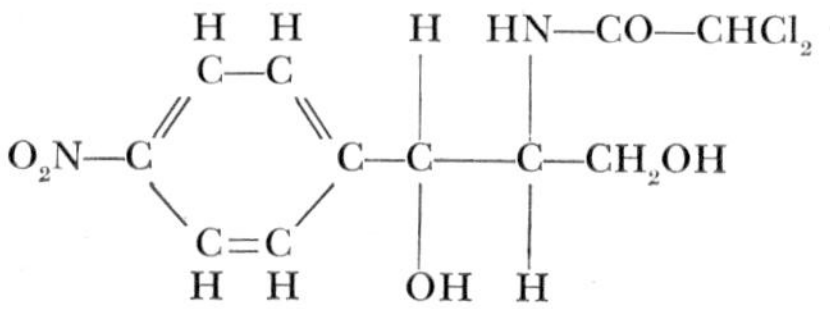

Chloromycetin (chloramphenicol)

The structural designation for Chloromycetin is D-(—)*threo*-2-dichloroacetamido-1-*p*-nitrophenyl-1,3-propanediol, which is the basis for the shorter term chloramphenicol.

Probably the most distinctive feature of its structure is the nitro group; it is, as far as the reviewers know, the first organic compound of biological origin that contains a nitro group. It is not unthinkable, of course, that organic nitro compounds exist, temporarily at least, in nitrification processes. The presence of chlorine is also noteworthy, although a number of mold products containing nonionic chlorine have already been isolated. Since it has two asymmetric carbons, there are three other isomers, but none of these has significant antibiotic activity.

Chloromycetin is a neutral compound that crystallizes as colorless needles or elongated plates and melts sharply at 150.1°C. It is optically active, $[\alpha]_D^{25} = -25.5°$ in ethyl acetate, and in water, it has a characteristic single absorption maximum at 278 mμ. It is not very soluble in water, 0.25 g per 100 ml, but is readily dissolved in methanol, ethanol, butanol, acetone, ethyl acetate, and several other organic solvents.

Activity[4c,64,106a,168a,180,240,304]

Chloromycetin is relatively inactive against Gram-positive bacteria, e.g., *S. aureus,* but is very active against Gram-negative bacteria, e.g., *E. coli,* and thus attacks many of the bacteria not affected by penicillin. It is very potent against bacteria producing certain intestinal diseases, e.g., typhoid and dysentery, and has given some promise in the treatment of brucellosis in humans. Its most notable characteristic, however, is its potency against certain rickettsial diseases, e.g., epidemic typhus, scrub typhus, murine typhus, and Rocky Mountain spotted fever. It has antiviral activity against psittacosis and lymphogranuloma, but is ineffective against a number of other virus diseases. It is only moderately active against the human tubercle bacillus and is inactive against actinomycetes, yeasts, fungi, and the malaria parasite. Cross resistance to Aureomycin and Terramycin has been noted in some organisms resistant to Chloromycetin.

TOXICITY AND METABOLISM[91,106b,119,125,168a,180,238,239,240]

Chloromycetin has a rather low toxicity. The intravenous LD_{50} for mice is about 245 mg per kg of body weight. Oral administration of 200 mg per kg a day to dogs over a period of 4 months had no cumulative toxic effect on organs or tissues.

Certain blood disorders (e.g., aplastic anemia, in which the bone marrow fails to produce red blood corpuscles with possible fatal results) have been observed in connection with the prolonged use of Chloromycetin. This has prompted the Food and Drug Administration to require a warning to be printed on the label as follows: "Blood dyscrasias may be associated with intermittent or prolonged use. It is essential that adequate blood studies be made." The nitrobenzene structure in Chloromycetin has been regarded by some as the possible cause of this toxicity, but there is as yet no agreement on this point. Approximately 90% of the administered dose is excreted as an inactive nitro compound and 10% as unchanged Chloromycetin in the urine in 24 hours. Liver and kidney tissues form inactive nitro compounds from Chloromycetin and reduce the nitro group to give aryl amines. An enzyme preparation has been obtained from cells of *Proteus vulgaris* and *Bacillus subtilis* that hydrolyzes the amide linkage, thus liberating the corresponding amine and dichloroacetic acid. Both of these products were isolated from the enzymic digest and identified by chemical means.

Circulin

PRODUCTION

This antibiotic is produced by an aerobic spore former, designated as *Bacillus circulans;* this is the basis for the name circulin.[191,192] Production has been studied in vessels ranging from shaken flasks to 2,000-gal tanks. In a dextrin-yeast-ammonium sulfate-salts medium, about 2,100 units per ml were obtained in 3 days in flask fermentations.[192] An improved medium containing dextrin, ammonium sulfate, rolled oats, and salts gave 5,000 to 6,000 units (about 1 mg) per ml in 100-gal fermentors under optimal conditions of aeration.[194] Since the pure circulin sulfate has an activity of about 6,000 units per mg, the circulin content of the broth in 100-gal fermentations was about 1 g per l, which is a high yield.

Circulin is obtained from the broth by adsorption on Darco G-60 at pH 6.5 and elution with 25% aqueous tertiary butyl alcohol at pH 2.5 to 4.0. Chromatography on Darco G-60 with tertiary butyl alcohol has given fractions analyzing 5,500 to 6,500 units per mg.[207] Pilot-plant runs of the antibiotic have been processed by precipitation with a sulfonated type of wetting agent, extraction with acidified acetone, and a second precipitation at higher concentrations of acetone. After solution in water, the product was lyophilized from the frozen state and gave a powder of about 65% purity. The recovery from beer to powder[194] was about 44%.

Chemical Properties

Circulin is a basic polypeptide that has been isolated as the sulfate, hydrochloride, picrate, helianthate and reineckate. The sulfate is a white, amorphous solid that melts at 226° to 228°C, is levorotatory with a specific rotation of —61.6°, and by paper chromatography behaves as a homogeneous substance. On hydrolysis, it yields D-leucine (10%), L-threonine (12%), L-α,γ-diaminobutyric acid (55 to 65%), and an optically active fatty acid, isomeric with pelargonic acid (10%). From these data, circulin appears to contain one molecule each of leucine, threonine, and fatty acid and five molecules of α,γ-diaminobutyric acid; it thus would have a minimum molecular weight of 854. It does not contain any free carboxyl groups and is apparently cyclic in character. The γ-amino groups of the diaminobutyric acid are all free and impart to the peptide its basic property. It is slowly inactivated by lipase (80% in 5 days); therefore, the fatty acid is believed to be attached through the hydroxyl group of threonine. (In the polymyxins, the fatty-acid residue is assumed to be linked to one of the five γ-amino groups of the diamino acid as an acyl derivative.)

Since circulin bears such a striking resemblance to the polymyxins, a comparison was made with respect to the action of lipase on circulin and on polymyxins A, B, and D. Only polymyxin B was inactivated by lipase and then more slowly than circulin. Polymyxin A, which contains the same amino acids as circulin, remained stable to lipase.[207] Because of its strong resemblance to the polymyxins, the question naturally arises as to whether circulin should not be called a polymyxin.

Activity and Toxicity

Circulin, like the polymyxins, is strongly active against most Gram-negative bacteria and relatively inactive against Gram-positive microorganisms. An exception to Gram-negative activity is reported by Bliss and Todd,[37] who found that circulin and polymyxins B and D had little or no activity against their strains of *Proteus*. Circulin and polymyxin B were about equally effective in protecting mice against infection with *Klebsiella pneumoniae* and both were much more potent than streptomycin, Aureomycin, or Chloromycetin.[37] Circulin is absorbed from the digestive tract in dogs, but not in mice.[265]

The LD_{50} dose of circulin for mice is 10 mg per kg by intravenous administration and 77 mg per kg by the subcutaneous route. It is much more toxic than polymyxin D, but seems to be less injurious to the kidneys.

Erythromycin

From broths of *Streptomyces erythreus* fermentations, research workers of Eli Lilly and Company have obtained a new antibiotic which acts on many Gram-positive bacteria, the typhus rickettsiae, some large viruses, and a few Gram-negative microorganisms. It thus covers the range of penicillin and part of that of the "broad spectrum" antibiotics. It is being produced commercially by Eli Lilly and Company under the trade name "Ilotycin,"[182a] by Abbott Laboratories as "Erythrocin,"[10b] and by the Upjohn Company as erythromycin, which is its generic name.

It can be produced by submerged fermentation in a soybean-corn steep-glucose medium maintained at 26° to 30°C. Yields of about 350 μg per ml have been obtained in a typical fermentation in 100 hours. Erythromycin production began at about 20 hours and continued uniformly to the end of the fermentation. The glucose concentration fell steadily during the first 60 hours and then leveled off. The pH of the medium rose from 6.4 at the start to about 7.2 in 90 hours. Erythromycin is obtained from the filtered broth by extraction at pH 9.0 with amyl acetate, transferring the antibiotic into water at pH 4.0, and concentration of the aqueous solution to about 10 mg of erythromycin per ml. If the aqueous solution is adjusted to pH 9.0, the free base separates in crystalline form. It can be recrystallized from a water-acetone (2:1) solution.[1,53a] A second clinically useful antibiotic, erythromycin B has been found in the broth by Pettinga, et al.[208a] A transfer of the culture has been deposited with the Northern Regional Research Laboratory, Peoria, Illinois, under the number NRRL 2338.[53a]

Chemical Properties and Composition[10b,53a,64a,106a,128c,182a]

Erythromycin is a basic compound which readily forms salts with acids. It has been used in the form of the crystalline base and as the hydrochloride. It is extremely soluble in alcohols, acetone, chloroform, and ethyl acetate; moderately soluble in ether; and less soluble in water. It is stable in solution at pH 6 to 8, but below or above this range, it loses its activity in 24 hours.

It is levorotatory with a specific rotation of —78° and exhibits a single weak broad absorption band at 280 mμ in the ultraviolet spectrum. From the elementary composition, an empirical formula $C_{37}H_{65}NO_{13}$ has been calculated. The molecular weight calculated from this formula, 732, appears probable by comparison with that obtained from titration data, 725. On acid degradation, erythromycin yields propionaldehyde, propionic acid, an amino sugar, desosamine (3-dimethylamino-4-desoxy-5-methyl aldopentose, $C_8H_{17}NO_3$), a second sugar, cladnose ($C_8H_{16}O_4$) and other unidentified products.

Activity[1,10b,126a,129a,148a,179c,210b]

Erythromycin has been shown to be active against Gram-positive bacteria and to some extent against a few Gram-negatives. It also has potency against mycobacteria, typhus rickettsiae, and against some of the larger viruses. It is not effective against *E. coli, A. aerogenes,* and members of the proteus, salmonella, and shigella groups of bacteria, nor against yeasts and molds. It is active against many organisms that have become resistant to penicillin and streptomycin, though such organisms may become resistant to erythromycin also. The situation seems to be a race between development of resistance by bacteria and development of new antibiotics by man. Erythromycin is administered by mouth in doses of 200 to 500 mg at 4 to 6 hour intervals.

Toxicity[7a,10b]

Careful testing preliminary to clinical use indicates a low toxicity for erythromycin. The LD_{50} for mice, rats, rabbits, and hamsters was above 3,000 mg per kg in oral administration, and for mice, above 2,500 mg per kg by subcutaneous injection.

In the seventy cases treated to date of writing side effects were rare and seemed to occur where the single doses (0.5 g) had been exceeded. Nausea, vomiting, and diarrhea were observed in occasional patients. Producers, however, give warning of the necessity of watching carefully for side effects of this new drug in cases where continued or repeated courses of erythromycin are administered.

Fumagillin

An unusual antibiotic, which when first discovered[127c] in 1949 was notable only for its activity against the bacteriophage of *Staphylococcus aureus,* but which has since then been raised to considerable importance when it was found by McCowen et al.[179b,179d] that it is a potent amoebicide. This antibiotic was called "H_3" when first discovered, because it was obtained from *Aspergillus fumigatus* H_3; but it was later given the name "fumagillin," under which designation it is being produced commercially. Fumagillin is apparently identical with an antibiotic reported under the odd name of phagopedin sigma. It was obtained from a mold later identified as *Aspergillus fumigatus.*[16a]

Production

Pilot-plant production of the antibiotic has been described by Eble and Hanson.[90a] Fifteen hundred gallons of dextrin-corn steep medium were seeded with a 5% volume of a 48-hour vegetative culture of *A. fumigatus* H_3 that had been grown in the same medium. After fermentation for 42 hours, the beer assayed 112 μg per ml of fumagillin. It was filtered with Dicalite, defatted with Skellysolve

B, extracted with chloroform, and the extract concentrated to dryness. The residual sirup was dissolved in acetone. Of the 1062 g of solids contained in this solution, approximately 210 g were fumagillin.

Crystalline fumagillin was obtained from the solution by concentrating to a thick suspension at room temperature under a nitrogen atmosphere, cooling the suspension under nitrogen for 18 hours at $-30°C$, centrifuging, and washing the crystalline substance five times with a total of 7,590 ml of tertiary butyl alcohol. It was then recrystallized from 3,000 ml methanol and 800 ml water, and 19.8 g of light-yellow crystals obtained.

CHEMICAL NATURE AND PROPERTIES

Eble and Hanson[90a] have determined the elementary composition (C,H,O,); melting point 189° to 194°C; molecular weight 436; optical rotation $[\alpha]_D^{25} = -26.6°$ (*c*, 0.25 in methanol); ultraviolet and infrared spectra of the compound and the chemical properties of some of its derivatives, e.g., methyl ester, bromine addition product, and 2,4-dinitrophenylhydrazone. From all these data, they reached the tentative conclusion that it is a monobasic acid with four conjugated double bonds, the remaining oxygen being partly accounted for as alkoxyl, dicarboxyl, or α-hydroxycarbonyl groups. The empirical formula proposed is $C_{27}H_{36}O_7$. Schenck et al.[233a] have obtained decatetraenedioic acid, $HOOC-(CH{=}CH)_4COOH$, from fumagillin by alkaline degradation and conclude that the antibiotic is a monoester of this acid, $[C_{16-17}H_{25-27}O_3]-O-CO-(CH{=}CH)_4COOH$.

ACTIVITY

Discovered as active against the bacteriophage of *S. aureus*, it was tested for antibacterial, antifungal, and antiviral potencies and had none of these. It sprang into considerable importance as an antibiotic when it was shown to be a potent amoebacide. McCowen et al.[179a] cleared rats of *Endamoeba histolytica* in 2 days by a total dosage of 11 mg per kg, and Hrenoff and Nakamura[140a] cured monkeys of amoebiasis by dosages of 50 to 125 mg per kg a day. This activity seems to be specific for amoebae; the antibiotic had no effect on trypanosomes, spirochetes, and other parasites tested.

TOXICITY

The toxicity of fumagillin is low. Eble and Hanson[90a] found the LD_{50} for mice to be 800 mg per kg in subcutaneous administration, while 2,000 mg per kg were tolerated orally. Hrenoff and Nakamura[140a] found no toxicity in their experiments with monkeys, although bromosulfein tests indicated that a dosage of 125 mg per kg a day might cause liver damage.

Magnamycin (Carbomycin)

Magnamycin is the trade-mark name and carbomycin the generic name of a new antibiotic, put on the market by Chas. Pfizer and Co., Inc.[256a] It is produced by a streptomyces called *Streptomyces halstedii* and can be obtained from the clarified broth (100 μg Magnamycin per ml) by extraction with methyl isobutyl ketone. After concentrating the extract *in vacuo,* the antibiotic is transferred into water by extracting at pH 2.0. The aqueous solution is washed with benzene, adjusted to pH 6.5, and then extracted with ether. The ether phase is dried and evaporated to give crystalline Magnamycin. A second crystallization from 50% methanol gave a product of 99.2% purity.

Chemical Properties[88a,222a,267d]

Magnamycin is a monobasic compound, having the tentative empirical formula $C_{41-42}H_{67-69}NO_{16}$ and a molecular weight of approximately 860. The homogeneity of the base has been established by solubility analysis, countercurrent distribution, and paper chromatography. It is levorotatory, $[\alpha]_D^{25} = -58.6°$ (*c,* 1% in chloroform), and melts with decomposition at 212° to 214°C. The base is readily soluble in organic solvents, but is almost insoluble in water and hexane. It forms stable, crystalline salts of which the hydrochloride is water soluble. The ultraviolet absorption spectrum shows a high, broad peak at 238mμ and a low, weak peak at 327 mμ.

Alkaline degradation has given 1 mole each of acetic acid, isovaleric acid, and dimethylamine. Mild acid hydrolysis yields the isovaleric acid ester of a sugar and an unidentified crystalline substance, $C_{29-30}H_{47-49}NO_{12}$. The sugar, called mycarose, is an unusual, branched-chain didesoxyaldohexose, $C_7H_{14}O_4$. It is apparently much like the sugar obtained from erythromycin. In chemical composition, physical properties, and microbial activity, erythromycin and Magnamycin are strikingly similar. When the structural chemistry of the two antibiotics has been worked out, these analogies will probably be made clear. The situation is remindful of that which existed before the structures of Aureomycin and Terramycin were unravelled.

Activity and Toxicity[96a,96b,111a,112a,287b,299a]

Magnamycin acts mainly on Gram-positive bacteria and has little or no activity on the Gram-negatives. Like erythromycin, it acts on rickettsia and large viruses. Microorganisms resistant to erythromycin also show some cross-resistance to Magnamycin and vice versa. Magnamycin has no cross-resistance with the older commercial antibiotics and should prove valuable in the treatment of the many cases now resistant to these antibiotics. It is administered to patients by mouth in divided doses totaling 1 to 2 g daily.

The toxicity of Magnamycin hydrochloride for mice by intravenous administration is given as LD_{50}, 550 mg per kg of body weight.

Neomycin

Source

Neomycin is an antibiotic recently discovered by Waksman and Lechevalier;[277] it is being manufactured by several companies. It is produced by the discoverers' *Streptomyces* sp. 3535, a microorganism closely related to *Streptomyces fradiae.* Neomycin consists of several components; therefore, it is often spoken of as the neomycin complex.[253] It is active against both Gram-positive and Gram-negative bacteria.

Production

Antibiotic production in shaken flasks, amounting to 900 units per ml,[278] was obtained in a soya peptone-glucose-meat extract-sodium chloride-zinc sulfate medium in 144 hours. Peptone and zinc were particularly effective in producing high yields. The medium was adjusted at a pH of 6.5 to 6.8 and incubated at 26° to 28°C, with continuous shaking of the flasks. In a 6,000-l fermentation, 2.25 g per l have been reported in 90 hours.[162a] One unit of neomycin is the minimum amount of antibiotic in 1 ml of nutrient agar that is required to inhibit the growth of *E. coli* ATCC 9637. Assuming an average of 250 units per mg for the mixed antibiotics, 900 units per ml on a weight basis is 3.6 g per l which is a large yield.

Purification and concentration of the flask product was effected by adsorption on and successive elution from Decalso and Darco G-60, precipitation as the picrate, and conversion to the hydrochloride to give a substance having a potency of 212 units per mg.[253] Purification of the 6,000-gal batch was accomplished by acidifying the broth with sulfuric acid to pH 2.7, filtering off the mycelial cake, adjusting the filtrate to pH 8.0, and adsorbing on Darco G-60. The neomycin was eluted from the carbon with 10% acetone at pH 2.0 and precipitated with more acetone. The crude neomycin sulfate was further purified by chromatography through a carbon column. The neomycin sulfate thus obtained appeared homogeneous by paper chromatography and solvent-distribution tests. It analyzed 1,000 μg per mg by the streptomycin *B. subtilis* assay.

Further purification of neomycin has been effected by precipitation with picric acid, chromatography through alumina, and crystallization as salts of different sulfonic acids, e.g., *p*- (*p′*-hydroxyphenylazo)-benzene sulfonic acid. A crystalline sulfonate, designated neomycin A, has been obtained from crude neomycin by Peck et al.[204] Recrystallization of the sulfonate from aqueous methanol gave a product with constant properties. The hydrochloride was prepared from the sulfonate by precipitation from butanol-hydrochloric acid with acetone. It had an activity of 1,700 units per mg by a cup diffusion assay with *B. subtilis* and 50 units per mg by *E. coli* turbidimetric assay.[300]

A second component, believed to make up much of the activity of the broth and designated neomycin B, has been obtained as the crystalline helianthate and the *p*- (*p*-hydroxyphenylazo)-benzene sulfonate by Regna and Murphy.[223] Sulfates prepared from the crystalline dye salts behaved as homogeneous substances when subjected to paper chromatography and had an activity of 220 units per mg by turbidimetric assay with *B. subtilis* and 255 units by *K. pneumoniae* turbidimetric assay.

Two homogenous antibiotics, neomycins B and C, have been obtained by Dutcher et al.[87] The B compound appears to be the same as the neomycin B of Regna and Murphy. Neomycins B and C were isolated by chromatography of the hydrochlorides in 80% methanol over alumina. Crystalline reineckates, picrolonates and *p*- (*p*-hydroxyphenylazo)-benzene sulfonates have been obtained of both components.

The potencies (units per milligram) of neomycins A, B, and C differ greatly, depending on the method of assay, even with the same test organism. By cup diffusion assay, A is two to three times as potent as B or C, but by turbidimetric assay, it is only about one tenth as active.[78,127b] This shows that A diffuses readily through agar whereas B and C do not. The difference in diffusion is probably related to the smaller size of the A molecule. For comparison of potency, the turbidimetric method gives a truer picture as then there is no problem of the antibiotics coming in contact with the test organism.

Swart, Lechevalier, and Waksman[253a] have examined five preparations of neomycin, one of their own and four made by antibiotics producers. These samples were tested in a Craig countercurrent apparatus and only one preparation, neomycin A, appeared to be fairly homogeneous. The others seemed to consist of three to four main components and several small fractions.

Contradictory results obtained by different investigators probably arise from the lack of a homogeneous standard, different methods of assay, and different systems of purification. The use of different methods is really desirable, as what appears homogeneous by one method may show up as a mixture by another procedure. The confusion that now exists will probably be cleared up as better methods of fractionation are developed and the chemistry of the different components is ascertained.

A quite different antibiotic, fradicin, is also produced in the neomycin fermentation.[254] The distinguishing feature of this antibiotic is its activity against fungi. Up to 69% of the antibiotic activity of the broth may consist of the antifungal agent if fermentation conditions are regulated to favor its production. Low aeration, increased amounts of soya peptone, and addition of glucose to the medium increased the production of fradicin. It is obtained from the broth by extraction with butanol, purified through ethanol and acetone, separated as oil from petroleum ether and then lyophilized to give a tan-colored powder. Fradicin has been obtained in crystalline form by Hickey and Hidy[132] who, on the basis of elementary-composition and molecular-weight determinations, proposed the empirical formula $C_{30}H_{34}N_4O_4$. Fradicin is a weak base, which has a specific rotation of $+65°$ and an ultraviolet-absorption spectrum with a prominent peak at 293 mμ and several other smaller ones. Inhibition of several pathogenic fungi by the crystalline compound required 1 to 10 μg per ml. Acute toxicity tests by intraperitoneal injection of mice gave an LD_{50} value of 4 mg per kg of body weight.

Chemical Nature and Properties

Although the elementary composition of neomycin has been determined a number of times, no empirical formula has been seriously considered for any of the components. The reluctance of investigators to offer empirical formulas is perhaps based on doubts as to the homogeneity of the different preparations. Dutcher and Donin[86a] suggested that neomycins B and C appeared to have the empirical formula $C_{29}H_{58}N_8O_{16}$, but the evidence for this formula does not seem to be complete or convincing.

All the neomycins are basic compounds and form salts with various acids, e.g., dye sulfonate, reineckate, hydrochloride, and sulfate. The neomycins contain nitrogen, all of which is present as amino nitrogen, since it is all removed by the action of nitrous acid. The ninhydrin test is positive and the Molisch and carbazole tests for carbohydrate residues are positive with neomycins B and C, but negative with neomycin A. Fehling's and Tollen's tests for reducing substances are negative with the original neomycins B and C, but after hydrolysis, the tests are positive. All of the compounds are dextrorotatory, but the specific rotation varies, being +83° for A, +80° for B and +54° to +58° for C. The neomycins are extremely stable to alkali, e.g., refluxing with excess barium hydroxide for 18 hours did not lessen the activity of a neomycin preparation. Neomycin A is very stable to acid, but B and C lose much of their activity on heating with acid.

Neomycin has been degraded with acid by Leach and Teeter[163a] to yield a crystalline base, designated neamine. The base still has some antibiotic potency, about 6% of that of the original neomycin. In a later paper, Leach and Teeter[163a] compared neamine with neomycin A and found the two to be identical. Only a month later, Dutcher and Donin[86b] reported that a methanolysis product which they had previously obtained[87] from neomycins B and C was the same as neomycin A.

By drastic hydrolysis of neomycin A, Kuehl et al.[159a] obtained a degradation product, $C_6H_{14}N_2O_3$, which they determined to be 1,3-diamino-4,5,6-trihydroxycyclohexane and established its structure unequivocally by synthesis. Leach and Teeter obtained the same cyclohexane from neamine. The question naturally arises as to what other component, if any, is contained in neomycin A, neamine, and the methanolysis product.

Dutcher et al.[87] reported two other hydrolysis products from neomycins B and C. One of these appeared to be a diaminodesoxyhexose and the other had the properties of a pentose. The presence of three components in neomycin suggests a pattern similar to that found in streptomycin. The cyclohexane base of neomycin is closely related to the streptidine of streptomycin, the pentose of neomycin may correspond to the streptose of streptomycin and the diaminodesoxyhexose of the new antibiotic may parallel the N-methylglucosamine of the older one. Bacterial spectra and development of bacterial resistance lend some support to the chemical data pointing to a relationship between the two antibiotics, but final decision on the extent of this relationship must wait until the structure of the neomycins has been determined.

Activity

Neomycin is active against numerous Gram-positive and Gram-negative bacteria, including the tubercle organism.[119a,267b,270,275,276,277] Bacteria do not develop resistance to neomycin as readily as to streptomycin and strains that are resistant to streptomycin are destroyed by neomycin. Resistance to neomycin can be developed slowly and such strains seem to have some resistance to streptomycin, but the reverse effect does not seem to occur. On a weight basis, neomycin and streptomycin have about the same activity.[287]

In vivo tests (mice, chick embryo, guinea pigs) have shown it to be effective in combatting infections with *S. aureus, S. schottmuelleri, S. pullorum, E. typhosa, K. pneumoniae, P. vulgaris,* and *M. tuberculosis.*[137,152,217,246] In humans, many cases of urinary and other infections, refractory to other antibiotics, have been treated successfully with neomycin.[83,162,156f,194a] In the treatment of tuberculosis, neomycin seems to be inferior to streptomycin. However, it may be useful in cases that are resistant to streptomycin.[150a,206b]

Toxicity

Neomycin is a toxic substance, but there still seems to be some question as to how great is its toxicity. Loss of weight and kidney damage in guinea pigs treated with neomycin for tuberculosis have been reported.[152,246] Kidney damage[267c] and progressive deafness[56a] in humans after neomycin therapy have been observed. Preparations containing about 200 units per mg had an LD_{50} of about 50 mg per kg when injected intravenously into mice.[137] By the subcutaneous route, the hydrochlorides of pure neomycins B and C had a mouse toxicity of LD_{50} amounting to 220 mg per kg and 290 mg per kg, respectively.[127a]

Polymyxins

Source

Polymyxin is a generic name adopted to indicate antibiotics obtained from strains of *Bacillus polymyxa.* To date, there are at least five such antibiotics, closely related in respect to producing organisms, activity, and chemical composition. These are named polymyxin A, B, C, D, and E. Originally, polymyxin A was known as aerosporin because it was produced by a bacterium identified as *Bacillus aerosporus.* This organism was later shown to be a strain of *B. polymyxa* and, at a conference of workers in this field, the more appropriate and taxonomically derived term, polymyxin, was agreed upon.[50]

Three groups of workers reported independently the existence of antibiotics produced by *B. polymyxa:* Ainsworth, Brown, and Brownlee[3] of the Wellcome Physiological Research Laboratories, England; Stansly, Shepherd, and White[245] of American Cyanamid; and Benedict and Langlykke[27] of the Northern Regional Research Laboratory in the United States.

Production

Of the polymyxins, B and D have been produced in shallow stationary cultures, shaken flasks, and aerated pilot-plant tanks. The following brief statement applies to the production of polymyxin D in an aerated medium of 15 l volume.[244] The medium contained about 1.0% glucose, 2% $(NH_4)_2SO_4$, 0.5% yeast extract (Difco), 0.2% KH_2PO_4, 0.05% $MgSO_4 \cdot 7H_2O$, 0.005% NaCl and 0.001% $FeSO_4 \cdot 7H_2O$ and was adjusted to pH 7.9. In previous flask fermentations, 1.0% soybean in place of yeast extract gave equally good yields. The medium was aerated at the rate of 64 l per hour for 5 days at 25°C. The pH fell to 6.3 on the third day and then rose to 6.8. In fourteen runs, the polymyxin ranged from 163 to

358 (average 216) units per ml. Assuming 1,800 units per mg, the average concentration of polymyxin is only about 0.12 g per l which is rather low as commercial antibiotics go.

For isolation of polymyxin D, the beer was filtered through Hyflo Super-Cel, adsorbed on Darco G-60, and eluted with methanol-hydrochloric acid. The polymyxin hydrochloride was precipitated with acetone, washed, and dried in a vacuum. The product analyzed 1,100 to 1,400 units per mg and the over-all recovery was 30 to 60%. The crude polymyxin hydrochloride was further purified by extraction with butanol at pH 9.0 and converted back to the hydrochloride (potency 1,700 to 1,800 units per mg). A product of somewhat higher potency was obtained by partition chromatography.[237]

A very potent crystalline derivative of polymyxin B has been obtained by adsorption on cotton succinate, elution, precipitation with Polar Yellow solution, precipitation of the free base with triethylamine, and conversion to crystalline polymyxin B naphthalene-β-sulfonic acid. This compound was levorotatory, $[\alpha]_D = -63.3°$, and paper chromatography showed only a single component.[225]

The cultures producing polymyxins A and B also form small quantities of other as yet unidentified polymyxins. If a number of bacteria of the polymyxa and related types were thoroughly studied, it is probable that many more antibiotic polypeptides similar to the polymyxins would be found.

Chemical Nature and Properties[25,61,148,298]

The polymyxins are basic polypeptides combined with a fatty acid ($C_8H_{17}COOH$). They all contain L-threonine, the C_9 fatty acid and a new amino acid, L-α,γ-diaminobutyric acid. (Some of the D-form of this acid was obtained, but this may have been produced by racemization during hydrolysis.) They differ in respect to the other amino acids found in the polypeptide. Polymyxin A contains D-leucine, polymyxin B yields D-leucine and phenylalanine, polymyxin C gives only phenylalanine, and polymyxin D contains D-leucine and D-serine, but no phenylalanine. The composition of polymyxin D is 5 moles of L-α,γ-diaminobutyric acid, 3 moles of threonine, 1 mole of D-leucine, 1 mole of D-serine and 1 mole of fatty acid; which gives a molecular weight of about 1,150. The amino acids appear to be arranged in the form of a closed cycle. The fatty acid apparently is linked to one of the diaminobutyric acid residues through the γ-amino group in the side chain to form an acyl derivative. The γ-amino groups of the other four diaminobutyric residues are free and give basicity to the compound. The fatty acid has not been fully characterized by synthesis, but degradation studies indicate it is (+)-6-methyloctanoic acid, $CH_3CH_2CH(CH_3)CH_2CH_2CH_2CH_2COOH$. The acid appears to be identical in all the polypeptides, although the optical activity of the acid from polymyxin A,[61] $[\alpha]_D = +8.2°$, seems to differ from that obtained from polymyxin B,[225] $[\alpha]_D = +32°$.

The various polymyxins differ in other respects; *e.g.*, solubility in butanol, titration curves, absorption spectra, and chromatographic behavior. Their antibiotic properties are unaffected by digestion with pepsin, trypsin, or papain.

Activity and Toxicity[51,52,52a,145a,150b,214,308]

The polymyxins are distinctive in being strongly active against Gram-negative bacteria and largely inactive against Gram-positive organisms. They have been tested clinically against typhoid, gastroenteritis, urinary infections, and whooping cough. Polymyxin B appeared unusually promising against the organism, *Hemophilus pertussis,* associated with whooping cough, but unfortunately some proteinuria seems to accompany its use and the antibiotic appears less promising than formerly. All of the polymyxins appear to cause renal damage of varying severity and duration, but this is least with polymyxin B. Administration of methionine, or choline, simultaneously with polymyxin A, prevented proteinuria in rats and dogs. Methylating or acetylating the antibiotic accomplished the same result. However, the author did not test the modified product to see if it still had antibiotic activity. If activity is retained and toxicity is removed, modification should greatly extend the usefulness of the polymyxins. In spite of the danger of causing damage, they have proved very useful in the treatment of refractory infections, such as *Proteus* and *Pseudomonas,* which often remain unaffected by other therapeutic agents.

The mean LD_{50} for polymyxin A is 6.9 mg per kg intravenously and 13.9 mg per kg intraperitoneally. Polymyxin B has about the same toxicity as the A form, but D is only one half to one third as toxic as A and B.

Terramycin

SOURCE

Terramycin is an important commercial antibiotic manufactured by Chas. Pfizer and Co., Inc. It is produced by a new actinomycete, *Streptomyces rimosus,* so-named because of the cracked appearance of the colonies on agar plates.[104,241] A culture of the organism has been deposited with the Fermentation Division of the Northern Regional Research Laboratory, Peoria, Illinois, under the number NRRL-2234. Terramycin was discovered in the course of an extensive screening program that in many years of testing covered one hundred thousand soil samples from all parts of the earth.[293] It is evident that much effort and expense is needed to find a new and useful antibiotic. Of course, during this search, probably many other new, but not useful, antibiotics were found by the discoverers of Terramycin. Although there are many papers dealing with the antimicrobial action and clinical evaluation of the antibiotic, only a short paper[104] and the patent[241] give information on its production and isolation.

FIGURE 21C. *Filtration of Terramycin Fermentation Broth* (Courtesy–Chas. Pfizer & Co., Inc.)

PRODUCTION

A number of different media are given in the patent[241] as being suitable for production of Terramycin. A typical medium contains 3% soybean meal, 0.5% corn starch, 0.1% N-Z-Amine B (an enzymic digest of casein), 0.3% sodium nitrate and 0.5% calcium carbonate. It is adjusted to pH 7.0 and, after sterilization, is seeded with a 2.5% by volume inoculum of the organism. Incubation with aeration and agitation is carried out at 24° to 30°C for about 48 hours and then the broth is processed. Yields given in the patent are expressed as coli dilution units (CDU) and range from 280 CDU per ml in 400-gal tanks to 2,560 in 1-gal glass fermentors. In shaken flasks, 640 CDU per ml are reported. A coli dilution unit is the quantity of Terramycin in 1 ml that is required to inhibit the growth of *E. coli* for 18 hours under specified conditions of inoculum, medium, and temperature. It is not possible to express these figures on a weight basis, as the quantity of pure Terramycin in 1 CDU is not given in the patent. Judging from another paper[135] on the inhibition of *E. coli* by crystalline Terramycin in liquid medium, the yields must be well over 1 g per l. No information has been published as to chemical changes in the medium during fermentation except that the pH rises from 7 to 8.

Terramycin can be isolated from the broth by a number of methods. One of those given in the patent involves filtering off the mycelium, adsorbing the Terramycin at pH 7.0 on Norite, eluting with water saturated with butanol at pH 1.5, transferring first to butanol at pH 9 and then back to acidified water. On neutralizing to pH 7.5, a precipitate of crude Terramycin is formed. Crystalline Terramycin can be obtained by further purification and final crystallization from water or methanol as the free base, the hydrochloride, or the sodium salt. An effective means of purification is a Craig counter-current distribution with butanol and water at pH 3. The crystals of the free base are white and those of the hydrochloride and sodium salt are lemon yellow.

Equipment used in the filtration of Terramycin broth is shown in Figure 21c.

Crystalline Terramycin is dried in ovens such as that shown in Figure 21d.

A second antibiotic, rimocidin, is produced by *S. rimosus*.[74b,235]

FIGURE 21d. *Loading a Drying Oven with Crystalline Terramycin* (Courtesy–Chas. Pfizer & Co., Inc.)

It is retained in the mycelium, instead of being secreted into the medium, and is obtained by extraction of the mycelium with butanol. The crystalline sulfate has been prepared and has the following percentage composition: C, 57.65; H, 7.82; N, 1.81; S, 2.03. It is dextrorotatory, $[\alpha]_D = +75.2°$, melts at 151°C with decomposition, and shows ultraviolet absorption at 279, 291, 304, and 318 mμ. The infrared absorption band shows many maxima.

The most distinctive antibiotic feature of rimocidin is its antifungal activity and lack of antibacterial potency. Inhibition of a number of pathogenic fungi has been demonstrated at concentrations of 1 to 5 mg per ml. It has some, but not striking, activity against protozoa, e.g., *Endamoeba histolytica.* The mouse toxicity of rimocidin (LD_{50}) is 20 to 30 mg per kg of body weight injected intravenously.

CHEMICAL NATURE AND PROPERTIES[104,224,241,247]

As has already been pointed out (p. 302), Terramycin has the same structural formula as Aureomycin except that it contains no

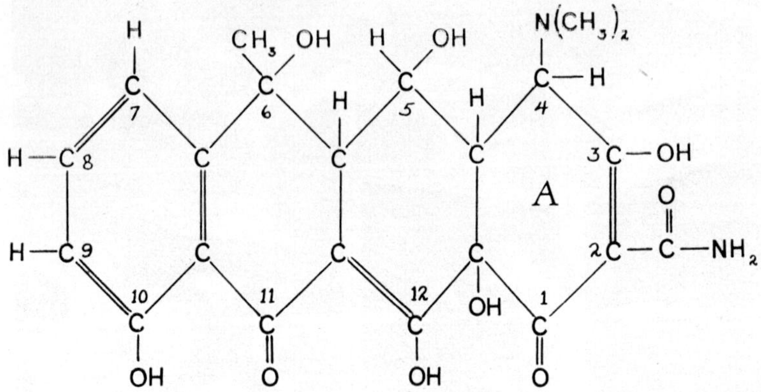

Terramycin

$C_{22}H_{24}N_2O_9$
M.W. 460.2

chlorine and has an additional hydroxyl group at 5. The finding of an antibiotic so closely related to Aureomycin and still not covered by the Aureomycin patent has been a most profitable discovery to one company and a corresponding loss to another. If one reflects on the many examples of closely related compounds that are produced by similar species or even by strains of the same species,[214c] it would not be surprising if other naphthacene antibiotics, for example, with a different number or distribution of chlorine or hydroxyl groups were discovered among the metabolic products of other streptomycetes.

Terramycin is an amphoteric compound and crystallizes as the dihydrate, the hydrochloride, and the sodium and potassium salts. From water, the hydrochloride comes out as platelets and from methanol as needles. It has a high specific rotation, −196°, in 0.1 N HCl. The free base melts at about 185°C, with decomposition. It is soluble in methanol, ethanol, acetone, and propylene glycol, and insoluble in ether and petroleum ether. In water, Terramycin is readily soluble at pH 1.2 (31.4 mg per ml) and at pH 9.0 (38.6 mg per ml). It is a very stable antibiotic. In water at pH 2.5 and 25°C, no detectable loss occurred in 30 days. At pH 7.0 and 37°C, 50% of the activity remained after 26 hours and at pH 10 and 37°C, the half life was 14 hours. In the dry state, the crystalline product can be stored at room temperature for long periods.

ACTIVITY[60a,135,145c,156d,166a,172,209,231c,241,306a]

Terramycin is active against a wide variety of microorganisms and disease-producing agents. It acts on most Gram-positive and Gram-negative bacteria, on rickettsiae, certain viruses, some spirochaetes, and certain protozoa. It does not appear to be effective, or is only slightly so, against the tubercle bacillus, bacteria of the proteus and pseudomonas types, yeasts, and fungi.

Terramycin is widely used in the treatment of pulmonary infections, including viral pneumonia, bacterial infections of the digestive and urinary tracts, typhus and other rickettsial fevers, the spirochete diseases syphilis and yaws, and the virus diseases lymphogranuloma venerum and granuloma inguinale. Promising results have been obtained in the treatment of brucellosis and malaria.

Since it is ineffective against proteus, pseudomonas, and monilia, these microorganisms may develop when their natural competitors, the Gram-negative bacteria, have been eliminated and thus bring about what is called a superinfection, that is, an infection superimposed on another. Such infections are likely to occur in the intestinal and urinary tracts. Microorganisms that are resistant to Terramycin are also resistant to Aureomycin and vice versa. Therefore, the use of either antibiotic in cases where the other is ineffective seems to be precluded.

Combinations of antibiotics are often used, but the combination may be less effective than adequate doses of the individual antibiotics alone. Useful combinations among penicillin, streptomycin, bacitracin, and neomycin have been noted. Members of this group may give bad results with Aureomycin, Chloromycetin, or Terramycin. The effect of combinations cannot be predicted. A report [145c] on the subject points out the many problems in making combinations and says "only if the organism proves resistant to single drugs by laboratory test or by adequate therapeutic trial should drug combinations be employed."

TOXICITY[104,172,241]

The toxicity of Terramycin is low, the LD_{50} for mice being about 200 mg per kg by intravenous injection. In clinical trials, there is no indication of injury to the kidney, liver, or bone marrow.

Tetracycline*

SOURCE[37b,65c,185a]

As mentioned previously (p. 302), Aureomycin (chlorotetracycline) and Terramycin (oxytetracycline) can be regarded as derivatives of a basic structure, tetracycline. The conversion of chlorotetracycline to tetracycline can be accomplished by simultaneous dechlorination and hydrogenation. Chlorotetracycline is dissolved in a suitable solvent (e.g., methyl Cellosolve), containing a palladium-charcoal catalyst and triethylamine, and hydrogen is passed in at atmospheric pressure and room temperature. The uptake of hydrogen is rapid and in 15 to 20 minutes is virtually complete. The reaction consumes about 1 mol of hydrogen per mol of chlorotetracycline. The triethylamine neutralizes the hydrochloric acid that is produced. After removal of the catalyst, the reaction mixture is poured into water and the free base of tetracycline crystallizes.

A later report[185a] by another group of workers contains the interesting information that tetracycline has been obtained by fermentation. The *Streptomyces* producing it was found in a sample of Texas soil and was isolated in the course of a search for organisms producing new antibiotics. To date the identity of the *Streptomyces* has not been published. The authors noted that the culture produced an antibiotic that resembled chlorotetracycline and oxytetracycline, but differed from these in certain respects and was evidently a different substance. High-yielding mutants of the organism were developed and improved means of purification were devised. To date no information has been published as to the medium, sequence of chemical changes during the fermentation, and yields of antibiotic. Tetracycline is also reported[37a] to be produced in small quantities by strains of *Streptomyces aureofaciens* simultaneously with chlorotetracycline. Probably the tetracycline culture produces also some chlorotetracycline, as cultures usually produce more than one antibiotic. Details regarding the recovery of tetracycline from the broth are not available but, because its properties are similar to those of chlorotetracycline and

* The trade names for tetracycline are: Achromycin (Lederle Laboratories Division of American Cyanamid Co.), Tetracyn (J. B. Roerig, Division of Chas. Pfizer and Co.), and Polycycline (Bristol Laboratories, Inc.).

oxytetracycline, probably the same general procedures could be used.

CHEMICAL NATURE AND PROPERTIES[37b,65c,185a]

Since tetracycline has the same basic structure as chlorotetracycline, but contains a hydrogen instead of chlorine, the molecular formula is $C_{22}H_{24}N_2O_8$ and the molecular weight is 444.2. As would be expected, some of its properties are similar and others different from those of chlorotetracycline. The specific rotation is −258° in 0.1 N HCl as compared to −240° for chlorotetracycline and −196° for oxytetracycline. The ultraviolet absorption spectrum differs markedly from that of chlorotetracycline, but is almost identical with that of oxytetracycline; strong peaks are present at 267 and 357 mμ. The infrared spectra of the three antibiotics likewise show similarities and differences. The three antibiotics can also be distinguished by countercurrent distribution and color tests. With *p*-dimethylaminobenzaldehyde in dilute aqueous solution, tetracycline gives an orange-yellow color, chlorotetracycline canary yellow, and oxytetracycline bluish green.

Another means of distinguishing the tetracyclines is by paper chromatography. Chlorotetracycline can be separated from the other two in a *n*-butanol-water system; and in *n*-amyl acetate saturated with water, tetracycline can be separated from oxytetracycline.

Tetracycline appears more stable than the chloro and hydroxy derivatives.

ACTIVITY AND TOXICITY[37a,72a,96c,156e,179a,185a,214a,299b,306b]

Tetracycline has a bacterial spectrum similar to that of chlorotetracycline and oxytetracycline and has much the same potency in experimental infections. Microorganisms made resistant to one tetracycline appear also resistant to the other two. Preliminary clinical trials in human patients show a higher tolerance level for tetracycline and an efficiency in counteracting infections apparently equal to that of the other tetracyclines.

The acute toxicity of tetracycline is about the same as that of the other tetracyclines. The LD_{50} for mice by intravenous injection is 170 mg per kg.

Tyrothricin (Gramicidin and Tyrocidine)

The literature on these antibiotics is so voluminous that only a few of the many papers can be referred to in this report. For more details, the reader should consult an extensive review by Hotchkiss[139] or a brief one by Neuberger.[196]

PRODUCTION

Tyrothricin is the oldest commercial antibiotic. It is produced by several firms, but, because of its limited use, does not bulk as large in trade as a number of other, more recent additions.

Gramicidin (sometimes called D after its discoverer, Dubos) and tyrocidine are separate antibiotics that have been obtained from tyrothricin by fractionation procedures. Both have been obtained in crystalline form and, for a long time, each has been regarded as homogeneous. However, recent work[23a,123] indicates that there are several components in both gramicidin and tyrocidine. A third crystalline product, gramicidin S, has been obtained, but it comes from a different organism from that producing gramicidin D and tyrocidine.

Bacillus brevis ATCC 8185 is the organism from which tyrothricin was obtained by Dubos and associates.[80,140] Another strain of *B. brevis* was used by Gause and Brazhnikova[116] for the production of gramicidin S. Many strains of *B. brevis* have been shown to produce antibiotics having some properties similar to those of gramicidin and tyrocidine; but they have not been sufficiently purified to indicate whether or not they are identical with gramicidin and tyrocidine. Probably they are all polypeptide mixtures and the separation and identification of the components of such mixtures is an exceedingly difficult and laborious undertaking.

Tyrothricin can be produced in shallow static media and also in submerged aerated cultures. Good yields of 0.5 to 0.9 g per l have been reported.[12a,139] Both synthetic and natural media have been used. The antibiotic can be readily obtained from the medium by precipitation at pH 4.8, extraction of the precipitate with alcohol, a second precipitation with salt solution, and fractionation of the precipitate (tyrothricin) with acetone-ether to give a soluble portion (gramicidin) and an insoluble portion (tyrocidine). The first portion can be crystallized from acetone to give gramicidin D and the second from methanol-hydrochloric acid to give tyrocidine hydrochloride. Gramicidin makes up 10 to 20% and tyrocidine about 40 to 60% of the tyrothricin product.[139]

CHEMICAL NATURE AND PROPERTIES[23a,23b,120,123,128a,132b,139,255,256]

The gramicidin complex (D) acts as a neutral polypeptide, apparently cyclic in nature and, on hydrolysis, gives five amino acids and ethanolamine. It has been calculated that there are three L-alanine, two glycine, six D-leucine, six L-tryptophan, five DL-valine, and one ethanolamine residues in the complex. The peptides, L-alanyl-D-leucine, L-alanyl-D-valine, D-leucyl-glycine, D-valyl-L-valine and L-valyl-D-valine have been isolated from partial hydrolysates. The data are not entirely consistent, but the discrepancies may be due to the fact that the gramicidin used, although carefully crystallized, was probably not homogeneous. Gramicidin D has been calculated to have a molecular weight of about 2,800 and a melting point of 228° to 230°C. When gramicidin D was fractionated by the Craig countercurrent-distribution method, the major component, A, melted at 227° to 228°C and a second, smaller component B melted at 258° to 259°C. Gramicidin A was found to contain somewhat more tryptophan than B. Evidence for the presence of at least two other components was obtained, but these did not crystallize sharply and appeared to be mixtures.

Crystalline tyrocidine is a mixture of polypeptides containing at least three components, A, B, and C. By extensive fractionation in the Craig countercurrent-distribution apparatus, the main component, A, was obtained in an apparently pure state and its amino acid content and molecular weight were determined. The amino acid residues found were: two of D-phenylalanine, and one each of L-phenylalanine, L-valine, L-tyrosine, L-leucine, L-proline, L-ornithine, L-glutamine, and L-asparagine. There are no free α-amino or α-carboxyl groups, which indicates that the peptide has a cyclic structure. The γ- and β-carboxyls of glutamic and aspartic acid, respectively, are present as the corresponding amides. The δ-amino group of ornithine is free and gives the peptide its basic property. The molecular weight 1,270, calculated from the amino acid content, agrees well with that (1,300) obtained by an independent method. Tyrocidine A hydrochloride has a melting point of 240° to 242°C and a specific levorotation of $-111°$.

The chemical character of gramicidin S is now thoroughly understood due to the work of Synge. The crystalline hydrochloride has been obtained from the cells by precipitation with acid and extraction with ethanol and the free peptide has also been crystallized. It is a cyclopentapeptide (or possibly a decapeptide) with one residue each of the amino acids named, in the following sequence: -L-valyl-L-ornithyl-L-leucyl-D-phenylalanyl-L-prolyl-. Open-chain pentapeptides of the same structure as gramicidin S have been synthesized, but the antibiotic activity of these peptides was only about one hundredth of that of the natural compound. From this, it appears that the antibiotic potency of gramicidin S is associated with its cyclic structure. Later evidence shows that gramicidin S is a decapeptide having two basic groups (ornithine) and a molecular weight of 1,142. The decapeptide structure helps explain why the synthetic pentapeptide has so little potency.

Activity and Toxicity[110,115,131,206,220,231,283]

Apparently, gramicidin D is active almost exclusively against Gram-positive organisms, but it is not inactivated by serum, etc., and, therefore, it is active *in vivo* as well as *in vitro*. Tyrocidine has stronger action against a wider bacterial spectrum, but is inhibited by serum, etc., and is, therefore, inactive *in vivo*. Neither antibiotic has proved a success in subcutaneous, intramuscular, or oral administration.

Tyrocidine is the most toxic because of its high hemolytic activity. Gramicidin D is not rapidly and strongly hemolytic like tyrocidine, but it brings about a slow and delayed hemolysis that continues for many hours. Combination of gramicidin D with formaldehyde decreases its hemolytic activity by 80% while reducing its antibacterial activity about 50%.

Much less information is available on the activity and toxicity of gramicidin S. It is reported to be active against a number of Gram-positive bacteria and, like tyrocidine, is ineffective in the presence of serum. When administered intraperitoneally, 15 to 20 mg per kg of body weight was toxic to rats.

Tyrothricin is in commercial production for therapeutic use, but its producers warn against any parenteral administration of the drug because it is a systemic poison when introduced into the blood stream (LD_{50} is about 3.75 mg per kg for mice intravenously). Oral administration seems to be ineffective. It is designed for topical use only, at the site of the infection, where it can be brought into direct contact with the infecting organism. Superficial ulcers, abscesses, wounds, etc., and infections in body cavities not directly connected with the blood stream usually respond to this antibiotic when the invading organisms are Gram-positive. Presence also of Gram-negative bacteria may interfere with the treatment. Doses sufficient for bactericidal action are not toxic to body tissues, though there is some toxicity at higher levels. Tyrothricin has had some use in veterinary medicine, also.

Viomycin

After 3 years of testing, viomycin was approved as a new drug by the Food and Drug Administration early in 1953 and has been placed on the market by Parke, Davis and Co. and Chas. Pfizer and Co., Inc. Streptomyces producing this antibiotic were discovered independently in the laboratories of these two firms and named *Streptomyces floridae* and *Streptomyces puniceus* respectively. The first name indicates the source of the soil sample from which the culture was isolated and the second denotes the purplish-red color of the vegetative mycelium produced by the culture.[23,103]

Production[23,65b]

Factors affecting viomycin production in shaken-flask fermentations have been reported for *S. floridae.* The organism can use a number of carbohydrates, but commercial glucose gave the highest titre. Several commercial nitrogenous substances gave good yields of viomycin. A typical high-yielding medium consisted of 1% commercial glucose, 0.8% acid hydrolyzed casein, 0.3% yeast hydrolyzate, 0.5% sodium chloride, and 0.1% calcium carbonate. The medium was adjusted to pH values ranging from 6.0 to 8.0, but the initial pH had no effect on viomycin titres. In 500-ml Erlenmeyer flasks, 100 ml of the medium, after inoculation with a spore suspension, was incubated at 24° to 26°C, while being swirled on a rotary shaker making 160 revolutions per minute. Antibiotic formation began after 30 hours and rose rapidly to about 600 μg per ml (calculated as viomycin sulfate) after 66 hours of incubation. Little additional antibiotic was formed after this time, but apparently the viomycin was not destroyed even after 144 hours of incubation. Irrespective of the initial pH, the medium reached a pH of about 7.6 in 24 hours, fell slightly to about 6.8 at 48 hours, reached a peak of 8.5 at 85 hours, and then remained virtually constant to the end of the incubation, i.e., 144 hours. No data are given in this report on mycelial growth or changes in the glucose content and nitrogen components of the medium. The titres are good for a first report, but it is reasonable to assume that higher yields have been obtained by the time the fermentation reached the tank-production stage. Production of other antibiotics simultaneously with viomycin has not been reported, but this may be assumed to occur since *Streptomyces* cultures usually form more than one antibiotic.

Crystalline viomycin sulfate has been obtained by adsorption on and elution from carbon followed by successive precipitations from 50% methanol.

Chemical Nature[23,103,128d]

A tentative empirical formula, $C_{18}H_{31-33}N_9O_8$, has been proposed for viomycin free base. It is a strong organic base that forms crystalline salts, e.g., sulfate, hydrochloride, picrate, and reineckate. It is of polypeptide nature, contains one primary amino group, but no free α-amino carboxyl. It gives positive Sakaguchi, biuret, and ninhydrin tests, but negative tests for carbohydrate components. The ultraviolet absorption spectrum shows a strong maximum which shifts from 268 mμ in 0.1 N HCl to 282.5 in 0.1 N NaOH. Viomycin sulfate is levorotatory, but the specific rotation varies from $-19.6°$ to $-40.6°$, depending on the method of drying the sample and the pH of the solution. Viomycin is stable at room temperature in acid or neutral solutions for 24 hours or longer. At 100°C and alkaline pH, it rapidly loses its antibiotic potency.

After acid hydrolysis and paper chromatography, five major and three minor ninhydrin spots were found. By passage through ion-exchange columns four of these ninhydrin positive components were isolated and three of them identified. These are L-serine, α,β-diaminopropionic acid, and β-lysine. The fourth component is a guanidino compound (strong Sakaguchi test) that has been obtained as a crystalline derivative of *p*-hydroxyazobenzene-*p*-sulfonic acid, but not characterized further. In addition to the amino acids, hydrolysis of viomycin liberates carbon dioxide, ammonia, and urea. The occurrence of two such unusual components as diaminopropionic acid and β-lysine illustrates the unique character of polypeptide antibiotics.

Activity[4a,10c,94,103,136,151a,209a,245a]

Viomycin is strongly active against the tuberculosis organism, has only low activity against Gram-positive or Gram-negative bacteria, and shows no inhibitory effect on fungi. Resistance to viomycin is acquired by the tubercle bacillus about as readily as resistance to streptomycin, but bacteria that have become resistant to streptomycin and isoniazid are still sensitive to viomycin. As in the case of streptomycin, emergence of strains resistant to viomycin is much retarded when it is given together with another therapeutic agent, e.g., *p*-aminosalicylic acid. However, since viomycin is less effective and apparently more toxic than streptomycin, it is valuable as an adjunct to streptomycin therapy rather than as the agent of first choice.

Viomycin is best administered intramuscularly, 2 g or less a day, in two divided doses at 12-hour intervals, twice weekly.

TOXICITY[10c,128c,203,209a,235c,292]

The acute LD_{50} for mice is given as 241 mg per kg by intravenous administration and 1,381 mg by subcutaneous route. Electrolyte imbalance, disturbance in renal function, eosinophilia, dizziness, and deafness have been noted rather frequently in the treatment of human patients, but these symptoms are usually mild or of short duration. However, as the drug has not been extensively used, clinicians advise that careful attention should be given to needs of the individual patient, size of dose, period of treatment, and development of unfavorable symptoms.

COMPARISON OF COMMERCIAL ANTIBIOTICS

Comparison of the potencies of different commercial antibiotics against specific organisms cannot be made with any assurance from the early reports, because different investigators used different methods of assay (e.g., cup, agar streak, dilution, turbidity, etc.) and also because they used different strains of the same organisms. Such miscellaneous data have a very limited value for purposes of comparison. It is also desirable to test the antibiotics against organisms freshly isolated from human infections and against stock cultures in laboratory collections.

Recently more useful data have appeared in the literature. As reported in these papers,[37,99,100,101,142,143,165,195,197e,225a,285,288,299] several commercial antibiotics have been tested simultaneously against selected organisms under identical conditions by a single group of investigators. These tests show that some antibiotics are definitely more potent than others against certain pathogens. In other papers, many strains, both laboratory stocks and new isolates, have been tested against several antibiotics in order to discover the variations among the strains. The variations are often great and indicate that types made resistant by previous exposure to the antibiotics are appearing.

Continuation of research along these lines should be of great value to the clinician who is trying to select the most suitable antibiotic for treatment of a particular disease.

NONTHERAPEUTIC USES FOR ANTIBIOTICS

A recent and unexpected development in the field of antibiotics is the stimulatory effect these substances have on the growth of young animals. This effect was first noted by Moore et al.[186a] and has since been reported by many investigators.[9,37d,47,56,65a,73,124,143a,156g,170a,171,214b,231a,232a,242a,248,249a,250,297] When fed in low concentrations antibiotics usually increase the growth of chicks, pigs, and calves by 10 to 20% in the first few weeks of life and in some experiments the increase in weight has been even greater. Utilization of the ration also appears to be more efficient in the presence of antibiotics. The mortality of young animals is also usually much reduced and the weak and unthrifty individuals seem to benefit most from the added antibiotics. As a result of these experiments, antibiotics are usually added to mixed feeds, particularly to those for poultry. Concentrates are manufactured by the antibiotics producer and sold to the mixed feed dealers, who incorporate these into the feed to give an antibiotic content of 1 to 10 g per ton of feed. There are many such concentrates on sale under various trade names. The tonnage of mixed feeds for poultry and swine in the United States is about 25 million tons a year. Much of this is now supplemented with antibiotics, so it is evident that the outlet for antibiotics in this field must amount to millions of dollars annually.

It is not clear how antibiotics act to produce the increased growth of animals, but the weight of evidence is in favor of the view that they eliminate harmful bacteria or reduce the number of microorganisms in the intestinal tract that compete with the animal for essential nutrients. Strong support for this interpretation is found in the evidence that germ-free animals give no response to the addition of antibiotics to the ration.

The effect of antibiotics on the growth of plants and control of plant diseases by means of antibiotics is being actively investigated and encouraging results have been obtained. The most widely-tested antibiotics are Actidione for fungal diseases and streptomycin for plant diseases of bacterial origin. The literature is now quite extensive and several reviews on it have been written.[7b,163b,197a,286a] An increased growth of plants under aseptic conditions in the presence of certain antibiotics, *e.g.*, penicillin and Terramycin, is an unexpected and puzzling result.[197b]

Other nontherapeutic uses of antibiotics, such as the preservation of certain foods and control of industrial infections, are subjects of current research.[235d]

NONCOMMERCIAL ANTIBIOTICS

In addition to the antibiotics that have been discussed in detail in the preceding sections of this chapter, there are many interesting and well-defined antibiotics which because they appear to lack commercial possibilities do not warrant extended discussion here. However, a summary of important facts about some of these antibiotics has been made and this information is given in Table 36. In making this selection, only antibiotics that are fairly well defined both from a chemical and a biological point of view have been included. Readers may disagree with the selections as being too inclusive or too exclusive. If they have erred, the authors are inclined to think that the fault has been in the direction of too extensive a list rather than one that is too limited.

In compiling this table, it has not been possible, unfortunately, to give complete information on the individual compounds. Many important facts have had to be omitted because of space limitations imposed by a tabular arrangement. It is believed that the conciseness and convenience of the table arrangement is to be preferred to a descriptive statement concerning each antibiotic.

In selecting the references, an effort has been made to include the first mention of the antibiotic, papers dealing with information given in the several columns, and the last paper on the compound observed at the time the table was compiled.

Table 36 indicates that the actinomycetes and the fungi are the chief sources of antibiotics. In recent years, the trend has been more strongly in the direction of the actinomycetes than toward the other sources. Bacteria do not play a very prominent part in the antibiotic program. Most of the antibiotics produced by bacteria are polypeptides and so far none of these has proved to be safe for parenteral use. Restriction to topical use has placed a severe limitation on their value.

TABLE 36. SUMMARY OF MAIN FACTS ABOUT OTHER WELL-DEFINED ANTIBIOTICS

Antibiotic	Source	Formula or chemical nature	Activity[a]	Toxicity, LD_{50}[b] mg per kg	References
Actinomycin A, B, C	*Actinomyces antibioticus,* etc.	Chromopeptides. A and B yield L-methylvaline, D-valine, sarcosine, L-proline and L-threonine; quinone (?). C gives D-isoleucine or D-alloisoleucine in place of D-valine.	G+ Y G− Y (?) F Y	0.5 (Ip or Sc)	48, 74, 157, 272, 279, 280
Actinorubin	*Actinomyces* strain A105	$C_6H_{14}O_2N_3$ or $C_9H_{22}O_4N_5$ (?)	G+ Y G− Y	36 (Ip)	149, 156, 187
Antimycin A	*Streptomyces* (unidentified)	$C_{28}H_{40}O_9N_2$. Yields: L-threonine, 3-aminosalicylic acid, L-methylethylacetic acid, and a $C_{11}H_{20}O_3$ acid.	F Y	25 (Sc)	2, 85, 156b, 164, 233b, 257a
Aspergillic acid	*Aspergillus flavus*	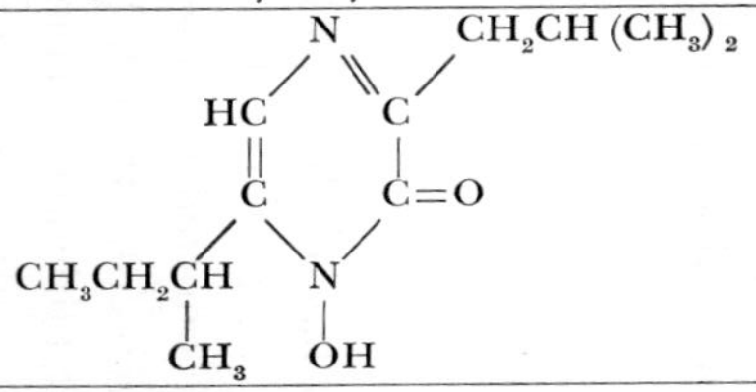	G+ Y G− Y F Y My Y	48.5 (Iv)	84, 86, 170, 204a, 219, 296, 302
Aureothricin (cf. thiolutin)	*Aspergillus farcinicus*	$C_9H_{10}N_2O_2S_2$. Yellow crystals. It is the propionamido derivative of the base $C_6H_6N_2OS_2$	G+ Y G− Y My Y	500 (Sc)	61a, 173, 262
Ayfivin (same as bacitracin A, B, and C)	*Bacillus licheniformis* strain A-5	A mixture of polypeptides. Yields aspartic acid, glutamic acid, histidine, ornithine, lysine, cystine, phenylalanine, leucine and isoleucine	G+ Y G− Slight	ca 500 (Sc)	15, 16, 132a, 197, 236
Borrelidin	*Streptomyces rochei*	$C_{28}H_{43}O_6N$ An acid with conjugated double bonds. Yields crystalline methyl and *p*-nitrobenzyl esters and other derivatives	*Sarcina lutea* Y Most bacteria N *Borellia* Y Enhances penicillin	74.7 (Sc) 39.0 (Iv)	29, 53, 126, 233c

[a] Activity against Gram-positive bacteria = G+, Gram-negative bacteria = G−, filamentous fungi = F, mycobacteria = My; Y = yes, N = no.

[b] LD_{50} = dose killing 50% of animals (mice unless otherwise stated); Ip = intraperitoneally, Sc = subcutaneously, Iv = intravenously.

Candidulin	*Aspergillus candidus*	$C_{11}H_{15}O_3N$	G+ (?) G− (?) My Y	ca. 250 (Sc)	243, 243a
Chetomin	*Chaetomium cochliodes,* etc.	$C_{16}H_{17}O_4N_3S_2$. Contains indole, four active hydrogens, methylimide and other groups, but no basic or acidic groups	G+ Y G− Y	"not very toxic"	117, 118, 269
Citrinin	*Penicillium citrinen,* etc. *Aspergillus niveus*	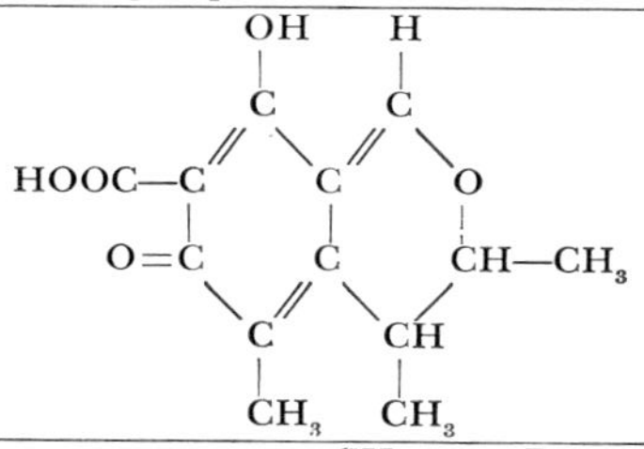	G+ Y G− N My (?) Protozoa Y	35 (Sc, Ip)	4b, 60, 71, 121, 127a, 147a, 216, 285
Enniatin (A, B, C) (Lateritiin, avenacin, fructigenin, sambucinin)	*Fusarium orthoceras* var. *enniatinum* etc.	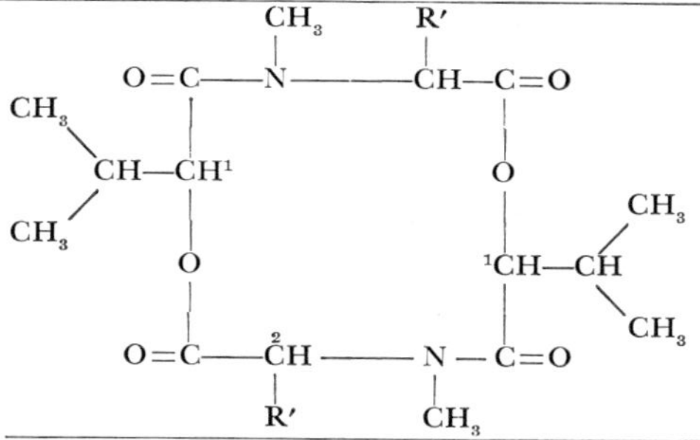 In Enniatin A: R′ = $CH(CH_3)C_2H_5$ In Enniatin B: R′ = $CH(CH_3)_2$ In Enniatin C: R′ = $CH_2CH(CH_3)_2$ [1]D-(−)-α-hydroxyisolvaleric acid 2 mols [2]*N*-methyl-L-isoleucine 2 mols (A), or *N*-methyl-L-valine 2 mols (B), or *N*-methyl-L-leucine 2 mols (C)	G+ Y G− Y	"Low"	67, 114, 210

TABLE 36 (*Continued*)

Antibiotic	Source	Formula or chemical nature	Activity[a]	Toxicity, LD_{50}[b] mg per kg	References
Fumigacin (Helvolic acid)	*Aspergillus fumigatus*	$C_{32}H_{44}O_8$ (?) (A keto monobasic acid)	G+ Y G− N	400 (Ip)	35, 63, 95, 185, 274
Fumigatin	*Aspergillus fumigatus*	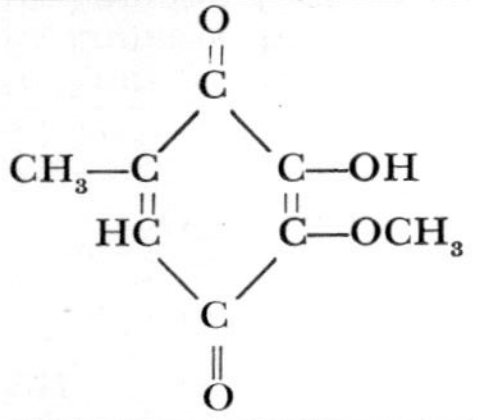	G+ Y G− N		11, 199, 271
	Aspergillus	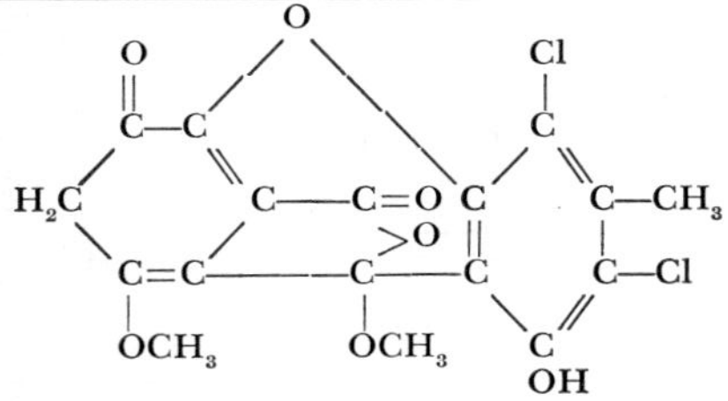 or 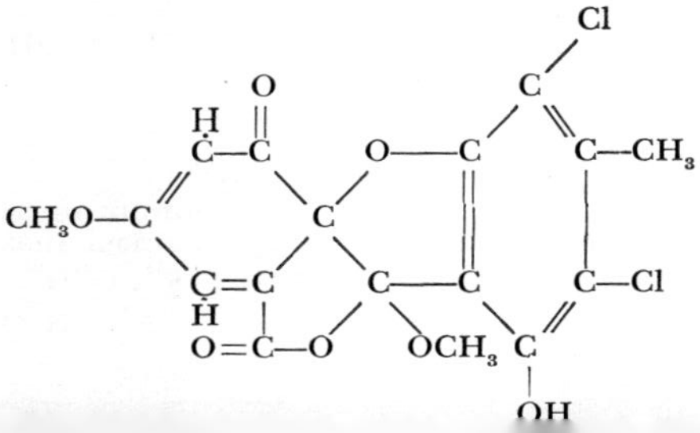	G+ Y G− N (?)		54, 174, 215, 227

Gladiolic acid	*Penicillium gladioli*	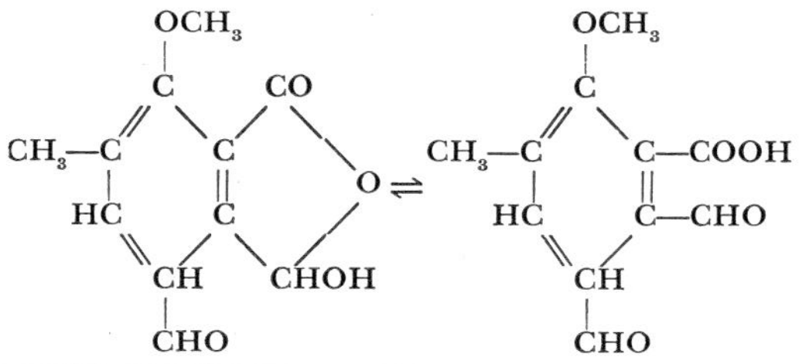	G+ Y G− N F Y		39, 41, 124a, 182
Gliotoxin	*Trichoderma lignorum (viride)* Aspergilli and penicillia	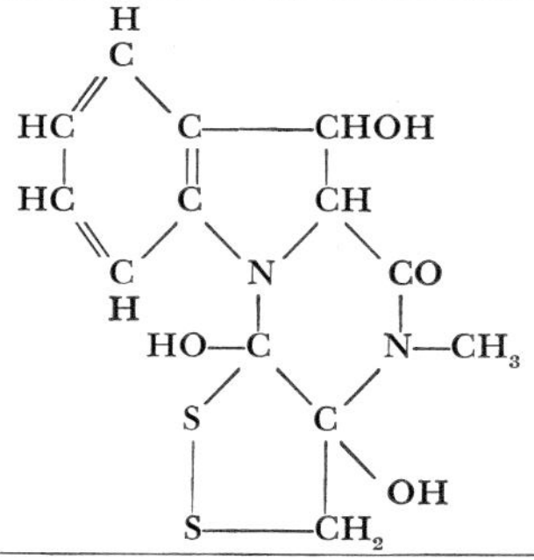	G+ Y G− N F Y My Y	19.5 (Iv)	38, 43, 88, 96, 147b, 156a, 185, 219, 233, 281, 286
Glutinosin	*Metarrhizium glutinosum*	$C_{48}H_{60}O_{16}$	G+ N (?) G− N F Y		40, 44, 46
Glyco-lipide	*Pseudomonas aeruginosa*	$C_{32}H_{58}O_{13} \cdot H_2O$ A conjugate of L-rhamnose (2 mols) and L-β-hydroxydecanoic acid (2 mols)	My Y	5 mg per mouse kills in 16 hours (Ip)	145
Griseolutein	*S. griseoluteus*	$C_{17}H_{15}O_6N_2$	G+ Y G− Y My Y	> 2000 (Sc)	197d, 259, 307a

TABLE 36 (*Continued*)

Antibiotic	Source	Formula or chemical nature	Activity[a]	Toxicity, LD_{50}[b] mg per kg	References
α-Hydroxyphenazine (hemipyocyanine)	*Pseudomonas pyocyanea* (aeruginosa)	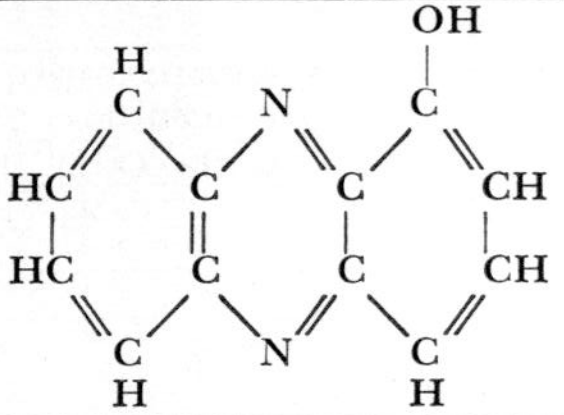	G+ Y (?) G− Y (?) F Y	Less toxic than pyocyanine	234, 249
Illudins M and S	*Clitocybe illudens*	$C_{15}H_{20}O_3$, M $C_{15}H_{22}O_4$, S	G+ Y G− Y My Y F Y	$LD_{100} = 15.6$	6
Iodinin	*Chromobacterium iodinium*	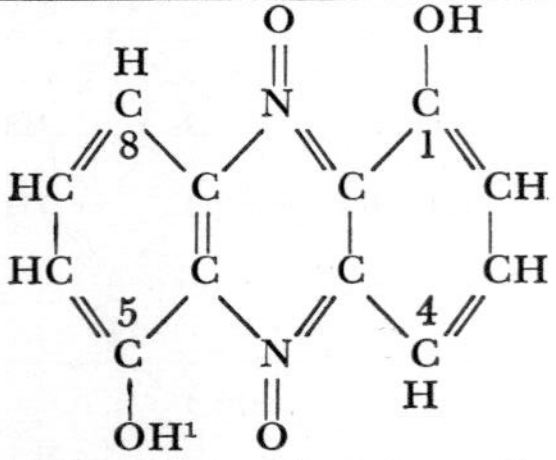 [1] OH is at either 5 or 8	G+ Y G− Slight		65, 181
	Fusarium	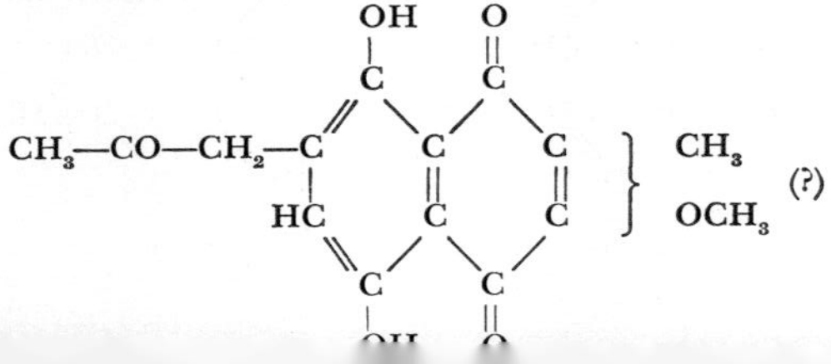	G+ Y G− N My Y	10 mg per mouse tolerated (Ip)	13, 14 231b

Kojic acid	*Aspergillus flavus,* etc.	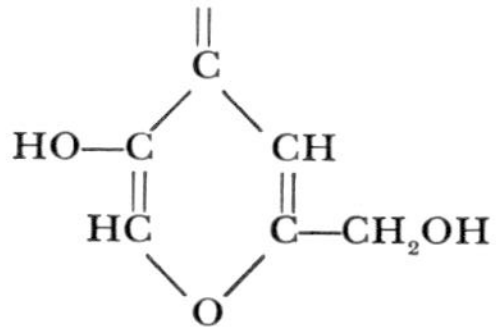	G+ Y (low) G− Y (low)	120-180 (Ip)	68, 146, 189
Lavendulin	Organism resembling *Streptomyces lavendulae*	$C_{49}H_{63}O_{18}N_{13}S_3$ (?) Lavendulin-helianthate	G+ Y G− Y	28 (Ip) 14 (Iv)	149, 156 187
Licheniformin	*Bacillus licheniformis*	Mixture of cyclic peptides. M.W. 3,800-4,000. Vary in potency and activity. Yield common amino acids on hydrolysis	G+ Y G− Y My Y	250 (Iv) 375 (Ip) 500 (Sc)	55
Marasmic acid	*Marasmius conigenus*	$C_{16}H_{20}O_4$ Unsaturated monobasic acid	G+ Y G− N F (?)	16-32 (Iv)	155
Mycophenolic acid	*Penicillium brevi-compactum,* etc.	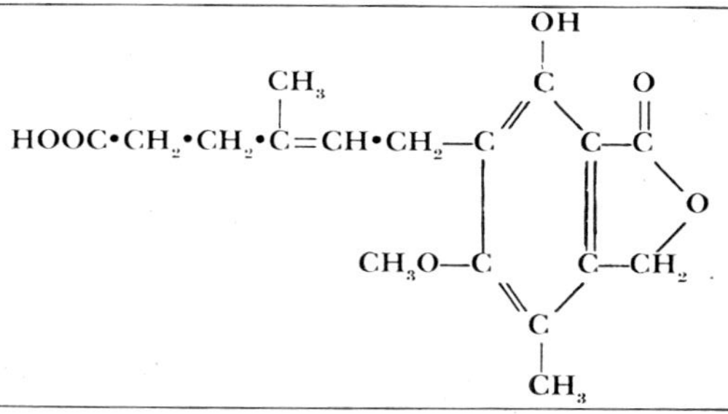	G+ Y G− N F Y	LD_{100} = 500 (Iv)	1a, 34, 36a, 106
Mycosubtilin	*Bacillus subtilis* 370	Polypeptide containing aspartic acid, tyrosine, and proline. Other amino acids uncertain.	G+ N (?) G− N F Y	25 lethal 12.5 not lethal (Sc)	284
Netropsin	*Streptomyces netropsis*	$C_{32}H_{48}O_4N_{18}$•4HCl. Gives guanidine base, $C_3H_5ON_3$, and monobasic acid, $C_{15}H_{20}O_3N_6$	G+ Y G− Y F Y	17 (Iv) 70 (Sc)	102

TABLE 36 (*Continued*)

Antibiotic	Source	Formula or chemical nature	Activity[a]	Toxicity, LD_{50}[b] mg per kg	References
Nisin	*Streptococcus lactis*, M354/07	Mixture of polypeptides. On hydrolysis gives lanthionine, cystathionine (or allocys.) and common amino acids. Cf. cinnamycin and subtilin	G+ Y	Between 20-30 (Iv, rabbit)	32, 33, 33a, 122a, 133, 134, 179
Patulin (clavacin clavatin claviformin expansin)	*Penicillium patulin*, etc. *Aspergillus clavatus*, etc.	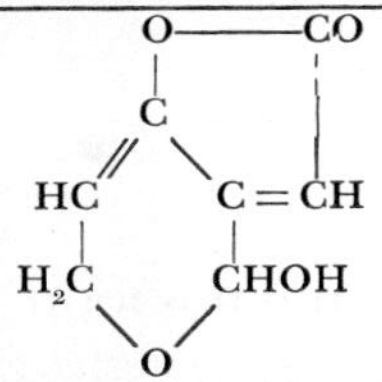	G+ Y G− Y F Y My Y	20-25 (Iv, Sc) 10 (Ip)	12, 28, 62, 74a, 146a, 193, 213, 266, 273, 305
Penicillic acid	*Penicillium puberulum*, etc. *Aspergillus ochraceus*	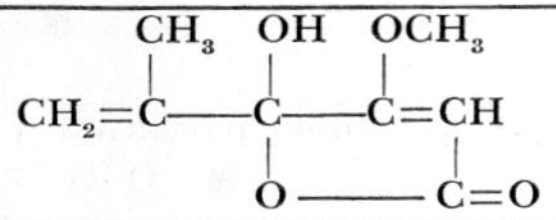	G+ Y G− Y My Y	110 (Sc) 250 (Iv)	4, 35a, 107, 190, 200, 221
Pleurotin	*Pleurotus griseus*	$C_{20}H_{22}O_5$	G+ Y G− N F Y	> 24 (Iv)	154, 229
Proactinomycin A, B, C	*Proactinomyces (Nocardia) gardneri*	A: $C_{27}H_{47}O_8N$ B: $C_{28}H_{49}O_8N$ C: $C_{24}H_{41}O_6N$	G+ Y G− (?)	A: 150 (Iv) B: 120 (Iv) C: 80 (Iv)	112, 176, 177
Prodigiosin	*Serratia marcescens*	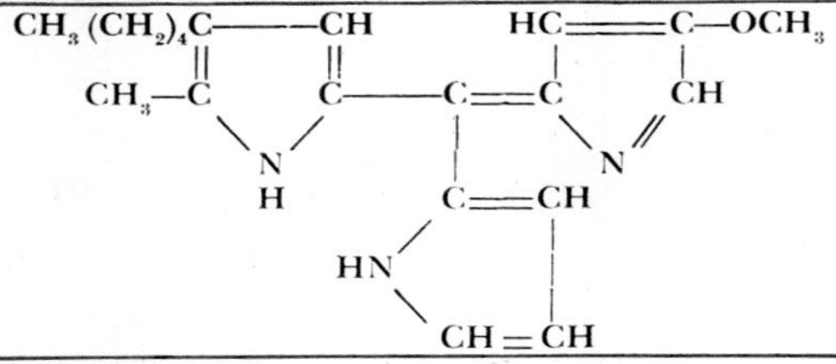	G+ Y (?) G− Y (?) F Y P Y		18a, 105, 160

Puberulic acid	*Penicillium puberulum,* etc.	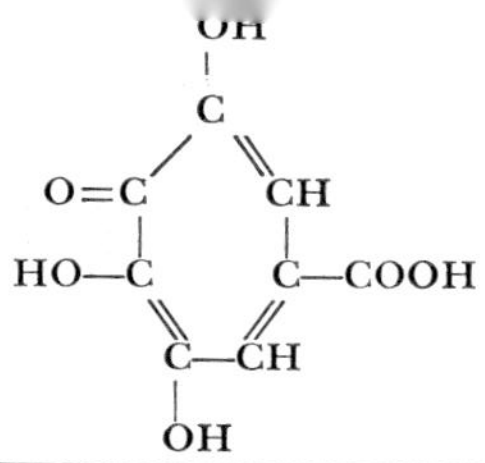	G+ Y G− N		19, 36, 69, 182
Puberulonic acid	*Penicillium aurantiovirens,* etc.	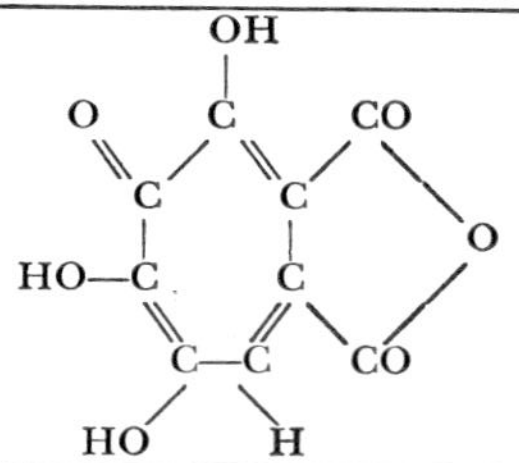	G+ Y G− N		16b, 19, 36, 69, 146b, 182
Pyo compounds	*Pseudomonas aeruginosa*	I_b, $C_{16}H_{21}NO$, 2-heptyl-4-quinolinol I_c, $C_{18}H_{25}NO$, 2-nonyl-4-quinolinol II, $C_{17}H_{23}NO_2$, ? III, $C_{18}H_{23}NO$, 2-(△′-nonenyl-)-4-quinolinol IV, $C_{16}H_{23}NO_3$, ?	G+ Y G− Slight	In carbonate 25-38 In fat about 75 (Ip)	129, 289a, 290, 291
Pyocyanine	*Pseudomonas aeruginosa,* etc.	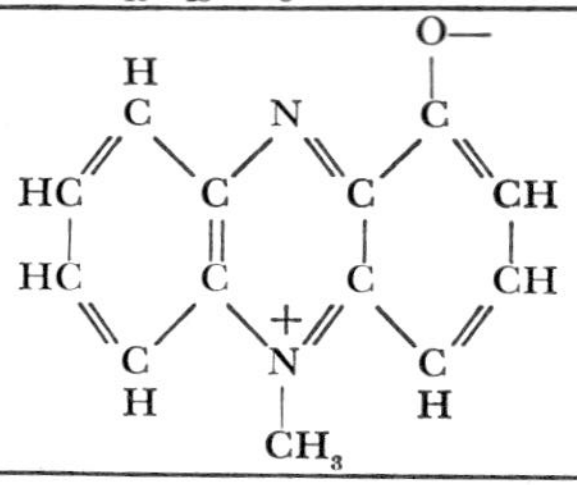	G+ Y G− YN F N	MLD is 100	198, 234, 249

TABLE 36 (*Continued*)

Antibiotic	Source	Formula or chemical nature	Activity[a]	Toxicity, LD_{50}[b] mg per kg	References
Pyolipic acid	*Pseudomonas pyocyanea*	A conjugate of L-β-hydroxydecanoic acid ($C_7H_{15}CHOH \cdot CH_2 \cdot COOH$) and L-rhamnose ($CH_3(CHOH)_4CHO$)	My Y		30, 31
Streptolin	*Streptomyces* (unidentified)	$C_{17}H_{31}O_8N_5 \cdot 3HX$ or $C_{24}H_{45}O_{11}N_7 \cdot 4HX$ Gives β-lysine,* and two other ninhydrin-positive compounds on hydrolysis.	G+ Y G− Y	5 (Iv)	161, 208, 228, 237a, 266a
Streptothricin	*Streptomyces lavendulae*	$C_{20}H_{34}O_9N_8 \cdot 3HX$ (hydrochloride, etc.) On hydrolysis gives β-lysine* and two other ninhydrin-positive products.	G+ Y G− Y My Y F Y	75 (Iv)	59, 109, 111, [illegible], 205, 218, 230, 252, 268, 301
Subtilin	*Bacillus subtilis*	Polypeptide. N.W. about 7,000. Yields eleven common amino acids, lanthionine and a new S-amino acid, probably: (Cf. cinnamycin and nisin) 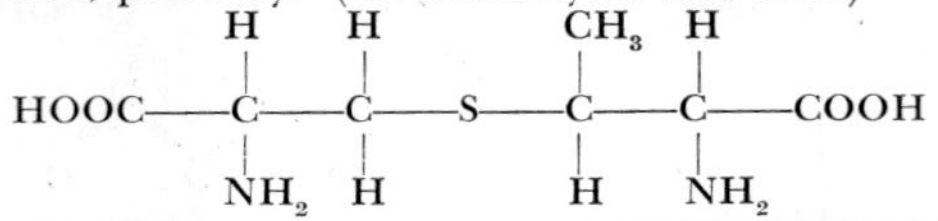	G+ Y G− Y My Y Viruses A&B of influenza	670 (Sc)	3a, 3b, 7, 47a, 56b, 97, 113, 144, 167, 168, 232, 251
Sulfactin	*Actinomyces* Sp. (possibly *roseus*)	$C_{38}H_{55}O_7N_{11}S_4$ or $C_{27}H_{40}O_5N_8S_3$	G+ Y G− N	149 (Ip)	150, 188
Thiolutin (cf. aureothricin)	*Streptomyces albus*	$C_8H_8N_2O_2S_2$. Yellow-orange neutral crystals. Acid hydrolysis gives acetic acid and a monoamine, $C_6H_6N_2OS_2$.	G+ Y G− Y My Y F Y P Y	25 (Sc)	61a, 235b, 257

* β-Lysine is $NH_2CH_2CH_2CH_2CH(NH_2)CH_2COOH$.

Toluquinone (5-methoxy-p-toluquinone)	*Coprinus simulus* *Lentinus degner*	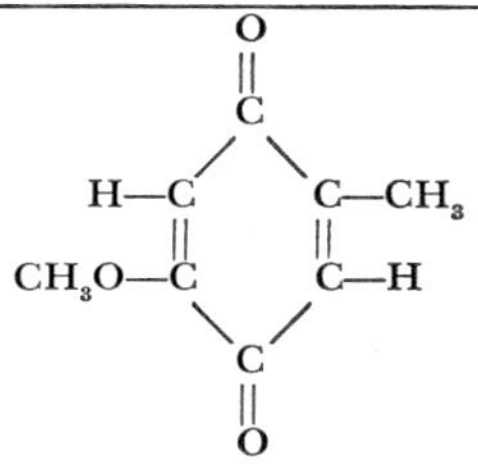	G+ Y G− Y F Y	25 (Iv)	5
Ustin, I, II, III	*Aspergillus ustus*	I, $C_{21}H_{17}O_6Cl_3$ II, $C_{21}H_{18}O_6Cl_2$ III, $C_{19}H_{15}O_5Cl_3$	G+ Y G− N My Y	6-8 mg tolerated by mice	77, 138, 216a
Viridin	*Trichoderma viride*	$C_{19}H_{16}O_6$. Contains one methoxy group. Two components are isomeric, but differ in properties. Very active against fungi.	G+ N G− N F Y		42, 45, 182, 267a
Xanthomycin	*Streptomyces* (unidentified)	$C_{23}H_{30}O_7N_3$•2HX. It is a methyloxylated quinone joined to a basic moiety. Gives methylamine, ethanolamine, etc., on degradation.	G+ Y G− Y	0.16 (Iv)	186, 220a, 258

TABLE 37. SUPPLEMENTARY TABLE OF WELL-DEFINED ANTIBIOTICS*

Antibotic	Source	Chemical Nature	Activity**	Toxicity,† LD_{50}, mg per kg	References
Actithiazic acid (Mycobacidin)	*Streptomyces virginiae, S. cinnamonensis*	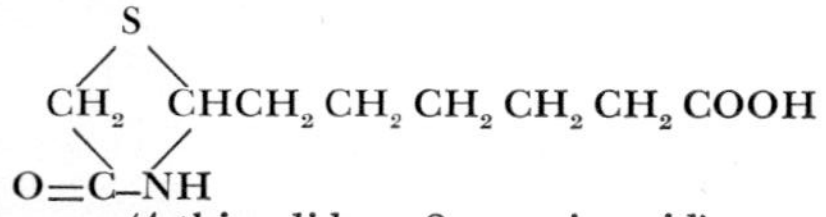 (4-thiazolidone-2-caproic acid)	My Y G+ N G− N	More than 1500 (Iv or Sc)	S17, S28, S36, S43, S46, S55, S56, S59
Alternaric acid	*Alternaria solani*	$C_{21}H_{30}O_8$ (Unsaturated dibasic acid)	F Y (low or no activity for bacteria)	Toxic to some plants	S8, S9, S26, S50a
Amicetin	*Streptomyces* sp.	$C_{29}H_{44}N_6O_9$. Yields cytosine, *p*-aminobenzoic acid, D-α-methyl serine and an unidentified product.	G+ Y G− Y My Y (marked)	90 (Iv) 600-700 (Sc)	S20, S20a, S45a
Bacillomycin	*B. subtilis*	Consists of several peptides. B component gives common amino acids probably high in glutamic.	G+ N G− N F Y	50-75 (Iv)	S40, S60, S65
Candicidin A, B, and C.	*Streptomyces griseus*	A: C = 62.9%; H = 9.6; N = 4.7 B: C = 57.8%; H = 9.9; N = 7.3	G+ N G− N My N F Y	Crude 663, A 277, B 159 } Sc Crude 79, A 47, B 53 } Iv	S38, S41
Catenulin	*Streptomyces* sp.	C = 50.2%; H = 5.3; N = 10.1 S = 7.9 (sulfate)	My Y	125 (Iv)	S19

* January 1951 to June 1953.

** Y = yes; N = no; G+ = Gram-positive bacteria; G− = Gram-negative bacteria; F = fungi; My = mycobacteria; P = protozoa.

† An asterisk (*) in this column indicates no data found in published papers.

Cephalosporins P_1, P_2, P_3, P_4, P_5, N	*Cephalosporium* unknown species	P_1, P_2, P_4 (acidic) crystallized P_1: $C = 65.8\%$; $H = 9.0$; $C—CH_3 = 12.1$, $acetyl = 16.8$, active $H = 0.6$. P_2: $C = 65.25$; $H = 8.5$; $C—CH_3 = 9.4$, $OCH_3 = 5.8$; $acetyl = 18.8$; active $H = 0.8$.	P { G+ Y, G− N N { G+ Y, G− Y	P_1, 500 (Iv)	S15, S48a, S53
Cinnamycin	*Streptomyces cinnamoneus*	Polypeptide. Contains common amino acids; also lanthionine and $C_7H_{14}N_2O_4S$ amino acid; cf. nisin and subtilin.	G+ Y My Y	*	S2
Cordycepin	*Cordyceps militaris*	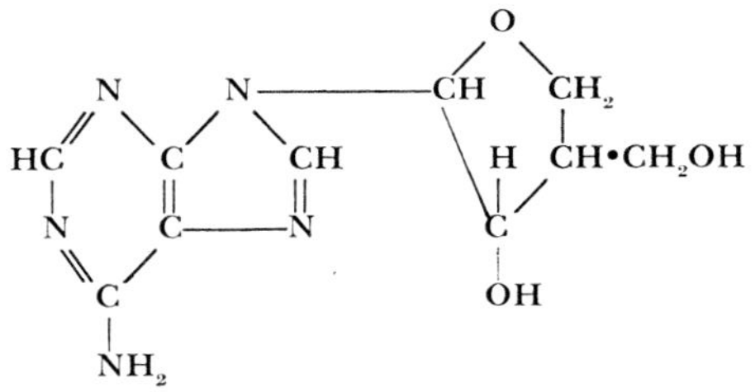	G+ Y My Y	"Of a low order"	S3, S18, S26
Exfoliatin	*Streptomyces exfoliatus*	$C_{27}H_{40}O_{16}Cl \cdot H_2O$ (?)	G+ Y G− N (except hemophilus)	500 (Sc)	S63
Fungistatin (antibiotic XG)	*Bacillus subtilis*	Polypeptide. Gives common amino acids.	F Y	90 (Iv)	S34, S41a, S51
Grisein	*Streptomyces griseus*	Probably $C_{40}H_{61}N_{10}O_{20}SFe$.	G+ Some G− " My " F N	None reported even with large doses	S39, S52

TABLE 37 (*Continued*)

Antibiotic	Source	Chemical Nature	Activity**	Toxicity,† LD_{50}, mg per kg	References
Griseofulvin	*Penicillium griseofulvum*	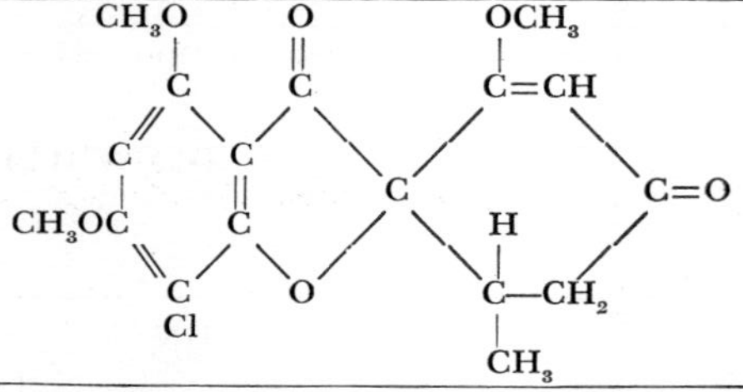	G+ N G− N F Y	Toxic to some plants	S6, S7, S27, S42, S47a, S49
Luteomycin	*Streptomyces* n. sp.	$C_{25}H_{29}NO_9 \cdot HCl$	G+ Y G− N	6 (Iv or Sc)	S30, S31, S48, S54
6-Methyl-1,4,—naphthoquinone	*Marasmius graminium*	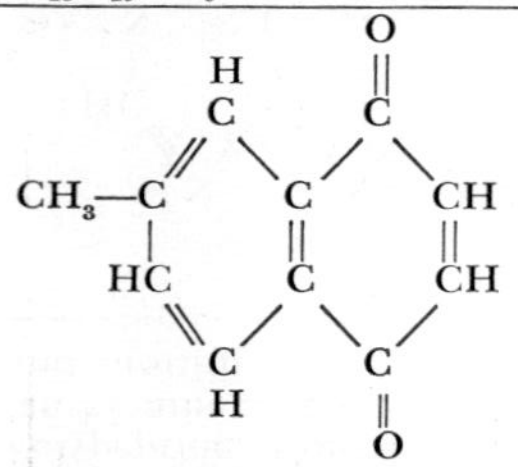	G+ Y	*	S1a
Mycomycin	*Nocardia acidophilus* n. sp.	HC≡C—C≡CCH=C=CHCH=CH—CH=CHCH$_2$COOH 3,5,7,8-tridecatetraene-10,12-diynoic acid.	G+ Y G− Y My Y F Y	Apparently nontoxic	S16, S37
Nigericin	*Streptomyces* unidentified	$C_{39}H_{69}O_{11}$ (monobasic acid)	G+ Y G− N My Some	2.5 (Ip)	S29

Nitrosporin	*Streptomyces nitrosporeus* n. sp.	$C_{20}H_{26}N_2O_6$ (?)	G+ Y G– Y	16 (Iv)	S62
Nocardamin	*Nocardia*	Proposed: 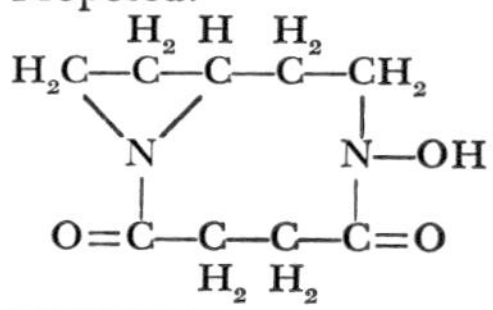	G+ "little" G– " My Y		S57, S58
Pikromycin (Cf. Proactinomycin in Table 36)	*Actinomyces* sp.	$C_{25}H_{43}NO_7$ (N present in dimethylamino group)	G+ Y	Lethal dose (Iv) 100-700	S13
Polypeptin	*Bacillus krzemieniewski*	Mixture of basic polypeptides. Main component resembles polymyxius in amino acid content.	G+ Y G– Y F Y	15	S31a, S35, S47
Puromycin	*Streptomyces alboniger*	$C_{22}H_{29}N_7O_5$ 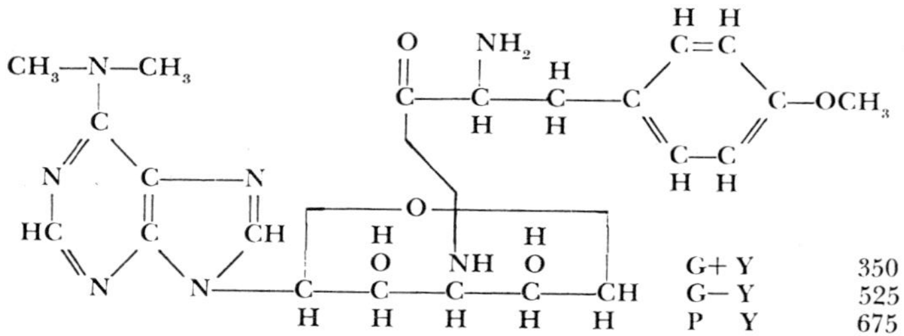	G+ Y G– Y P Y	350 (Iv) 525 (Ip) 675 (oral)	S1, S50, S64

TABLE 37 (*Continued*)

Antibiotic	Source	Chemical Nature	Activity**	Toxicity,† LD_{50}, mg per kg	References
Resistomycin	*Streptomyces resistomycificus*	$C_{23}H_{18}O_6$ "weakly acid" "4 active H atoms"	G+ Y My Y	*	S14
Rhodomycins (A and B)	*Streptomyces purpurascens*	$C_{20}H_{29}NO_7$ or $C_{40}H_{58}N_2O_{14}$	G+ Y G– slight F N	*	S10, S11, S12
Thioaurin	*Streptomyces* sp.	$C_{14}H_{12}O_4N_4S_4$ (?)	G+ Y G– Y	16 (Iv)	S5
Trichothecin	*Trichothecium roseum* Link	$C_{19}H_{24}O_5$ (an ester)	F Y	(for plant disease only)	S21, S22, S23
Vinactin A, B, C (Cf. viomycin)	*Actinomyces vinaceus*	Peptides. A and B give lysine, serine, alanine, glycine, glutamic acid, aspartic acid. C gives serine, sarcosine, glycine, and an unidentified amino acid.	G+ Y G– Y My Y Rickettsia Y	*	S45, S61
X-206 X-464 X-537A	*Streptomyces* unidentified species	$C_{46-47}H_{80-82}O_{13}$ $C_{25}H_{40}O_7$ $C_{34}H_{52}O_8$ Optically active organic acids	All alike G+ Y G– N F N	11 (Sc) 2.5 (Ip) 40 (Ip)	S4

In spite of the fact that only antibiotics for which there are considerable chemical data have been included in Table 36, a study of these data shows that there is a dearth of knowledge on the chemical nature of these substances. Structural formulas are known for less than half of the compounds. It is difficult, costly, and time consuming to work out the structure of antibiotics, but the picture will not be clear until the structure is known. Only then can a proper understanding of the chemotherapeutic action of the antibiotics be obtained. The chemical structures that are known are so varied that only a beginning can be made toward developing a theory as to the type of structure a compound must have in order to show biological activity. The mode of action of antibiotics is probably more obscure even than their structural specificity. In either field there is abundant opportunity for fundamental work.

A related and equally obscure feature of antibiotics is the reason for their toxicity. The LD_{50} figures range from 0.16 mg per kg for xanthomycin to several hundred times that figure for a number of the others. It is not unthinkable that some of the compounds now regarded as much too toxic may prove useful against certain infections when administered under carefully controlled conditions. The relative nontoxicity of penicillin has led the medical profession to expect effective antibiotics of low toxicity. Such compounds are rare and it may be worth while to go back and restudy some of those that have been discarded as well as to continue the search for new compounds with antibiotic activity.

ACKNOWLEDGMENTS

The authors are indebted to the scientific staffs of many commercial producers of antibiotics for bibliographies and unpublished information, for reading the sections of this chapter concerned with their special products, and for helpful criticism of the contents of these sections. Credit should go to the following individuals and companies for this assistance: R. D. Coghill and J. C. Sylvester, Abbott Laboratories; R. G. Shepherd, American Cyanamid Co.; D. S. Searle, Burroughs Wellcome and Co. (U.S.A.), Inc.; J. Martin and H. E. Stavely, Commercial Solvents Corp.; J. M. McGuire and G. W. Probst, Eli Lilly and Co.; N. Bohonos, Lederle Laboratories Division, American Cyanamid Co.; N. Sjolander, Heyden Chemical Corp.; K. Folkers, R. L. Peck, and H. B. Woodruff, Merck and Co.,

Inc.; J. Ehrlich, Parke, Davis and Co.; G. St. Clair, S. B. Penick and Co.; A. C. Finlay, J. H. Kane, P. P. Regna, and E. R. Weyer, Chas. Pfizer and Co., Inc.; R. Donovick, J. D. Dutcher, A. F. Langlykke, and O. Wintersteiner, E. R. Squibb and Sons; G. F. Cartland, D. R. Colingsworth, J. H. Ford, and D. H. Peterson, The Upjohn Co.; K. L. Howard, Wallerstein Co., Inc. The authors are also indebted to several other persons for helpful information; *viz.*, R. E. Feeney, J. C. Lewis, H. Umezawa, S. A. Waksman, and H. Welch. The individuals to whom we have made acknowledgments are, however, in no way responsible for the statements or interpretations contained in this chapter. While we have made strenuous efforts to avoid errors of omission and commission, there are undoubtedly some of each kind in this review. For any such errors the authors bespeak the indulgence of readers and offer in advance their apologies to the investigators who may become victims of these errors.

BIBLIOGRAPHY

1. Abbott Laboratories, *Personal communication.*
1a. Abraham, E. P. *Biochem. J.*, **39**, 398 (1945).
2. Ahmad, K., H. G. Schneider, and F. M. Strong, *Arch. Biochem.*, **28**, 281 (1950).
3. Ainsworth, G. C., A. M. Brown, and G. Brownlee, *Nature*, **160**, 263 (1947).
3a. Alderton, G., *J. Am. Chem. Soc.*, **75**, 2391 (1953).
3b. Alderton, G., and H. L. Fevold, *J. Am. Chem. Soc.*, **73**, 463 (1951).
4. Alsberg, C. L., and O. F. Black, *U. S. Dept. Agr. Bur. Plant Ind., Bull. 270* (1913).
4a. Amberson, J. A., Transactions, 11th Conference on the Chemotherapy of Tuberculosis, *U. S. Veterans Administration 286* (1952).
4b. Ambrose, A. M., and F. DeEds, *J. Pharmacol. Exptl. Therap.*, **88**, 173 (1946).
4c. Ambrus, J. L., C. M. Ambrus, C. N. Sideri, and J. W. E. Harrisson, *Antibiotics and Chemotherapy*, **3**, 16 (1953).
4d. American Cyanamid Co., British Patent 672,510 (1952).
5. Anchel, M., A. Hervey, F. Kavanagh, J. Polatnick, and W. J. Robbins, *Proc. Natl. Acad. Sci. U. S.*, **34**, 498 (1948).
6. Anchel, M., A. Hervey, and W. J. Robbins, *Proc. Natl. Acad. Sci. U. S.*, **36**, 300 (1950); **38**, 927 (1952).

7. Anderson, H. H., G. G. Villela, E L. Hansen, and R. K. Reed, *Science,* **103**, 419 (1946).
7a. Anderson, R. C., P. N. Harris, and K. K. Chen, *J. Am. Pharm. Assoc.,* **41**, 555 (1952).
7b. Anderson, H. W., and D. Gottlieb, *Econ. Bot.,* **6**, 294 (1952).
8. Anker, H. S., B. A. Johnson, J. Goldberg, and F. L. Meleney, *J. Bact.,* **55**, 249 (1948).
9. Anon., *Bi-Con Feed Supplements,* New York, Chas. Pfizer and Co., 1950.
10. Anon., *Bacitracin, A New Antibiotic,* Commercial Solvents Corp., 1949.
10a. Anon., Circ., Upjohn Co., 1952.
10b. Anon., *Erythrocin,* Abbott Laboratories, 1952.
10c. Anon., Circular, Chas. Pfizer and Co., 1953.
11. Anslow, W. K., and H. Raistrick, *Biochem. J.,* **32**, 687 (1938).
12. Anslow, W. K., H. Raistrick, and G. Smith, *J. Soc. Chem. Ind.,* **62**, 236 (1943).
13. Arnstein, H. R. V., and A. H. Cook, *J. Chem. Soc.,* **1947**, 1021.
14. Arnstein, H. R. V., A. H. Cook, and M. S. Lacey, *Brit. J. Exptl. Path.,* **27**, 349 (1946).
15. Arriagada, A., H. W. Florey, M. A. Jennings, and I. G. Wallmark, *Brit. J. Exptl. Path.,* **30**, 458 (1949).
16. Arriagada, A., M. C. Savage, E. P. Abraham, N. G. Heatley, and A. E. Sharp, *Brit. J. Exptl. Path.,* **30**, 425 (1949).
16a. Asheshov, I. N., F. Strelitz, and F. A. Hall, *Antibiotics and Chemotherapy,* **2**, 361 (1952).
16b. Aulin-Erdtman, G., *Acta Skand.,* **5**, 301 (1951).
17. Bachman, M. C., *J. Clin. Invest.,* **28**, 864 (1949).
18. Bailey, J. H., and C. J. Cavallito, *Ann. Rev. Microbiol.,* **2**, 143 (1948).
18a. Baker, B. R., and R. E. Schaub., *J. Am. Chem. Soc.,* **75**, 3864 (1953).
18b. Balamuth, W., and M. M. Brent, *Proc. Soc. Exptl. Biol. Med.,* **75**, 374 (1950).
19. Barger, G., and O. Dorrer, *Biochem. J.,* **28**, 11 (1934).
19a. Baron, A. L., *Handbook of Antibiotics,* New York, Reinhold, 1950.
20. Barry, G. T., J. C. Gregory, and L. C. Craig, *J. Biol. Chem.,* **175**, 485 (1948).
21. Bartz, Q. R., *J. Biol. Chem.,* **172**, 445 (1948).
22. Bartz, Q. R., U. S. Patent 2,483,871 (1949).

23. Bartz, Q. R., J. Ehrlich, J. D. Mold, M. A. Penner, and R. M. Smith, *Am. Rev. Tuberc.*, **63**, 4 (1951).
23a. Battersby, A. R., and L. C. Craig, *J Am. Chem. Soc.*, **73**, 1887 (1951).
23b. Battersby, A. R., and L. C. Craig, *J. Am. Chem. Soc.*, **74**, 4019, 4023 (1952).
24. Bell, J. A., M. Pittman, and B. J. Olson, *Public Health Repts.*, **64**, 589 (1949).
25. Bell, P. H., J. F. Bone, J. P. English, C. E. Fellows, K. S. Howard, M. M. Rogers, R. G. Shepherd, and R. Winterbottom, *Ann. N. Y. Acad. Sci.*, **51**, 897 (1949).
26. Benedict, R. G., and A. F. Langlykke, *Ann. Rev. Microbiol.*, **1**, 193 (1947).
27. Benedict, R. G., and A. F. Langlykke, *J. Bact.*, **54**, 24 (1947).
28. Bergel, F., A. L. Morrison, A. R. Moss, R. Klein, H. Rinderknecht, and J. L. Ward, *Nature,* **152**, 750 (1943).
29. Berger, J., L. M. Jampolsky, and M. W. Goldberg, *Arch. Biochem.*, **22**, 476 (1949).
30. Bergström, S., H. Theorell, and H. Davide, *Arch. Biochem.*, **10**, 165 (1946).
31. Bergström, S., H. Theorell, and H. Davide, *Arkiv. Kemi Mineral Geol.*, (A) **23** (13), 1 (1946).
32. Berridge, N. J., *Lancet,* **253**, 5 (1947).
33. Berridge, N. J., *Biochem. J.*, **45**, 486 (1949).
33a. Berridge, N. J., G. G. F. Newton, and E. P. Abraham, *Biochem. J.*, **52**, 529 (1952).
34. Birkinshaw, J. H., A. Bracken, E. N. Morgan, and H. Raistrick, *Biochem. J.*, **43**, 216 (1948).
35. Birkinshaw, J. H., A. Bracken, and H. Raistrick, *Biochem. J.*, **39**, 70 (1945).
35a. Birkinshaw, J. H., A. E. Oxford, and H. Raistrick, *Biochem. J.*, **30**, 394 (1936).
36. Birkinshaw, J. H., and H. Raistrick, *Biochem. J.*, **26**, 441 (1932).
36a. Birkinshaw, J. H., H. Raistrick, and D. J. Ross, *Biochem. J.*, **50**, 630 (1952).
37. Bliss, E. A., and P. H. Todd, *J. Bact.*, **58**, 61 (1949).
37a. Bohonos, N., A. C. Dornbush, L. I. Feldman, J. H. Martin, E. Pelcak, and J. H. Williams, *Antibiotics Annual,* 49, 1953-54, New York, Medical Encyclopedia Inc. Also personal communication from N. Bohonos.
37b. Boothe, J. H., J. Morton II, J. P. Petisi, R. G. Wilkinson, and J. H. Williams, *J. Am. Chem. Soc.*, **75**, 4621 (1953).

Complete paper in *Antibiotics Annual,* 46, 1953-54, New York, Medical Encyclopedia Inc.

37c. Brainerd, H. D., H. B. Bruyn, G. Meiklejohn, and L. O'Gara, *Antibiotics and Chemotherapy,* **1**, 447 (1951).

37d. Braude, R., H. D. Wallace, and T. J. Cunha, *Antibiotics and Chemotherapy,* **3**, 271 (1953).

38. Brian, P. W., *Nature,* **154**, 667 (1944).

39. Brian, P. W., P. J. Curtis, J. F. Grove, H. G. Hemming, and J. C. McGowan, *Nature,* **157**, 697 (1946).

40. Brian, P. W., P. J. Curtis, and H. G. Hemming, *Proc. Roy. Soc. (London)* (B), **135**, 106 (1947).

41. Brian, P. W., P. J. Curtis, and H. G. Hemming, *J. Gen. Microbiol.,* **2**, 341 (1948).

42. Brian, P. W., P. J. Curtis, H. G. Hemming, and J. C. McGowan, *Ann. Applied Biol.,* **33**, 190 (1946).

43. Brian, P. W., and H. G. Hemming, *Ann. Applied Biol.,* **32**, 214 (1945).

44. Brian, P. W., H. G. Hemming, and E. G. Jefferys, *Mycologia,* **40**, 363 (1948).

45. Brian, P. W., and J. C. McGowan, *Nature,* **156**, 144 (1945).

46. Brian, P. W., and J. C. McGowan, *Nature,* **157**, 334 (1946).

47. Briggs, G. M., *Trans. Am. Assoc. Cereal Chem.,* **10**, 31 (1952).

47a. Brink, N. G., J. Mayfield, and K. Folkers, *J. Am. Chem. Soc.,* **73**, 330 (1951).

48. Brockman, H., N. Grubhofer, W. Kass, and H. Kalbe, *Chem. Ber.,* **84**, 260 (1951).

49. Broschard, R. W., A. C. Dornbush, S. Gordon, B. L. Hutchings, A. R. Kohler, G. Krupka, S. Kushner, D. V. Lefemine, and C. Pidacks, *Science,* **109**, 199 (1949).

50. Brownlee, G., *Ann. N. Y. Acad. Sci.,* **51**, 875 (1949).

51. Brownlee, G., *Ann. N. Y. Acad. Sci.,* **51**, 998 (1949).

52. Brownlee, G., S. R. M. Bushby, and E. I. Short, *Ann. N. Y. Acad. Sci.,* **51**, 891 (1949).

52a. Brownlee, G., S. R. M. Bushby, and E. I. Short, *Brit. J. Pharmacol. Chem.,* **7**, 170 (1952).

53. Buck, M., A. C. Farr, and R. J. Schnitzer, *Trans. N. Y. Acad. Sci.,* (2) **11**, 207 (1949).

53a. Bunch, R. L., and J. M. McGuire, U. S. Patent 2,653,899 (1953).

54. Calam, C. T., P. W. Clutterbuck, A. E. Oxford, and H. Raistrick, *Biochem. J.,* **41**, 458 (1947).

55. Callow, R. K., and T. S. Work, *Biochem. J.,* **51**, 558 (1952).

56. Carpenter, L. E., *Arch. Biochem.*, **27**, 469 (1950).
56a. Carr, D. T., K. H. Pfuetze, H. A. Brown, B. E. Douglass, and A. G. Karlson, *Am. Rev. Tuberc.*, **63**, 427 (1951).
56b. Carson, J. F., *J. Am. Chem. Soc.*, **74**, 1480 (1952).
57. Carter, H. E., and J. H. Ford, *Ann. Rev. Biochem.*, **19**, 487 (1950).
58. Carter, H. E., D. Gottlieb, and H. W. Anderson, *Science*, **107**, 113 (1948).
59. Carter, H. E., W. R. Hearn, E. M. Lansford, A. C. Page, N. P. Salsman, D. Shapiro, and W. R. Taylor, *J. Am. Chem. Soc.*, **74**, 3704 (1952).
60. Cartwright, N. J., A. Robertson, and W. B. Whalley, *J. Chem. Soc.*, **1949**, 1563.
60a. Castenada, M. R., and G. G. Ibarra, *Antibiotics and Chemotherapy*, **2**, 86 (1952).
61. Catch, J. R., T. S. G. Jones, and S. Wilkinson, *Ann. N. Y. Acad. Sci.*, **51**, 917 (1949).
61a. Celmer, W. D., F. W. Tanner, M. Harfenist, T. M. Lees, and I. A. Solomons, *J. Am. Chem. Soc.*, **74**, 6304 (1952).
62. Chain, E. B., H. W. Florey, and M. A. Jennings, *Brit. J. Exptl. Path.*, **23**, 202 (1942).
63. Chain, E., H. W. Florey, M. A. Jennings, and T. I. Williams, *Brit. J. Exptl. Path.*, **24**, 108 (1943).
64. Chittenden, G. A., E. A. Sharp, E. C. Vonder Heide, A. C. Bratton, A. J. Glazko, and F. D. Stimpert, *J. Urol.*, **62**, 771 (1949).
64a. Clark, R. K., *Antibiotics and Chemotherapy*, **3**, 663 (1953).
65. Clemo, G. R., and A. F. Daglish, *Nature*, **162**, 776 (1948).
65a. Coates, M. E., C. D. Dickinson, G. F. Harrison, S. K. Kon, J. W. G. Porter, S. H. Cummins, and W. F. J. Cuthbertson, *J. Sci. Food Agr.*, **3**, 43 (1952).
65b. Coffey, G. L., J. E. Oyaas, and J. Ehrlich, *Antibiotics and Chemotherapy*, **1**, 203 (1951).
65c. Conover, L. H., W. T. Moreland, A. R. English, C. R. Stephens, and F. J. Pilgrim, *J. Am. Chem. Soc.*, **75**, 4622 (1953).
66. Controulis, J., M. C. Rebstock, and H. M. Crooks, *J. Am. Chem. Soc.*, **71**, 2463 (1949).
67. Cook, A. H., S. F. Cox, and T. H. Farmer, *J. Chem., Soc.*, **1949**, 1022.
68. Cook, A. H., and M. S. Lacey, *Nature*, **155**, 790 (1945).
69. Corbett, R. E., A. W. Johnson, and A. R. Todd, *J. Chem. Soc.*, **1950**, 6.

70. Craig, L. C., J. D. Gregory, and G. T. Barry, *J. Clin. Invest.*, **28**, 1014 (1949).
70a. Craig, L. C., W. Hausman, and J. R. Weisiger, *J. Biol. Chem.*, **199**, 865 (1952).
70b. Craig, L. C., J. R. Weisiger, W. Hausman, and E. J. Harfenist, *J. Biol. Chem.*, **199**, 259 (1952).
71. Cram, D. J., *J. Am. Chem. Soc.*, **70**, 440 (1948).
72. Crooks, H. M., M. C. Rebstock, J. Controulis, and Q. R. Bartz, U. S. Patents 2,483,884 and 2,483,885 (1949).
72a. Cunningham, R. W., L. R. Hines, E. H. Stokey, R. E. Vessey, and N. N. Yuda, *Antibiotics Annual*, 63, 1953-54, New York, Medical Encyclopedia Inc.
73. Cutherbertson, W. F. J., *J. Sci. Food Agr.*, **3**, 49 (1952).
74. Dalgliesh, C. E., A. W. Johnson, A. R. Todd, and L. C. Vining, *J. Chem. Soc.*, **1950**, 294.
74a. Darken, M. A., and N. O. Sjolander, *Antibiotics and Chemotherapy*, **1**, 573 (1951).
74b. Davisson, J. W., F. W. Tanner, A. C. Finlay, and I. A. Solomons, *Antibiotics and Chemotherapy*, **1**, 289 (1951).
75. Debono, J. E. *Lancet*, **257**, 326 (1949).
76. Derzavis, J. L., J. S. Rice, and L. S. Leland, *J. Am. Med. Assoc.*, **141**, 191 (1949).
77. Doering, W. E., R. J. Dubos, D. S. Noyce, and R. Dreyfus, *J. Am. Chem. Soc.*, **68**, 725 (1946).
78. Donovick, R., *Personal communication.*
79. Dornbush, A. C., and E. J. Pelcak, *Ann. N. Y. Acad. Sci.*, **51**, 218 (1949).
80. Dubos, R. J., *J. Exptl. Med.*, **70**, 1 (1939).
81. Duggar, B. M., *Ann. N. Y. Acad. Sci.*, **51**, 177 (1948).
82. Duggar, B. M., U. S. Patent 2,482,055 (1949).
83. Duncan, G. G., C. F. Clancy, J. R. Wolgamot, and B. Beidleman, *J. Am. Med. Assoc.*, **145**, 75 (1951).
84. Dunn, G., J. J. Gallagher, G. T. Newbold, and F. S. Spring, *J. Chem. Soc.* (Suppl. Issue No. 1), S126 (1949); 2091 (1949).
85. Dunshee, B. R., C. Leben, G. W. Keitt, and F. M. Strong, *J. Am. Chem. Soc.*, **71**, 2436 (1949).
86. Dutcher, J. D., *J. Biol. Chem.*, **171**, 321, 341 (1947).
86a. Dutcher, J. D., and M. D. Donin, *J. Am. Chem. Soc.*, **74**, 3420 (1952).
87. Dutcher, J. D., N. Hosanky, M. N. Donin, and O. Wintersteiner, *J. Am. Chem. Soc.*, **73**, 1384 (1951).
88. Dutcher, J. D., J. R. Johnson, and W. F. Bruce, *J. Am. Chem. Soc.*, **67**, 1736 (1945).

88a. Dutcher, J. D., J. Vandeputte, S. Fox, and L. J. Heuser, *Antibiotics and Chemotherapy,* **3**, 910 (1953).
89. Eagle, H., A. D. Musselman, and R. Fleischman, *J. Bact.,* **55**, 347 (1948).
90. Eagle, H., and R. Fleischman, *Proc. Soc. Exptl. Biol. Med.,* **68**, 415 (1948).
90a. Eble, T. E., and F. R. Hanson, *Antibiotics and Chemotherapy,* **1**, 54 (1951).
91. Ehrlich, J., Q. R. Bartz, R. M. Smith, D. A. Joslyn, and P. R. Burkholder, *Science,* **106**, 417 (1947).
92. Ehrlich, J., D. Gottlieb, P. R. Burkholder, L. E. Anderson, and T. G. Pridham, *J. Bact.,* **56**, 467 (1948).
93. Ehrlich, J., R. M. Smith, and M. A. Penner, U. S. Patent 2,483,892 (1949).
94. Ehrlich, J., R. M. Smith, M. A. Penner, L. E. Anderson, and A. C. Bratton, *Am. Rev. Tuberc.,* **63**, 7 (1951).
95. Elliot, W. H., P. A. Katzman, S. A. Thayer, and E. A. Doisy, *Federation Proc.,* **6**, 250 (1947).
96. Elvidge, J. A., and F. S. Spring, *J. Chem. Soc.,* **1949**, S135.
96a. English, A. R., M. F. Field, S. R. Szendy, N. J. Tagliani, and R. A. Fitts, *Antibiotics and Chemotherapy,* **2**, 678 (1952).
96b. English, A. R., H. E. Mullady, and R. A. Fitts, *Antibiotics and Chemotherapy,* **3**, 94 (1953).
96c. English, A. R., S. Y. P'an, J. F. Gardocki, and W. A. Wright, *Antibiotics Annual,* 70, 1953-54, New York, Medical Encyclopedia Inc.
97. Fevold, H. L., K. P. Dimick, and A. A. Klose, *Arch. Biochem.,* **18**, 27 (1948).
98. Finland, M., H. S. Collins, T. M. Gocke, and E. B. Wells, *Ann. Internal Med.,* **31**, 39 (1949).
99. Finland, M., P. F. Frank, and C. Wilcox, *Am. J. Clin. Path.,* **20**, 325 (1950).
100. Finland, M., and C. Wilcox, *Am. J. Clin. Path.,* **20**, 335 (1950).
101. Finland, M., C. Wilcox, and P. F. Frank, *Am. J. Clin. Path.,* **20**, 208 (1950).
102. Finlay, A. C., F. A. Hochstein, B. A. Sobin, and F. X. Murphy, *J. Am. Chem. Soc.,* **73**, 341 (1951).
103. Finlay, A. C., G. L. Hobby, F. Hochstein, T. M. Lees, T. F. Lenert, J. A. Means, S. Y. P'an, P. P. Regna, J. B. Routien, B. A. Sobin, K. B. Tate, and J. H. Kane, *Am. Rev. Tuberc.,* **63**, 1 (1951).
104. Finlay, A. C., G. L. Hobby, S. Y. P'an, P. P. Regna, J. B.

Routien, D. B. Seeley, G. M. Shull, B. A. Sobin, I. A. Solomons, J. W. Vinson, and J. H. Kane, *Science,* **111,** 85 (1950).

105. Florey, H. W., E. Chain, N. G. Heatley, M. A. Jennings, A. G. Sanders, E. P. Abraham, and M. E. Florey, *Antibiotics,* 2 vols., New York, Oxford University Press, 1949.

106. Florey, H. W., K. Gilliver, M. A. Jennings, and A. G. Sanders, *Lancet,* **250,** 46 (1946).

106a. Flynn, E. H., M. V. Sigal, and P. F. Wiley, *Abstracts Am. Chem. Soc.,* 124th Meeting, Chicago, 47-0 (1953).

106b. Food and Drug Administration, FSA-E 53, 7236, August 14, 1952.

107. Ford, J. H., A. R. Johnson, and J. W. Hinman, *J. Am. Chem. Soc.,* **72,** 4529 (1950).

108. Ford, J. H., and B. E. Leach, *J. Am. Chem. Soc.,* **70,** 1223 (1948).

109. Foster, J. W., and H. B. Woodruff, *Arch. Biochem.,* **3,** 241 (1943).

110. Fraenkel-Conrat, H., J. C. Lewis, K. P. Dimick, B. Edwards, H. C. Reitz, R. E. Ferrel, B. A. Brandon, and H. S. Olcott, *Proc. Soc. Exptl. Biol. Med.,* **63,** 302 (1946).

111. Fried, J., and O. Wintersteiner, *Science,* **101,** 613 (1945).

111a. Fusillo, M. H., H. E. Noyes, E. J. Pulaski, and J. Y. S. Tom, *Antibiotics and Chemotherapy,* **3,** 581 (1953).

111b. Fusillo, M. H., M. J. Romansky, and D. M. Kuhns, *Antibiotics and Chemotherapy,* **3,** 35 (1953).

112. Gardner, A. D., and E. Chain, *Brit. J. Exptl. Path.,* **23,** 123 (1942).

112a. Gardocki, J. F., S. Y. P'an, A. L. Rappuzi, G. M. Fanelli, and E. K. Timmins, *Antibiotics and Chemotherapy,* **3,** 55 (1953).

113. Garibaldi, J. A., and R. E. Feeney, *Ind. Eng. Chem.,* **41,** 432 (1949).

114. Gäumann, E., S. Roth, L. Ettlinger, P. A. Plattner, and U. Nager, *Experientia,* **3,** 202 (1947).

115. Gause, G. F., *Lancet,* **251,** 46 (1946).

116. Gause, G. F., and M. G. Brazhnikova, *Nature,* **154,** 703 (1944).

117. Geiger, W. B., *Arch. Biochem.,* **21,** 125 (1949).

118. Geiger, W. B., J. E. Conn, and S. A. Waksman, *J. Bact.,* **48,** 531 (1944).

118a. Georg, L. K., L. Ajello, and M. A. Gordon, *Science,* **114,** 387 (1951).

118b. Gernez-Rieux, C., A. Taquet, and C. Chenet, *Ann. inst. Pasteur,* **81**, 158 (1951).
119. Glazko, A. J., L. M. Wolf, W. A. Dill, and A. C. Bratton, *J. Pharmacol. Exptl. Therap.,* **96**, 445 (1949).
119a. Goldin, M., *Antibiotics and Chemotherapy,* **3**, 881 (1953).
120. Gordon, A. H., A. J. P. Martin, and R. L. M. Synge, *Biochem. J.,* **37**, 86 (1943).
121. Gore, T. S., T. B. Panse, and K. Venkataraman, *Nature,* **157**, 333 (1946); *J. Am. Chem. Soc.,* **70**, 2287 (1948).
122. Gottlieb, D., P. K. Bhattacharyya, H. W. Anderson, and H. E. Carter, *J. Bact.,* **55**, 409 (1948).
122a. Gowans, J. L., N. Smith, and H. W. Florey, *Brit. J. Pharmacol.,* **7**, 438 (1952).
122b. Green, S. R., and P. P. Gray, *Arch. Biochem. Biophys.,* **32**, 59 (1951).
123. Gregory, J. D., and L. C. Craig, *J. Biol. Chem.,* **172**, 839 (1948).
123a. Griggs, J. F., *Antibiotics and Chemotherapy,* **2**, 290 (1952).
124. Groschke, A. C., and R. J. Evans, *Poultry Sci.,* **29**, 616 (1950).
124a. Grove, J. F., *Biochem. J.,* **50**, 648 (1952).
125. Gruhzit, O. M., R. A. Fisken, T. F. Reutner, and E. Martino, *J. Clin. Invest.,* **28**, 943 (1949).
126. Grunberg, E., D. Eldridge, G. Soo-Hoo, and D. R. Kelley, *Trans. N. Y. Acad. Sci.,* **11**, 210 (1949).
126a. Haight, T. H., and M. Finland, *Proc. Soc. Exptl. Biol. Med.,* **81**, 175, 183, 188 (1952).
127. Hailman, H. F., The Upjohn Co., *Personal communication.*
127a. Hamada, Y., H. Fujitani, K. Okamoto, and S. Konishi, *J. Antibiotics (Japan),* **5**, 541 (1952).
127b. Hamre, D. M., F. E. Pansy, D. N. Lapedes, D. Perlman, A. P. Bayan, and R. Donovick, *Antibiotics and Chemotherapy,* **2**, 135 (1952).
127c. Hanson, F. R., and T. E. Eble, *J. Bact.,* **58**, 527 (1949); U. S. Patent 2,652,356 (1953).
128. Harned, B. K., R. W. Cunningham, M. C. Clark, R. Cosgrove, C. H. Hine, W. J. McCauley, E. Stokey, R. E. Vessey, N. N. Yuda, and Y. SubbaRow, *Ann. N. Y. Acad. Sci.,* **51**, 182 (1948).
128a. Harris, J. I., and T. S. Work, *Biochem. J.,* **46**, 582 (1950).
128b. Hart, P. D., and B. Moss, *J. Gen. Microbiol.,* **4**, 244 (1950).
128c. Hasbrouck, R. B., and F. C. Garven, *Antibiotics and Chemotherapy,* **3**, 1040 (1953).

128d. Haskell, T. H., S. A. Fusari, R. P. Frohardt, and Q. R. Bartz, *J. Am. Chem. Soc.,* **74**, 599 (1952).

128e. Hasselmann-Kahlert, M., *Antibiotics and Chemotherapy,* **2**, 290 (1952).

129. Hays, E. E., I. C. Wells, P. A. Katzman, C. K. Cain, F. A. Jacobs, S. A. Thayer, E. A. Doisy, W. L. Gaby, E. C. Roberts, R. D. Muir, C. J. Carroll, L. R. Jones, and N. J. Wade, *J. Biol. Chem.,* **159**, 725 (1945).

129a. Heilman, F. R., W. E. Herrell, W. E. Wellman, and J. E. Geraci, *Proc. Staff Meet. Mayo Clinic,* **27**, 285 (1952).

130. Herrell, W. E., *Ann. Rev. Microbiol.,* **4**, 101 (1950).

131. Herrell, W. E., and D. Heilman, *Am. J. Med. Sci.,* **205**, 157 (1943).

132. Hickey, R. J., and P. H. Hidy, *Science,* **113**, 361 (1951).

132a. Hills, G. M., F. C. Belton, and E. D. Blatchley, *Brit. J. Exptl. Path.,* **30**, 427 (1949).

132b. Hinman, J. W., E. L. Caron, and H. N. Christensen, *J. Am. Chem. Soc.,* **72**, 1620 (1950).

133. Hirsch, A., *J. Gen. Microbiol.,* **4**, 70 (1950).

134. Hirsch, A., and A. T. R. Mattick, *Lancet,* **257**, 190 (1949).

135. Hobby, G. L., N. Dougherty, T. F. Lenert, E. Hudders, and M. Kiseluk, *Proc. Soc. Exptl. Biol. Med.,* **73**, 503 (1950).

136. Hobby, G. L., T. F. Lenert, M. Donikian, and D. Pikula, *Am. Rev. Tuberc.,* **63**, 17 (1951).

137. Hobby, G. L., T. F. Lenert, and N. Dougherty, *Ann. N. Y. Acad. Sci.,* **52**, 775 (1949).

138. Hogeboom, G. H., and L. C. Craig, *J. Biol. Chem.,* **162**, 363 (1946).

139. Hotchkiss, R. D., *Advances in Enzymol.,* **4**, 153 (1944).

140. Hotchkiss, R. D., and R. J. Dubos, *J. Exptl. Med.,* **73**, 629 (1941); *J. Biol. Chem.,* **136**, 803 (1940).

140a. Hrenoff, A. K., and M. Nakamura, *Proc. Soc. Exptl. Biol. Med.,* **77**, 162 (1951).

141. Hutchison, D., E. A. Swart, and S. A. Waksman, *Arch. Biochem.,* **22**, 16 (1949).

142. Jackson, G. G., T. M. Gocke, H. S. Collins, and M. Finland, *J. Infectious Diseases,* **87**, 63 (1950).

143. Jackson, G. G., T. M. Gocke, C. Wilcox, and M. Finland, *Am. J. Clin. Path.,* **20**, 218 (1950).

143a. Inskeep, G. C., R. E. Bennett, J. F. Dudley, and M. W. Shepard, *Ind. Eng. Chem.,* **43**, 1488 (1951).

144. Jansen, E. F., and D. J. Hirschmann, *Arch. Biochem.,* **4**, 297 (1944).

145. Jarvis, F. G., and M. J. Johnson, *J. Am. Chem. Soc.,* **71,** 4124 (1949).
145a. Jawetz, E., *Arch. Internal Med.,* **89,** 90 (1952).
145b. Jawetz, E., and J. B. Gunnison, *Antibiotics and Chemotherapy,* **2,** 243 (1952).
145c. Jawetz, E., and J. B. Gunnison, *J. Am. Med. Assoc.,* **150,** 693 (1952).
146. Jennings, M. A., and T. I. Williams, *Nature,* **155,** 302 (1945).
146a. Jirovec. O., *Experientia,* **5,** 74 (1949).
146b. Johnson, A. W., N. Sheppard, and A. R. Todd, *J. Chem. Soc.,* **1951,** 1139.
147. Johnson, B. A., H. Anker, and F. L. Meleney, *Science,* **102,** 376 (1945).
147a. Johnson, D. H., A. Robertson, and W. B. Whalley, *J. Chem. Soc.,* **1950,** 2971.
147b. Johnson, J. R., and J. H. Andreen, *J. Am. Chem. Soc.,* **72,** 2862 (1950).
148. Jones, T. S. G., *Ann. N. Y. Acad. Sci.,* **51,** 909 (1949).
148a. Josselyn, L. E., and J. C. Sylvester, *Antibiotics and Chemotherapy,* **3,** 63 (1953).
149. Junowicz-Kocholaty, R., and W. Kocholaty, *J. Biol. Chem.,* **168,** 757 (1947).
150. Junowicz-Kocholaty, R., W. Kocholaty, and A. Kelner, *J. Biol. Chem.,* **168,** 765 (1947).
150a. Kadison, E. R., I. F. Volini, S. J. Hoffman, and O. Felsenfeld, *J. Am. Med. Assoc.,* **145,** 1307 (1951).
150b. Kagan, B. M., D. Krevsky, A. Milzer, and M. Locke, *J. Lab. Clin. Med.,* **37,** 402 (1951).
151. Karel, L., and E. S. Roach, *A Dictionary of Antibiosis,* New York, Columbia University Press, 1951.
151a. Karlson, A. G., and J. H. Gainer, *Am. Rev. Tuberc.,* **63,** 36 (1951).
152. Karlson, A. G., J. H. Gainer, and W. H. Feldman, *Diseases of the Chest,* **17,** 493 (1950).
153. Kavanagh, F., *Advances in Enzymol.,* **7,** 461 (1947).
154. Kavanagh, F., *Arch. Biochem.,* **15,** 95 (1947).
155. Kavanagh, F., A. Hervey, and W. J. Robbins, *Proc. Natl. Acad. Sci. U. S.,* **35,** 343 (1949).
156. Kellner, A., and H. E. Morton, *J. Bact.,* **53,** 695 (1947).
156a. Kenner, B. A., and F. J. Murray, *Antibiotics and Chemotherapy,* **1,** 509 (1951).
156b. Kido, G. S., and E. Spyhalski, *Science,* **112,** 172 (1950).
156c. Kielova-Rodova, H., *Experientia,* **5,** 242 (1949).

156d. Kirby, W. M. M., *Ann. Rev. Microbiol.*, **6**, 37 (1952).

156e. Kiser, J. S., G. C. DeMello, V. Eve, H. Lindh, L. Malone, F. Popken, A. Schurr, and M. K. Waters, *Antibiotics Annual*, 56, 1953-54, New York, Medical Encyclopedia Inc.

156f. Knight, V., R. C. Hardy, and J. Negrin, *J. Am. Med. Assoc.*, **149**, 1395 (1952).

156g. Knodt, C. B., *Antibiotics and Chemotherapy*, **3**, 442 (1953).

157. Kocholaty, W., R. Junowicz-Kocholaty, and A. Kelner, *Arch. Biochem.*, **17**, 191 (1948).

158. Kornfeld, E. C., and R. G. Jones, *Science*, **108**, 437 (1948).

159. Kornfeld, E. C., R. G. Jones, and T. V. Parke, *J. Am. Chem. Soc.*, **71**, 150 (1949).

159a. Kuehl, F. A., M. N. Bishop, and K. Folkers, *J. Am. Chem. Soc.*, **73**, 881 (1951).

160. Lack, A., *Proc. Soc. Exptl. Biol. Med.*, **72**, 656 (1949).

161. Larson, L. M., H. Sternberg, and W. H. Peterson, *J. Am. Chem. Soc.*, **75**, 2036 (1953).

162. Lazar, A. M., and J. Fishman, *Eye, Ear, Nose, Throat Monthly*, **29**, 484 (1950).

162a. Leach, B. E., W. E. DeVries, H. A. Nelson, W. G. Jackson, and J. S. Evans, *J. Am. Chem. Soc.*, **73**, 2797 (1951).

163. Leach, B. E., J. H. Ford, and A. J. Whiffen, *J. Am. Chem. Soc.*, **69**, 474 (1947).

163a. Leach, B. E., and C. M. Teeters, *J. Am. Chem. Soc.*, **74**, 3187 (1952).

163b. Leben, C., *Abstracts Am. Chem. Soc.*, 124th Meeting, Chicago, 29A (1953).

164. Leben, C., and G. W. Keitt, *Phytopath.*, **38**, 899 (1948); **39**, 529 (1949).

165. Leberman, P., P. F. Smith, and H. E. Morton, *J. Urol.*, **64**, 167 (1950).

165a. Lee, S. B., *Ind. Eng. Chem.*, **42**, 1672 (1950).

165b. Legator, M., and D. Gottlieb, *Antibiotics and Chemotherapy*, **3**, 809 (1953).

166. Lennette, E. H., G. Meiklejohn, and H. M. Thelen, *Ann. N. Y. Acad. Sci.*, **51**, 331 (1948).

166a. Levaditi, C., and A. Vaisman, *Antibiotics and Chemotherapy*, **1**, 425 (1951).

167. Lewis, J. C., and N. S. Snell, *J. Am. Chem. Soc.*, **73**, 4812 (1951).

168. Lewis, J. C., R. E. Feeney, J. A. Garibaldi, H. D. Michener, D. J. Hirschmann, D. H. Traufler, A. F. Langlykke, H. D.

Lightbody, J. J. Stubbs, and H. Humfeld, *Arch. Biochem.,* **14**, 415 (1947).

168a. Lewis, C. N., L. E. Putnam, F. D. Hendricks, I. Kerlan, and H. Welch, *Antibiotics and Chemotherapy,* **2**, 601 (1952).

169. Long, L. M., and H. D. Troutman, *J. Am. Chem. Soc.,* **71**, 2469, 2473 (1949).

170. Lott, W. A., and E. Shaw, *J. Am. Chem. Soc.,* **71**, 70 (1949).

170a. Luckey, T. D., *Studies on the Growth Effect of Antibiotics in Germ-Free Animals,* Lebund Institute, University of Notre Dame, Notre Dame, Ind., 1952.

171. Many authors, Abstracts of Papers, 42nd Annual Meeting Am. Soc. Animal Production, *J. Animal Sci.,* **9**, 646-671 (1950).

172. Many authors, Conference on Terramycin, *Ann. N. Y. Acad. Sci.,* **53**, 223-459 (1950).

173. Maeda, K., *J. Antibiotics,* **2**, 795 (1949).

174. Marcus, S., *Biochem. J.,* **41**, 462 (1947).

175. Markunas, P. C., R. E. Bennett, and M. C. Bachman, *Personal communication.*

176. Marston, R. Q., *Brit. J. Exptl. Path.,* **30**, 398 (1949).

177. Marston, R. Q., and H. W. Florey, *Brit. J. Exptl. Path.,* **30**, 407 (1949).

178. Martin, J., *Personal communication.*

179. Mattick, A. T. R., and A. Hirsch, *Lancet,* **253**, 5 (1947).

179a. Maynard, A. J., C. Andriola, and A. Prigot, *Antibiotics Annual,* 102, 1953-54, New York, Medical Encyclopedia Inc.

179b. McCowen, M. C., M. E. Callender, and J. T. Lawlis, *Science,* **113**, 202 (1951).

179c. McCowen, M. C., M. E. Callender, J. F. Lawlis, and M. C. Brandt, *Am. J. Trop. Med. and Hyg.,* **2**, 212 (1953).

179d. McCowen, M. C., and J. F. Lawlis, *J. Parasitol. Suppl.,* **36**, 25 (1950).

180. McLean, I. W., J. L. Schwab, A. B. Hillegas, and A. S. Schlingman, *J. Clin. Invest.,* **28**, 953 (1949).

181. McIlwain, H., *Biochem. J.,* **37**, 265 (1943).

182. McGowan, J. C., *Chemistry and Industry,* **66**, 205 (1947).

182a. McGuire, J. M., R. L. Bunch, R. C. Anderson, H. E. Boaz, E. H. Flynn, H. M. Powell, and J. W. Smith, *Antibiotics and Chemotherapy,* **2**, 281 (1952).

182b. McVay, L. V., and D. H. Sprunt, *Southern Med. J.,* **45**, 183 (1952).

183. Meleney, F. L., and B. Johnson, *J. Am. Med. Assoc.,* **133**, 675 (1947).

184. Meleney, F. L., and B. Johnson, *Am. J. Med.,* **7**, 794 (1949).

185. Menzel, A. E. O., O. Wintersteiner, and J. C. Hoogerheide, *J. Biol. Chem.*, **152**, 419 (1944).
185a. Minieri, P. P., M. C. Firman, A. G. Mistretta, A. Abbey, C. E. Bricker, N. E. Rigler, and H. Sokol, *Antibiotics Annual*, 81, 1953-54, New York, Medical Encyclopedia Inc.
186. Mold, J. D., and Q. R. Bartz, *J. Am. Chem. Soc.*, **72**, 1847 (1950).
186a. Moore, P. R., A. Evenson, T. D. Luckey, E. McCoy, C. A. Elvehjem, and E. B. Hart, *J. Biol. Chem.*, **165**, 437 (1946).
187. Morton, H. E., *Proc. Soc. Exptl. Biol. Med.*, **64**, 327 (1947).
188. Morton, H. E., *Proc. Soc. Exptl. Biol. Med.*, **66**, 345 (1947).
189. Morton, H. E., W. Kocholaty, R. Junowicz-Kocholaty, and A. Kelner, *J. Bact.*, **50**, 579 (1945).
190. Murnaghan, M. F., *J. Pharmacol. Exptl. Therap.*, **88**, 119 (1946).
191. Murray, F. J., and P. A. Tetrault, *Proc. Soc. Am. Bact.*, **1**, 20 (1948).
192. Murray, F. J., P. A. Tetrault, O. W. Kaufmann, H. Koffler, D. H. Peterson, and D. R. Colingsworth, *J. Bact.*, **57**, 305 (1949).
193. Nauta, W. T., H. K. Oosterhuis, A. C. van der Linden, P. van Duyn, and J. W. Dienske, *Rec. trav. chim.*, **65**, 865 (1946).
194. Nelson, H. A., C. DeBoer, and W. H. Devries, *Ind. Eng. Chem.*, **42**, 1259 (1950).
194a. Nesbit, R. M., A. I. Dodson, and C. C. MacKinney, *Antibiotics and Chemotherapy*, **2**, 447 (1952).
195. Neter, E., and G. A. Gorzynski, *Proc. Soc. Exptl. Biol. Med.*, **74**, 328 (1950).
196. Neuberger, A., *Advances in Protein Chem.*, **4**, 367 (1948).
197. Newton, G. G. F., and E. P. Abraham, *Biochem. J.*, **47**, 257 (1950).
197a. Nickell, L. G., *Antibiotics and Chemotherapy*, **3**, 449 (1953).
197b. Nickell, L. G., and A. C. Finlay, *Abstracts Am. Chem. Soc.*, 124th Meeting, Chicago, 28A (1953).
197c. Niedercorn, J. G., U. S. Patent 2,609,329 (1952).
197d. Ogata, Y., K. Nitta, S. Yamazaki, O. Taya, T. Takeuchi, and H. Unezawa, *J. Antibiotics (Japan)*, **6**, 139 (1953).
197e. Olitsky, I., and G. W. Bierman, *Antibiotics and Chemotherapy*, **2**, 344 (1952).
198. Oxford, A. E., *Ann. Rev. Biochem.*, **14**, 749 (1945).
199. Oxford, A. E., and H. Raistrick, *Chemistry and Industry*, **61**, 128 (1942).

200. Oxford, A. E., H. Raistrick, and G. Smith, *Chemistry and Industry,* **61**, 22 (1942).
201. Oyaas, J. E., J. Ehrlich, and R. M. Smith, *Ind. Eng. Chem.,* **42**, 1775 (1950).
201a. Paine, T. F., *Antibiotics and Chemotherapy,* **2**, 653 (1952).
202. Paine, T. F., H. S. Collins, and M. Finland, *J. Bact.,* **56**, 489 (1948).
203. P'an, S. Y., T. V. Halley, J. C. Reilly, and A. M. Pekich, *Am. Rev. Tuberc.,* **63**, 44 (1951).
203a. Pansy, F. E., P. Kann, J. F. Pagano, and R. Donovick, *Proc. Soc. Exptl. Biol. Med.,* **75**, 618 (1950).
203b. Pappenfort, R. B., and E. S. Schnall, *Ann. Internal Med.,* **33**, 729 (1951).
204. Peck, R. L., C. E. Hoffhine, P. Gale, and K. Folkers, *J. Am. Chem. Soc.,* **71**, 2590 (1949); **75**, 1018 (1953).
204a. Peck, R. L., and J. E. Lyons, *Ann. Rev. Biochem.,* **20**, 367 (1951).
205. Peck, R. L., A. Walti, R. P. Graber, E. H. Flynn, C. E. Hoffhine, V. Allfrey, and K. Folkers, *J. Am. Chem. Soc.,* **68**, 772 (1946).
206. Penick, S. B., and Co., *Annotated Bibliography of Tyrothricin,* New York, Research Division, S. B. Penick and Co., 1947.
206a. Perlman, D., W. E. Brown, and S. B. Lee, *Ind. Eng. Chem.,* **44**, 1996 (1952).
206b. Perry, T. L., *Am. Rev. Tuberc.,* **65**, 325 (1952).
207. Peterson, D. H., and L. M. Reineke, *J. Biol. Chem.,* **181**, 95 (1949).
208. Peterson, D. H., D. R. Colingsworth, L. M. Reineke, and C. DeBoer, *J. Am. Chem. Soc.,* **69**, 3145 (1947).
208a. Pettinga, C. W., W. M. Stark, and F. R. Van Abeele, *Abstracts Am. Chem. Soc.,* 124th Meeting, Chicago, 47-O (1953).
209. Pfizer, Chas., and Co., *Terramycin, A summary report with reference to literature,* Brooklyn, Chas. Pfizer and Co., 1950.
209a. Pidacks, C., and E. E. Starbird, U. S. Patent 2,586,766 (1952).
209b. Pitts, F. W., E. T. O'Dell, W. E. Dye, F. J. Hughes, and C. W. Tempel, *Transactions, 11th Conference on Chemotherapy of Tuberculosis, U. S. Veterans Administration 270* (1952).
210. Plattner, P. A., and U. Nager, *Helv. Chim. Acta,* **31**, 594, 665, 2203 (1948).
210a. Perath, J., *Acta Chem. Skand.,* **6**, 1237 (1952).
210b. Powell, H. M., W. M. Beniece, R. C. Pettinger, R. L. Stone,

and C. G. Culbertson, *Antibiotics and Chemotherapy,* **3**, 165 (1953).
211. Pratt, R., and J. Dufrenoy, *Antibiotics,* Philadelphia, Lippincott, 1949.
212. Price, C. W., W. A. Randall, and H. Welch, *Ann. N. Y. Acad. Sci.,* **51**, 211 (1948).
213. Puetzer, B., C. H. Nield, and R. H. Barry, *J. Am. Chem. Soc.,* **67**, 832 (1945).
214. Pulaski, E. J., H. J. Baker, M. J. Rosenberg, and J. F. Connell, *J. Clin. Invest.,* **28**, 1028 (1949).
214a. Putnam, L. E., F. D. Hendricks, and H. Welch, *Antibiotics Annual,* 88, 1953-54, New York, Medical Encyclopedia Inc.
214b. Quinn, L. Y., M. D. Lane, G. C. Ashton, H. M. Maddock, and D. V. Catron, *Antibiotics and Chemotherapy,* **3**, 622 (1953).
214c. Raistrick, H., *Proc. Roy Soc. (London),* A **199**, 141 (1949).
215. Raistrick, H., and G. Smith, *Biochem. J.,* **30**, 1315 (1936).
216. Raistrick, H., and G. Smith, *Chemistry and Industry,* **60**, 828 (1941).
216a. Raistrick, H., and C. E. Stickings, *Biochem. J.,* **48**, 53 (1951).
217. Rake, G., *Ann. N. Y. Acad. Sci.,* **52**, 765 (1949).
218. Rake, G., D. M. Hamre, F. Kavanagh, W. L. Koerber, and R. Donovick, *Am. J. Med. Sci.,* **210**, 61 (1945).
219. Rake, G., C. M. McKee, D. M. Hamre, and C. L. Houck, *J. Immunol.,* **48**, 271 (1944).
220. Rammelkamp, C. H., and L. Weinstein, *Proc. Soc. Exptl. Biol. Med.,* **48**, 211 (1941).
220a. Rao, K. V., and W. H. Peterson, *Abstracts Am. Chem. Soc.,* 121st Meeting, Milwaukee, 8C (1952); *J. Am. Chem. Soc.,* **76**, 1335, 1338 (1954).
221. Raphael, R. A., *J. Chem. Soc.,* **1947**, 805.
222. Rebstock, M. C., H. M. Crooks, J. Controulis, and Q. R. Bartz, *J. Am. Chem. Soc.,* **71**, 2458 (1949).
222a. Regna, P. P., F. A. Hochstein, R. L. Wagner, and R. B. Woodward, *J. Am. Chem. Soc.,* **75**, 4625 (1953).
223. Regna, P. P., and F. X. Murphy, *J. Am. Chem. Soc.,* **72**, 1045 (1950).
224. Regna, P. P., and I. A. Solomons, *Ann. N. Y. Acad. Sci.,* **53**, 229 (1950).
225. Regna, P. P., I. A. Solomons, B. K. Forscher, and A. E. Timreck, *J. Clin. Invest.,* **28**, 1022 (1949).
225a. Reid, J. D., M. M. Jones, and E. C. Bryce, *Antibiotics and Chemotherapy,* **2**, 357 (1952).
226. Rieveschel, G., Jr., *Personal communication.*

227. Rinderknecht, H., J. L. Ward, F. Bergel, and A. L. Morrison, *Biochem. J.*, **41**, 463 (1947).
228. Rivett, R. W., and W. H. Peterson, *J. Am. Chem. Soc.*, **69**, 3006 (1947).
229. Robbins, W. J., F. Kavanagh, and A. Hervey, *Proc. Natl. Acad. Sci. U. S.*, **33**, 176 (1947).
230. Robinson, H. J., O. E. Graessle, and D. G. Smith, *Science*, **99**, 540 (1944).
231. Robinson, H. J., and H. Molitor, *J. Pharmacol. Exptl. Therap.*, **74**, 75 (1942).
231a. Romoser, G. L., M. S. Shorb, and G. F. Combs, *Poultry Sci.*, **31**, 932 (1952).
231b. Ruelius, H. W., and A. Gauhe, *Ann. Chem., Justus Liebig's*, **569**, 38 (1950).
231c. Ruiz-Sanchez, F., J. Casillas, M. Paredes, E. J. Valazuez, and R. Riebeling, *Antibiotics and Chemotherapy*, **2**, 50 (1952).
232. Salle, A. J., and G. J. Jann, *J. Bact.*, **54**, 269 (1947); *J. Clin. Invest.*, **28**, 1036 (1949).
232a. Sauberlich, W. E., *J. Nutrition*, **46**, 99 (1952).
233. Schatz, A., and S. A. Waksman, *Proc. Soc. Exptl. Biol. Med.*, **57**, 244 (1944).
233a. Schenck, J. R., M. P. Hargie, D. S. Tarbell, and P. Hoffman, *J. Am. Chem. Soc.*, **75**, 2274 (1953).
233b. Schneider, H. G., G. M. Tener, and F. M. Strong, *Arch. Biochem. Biophys.*, **37**, 147 (1952).
233c. Schnitzer, R. J., M. Buck, and A. C. Farr, *J. Clin. Invest.*, **28**, 1047 (1949).
234. Schoenthal, R., *Brit. J. Exptl. Path.*, **22**, 137 (1941).
235. Scudi, J. V., and W. Antopol, *Proc. Soc. Exptl. Biol. Med.*, **64**, 503 (1947).
235a. Seneca, H., J. H. Kane, and J. Rockenbach, *Antibiotics and Chemotherapy*, **2**, 435 (1952).
235b. Seneca, H., J. H. Kane, and J. Rockenbach, *Antibiotics and Chemotherapy*, **2**, 357 (1952).
235c. Several authors, *Quarterly Progress Report on Chemotherapy of Tuberculosis*, U. S. Veterans Administration (October, 1952).
235d. Several authors, *Abstracts Am. Chem. Soc.*, 124th Meeting, Chicago, 24-31A (1953).
236. Sharp, V. E., A. Arriagada, G. G. F. Newton, and E. P. Abraham, *Brit. J. Exptl. Path.*, **30**, 444 (1949).
237. Shepherd, R. G., P. G. Stansly, R. Winterbottom, J. P.

English, C. E. Fellows, N. H. Ananenko, and G. L. Guillet, *J. Am. Chem. Soc.*, **70**, 3771 (1948).
237a. Smissman, E. E., R. W. Sharpe, B. F. Aycock, E. E. van Tamelen, and W. H. Peterson, *J. Am. Chem. Soc.*, **75**, 2029 (1953).
238. Smith, G. N., C. S. Worrel, and B. L. Lilligren, *Science*, **110**, 297 (1949).
239. Smith, G. N., C. S. Worrel, and A. L. Swanson, *J. Bact.*, **58**, 803 (1949).
240. Smith, R. M., D. A. Joslyn, O. M. Gruhzit, I. W. McLean, M. A. Penner, and J. Ehrlich, *J. Bact.*, **55**, 425 (1948).
241. Sobin, B. A., A. C. Finlay, and J. H. Kane, U. S. Patent 2,516,080 (1950).
242. Spink, W. W., A. I. Braude, M. R. Castaneda, and R. S. Groytia, *J. Am. Med. Assoc.*, **138**, 1145 (1948).
242a. Staff Survey, *Agr. Food Chem.*, **1**, 1096 (1953).
243. Stansly, P. G., and N. H. Ananenko, *Arch. Biochem.*, **23**, 256 (1949).
243a. Stansly, P. G., and N. H. Ananenko, *J. Clin. Invest.*, **28**, 1047 (1949).
244. Stansly, P. G., M. E. Schlosser, N. H. Ananenko, and M. H. Cook, *J. Bact.*, **55**, 573 (1948).
245. Stansly, P. G., R. G. Shepherd, and H. J. White, *Bull. Johns Hopkins Hosp.*, **81**, 43 (1947).
245a. Steenken, W., and E. Wolinsky, *Am. Rev. Tuberc.*, **63**, 30 (1951).
246. Steenken, W., E. Wolinsky, and B. J. Bolinger, *Am. Rev. Tuberc.*, **62**, 300 (1950).
247. Stephens, C. R., L. H. Conover, F. A. Hochstein, P. P. Regna, F. J. Pilgrim, K. J. Brunings, and R. B. Woodward, *J. Am. Chem. Soc.*, **74**, 4976 (1952).
248. Stern, J. R., and J. McGinnis, *Arch. Biochem.*, **28**, 364 (1950).
249. Stokes, J. L., R. L. Peck, and C. R. Woodward, *Proc. Soc. Exptl. Biol. Med.*, **51**, 126 (1942).
249a. Stokstad, E. L. R., *Antibiotics and Chemotherapy*, **3**, 434 (1953).
250. Stokstad, E. L. R., and T. H. Jukes, *Proc. Soc. Exptl. Biol. Med.*, **73**, 523 (1949).
251. Stubbs, J. J., R. E. Feeney, J. C. Lewis, I. C. Feustel, H. D. Lightbody, and J. A. Garibaldi, *Arch. Biochem.*, **14**, 427 (1947).
252. Swart, E. A., *J. Am. Chem. Soc.*, **71**, 2942 (1949).
253. Swart, E. A., D. Hutchison, and S. A. Waksman, *Arch. Biochem.*, **24**, 92 (1949).

253a. Swart, E. A., H. A. Lechevalier, and S. A. Waksman, *J. Am. Chem. Soc.,* **73**, 3253 (1951).
254. Swart, E. A., A. H. Romano, and S. A. Waksman, *Proc. Soc. Exptl. Biol. Med.,* **73**, 376 (1950).
255. Synge, R. L. M., *Biochem. J.,* **39**, 363 (1945).
256. Synge, R. L. M., *Biochem. J.,* **44**, 542 (1949).
256a. Tanner, F. W., A. R. English, T. M. Lees, and J. B. Routien, *Antibiotics and Chemotherapy,* **2**, 441 (1952).
257. Tanner, F. W., J. A. Means, and J. W. Davisson, *Abstracts Am. Chem. Soc.,* 118th Meeting, 18A (1950).
257a. Tener, G. M., E. E. Van Tamelen, and F. M. Strong, *J. Am. Chem. Soc.,* **75**, 3623 (1953).
258. Thorne, C. B., and W. H. Peterson, *J. Biol. Chem.,* **176**, 413 (1948).
259. Umezawa, H., S. Hayano, K. Maeda, Y. Ogata, and Y. Okami, *J. Antibiotics,* **4**, 37 (1951).
260. Umezawa, H., R. Kametani, T. Osato, K. Takeda, H. Kanari, R. Utahara, A. Kawahara, and R. Wada, *J. Antibiotics,* **2**, Supplement B 100 (1950).
261. Umezawa, H., and K. Maeda, *J. Antibiotics,* **3**, 41 (1950).
262. Umezawa, H., K. Maeda, and H. Kosaka, *Jap. Med. J.,* **1**, 512 (1948).
263. Umezawa, H., T. Tazaki, and S. Fukuyama, *J. Antibiotics,* **2**, Supplement B 91 (1949).
264. Umezawa, H., T. Tazaki, H. Kanari, Y. Okami, and S. Fukuyama, *Jap. Med. J.,* **1**, 358 (1949).
265. Vander Brook, M. J., and M. T. Richmond, *J. Clin. Invest.,* **28**, 1032 (1949).
265a. Van Dyck, P., and P. De Somer, *Antibiotics and Chemotherapy,* **2**, 184 (1952).
266. Van Luijk, A., *Med. Phytopathol. Lab. W. C. Scholten,* **14**, 43 (1938).
266a. Van Tamelen, E. E., and E. E. Smissman, *J. Am. Chem. Soc.,* **74**, 3713 (1952); **75**, 2031 (1953).
267. Vaughn, J. R., and Klomparens, *Phytopath.,* **42**, 22 (1952).
267a. Vischer, E. B., S. R. Howland, and H. Raudnitz, *Nature,* **165**, 528 (1950).
267b. Waisbren, B. A., and W. W. Spink, *Proc. Soc. Exptl. Biol. Med.,* **74**, 35 (1950).
267c. Waisbren, B. A., and W. W. Spink, *Ann. Internal Med.,* **33**, 1099 (1950).
267d. Wagner, R. L., F. A. Hochstein, K. Murai, N. Messina, and P. P. Regna, *J. Am. Chem. Soc.,* **75**, 4684 (1953).

268. Waksman, S. A., *J. Bact.*, **46**, 299 (1843).
269. Waksman, S. A., and E. Bugie, *J. Bact.*, **48**, 527 (1944).
270. Waksman, S. A., J. Frankel, and O. Graessle, *J. Bact.*, **58**, 229 (1949).
271. Waksman, S. A., and W. B. Geiger, *J. Bact.*, **47**, 391 (1944).
272. Waksman, S. A., W. B. Geiger, and D. M. Reynolds, *Proc. Natl. Acad. Sci.*, **32**, 117 (1946).
273. Waksman, S. A., E. S. Horning, and E. L. Spencer, *Science*, **96**, 202 (1942).
274. Waksman, S. A., E. S. Horning, and E. L. Spencer, *J. Bact.*, **45**, 233 (1943).
275. Waksman, S. A., D. Hutchison, and E. Katz, *Am. Rev. Tuberc.*, **60**, 78 (1949).
276. Waksman, S. A., E. Katz, and H. Lechevalier, *J. Lab. Clin. Med.*, **36**, 93 (1950).
277. Waksman, S. A., and H. A. Lechevalier, *Science*, **109**, 305 (1949).
278. Waksman, S. A., H. A. Lechevalier, and D. A. Harris, *J. Clin. Invest.*, **28**, 934 (1949).
279. Waksman, S. A., and M. Tishler, *J. Biol. Chem.*, **142**, 519 (1942).
280. Waksman, S. A., and H. B. Woodruff, *Proc. Soc. Exptl. Biol. Med.*, **45**, 609 (1940).
281. Waksman, S. A., and H. B. Woodruff, *J. Bact.*, **44**, 373 (1942).
282. Waller, C. W., B. L. Hutchings, R. W. Broschard, A. A. Goldman, W. J. Stein, C. F. Wolf, and J. H. Williams, *J. Am. Chem. Soc.*, **74**, 4981 (1952).
283. Wallerstein Co., *Tyrothricin Bibliography*, New York, Wallerstein Laboratories Inc., 1952
284. Walton, R. B., and H. B. Woodruff, *J. Clin. Invest.*, **28**, 924 (1949).
285. Warren, H. H., G. Dougherty, and E. S. Wallis, *J. Am. Chem. Soc.*, **71**, 3422 (1949).
286. Weindling, R., and O. H. Emerson, *Phytopathology*, **26**, 1068 (1936).
286a. Weindling, R., H. Katznelson, and H. P. Beale, *Ann. Rev. Microbiol.*, **4**, 247 (1950).
287. Weiss, D., and S. A. Waksman, *Proc. Natl. Acad. Sci.*, **36**, 293 (1950).
287a. Welch, H., *Antibiotics and Chemotherapy*, **2**, 279 (1952).
287b. Welch, H., W. A. Randall, R. J. Reedy, and J. Kramer, *Antibiotics and Chemotherapy*, **2**, 693 (1952).

288. Welch, H., R. J. Reedy, and S. W. Wolfson, *J. Lab. Clin. Med.,* **35**, 663 (1950).
289. Wells, E. B., C. Shih-Man, G. G. Jackson, and M. Finland, *J. Pediatrics,* **36**, 752 (1950).
289a. Wells, I. C., *J. Biol. Chem.,* **196**, 331 (1952).
290. Wells, I. C., W. H. Elliott, S. A. Thayer, and E. A. Doisy, *Federation Proc.,* **7**, 198 (1948).
291. Wells, I. C., E. E. Hays, N. J. Wade, W. L. Gaby, C. J. Carroll, L. R. Jones, and E. A. Doisy, *J. Biol. Chem.,* **167**, 53 (1947).
292. Werner, C. A., R. Tompsett, C. Muschenheim, and W. McDermott, *Am. Rev. Tuberc.,* **63**, 49 (1951).
292a. Werner, C. A., C. Adams, and R. DuBois, *Proc. Soc. Exptl. Biol. Med.,* **76**, 292 (1951).
293. Weyer, E. R., *J. Am. Pharm. Assoc.,* Practical Pharmacy Edition, **11**, 230 (1950).
294. Whiffen, A. J., *J. Bact.,* **56**, 283 (1948).
295. Whiffen, A. J., N. Bohonos, and R. L. Emerson, *J. Bact.,* **52**, 610 (1946).
296. White, E. C., and J. H. Hill, *J. Bact.,* **45**, 433 (1943).
297. Whitehill, A. R., J. J. Oleson, and B. L. Hutchings, *Proc. Soc. Exptl. Biol. Med.,* **74**, 11 (1950).
298. Wilkinson, S., *Nature,* **164**, 622 (1949).
299. Wintersteiner, O., and J. D. Dutcher, *Ann. Rev. Biochem.,* **18**, 559 (1949).
299a. Wong, S. C., C. G. James, and A. Finlay, *Antibiotics and Chemotherapy,* **3**, 741 (1953).
299b. Wood, W. S., G. P. Kipnis, and H. F. Dowling, *Antibiotics Annual,* 98, 1953-54, New York, Medical Encyclopedia Inc.
300. Woodruff, H. B., *Personal communication.*
301. Woodruff, H. B., and J. W. Foster, *Proc. Soc. Exptl. Biol. Med.,* **57**, 88 (1944).
302. Woodward, C. R., *J. Bact.,* **54**, 375 (1947).
303. Woodward, T. E., *Ann. Internal Med.,* **31**, 53 (1949).
304. Woodward, T. E., W. T. Raby, W. Eppes, W. A. Holbrook, and J. A. Hightower, *J. Am. Med. Assoc.,* **139**, 830 (1949).
305. Woodward, R. B., and G. Singh, *J. Am. Chem. Soc.,* **72**, 1428 (1950).
305a. Work, T. S., *Ann. Rev. Biochem.,* **21**, 434 (1952).
306. Wright, L. T., M. Sanders, M. A. Logan, A. Prigot, and L. M. Hill, *Ann. N. Y. Acad. Sci.,* **51**, 318 (1948).
306a. Wright, L. T., J. C. Whitaker, R. S. Wilkinson, and M. S. Beinfield, *Antibiotics and Chemotherapy,* **1**, 193 (1951).

306b. Wright, S. S., E. M. Purcell, B. D. Love, Jr., T. W. Mou, E. H. Kass, and M. Finland, *Antibiotics Annual,* 92, 1953-54, New York, Medical Encyclopedia Inc.
307. Yagishita, K., T. Osato, R. Utahara, and H. Umezawa, *J. Antibiotics,* **4**, Supplement A 48 (1951).
307a. Yagishita, K., R. Utahara, M. Ueda, T. Osato, and H. Umezawa, *J. Antibiotics (Japan),* **6**, 113 (1953).
308. Yow, E. M., *J. Am. Med. Assoc.,* **149**, 1184 (1952).

BIBLIOGRAPHY FOR SUPPLEMENTARY TABLE

S1. Baker, B. R., and R. E. Schaub, *J. Am. Chem. Soc.,* **75**, 3864 (1953).
S1a. Bendz, G., *Acta Chem. Scand.,* **2**, 192 (1948); **5**, 489 (1951).
S2. Benedict, R. G., W. Dvonch, O. L. Shotwell, T. G. Pridham, and L. A. Lindenfelser, *Antibiotics and Chemotherapy,* **2**, 591 (1952).
S3. Bentley, H. R., K. G. Cunningham, and F. S. Spring, *J. Chem. Soc.,* **1951**, 2301.
S4. Berger, J., A. I. Rachin, W. E. Scott, L. H. Sternbach, and M. W. Goldberg, *J. Am. Chem. Soc.,* **73**, 2295 (1951).
S5. Bolhofer, W. A., R. A. Machlowitz, and J. Charney, *Antibiotics and Chemotherapy,* **3**, 382 (1953).
S6. Brian, P. W., *Ann. Bot.,* **13**, 59 (1949).
S7. Brian, P. W., P. J. Curtis, and H. G. Hemming, *Trans. Brit. Mycolog. Soc.,* **29**, 173 (1946).
S8. Brian, P. W., P. J. Curtis, H. G. Hemming, E. G. Jeffreys, C. H. Unwin, and J. M. Wright, *J. Gen. Microbiol.,* **5**, 619 (1951).
S9. Brian, P. W., P. J. Curtis, H. G. Hemming, C. H. Unwin, and J. M. Wright, *Nature,* **164**, 534 (1949).
S10. Brockman, H., and K. Bauer, *Naturwiss.,* **37**, 492 (1950).
S11. Brockman, H., K. Bauer, and I. Borchers, *Chem. Ber.,* **84**, 700 (1951).
S12. Brockman, H., and I. Borchers, *Chem. Ber.,* **86**, 261 (1953).
S13. Brockman, H., and W. Henkel, *Chem. Ber.,* **84**, 284 (1951); **85**, 426 (1952).
S14. Brockman, H., and G. Schmidt-Kastner, *Naturwiss.,* **38**, 479 (1951).
S15. Burton, H. S., and E. P. Abraham, *Biochem. J.,* **50**, 168, (1951).
S16. Celmer, W. D., and I. A. Solomons, *J. Am. Chem. Soc.,* **74**, 1870, 2245, 3838 (1952); **75**, 1372 (1953).

S17. Clark, R. K., Jr., and J. R. Schenck, *Arch. Biochem.*, **40**, 270 (1952).

S18. Cunningham, K. G., S. A. Hutchinson, W. Manson, and F. S. Spring, *J. Chem. Soc.*, **1951**, 2299.

S19. Davisson, J. W., I. A. Solomons, and T. M. Lees, *Antibiotics and Chemotherapy*, **2**, 460 (1952).

S20. DeBoer, C., E. L. Caron, and J. W. Hinman, *J. Am. Chem. Soc.*, **75**, 499, 5864 (1953).

S20a. Flynn, E. H., J. W. Hinman, E. L. Caron, and D. O. Woolf, Jr., *J. Am. Chem. Soc.*, **75**, 5867 (1953).

S21. Freeman, G. G., and J. E. Gill, *Nature*, **166**, 698 (1950).

S22. Freeman, G. G., and R. I. Morrison, *Biochem. J.*, **44**, 1 (1949).

S23. Freeman, G. G., and R. I. Morrison, *J. Gen. Microbiol.*, **3**, 60 (1949).

S24. Groupe, V., L. H. Pugh, and A. S. Levine, *Proc. Soc. Exp. Biol. Med.*, **80**, 710 (1952).

S25. Grove, J. F., *J. Chem. Soc.*, 4056 (1952).

S26. Grove, J. F., and associates, four papers (I to IV), *J. Chem. Soc.*, 3949-4002 (1952).

S27. Grove, J. F., D. Ismay, J. MacMillan, T. P. C. Mulholland, and M. A. T. Rogers, *Chemistry and Industry*, **1951**, 219.

S28. Grundy, W. E., A. L. Whitman, E. G. Rdzok, E. J. Rdzok, M. E. Hanes, and J. C. Sylvester, *Antibiotics and Chemotherapy*, **2**, 399 (1952).

S29. Harned, R. L., P. H. Hidy, C. J. Corum, and K. L. Jones, *Antibiotics and Chemotherapy*, **1**, 594 (1951).

S30. Hata, T., T. Higuchi, Y. Sano, and K. Sowachika, *J. Antibiotics* (Japan), **3**, 313 (1950).

S31. Hata, T., N. Ohki, and T. Higuchi, *J. Antibiotics* (Japan), **5**, 529 (1952).

S31a. Hausman, W., and L. C. Craig, *J. Biol. Chem.*, **198**, 405 (1952).

S32. Heatley, N. G., J. L. Gowans, H. W. Florey, and A. G. Sanders, *Brit. J. Exptl. Path.*, **33**, 105 (1952).

S33. Hickey, R. J., C. J. Corum, P. H. Hidy, I. R. Cohen, U. F. B. Nager, and E. Kropp, *Antibiotics and Chemotherapy*, **2**, 472 (1952).

S34. Hobby, G. L., P. P. Regna, N. Dougherty, and W. E. Steig, *J. Clin. Invest.*, **28**, 927 (1949).

S35. Howell, S. F., *J. Biol. Chem.*, **186**, 863 (1950).

S36. Hwang, K., *Antibiotics and Chemotherapy*, **2**, 453 (1952).

S37. Johnson, E. A., and K. L. Burdon, *J. Bact.*, **54**, 281 (1949).

S38. Kligman, A. M., and F. S. Lewis, *Proc. Soc. Exp. Biol. Med.,* **82**, 339 (1953).
S39. Kuehl, F. A., Jr., M. N. Bishop, L. Chaiet, and K. Folkers, *J. Am. Chem. Soc.,* **73**, 1770 (1951).
S40. Landy, M., G. H. Warren, S. B. Rosenman, and L. G. Colio, *Proc. Soc. Exp. Biol. Med.,* **67**, 539 (1948).
S41. Lechevalier, H., R. F. Acker, C. T. Corke, C. F. Haenseler, and S. A. Waksman, *Mycologia,* **45**, 155 (1953).
S41a. Lewis, G. M., M. E. Hopper, and S. Shultz, *Arch. Dermat. Syph.,* **54**, 300 (1946).
S42. MacMillan, J., *Chemistry and Industry,* **1951**, 719.
S43. Maeda, K., Y. Okami, H. Kosaka, O. Taya, and H. Umezawa, *J. Antibiotics* (Japan), **5**, 572 (1952).
S44. Markham, N. P., N. G. Heatley, A. G. Sanders, and H. W. Florey, *Brit. J. Exptl. Path.,* **32**, 136 (1951).
S45. Mayer, R. L., *Proc. 12th Cong. Pure and Applied Chemistry,* New York, September, 1951, p. 283.
S45a. McCormick, M., and M. M. Hoehn, *Antibiotics and Chemotherapy,* **3**, 581 (1953).
S46. McLamore, W. M., W. D. Celmer, V. V. Bogert, F. C. Pennington, and I. A. Solomons, *J. Am. Chem. Soc.,* **74**, 2946 (1952).
S47. McLeod, C., *J. Bact.,* **56**, 749 (1948).
S47a. Mulholland, T. P. C., *J. Chem. Soc.,* **1952**, 3987, 3994.
S48. Nakase, Y., and T. Hata, *J. Antibiotics* (Japan), **5**, 542 (1952).
S48a. Newton, G. G. F., and E. P. Abraham, *Nature,* **172**, 395 (1953).
S49. Oxford, A. E., H. Raistrick, and P. Simonart, *Biochem. J.,* **33**, 240 (1939).
S50. Porter, J. N., R. I. Hewitt, C. W. Hesseltine, G. Krupka, J. A. Lowery, W. S. Wallace, N. Bohonos, and J. N. Williams, *Antibiotics and Chemotherapy,* **2**, 409 (1952).
S50a. Pound, and M. A. Stahmann, *Phytopathology,* **41**, 1104 (1951).
S51. Regna, P. P., R. A. Carboni, and W. E. Steig, *Am. Chem. Soc. Meeting in Miniature,* Brooklyn, March 17, 1950.
S52. Reynolds, D. M., and S. A. Waksman, *J. Bact.,* **55**, 739 (1948).
S53. Ritchie, E. C., N. Smith, and H. W. Florey, *Brit. J. Pharmacol.,* **6**, 439 (1951).
S54. Sano, Y., *J. Antibiotics* (Japan), **5**, 535 (1952).
S55. Schenck, J. R., and A. F. De Rose, *Arch. Biochem.,* **40**, 263 (1952).
S56. Sobin, B. A., *J. Am. Chem. Soc.,* **74**, 2947 (1952).

S57. Stoll, A., A. Brack, and J. Renz, *Schweiz. Z. Path. u. Bakt.*, **14**, 225 (1951).

S58. Stoll, A., J. Renz, and A. Brack, *Helv. Chim. Acta,* **34**, 862 (1951).

S59. Tejera, E., E. J. Backus, M. Dann, C. D. Ervin, A. J. Shakofski, S. O. Thomas, N. Bohonos, and J. H. Williams, *Antibiotics and Chemotherapy,* **2**, 333 (1952).

S60. Tint, H., and W. Reiss, *J. Biol. Chem.*, **190**, 133 (1951).

S61. Townley, R. W., R. P. Mull, and C. R. Scholz, *Proc. 12th Intern. Cong. Pure and Applied Chemistry,* New York, September 1951, p. 284.

S62. Umezawa, H., and T. Takeuchi, *J. Antibiotics* (Japan), **5**, 270 (1952).

S63. Umezawa, H., S. Takahashi, T. Takeuchi, K. Maeda, and Y. Okami, *J. Antibiotics* (Japan), **5**, 466 (1952).

S64. Waller, C. W., P. W. Fryth, B. L. Hutchings, and J. H. Williams, *J. Am. Chem. Soc.*, **75**, 2025 (1953).

S65. Williams, R. J., and H. Kirby, *Science,* **107**, 481 (1948).

Part VI. MISCELLANEOUS

CHAPTER 10

MISCELLANEOUS FERMENTATIONS

Alfred R. Stanley and R. J. Hickey

DEXTRAN FERMENTATION

Dextran is a microbiologically produced polysaccharide well known around sugar refineries and present in very small amounts in practically all commercial sucrose. This compound has been used on a limited scale for some time as a stabilizer in certain food products, such as sirups and ice creams, and has recently attained a leading position in the field of blood extenders for use in place of blood plasma in cases of shock and loss of blood.

Historical

As recorded by Browne,[10] the first report on dextran goes back to 1822. In this year, Vanquelin sent four bottles of cane juice from Martinique to France, but on arrival, it was found that the juice had changed to a thick mucilage. Vanquelin studied this gummy mass, but could not determine its composition, probably because of a lack of analytical procedures. Pasteur, Durin, Scheibler, and many others studied and described this and similar gummy fermentations of both cane and beet sugar and, in 1861, Pasteur[69] described two "sorts" of viscous fermentations due to two distinct "organized ferments."

Durin,[17,18] was convinced that this gum was composed of cellulose, but its true composition, consisting only of glucose molecules, was first discovered by Scheibler.[76] He named it dextran because it was similar to dextrin, but was in error in believing the substance to occur naturally in cells of the sugar beet.

The causal organism of this fermentation was first named by Cienkowski[13] as *Ascococcus mesenteroides,* believing the slime to be due to a true zoogleal formation. Van Tieghem[107] gave the first adequate description of the organism noting its similarity to the genus *Nostoc* of the green algae and so he named it *Leuconostoc mesenteroides.*

Several other workers, including Bechamp,[6] Daumichen,[16] Brautigam,[9] and Happ[33] studied the fermentation from different points of view, but with no particularly outstanding results.

Pure cultures of *Leuconostoc mesenteroides* were first isolated by Leisenberg and Zopf[55] and strains from different sources showed no marked variation. Two other isolants were named *L. opalanitza* and *L. aller* by Zettnow.[118] Another strain named *Lactococcus dextranicus* was isolated by Beijerinck[7] in 1914. It was described as a lactic acid-producing coccus which formed dextran from only sucrose.

A very careful study of the genus *Leuconostoc* was conducted by Hucker and Pederson[45] who classified three distinct species, namely *L. mesenteroides, L. dextranicus,* and *L. citrovorus.* This classification is based primarily on reactions in a variety of carbohydrates and polyhydric alcohols. From this classification, it is quite certain that the organisms described by Liesenberg and Zopf[55] and by Zettnow[118] were of the species *mesenteroides,* whereas those described by Beijerinck and later by Fernbach, Schoen, and Hagiwara[25,26] were of the *dextranicus* type.

More detailed reviews covering the period up to the early years of this century are those of Lafar,[51] from the microbiological point of view, and of Lippmann,[56] from the viewpoint of a sugar chemist.

Raw Materials

Sucrose is the principal raw material for dextran formation. It may be used in the form of molasses, as noted by Bauer[5] and

Owen,[68] or as refined sucrose with which most of the work has been done. In addition, plant juices have served as the raw material in several studies, for example, aquamiel,[72] sugar-beet juice,[8,102] wine,[109] and sugar-cane juice.[114] Most investigators report that sucrose is the only carbohydrate from which they were able to obtain dextran, but Leisenberg and Zopf[55] reported the production of dextran from both sucrose and dextrose. The work of the last 20 years does not, however, tend to substantiate this.[8a]

In 1931, Tarr and Hibbert[101] reported that, with four strains of *L. mesenteroides* studied, pronounced precipitation occurred in fermented sucrose media treated with 5 volumes of 95% alcohol. Only slight precipitation was found in glucose media with the two strains received from Holland when treated in the same manner. They conclude from this that "It seems probable that only very active cultures of *L. mesenteroides* are capable of forming dextran from glucose."

The production of dextran from refined cane sugar has been shown to be stimulated by certain crude plant products, including raw sugar,[11] maple sugar,[80] and an extract of figs.[108]

The conversion of dextrin, Lintner's soluble starch, certain hydrolysates of amylopectin, and of crystalline amylose into dextran has been described.[40a,41] The acetic acid bacteria are involved in this conversion, whereas the lactic acid types are used with sucrose.

Fermentation Mechanism

The mechanism of the fermentation producing dextran from sucrose is not too well understood.

Kagan, Lyatker, and Tsvasman,[49] in 1942, stated that *L. mesenteroides* was capable of binding inorganic phosphorus in the presence of sucrose, but not of glucose or fructose. In sucrose, then, the Cori ester is formed, as is shown by its accumulation at the expense of inorganic phosphorus. Later, Hehre[39,40a] concluded from his studies that phosphorylated sugar does not enter into this synthesis of dextran.

Leibowitz and Hestrin[52] stated in 1945 that the Cori ester does not enter into this reaction nor is sucrose acted on by the ordinary phosphorylase. They concluded that, while the exact mechanism is not certain, the process is direct and involves a primary hydrolysis.

In 1950, Hassid and Doudoroff[35] stated that the reaction involves the substitution of a 1,6-glucosidic linkage for the glucose-fructose bond in the sucrose molecule. Evidently only the glucose portion of sucrose is used for formation of the dextran molecule. Neither glucose alone nor glucose-fructose mixtures will yield dextran. Sucrose is considered essential.[8a]

A reaction for the formation of dextran from sucrose by the action of dextransucrase appears to be as follows:[40a]

In addition to the 1,6-glucosidic linkage in dextran, some 1,4-maltosidic cross-linkage is also found.[8a,40a]

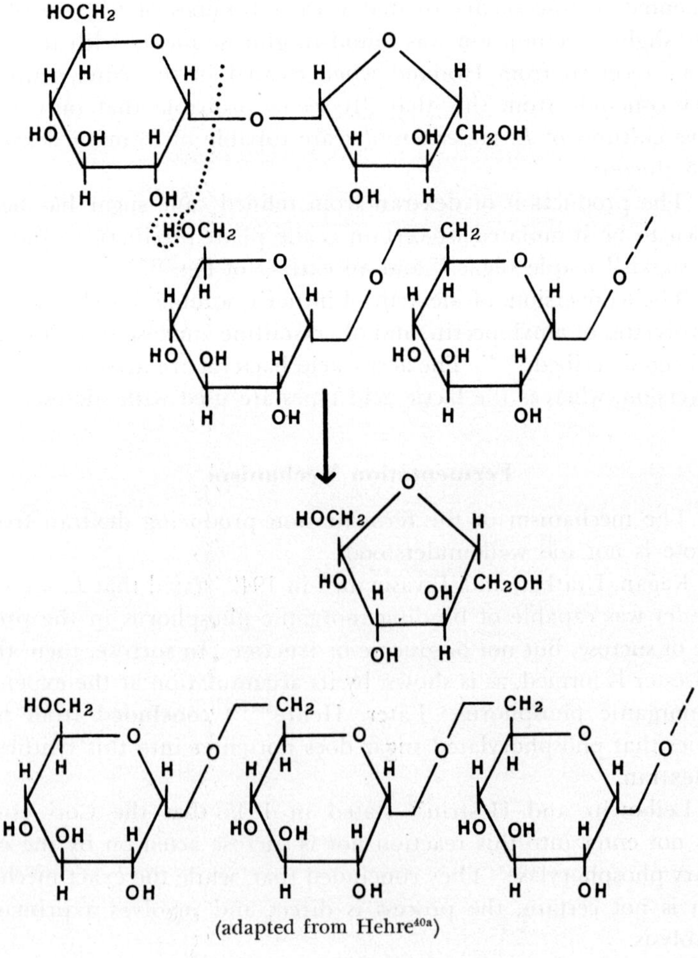

(adapted from Hehre[40a])

Cultures and Media

Several different media and modifications have been described for use with *Leuconostoc mesenteroides*. Tarr and Hibbert,[101] in 1931, suggested 10% sucrose, 0.5% KH_2PO_4, 0.5% NaCl, and 0.5% peptone. A recent report[8a] indicates that *L. mesenteroides* NRRL B-512 is generally employed for the preparation of clinical dextran of commerce.

The use of maple sugar as a part of the sucrose was suggested[80] in 1938. This undoubtedly contributed a number of growth factors.

Yeast extract was substituted for peptone, and ammonium and magnesium sulfates were added by Hassid and Barker,[34] in 1940. Their medium was as follows: 600 g sucrose, 15.0 g yeast extract, 30 g K_2HPO_4, 1.2 g $MgSO_4 \cdot 7H_2O$, 3.6 g $(NH_4)_2SO_4$, and 6 l distilled water.

This medium was modified by Jeanes, Wilham and Miers[48] who added sodium chloride in 0.1% concentration. Other variations are composed of the previously listed ingredients, but in different proportions. Calcium carbonate has sometimes been added to media to control the pH.[48]

Suggested substitutes for peptone and yeast extract are corn steep liquor,[68,111] malt sprouts,[111] proteolyzed wheat grain,[68] and hydrolyzed juice of *Agave atrovireus*.[72]

Stacey and Swift,[79] in 1948, described a somewhat different combination as follows: 4.9 g microcosmic salt, 0.5 g *p*-aminobenzoic acid, 1.0 g potassium hydrogen phosphate, 0.5 g potassium chloride, 0.5 g hydrated magnesium sulfate, 0.01 g hydrated ferrous sulfate, 0.01 g peptone, 100 g sucrose, and 1,000 ml distilled water.

The production of dextran by other bacteria was reported by Hehre and coworkers.[41,42] When growing *Acetobacter viscosum* or *A. capsulatum,* they used a medium consisting of 5% dextrin or other carbohydrate and 0.5% yeast extract, while the broth for streptococci contained 5% sucrose, 1% tryptose peptone, 0.2% anhydrous sodium phosphate, and 0.5% sodium chloride.

Incubation temperatures have varied somewhat between investigators. *Leuconostoc* strains have been grown at either 20° to 25°C or at 30°C and until 1948 the fermentation period was extended for 5 to 10 days. Jeanes, Wilham, and Miers[48] and Sanchez-Marroquin and Arciniega[72] reduced this time to between 24 and 26 hours for a high-viscosity dextran. *Acetobacter* fermentations were run at 25°C for 1 week, whereas the streptococci were incubated at 37°C for 5 days.

One of the fungi (*Botrytis cinerea*) has also been described as producing dextran from sucrose.[109] This organism has given some trouble in wine making.

Growth factors required by the various species of *Leuconostoc* have been studied by a number of investigators,[73,74,100] but the only studies relating these factors to dextran production are those by Carlson and coworkers.[11,12,115,116] They found that all strains of *Leuconostoc* studied required thiamine, nicotinic acid, and pantothenic acid. Requirements for folic acid and riboflavin varied with the strains and while biotin was required in a glucose medium, it was not required in the presence of sucrose. *p*-Aminobenzoic acid was not required for dextran synthesis by the strains studied and the stimulation afforded by raw cane sugar could not be identified with any of the vitamins, or their combinations, studied.

The production of dextran by cell-free enzymes has been described by Hehre,[38b,40a] Stacey,[78d] and Koepsell and Tsuchiya.[49a] According to the last authors, this enzyme, dextransucrase, can be produced in large quantity in a corn steep liquor-mineral salts medium. The molecular size of the resulting dextran can be varied by changing the conditions of the reaction. Large amounts of fructose are present in the reaction solution when dextran formation is completed. Enzymic synthesis of dextran, levan, and other polysaccharides was reviewed in 1951 by Hehre.[40a]

Dextran Recovery

It was recognized at least as early as 1903 that dextran could be precipitated by alcohol.[71] The proportions used have varied, however, from 1 to 5 volumes as compared with the dextran solution.

One volume of alcohol per volume of dextran solution was used by Hassid and Barker,[34] Stacey and Swift,[79] Jeanes, Wilham, and Miers,[48] and Haworth and Stacey.[36] Jeanes, Wilham, and Miers[48] passed dextran beer through a supercentrifuge to remove the bacterial cells before precipitating the dextran, but no mention is made of such a procedure by the other investigators.

Stacey and Yound[80] and Daker and Stacey[14] used 2 volumes of alcohol for precipitation, whereas Tarr and Hibbert[101] used 5 volumes.

Fractional precipitation was utilized by Hehre and Neill,[42] in 1946, in a study of the polysaccharides produced by streptococci. Their samples were diluted 1:10 with 10% sodium acetate and 1.2 and 2.5 volumes of 95% alcohol were added to separate portions of each sample. The sample treated with 1.2 volumes of alcohol was used as an index of the dextran present, whereas the sample treated with 2.5 volumes was considered to be a test for either dextran or levan.

The only method described to date for the purification of dextran is that of reprecipitation.[48,79] Stacey and Swift[79] reprecipitated from water at different pH values until essentially ash-free dextran was obtained. Information on industrial methods was supplied in 1953.[8a]

Yields of dextran reported in the literature vary from 10%[25,26] to 40%[36] of the starting sucrose with most reports in the range of 25 to 35%.[48,79,80,101] The theoretical yield, assuming that only the glucose fraction of sucrose goes into the dextran molecule, would be approximately 47% of the sucrose.

The Structure of Dextran

As mentioned previously, the true nature of dextran, as a compound made up entirely of glucose units, was reported by Scheibler,[76] in 1869. Information on how these molecules were arranged, however, has appeared only in the last 20 years.

The complete methylation of dextran produced by one strain of *Leuconostoc mesenteroides* was reported by Fowler, Buckland, Branns, and Hibbert,[27] in 1937, to yield dimethyl, trimethyl, and tetramethyl methylglucosides in a ratio of 1:3:1. They conclude from these results that dextran is a polymer of pentaglucopyranose anhydride units, one being attached as a side chain and the remaining four most probably being connected in "linear chain union." Three linkages are of the 1:6 type, while the other two may be either 1:6 or 1:4. It is possible that the side chains are composed of more than one glucopyranose unit. These results were confirmed by Levi, Hawkins, and Hibbert,[54] in 1942, and again by Stacey and Swift,[79] in 1948.

Dextran produced by a strain of *Leuconastoc dextranicus* was reported by Fairhead, Hunter, and Hibbert,[24] in 1938, to show a very different result on methylation. No tetramethyl methylglucoside was formed. The product was principally the trimethyl derivative with about 10% of the dimethyl. This would indicate a very linear molecular arrangement with few side chains. Essentially the same results were reported by Peat, Schluchterer, and Stacey[70] the next year. Their results indicated a chain length of 200 to 550 glucose units. The same type of molecular arrangement was found by Daker and Stacey[15] for a dextran produced by *Betabacterium vermiforme* (Ward-Mayer). End-group determinations indicated a chain length of 25 glucose units whereas the osmotic pressure method of molecular-weight determination indicated a chain of about 500 units.

Uses of Dextran

Several ethers, esters, and mixed ether-esters of dextran were described by Stahly and Carlson,[82-99] in a series of patents, in 1940 and 1941. Certain of these are described as suitable for use in the production of lacquers. The sulfuric acid ester of dextran was described by Grönwall, Ingelman, and Mosimann,[32] in 1945, as having an anticoagulant potency comparable to that of heparin. Other derivatives which have been described include the acetate[112,113] and aryl carbamates.[117] The use of an acid ester as a cation-exchange medium was described by McIntire and Schenck,[63] in 1948.

Dextran itself is a good stabilizer for sugar sirups,[57,111] ice cream, and other confections and has been the basis for the operation of one company in the United States. It has also been suggested as a stabilizer for drilling muds for oil and gas wells.[4]

Interest in dextran was revived by the reports of Grönwall and Ingelman[29,30,31] on its use as a substitute for blood plasma in transfusions. The experience of investigators with dextran in Sweden and a number of reports of clinical investigations with this general type have been well summarized by Ingelman,[47] in 1949. In this report, he described the product used as a partially hydrolyzed dextran with molecular weights in the range 20,000 to 200,000. Hydrolysis was illustrated by an example in which an 8% solution of raw dextran was heated to 90°C for 20 minutes in a solution of

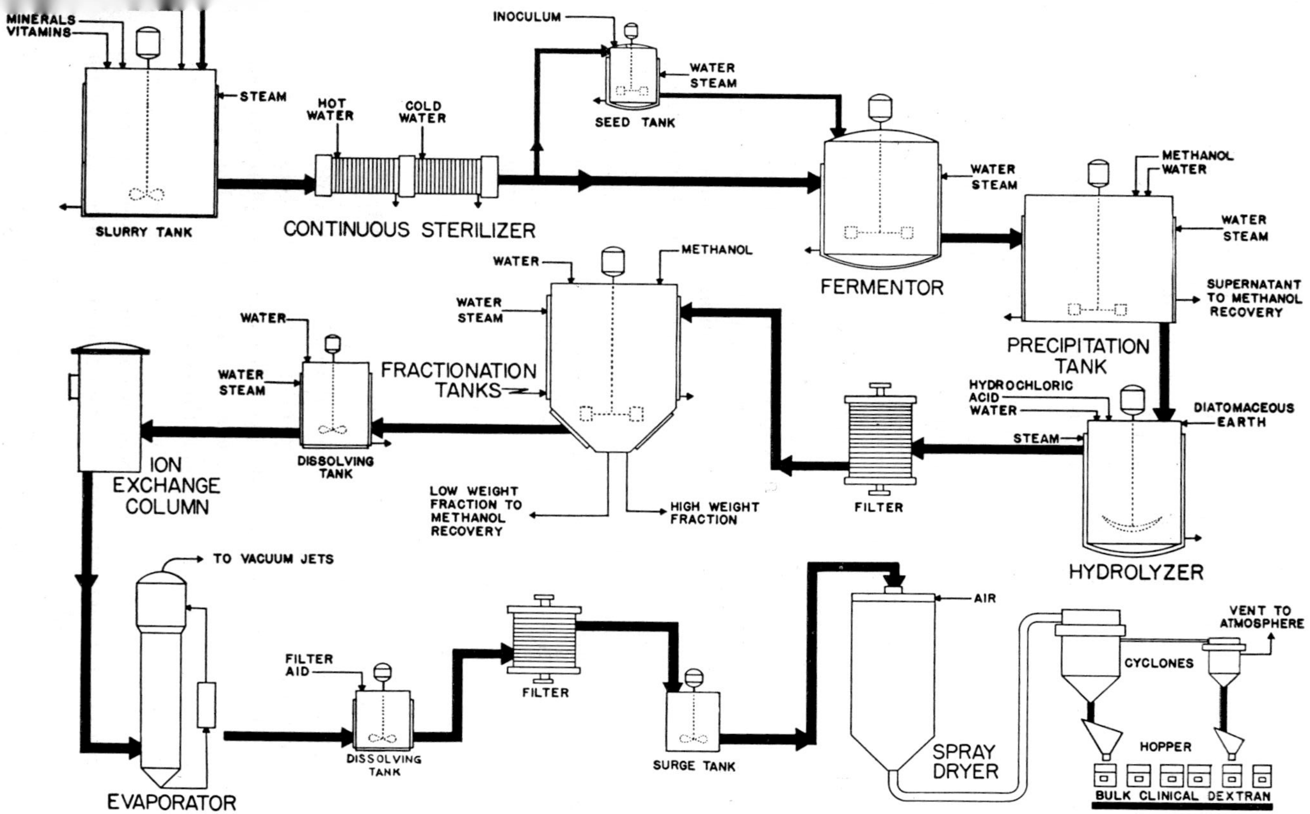

FIGURE 22. *Flow Diagram for the Clinical Dextran Process of Commercial Solvents Corporation* [Reprinted by permission from a paper by Bixler, Hines, McGhee, and Shurter, *Ind. Eng. Chem.*, **45**, 692 (1953)]

0.12 *N* hydrochoric acid. The hydrolysis was stopped by neutralization with sodium hydroxide and the dextran precipitated with alcohol and dried. Spray-drying is employed with success for commercial production of dextran.[8a] A neutral 6% (w/v) solution of this partially hydrolyzed dextran with a relative viscosity of 3.3 (compared to distilled water) may be sterilized by heating to 120°C for ½ hour. This type of a solution containing in addition 0.9% NaCl and passing certain specifications[8a] is used for injection.

Radioactive dextran employing C^{14} was prepared in 1952[78a] from sucrose which was photosynthesized by canna leaves which metabolized $C^{14}O_2$ supplied to them. Clinical dextran prepared in this manner was employed with success to follow the fate of injected material. Essentially all of the radioactivity was accounted for in expired CO_2 and in body excretions.

Recently,[1a,8a] it has been reported that four producers in the United States are supplying clinical dextran for use as a blood-plasma extender. Production methods, equipment, and application were described in some detail in 1953.[8a] The Armed Forces specifications for clinical dextran were included in this report.[8a] An average molecular weight of 75,000 ± 25,000 as determined by light-scattering was specified. The high molecular weight fraction (5 to 10%) should not exceed 200,000 and the low fraction (5 to 10%) should not be less than 25,000. A flow diagram for the production of clinical dextran at the plant of Commercial Solvents Corporation in Terre Haute, Indiana, is shown in Figure 22.

THE SYNTHESIS OF *l*-EPHEDRINE

It has been commonly observed in drug studies that activity and utility of a drug may be dependent on its steric configuration. Great differences may be observed between the pharmacologic activities of *d*- and *l*-isomers of the same compound. For example, the *l*-isomer of ephedrine is about thirty times more active than the *d*-isomer in pressor properties. Chemical syntheses of many of the optically active drugs have been accomplished, but synthesis of racemates has generally been involved and resolution is, to say the least, often troublesome and costly.

The alkaloid, *l*-ephedrine, has been successfully prepared by the unique method of Hildebrandt and Klavehn,[44] described in

1934, in which an asymmetric synthesis is accomplished fermentatively. An *l*-intermediate is prepared rather than the more common *dl*-intermediate.

Ephedrine was originally obtained from the Chinese plant, *ma huang*. The medicinal properties of the plant have been known to the Chinese for several thousand years. Ephedrine is a sympathomimetic amine which has been widely used for the relief of congestion in colds and in other disorders. The history and pharmacology of the drug have been discussed by Goodman and Gilman.[28]

The Hildebrandt and Klavehn synthesis[44] makes use of the observations of Neuberg[65,66] who found, in 1921 and later, that top yeast in a glucose medium will fermentatively convert benzaldehyde to *l*-1-phenyl-1-hydroxy-2-propanone:

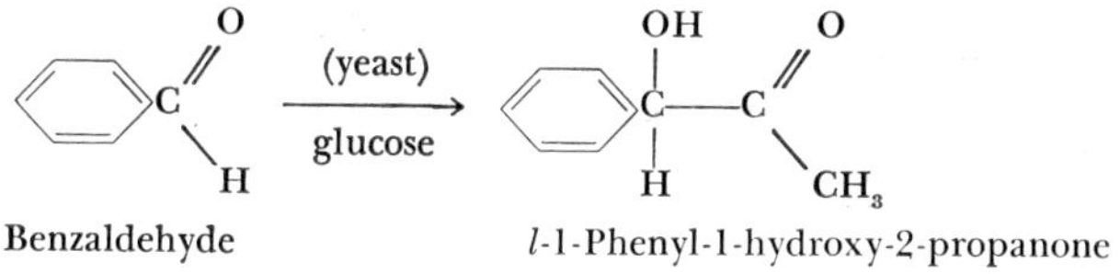

The *l*-keto alcohol thus obtained is then reductively condensed with monomethylamine by chemical means to yield primarily *l*-ephedrine, along with pseudoephedrine, which may be separated from the ephedrine.

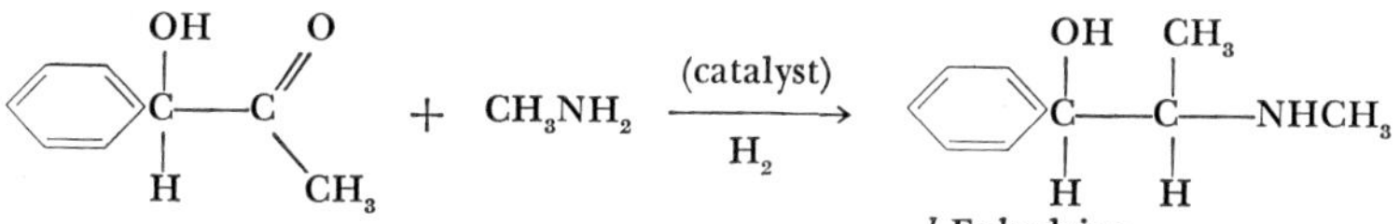

l-Ephedrine

The reduction is accomplished by means of aluminum amalgam or by catalytic hydrogenation using colloidal platinum.

The fermentative phase of the synthesis, according to Neuberg, would proceed, for example, by fermenting a mixture of 125 g of glucose sirup and 100 g of yeast in 2.5 l of solution for ½ hour. Following this, 10 g of benzaldehyde is added in increments with stirring. At the end of 3 days, the yeast is removed and, by ether extraction and subsequent isolation procedures, about 2.5 g of *l*-1-phenyl-1-hydroxy-2-propanone is obtained, along with significant amounts of benzyl alcohol and benzoic acid.

Hildebrandt and Klavehn[44] cite several examples of procedures for conversion of the ketoalcohol to ephedrine. In one procedure, to a solution of 100 g of *l*-1-phenyl-1-hydroxy-2-propanone in 200 ml of ether, 75 g of a 33% solution of monomethylamine is added. The mixture is agitated for about 30 minutes, during which time, an exothermic condensation occurs yielding the following imine:

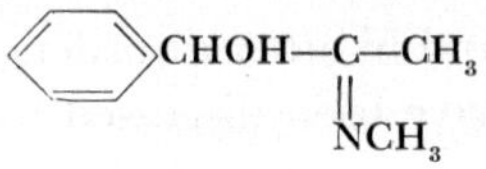

The mixture is then catalytically hydrogenated, using 70 ml of 1% colloidal platinum. The *l*-ephedrine is recovered as the crystalline hydrochloride after filtration and treatment with hydrochloric acid. The yield of final product was not given for this example; in a related example employing more impure starting material, about 110 g of ephedrine hydrochloride was obtained from 300 g of crude ketoalcohol oil. The fact that the imine is asymmetric undoubtedly affects the proportions of isomers obtained on reduction.

Studies on the mechanism of biosynthesis of *l*-1-phenyl-1-hydroxy-2-propanone (phenylacetylcarbinol) were reported by Smith and Hendlin[78c] in considerable detail in 1953.

It is to be expected that modifications of the original procedures are employed in actual practice.

THE ACTION OF MICROORGANISMS ON STEROIDS

Steroid reactions have been accomplished in the laboratory generally by purely chemical means. The fact that many different steroids, such as bile acids, sex hormones, and D-vitamins, are essential to biological processes indicates, of course, that enzymic reactions of steroids occur. Reactions brought about by microorganisms might be expected and some such reactions have been observed. Mildness of reaction conditions and lack of excessive oxidation by-products have been noted[77] as advantages of the biological oxidation over the chemical methods.

A few instances of microbiologic steroid conversions will be noted. Both oxidative and reductive reactions have been brought about microbiologically. The reactions have generally concerned oxidation of hydroxyls to carbonyls, or the reduction of carbonyls to hydroxyls. Recently oxidations have been described in which a hydroxyl has been substituted for hydrogen directly on the ring positions. The general subject was reviewed fairly extensively through 1945 by Fischer;[26a] direct introduction of oxygen on a steroid nucleus was not reported until about 1947.[50a]

Oxidative Reactions

In 1938, Mamoli[58] described the conversion of dehydroandrosterone to androstenedione by bacteriological oxidative means. The organism employed was probably *Flavobacterium helvolum*[59] or a very closely related organism. *Acetobacter pastorianum* has also been reported[50] to produce this reaction which was described further, in 1939, by Mamoli, Koch, and Teschen.[60]

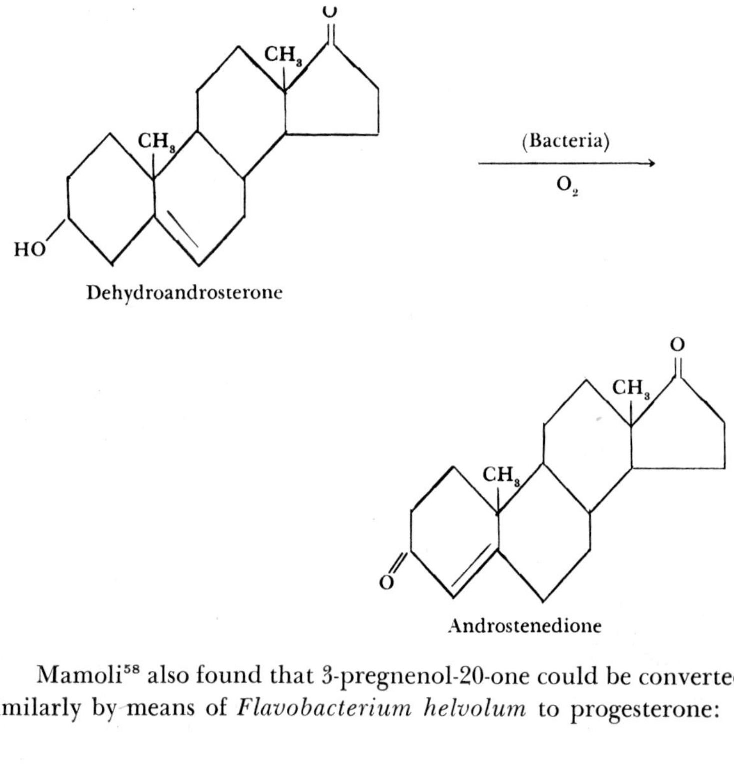

Mamoli[58] also found that 3-pregnenol-20-one could be converted similarly by means of *Flavobacterium helvolum* to progesterone:

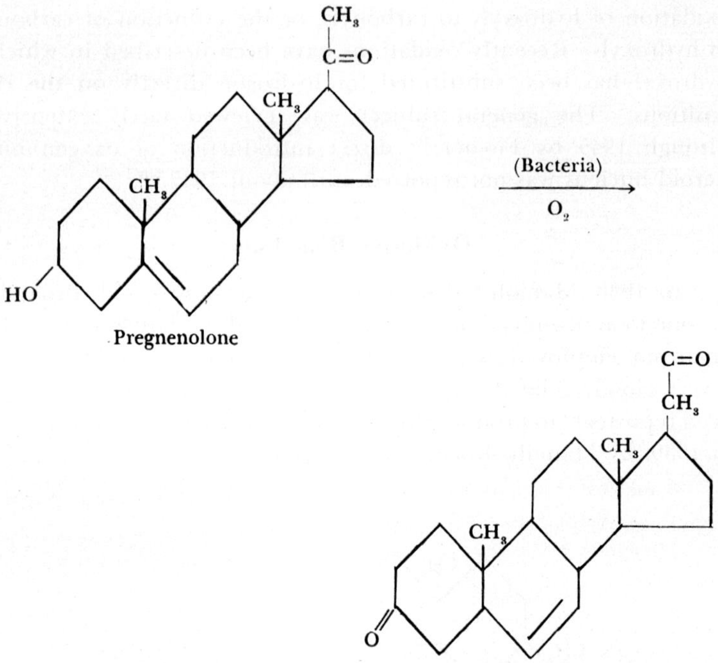

Pregnenolone

Progesterone

In addition, Mamoli[58] attempted to oxidize cholesterol by means of *Fl. helvolum,* but was unsuccessful. Subsequently it was found that both *Acetobacter xylinum*[50] and species of *Nocardia*[103] would oxidize cholesterol to △4-cholestenone:

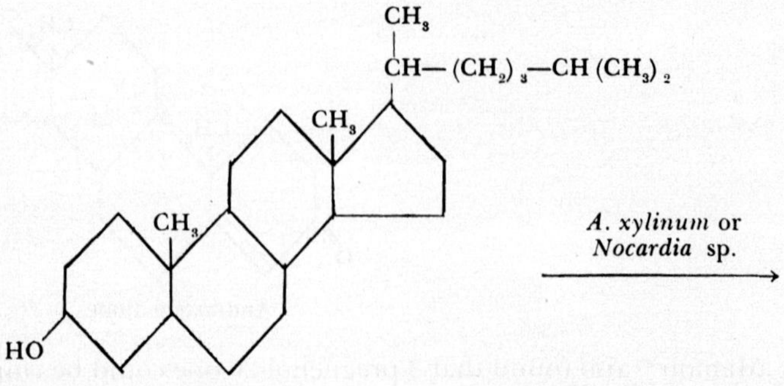

Cholesterol

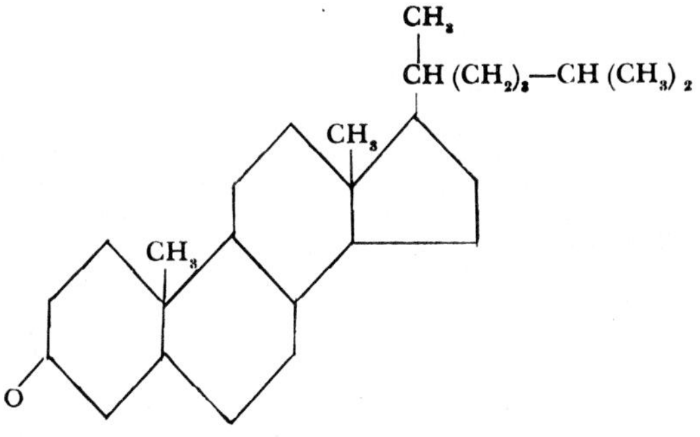

Cholestenone

For this reaction, Turfitt[103] employed a medium of inorganic salts, with the sterol as the only organic carbon source. Flasks inoculated with *Nocardia* were incubated statically for 1 month at 25°C prior to recovery of the product. Turfitt[104,105,106] has also reported other related investigations. In 1949, Schatz Savard, and Pintner[75] described studies with a number of *Nocardia* and other fungi, along with several bacteria. Both static and shake flasks were employed; more rapid metabolism was found in the shake flasks. The presence of calcium carbonate in the media as a buffer was advantageous to fermentative attack. Steroid conversions were not discussed.

Hughes and Schmidt[46] showed, in 1942, that a strain of *Alcaligenes faecalis* in nutrient medium, occasionally agitated, would oxidize desoxycholic, hyodesoxycholic, and lithocholic acids and also dehydroisoandrosterone to corresponding ketoderivatives. Prior studies of a similar nature[77,78] showed that *A. faecalis* would also oxidize cholic acid to ketocholanic acids. Hughes and Schmidt[46] also attempted the oxidations of estradiol and estriol under similar conditions, but were unsuccessful.

In 1941, Ercoli[21] observed that estradiol could be converted to estrone by the use of a heavy culture of *Flavobacterium dehydrogenans* (*Micrococcus dehydrogenans*). Other steroid reactions were also observed to be brought about by this organism.[19,20] Similarly, in 1944, Zimmerman and May[119] found that pseudodiphtheria bacilli were capable of converting estradiol to estrone and androstenediol to testosterone.

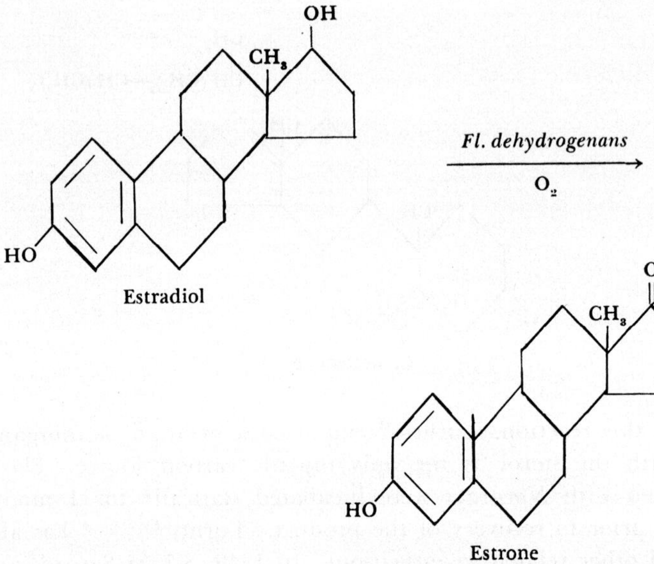

Similar findings were reported, in 1942, by Arnaudi[2] who showed that *Fl. dehydrogenans* would convert pregnenolone to progesterone, dehydroandrosterone to androstenedione, and estradiol to estrone, along with other reactions. In 1944, Ercoli and Molina[22,23] reported a new organism, *Fl. androstenedionicum,* which they found would convert androstenediol to androstenedione. Another organism, *Fl. carbonilicum,* behaved somewhat similarly;[64] it was found to oxidize androstenediol to a mixture of testosterone and androstenedione. The subject was studied further, in 1946, by Arnaudi.[3]

As a specific example of microbiological steroid oxidation, Mamoli[59] has claimed, in a 1944 patent, a method for conversion of 21-acetoxypregnenol-3-one-20 to desoxycorticosterone. In his procedure, the medium employed was composed of 60 ml sterile yeast water buffered with 10 ml each of 0.2 *M* Na_2HPO_4 and 0.2 *M* KH_2PO_4. To this medium 200 mg of pulverized 21-acetoxypregnenol-3-one-20 was added and the mixture was sterilized for 1 hour. The cooled, sterile medium was then inoculated with a culture of *Flavobacterium helvolum* ("*Bacterium Coryne*"), after which, agitation was employed for 6 days under an oxygen atmosphere. At the end of this period, the product was filtered, the residue was dissolved in acetone, and the bacterial residue was removed from the acetone extract by filtration. On removal of the acetone by evaporation, the residue was recrystallized with care from "diluted acetone." The first crystal crop of 54 mg was composed of unreacted 21-acetoxypregnenol-3-one-20. The mother liquor was evaporated and the residue was taken up in acetone. After diluting this acetone solution with ether, a 60 mg crop of crystals was obtained which was identified as desoxycorticosterone, M.P. 139° to 140°C. Other related dehydrogenation reactions were also described.

Additional reactions were set forth in a patent of Koester, Mamoli, and Vercellone.[50] Microorganisms employed by these investigators included both yeast and *Acetobacter*. In one example cited, *Acetobacter pastorianum* in surface culture converted androstenediol to testosterone. Similarly, androstenediol monopropionate-17 was oxidized to testosterone propionate.

An example was also cited[50] in which a Milan yeast was employed for both oxidation and reduction to convert dehydroandrosterone to testosterone:

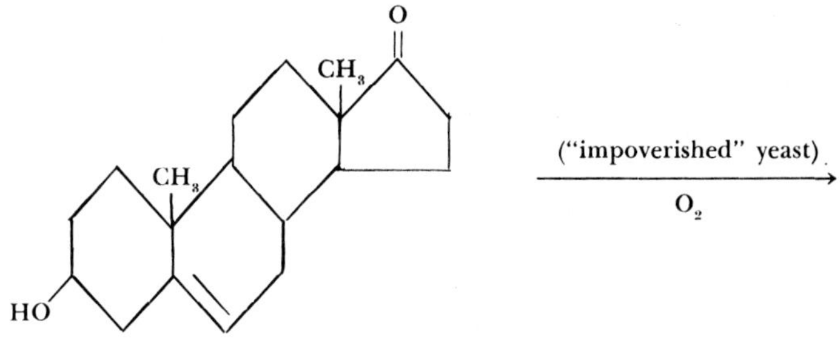

Dehydroandrosterone

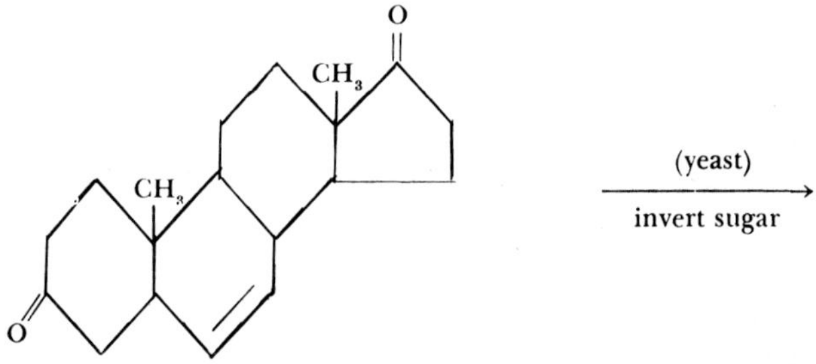

Androstenedione

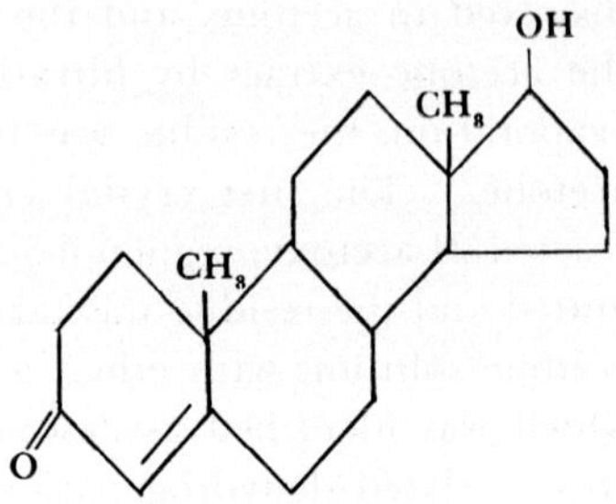

Testosterone

Experimentally, 8 g of yeast in 50 ml of phosphate buffer was shaken for 20 hours at 32°C in an oxygen atmosphere. This procedure produced an "impoverished" yeast. To the mixture, 200 mg of dehydroandrosterone, suspended in 30 ml of water, was added. The resulting mixture was then shaken for an additional 48 hours under oxygen. At this point, 25 g of invert sugar, dissolved in 150 ml of water, was added and the mixture was allowed to stand and ferment for 3 days at room temperature. From this mixture, 120 mg of testosterone was recovered.

In the last few years, there have been considerable activity and success in conducting biological oxidations of steroids for introducing oxygen at carbons of the steroid nucleus. In 1947 and 1949, Krámli and Horváth[50a] reported that 7-hydroxycholesterol was derived from cholesterol by microbiological oxidation with *Proactinomyces roseus* or by *Azotobacter* species. Perlman, Titus, and Fried[71a] have recently reported that an unidentified actinomycete was found which would convert progesterone in simple nutrient media to 16-α-hydroxyprogesterone in good yield. In addition, small amounts of pregnanol-16-α-dione-3,20 and a dihydroxyprogesterone were produced.

In 1949 and later, Hechter, Jacobson, Jeanloz, Marshall, Pincus, and Schenker[37,38] demonstrated that oxygen could be introduced at carbon 11 by perfusion of steroid-containing fluids through adrenal glands. This was, of course, a very highly selective oxidation and was quite remarkable from the organic chemist's point of view, since the 11-position is considered as highly inaccessible. Introduction of oxygen at carbon 11 in this manner was an important and unique step in the field of corticosteroid syntheses. An example of such an oxidation is as follows:

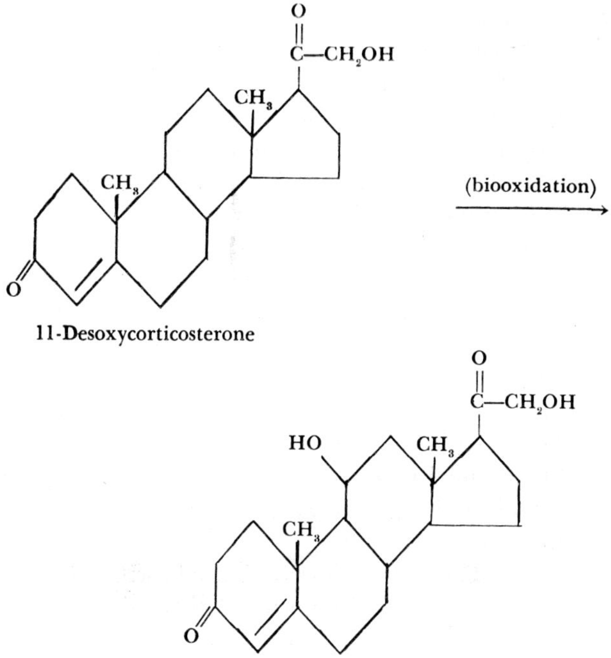

11-Desoxycorticosterone

Corticosterone

Other related investigations have been described, and further studies of adrenal biooxidations of steroids have been reported.[36a,38a,117a]

In 1952, Peterson and Murray,[71b,71c] working in the research laboratories of the Upjohn Company, Kalamazoo, Michigan, reported that they had been able to oxygenate steroids at carbon 11 by means of a fungal biooxidation, employing certain *Mucorales*. As an example, *Rhizopus arrhizus* was utilized effectively for the oxidation of progesterone to 11-α-hydroxyprogesterone in a medium containing lactalbumin digest, corn steep and glucose. The oxidation required 24 to 48 hours and resulted in yields of 10% and higher. Along with the monohydroxy compound, dihydroxy steroids were also isolated. Similar biooxidations at carbon 11 occurred with other steroids, including androstenedione, 11-desoxycorticosterone and 11-desoxy-17-hydroxycorticosterone. The 1952 patent of Murray and Peterson[71c] cites many examples of the general procedure, including variations in substrate and culture. Paper chromatographic analytical methods were described. These were based on the technique of Burton, Zaffaroni, and Keutmann.[10a] Additional paper chromatographic methods have been described by Shull, Sardinas, and Nubel.[78b]

A series of papers was presented subsequently in 1952[71g] and in 1953[18a,18b,63a,63b,71c,71h] which detailed a number of specific steroid transformations of the general type noted in the 1952 patent of Murray and Peterson.[71c] In addition to the microbiological oxidative transformations of steroids, many chemical transformations are also described in these papers. The purely chemical reactions of steroids are outside the scope of this book (see also Peterson[71b]).

In addition to *Rhizopus arrhizus* Fischer (ATCC 11145), which was a specific strain employed, *Rhizopus nigricans* Ehrb (ATCC 62276) has been used for oxygenation of steroids at carbon 11. For example, this second fungus was reported[63a] to convert 4,16-pregnadien-3,20-dione to 11-α-hydroxy-17 α-progesterone under oxidative conditions.

Oxidation at C-11 is not always obtained with these *Rhizopus* species. For example, it was reported[18a] that, although *Rhizopus nigricans* converted 11-desoxycorticosterone (or its acetate) to 11-epicorticosterone, *Rhizopus arrhizus* converted this substrate to 6-β-hydroxy-11-desoxycorticosterone. Introduction of oxygen into the ring by the second culture was accomplished at position 6 rather than at position 11, in contrast to *Rhizopus nigricans* which attacked position 11.

In 1952, Colingsworth, Brunner, and Haines[13a] reported that *Streptomyces fradiae* No. 3535 (Waksman) could also oxidize steroids at carbon 11. As an example, 5 g of 11-desoxy-17-hydroxycorticosterone (Reichstein's compound S) was dispersed in 15 l medium containing glucose, soybean meal, and distillers' solubles. This was incubated in rotary shake flasks with *S. fradiae* for 7 hours at 24°C. From this mixture 4.86 g of a neutral hormone concentrate was obtained which contained, according to bioassay, 140 mg of 17-hydroxycorticosterone (Kendall's compound F). Purification was accomplished by partition chromatography. A trace of 11-dehydro-17-hydroxycorticosterone was also detected.

In addition, according to Fried and associates,[27a] in 1952, certain members of the genus *Aspergillus* can also oxygenate at carbon 11. For example, *Aspergillus niger* (Wis. 72-2) was reported to attack progesterone, Reichstein's compound S, desoxycorticosterone, and 17α-hydroxyprogesterone in this manner. Progesterone, as an example, gave a mixture of 11α-hydroxyprogesterone in 35% yield and 6ξ, 11-α-dihydroxyprogesterone in 20% yield.

A patent was issued in 1953 to Shull, Kita, and Davisson* of Chas. Pfizer & Co., Inc., pertaining to the hydroxylation of steroids at the 11 position, using the genus *Curvularia,* or an enzyme preparation from these organisms. For example, *Curvularia lunata* (NRRL-2380) was grown aerobically for 22 hours in a medium consisting of salts, sucrose, and malt extract. The mycelium from 2 l of such a fermentation, for example, is separated by filtration. It is then washed and suspended in an equal volume of distilled water. Such a preparation was found to be capable of converting 0.5 g of 17-hydroxy-11-desoxycorticosterone (added in 20 ml ethanol solution) after 16 hours of aeration to 17-hydroxycorticosterone in 35% yield. Purification involved extraction with chloroform and chromatography on silica gel. In a similar manner, *C. pallescens* (NRRL-2381) brought about the conversion of desoxycorticosterone to corticosterone. A 28% yield was obtained.

Some degradations of steroids were also reported recently to be brought about by certain other fungi;[71d] side chains were removed. Among the fungi capable of doing this are *Gliocladium catenulatum, Penicillium lilacinum* Thom, *Aspergillus flavus,* and *Penicillium adametzi.* For example, progesterone may be degraded to androstenedione and other oxygenated steroids by *Gliocladium catenulatum.*

In view of the importance of cortisone and other adrenocortical steroids in medicine, some of these discoveries may be considered as major accomplishments in the advancement of biological science.

Reductive Reactions

The microbiological reductive reactions of steroids have generally involved the reduction of ketone groups to hydroxyl groups.

In 1937, Mamoli and Vercellone[61] described several microbiological reductions of ketosteroids for which yeast was used as the reducing organism. These investigators demonstrated, for ex-

* Shull, G. M., D. A. Kita, and J. W. Davisson, U. S. Patent 2,658,023 (1953). [*C.A.* **48**, 2332 (1954)].

ample, that △4-androstenedione could be reduced by the action of yeast to testosterone:

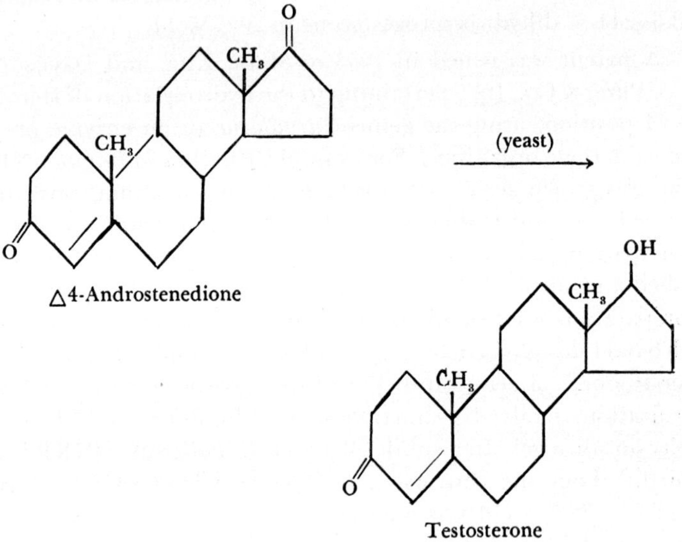

By the same general procedure, i.e., reduction by yeast at position 17, these investigators[62] also were able to reduce dehydroandrosterone to androstenediol:

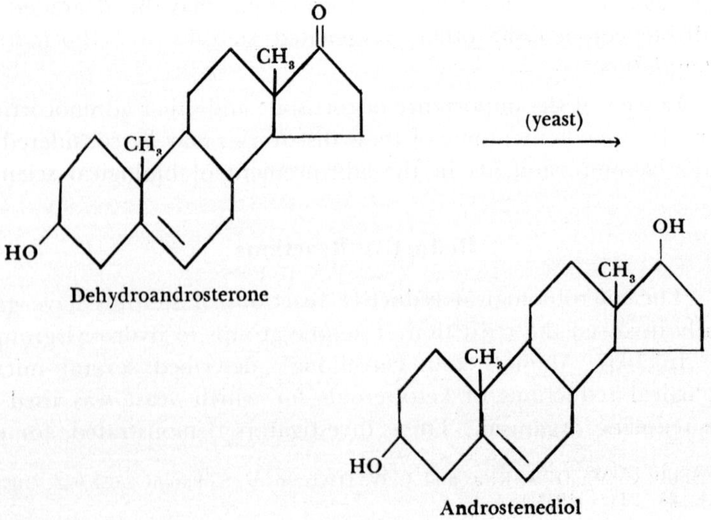

Vercellone and Mamoli[110] were able to obtain reduction of the 3-keto group as well as at position 17. This is seen from the fact that they could reduce androstanedione to isoandrostanediol, again by the use of yeast. There was no ring unsaturation in this example:

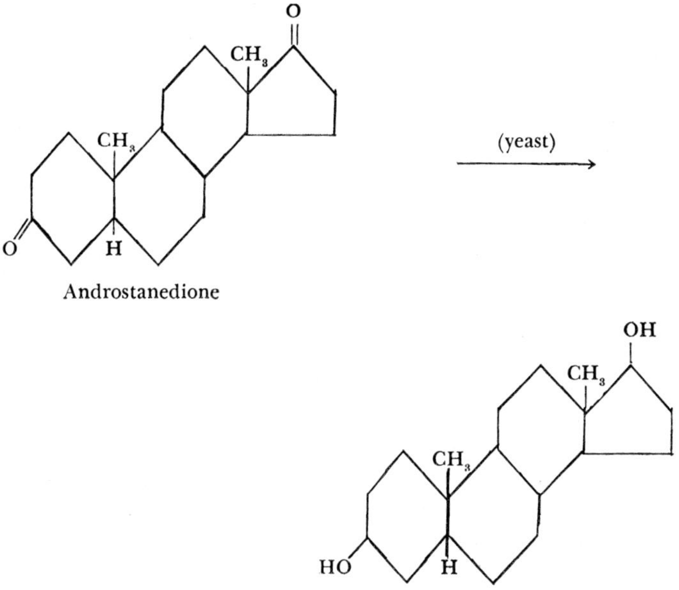

Androstanedione

Isoandrostanediol

The preceding material on the microbiological attack of steroids has been somewhat limited in comparison with the information available. For detailed information, the original papers should be studied. In 1953, Peterson[71b] reviewed the background and developments leading to the preparation of cortisone and hydrocortisone through microbiological approaches. Considerable additional references and information may be found in this report.

At the present time, it would be difficult or impossible to determine the extent of commercial utility of the various specific steroid reactions noted and to indicate their relative values. Reactions resulting in improved means for the economic preparation of cortisone and hydrocortisone would undoubtedly be potentially or actually of economic value. The results of investigations such as these should have considerable additional value in suggesting that certain other organic chemical reactions which are difficult to accomplish might be brought about by the proper application of microorganism or enzyme reactions.

BIBLIOGRAPHY

1. Anon., *Chem. Eng. News,* **29**, 4000 (1951).
1a. Anon., *Chem. Eng.,* **59**, No. 9, 215 (1952).
2. Arnaudi, C., *Boll. ist. sieroterap. milan,* **21**, 1 (1942).
3. Arnaudi, C., *Schweiz. Z. Path u. Bakt.,* **9**, 607 (1946).
4. Bailey, L. H., L. Owen, and L. Owen, Jr., U. S. Patent 2,360,327 (1944).
5. Bauer, E., *J. Chem. Soc.,* **44**, 105 (1883).
6. Bechamp, A., *Compt. rend.,* **93**, 78 (1881).
7. Beijerinck, M. W., *Fol. Mikrobiol.,* **1**, 377 (1912).
8. Belval, H., *Pub. inst. belge amelioration betterave,* **12**, 31 (1944); *C. A.,* **40**, 3631 (1946).
8a. Bixler, G. H., G. E. Hines, R. M. McGhee, and R. A. Shurter, *Ind. Eng. Chem.,* **45**, 692 (1953).
9. Brautigam, W., *Chem. Centr.,* **63**, 648 (1892).
10. Browne, C. A., Jr., *J. Am. Chem. Soc.,* **28**, 453 (1906).
10a. Burton, R. B., A. Zaffaroni, and E. H. Keutmann, *Science,* **110**, 442 (1949); **111**, 6 (1950).
11. Carlson, W., and G. L. Stahly, *J. Bact.,* **37**, 230 (1939).
12. Carlson, W., and V. Whiteside-Carlson, *Proc. Soc. Exp. Biol. Med.,* **71**, 416 (1949).
13. Cienkowski, L., *J. Chem. Soc.,* **38**, 334 (1880).
13a. Colingsworth, D. R., M. P. Brunner, and W. J. Haines, *J. Am. Chem. Soc.,* **74**, 2381 (1952).
14. Daker, W. D., and M. Stacey, *Biochem. J.,* **32**, 1946 (1938).
15. Daker, W. D., and M. Stacey, *J. Chem. Soc.,* **1939**, 585.
16. Daumichen, P., *Z. Verein der deut. Zucker-Ind.,* **40**, 701 (1890).
17. Durin, E., *Compt. rend.,* **83**, 128 (1876); Abstr. *J. Chem. Soc.,* **30**, 540 (1876).
18. Durin, E., *Compt. rend.,* **83**, 355 (1876); Abstr. *J. Chem. Soc.,* **31**, 106 (1877).
18a. Eppstein, S. H., P. D. Meister, D. H. Peterson, H. C. Murray, H. M. Leigh, D. A. Lyttle, L. M. Reineke, and A. Weintraub, *J. Am. Chem. Soc.,* **75**, 408 (1953).
18b. Eppstein, S. H., D. H. Peterson, H. M. Leigh, H. C. Murray, A. Weintraub, L. M. Reineke, and P. D. Meister, *J. Am. Chem. Soc.,* **75**, 421 (1953).
19. Ercoli, A., *Boll. sci. facolta chim. ind. Bologna,* **1940**, 279.

20. Ercoli, A., *Z. physiol. Chem.,* **270**, 266 (1941).
21. Ercoli, A., *Biochim. e. terap. sper.,* **28**, 215 (1941).
22. Ercoli, A., and L. Molina, *Boll. ist. sieroterap. milan.,* **23**, 158 (1944).
23. Ercoli, A., and L. Molina, *Boll. ist. sieroterap. milan.,* **23**, 175 (1944).
24. Fairhead, E. C., M. J. Hunter, and H. Hibbert, *Can. J. Research,* **16B**, 151 (1938).
25. Fernbach, A., M. Schoen, and S. Hagiwara, *Compt. rend. soc. biol.,* **92**, 1418 (1925).
26. Fernbach, A., M. Schoen, and S. Hagiwara, *Ann. Brass. Dist.,* **23**, 321 (1925); *C. A.,* **20**, 1643 (1926).
26a. Fischer, F. G., in *Newer Methods of Preparative Organic Chemistry,* New York, Interscience, 1948.
27. Fowler, F. L., I. K. Buckland, F. Brauns, and H. Hibbert, *Can. J. Research,* **15B**, 486 (1937).
27a. Fried, J., R. W. Thoma, J. R. Gerke, J. E. Herz, M. N. Donin, and D. Perlman, *J. Am. Chem. Soc.,* **74**, 3962 (1952).
28. Goodman, L., and A. Gilman, *The Pharmacological Basis of Therapeutics,* New York, Macmillan, 1943.
29. Grönwall, A., and B. Ingelman, *Acta Physiol. Scand.,* **7**, 97 (1944).
30. Grönwall, A., and B. Ingelman, *Nord. Med.,* **21**, 247 (1944); *C. A.,* **40**, 7400 (1946).
31. Grönwall, A., and B. Ingelman, *Nature,* **155**, 45 (1945).
32. Grönwall, A., B. Ingelman, and H. Mosimann, *Upsala Lakarefören Förh.,* **50**, 397 (1945); *Chimie & industrie,* **55**, 706 (1946); *C. A.,* **41**, 5213 (1947).
33. Happ, C., *Centr. Bakt. Parasitenk., I Abt.,* **14**, 175 (1893).
34. Hassid, W. Z., and H. A. Barker, *J. Biol. Chem.,* **134**, 163 (1940).
35. Hassid, W. Z., and M. Doudoroff, *Advances in Enzymology,* **10**, 123, 124, 140 (1950).
36. Haworth, W. N., and M. Stacey, British Patent 618,999 (1949).
36a. Hayano, M., and R. I. Dorfman, *Arch. Biochem.,* **36**, 237 (1952).
37. Hechter, O., R. P. Jacobson, R. Jeanloz, C. W. Marshall, G. Pincus, and V. Schenker, *J. Am. Chem. Soc.,* **71**, 3261 (1949).
38. Hechter, O., R. P. Jacobson, R. Jeanloz, C. W. Marshall, G. Pincus, and V. Schenker, *Arch. Biochem.,* **25**, 457 (1950).
38a. Hechter, O., R. P. Jacobson, R. Jeanloz, H. Levy, G. Pincus, and V. Schenker, *Proc. Am. Diabetes Assoc.,* **10**, 39 (1950).

38b. Hehre, E. J., *Science,* **93**, 237 (1941).
39. Hehre, E. J., *Proc. Soc. Exp. Biol. Med.,* **54**, 240 (1943).
40. Hehre, E. J., *Trans. N. Y. Acad. Sci.,* **10**, 188 (1948).
40a. Hehre, E. J., *Advances in Enzymology,* **11**, 297 (1951).
41. Hehre, E. J. and D. M. Hamilton, *Proc. Soc. Exp. Biol. Med.,* **71**, 336 (1949).
42. Hehre, E. J., and J. M. Neill, *J. Exptl. Med.,* **83**, 147 (1946).
43. Hehre, E. J., and J. Y. Sugg, *J. Exptl. Med.,* **75**, 339 (1942).
44. Hildebrandt, G., and W. Klavehn, U. S. Patent 1,956,950 (1934).
45. Hucker, G. J., and C. S. Pederson, *N. Y. Agr. Exp. Sta. Tech. Bull., 167,* 3 (1930).
46. Hughes, H. B., and L. H. Schmidt, *Proc. Soc. Exp. Biol. Med.,* **51**, 162, (1942).
47. Ingelman, B., *Upsala Lakareforen Förh.,* **54**, 107 (1949).
48. Jeanes, A., C. A. Wilham, and J. C. Miers, *J. Biol. Chem.,* **176**, 603 (1948).
49. Kagan, B. O., S. N. Lyatker, and E. M. Tsvasman, *Biokhimiya,* **7**, 93 (1942); *C. A.,* **37**, 4760 (1943).
49a. Koepsell, H. J., and H. M. Tsuchiya, *J. Bact.,* **63**, 293 (1952).
50. Koester, H., L. Mamoli, and A. Vercellone, U. S. Patent 2,236,574 (1941).
50a. Krämli, A., and J. Horváth, *Nature,* **160**, 639 (1947); **163**, 219, (1949).
51. Lafar, F., *Handbuch der technischen Mykologia,* 2 volumes Jena, 1904, 1907.
52. Liebowitz, J., and S. Hestrin, *Adv. in Enzymology,* **5**, 87 (1945).
53. Levi, I., W. L. Hawkins, and H. Hibbert, *J. Am. Chem. Soc.,* **64**, 1957 (1942).
54. Levi, I., W. L. Hawkins, and H. Hibbert, *J. Am. Chem. Soc.,* **64**, 1959 (1942).
55. Liesenberg, C., and W. Zopf, Quoted by Lippmann.[56]
56. Lippmann, E. O., Von., *Die Chemie der Zukerarten,* 3rd Ed., 427 (1904).
57. Mahoney, J. C., U. S. Patent 2,089,217 (1937).
58. Mamoli, L., *Ber.,* **71**, 2701 (1938).
59. Mamoli, L., U. S. Patent 2,341,110 (1944).
60. Mamoli, L., R. Koch, and H. Teschen, *Naturwissenschaften,* **27**, 319 (1939).
61. Mamoli, L., and A. Vercellone, *Ber.,* **70**, 470 (1937).
62. Mamoli, L., and A. Vercellone, *Z. physiol. Chem.,* **245**, 93 (1937).

63. McIntire, F. C., and J. R. Schenck, *J. Am. Chem. Soc.,* **70,** 1193 (1948).
63a. Meister, P. D., D. H. Peterson, H. C. Murray, S. H. Eppstein, L. M. Reineke, A. Weintraub, and H. M. Leigh, *J. Am. Chem. Soc.,* **75,** 55 (1953).
63b. Meister, P. D., D. H. Peterson, H. C. Murray, G. B. Spero, S. H. Eppstein, A. Weintraub, L. M. Reineke, and H. M. Leigh, *J. Am. Chem. Soc.,* **75,** 416 (1953).
64. Molina, L., and A. Ercoli, *Boll. ist. sieroterap. milan,* **23,** 164 (1944).
65. Neuberg, C., *Biochem. Z.,* **115,** 282 (1921).
66. Neuberg, C., *Biochem. Z.,* **128,** 611 (1922).
67. Nordström, L., and E. Hultin, *Svensk. Kem. Tid.,* **60,** 283 (1948); *C. A.,* **43,** 3050 (1949).
68. Owen, Wm. L., *Sugar,* **42,** No. 8, 28 (1948); *C. A.,* **42,** 8004 (1948).
69. Pasteur, L., Quoted *Chem. Rev.,* **3,** 403 (1926).
70. Peat, S., E. Schluchterer, and M. Stacey, *J. Chem. Soc.,* **1939,** 581.
71. Penrod, J., French Patent 339,228 (1905); Abstr. *J. Soc. Chem. Ind.,* **24,** 245 (1905).
71a. Perlman, D., E. Titus, and J. Fried, *J. Am. Chem. Soc.,* **74,** 2126 (1952).
71b. Peterson, D. H., *Research,* **6,** 309 (1953).
71c. Peterson, D. H., S. H. Eppstein, P. D. Meister, B. J. Magerlien, H. C. Murray, H. M. Leigh, A. Weintraub, and L. M. Reineke, *J. Am. Chem. Soc.,* **75,** 412 (1953).
71d. Peterson, D. H., S. H. Eppstein, P. D. Meister, H. C. Murray, H. M. Leigh, A. Weintraub, and L. M. Reineke, *J. Am. Chem. Soc.,* **75,** 5768 (1953).
71e. Peterson, D. H., and H. C. Murray, *J. Am. Chem. Soc.,* **74,** 1871 (1952).
71f. Peterson, D. H., and H. C. Murray, U. S. Patent 2,602,769 (1952).
71g. Peterson, D. H., H. C. Murray, S. H. Eppstein, L. M. Reineke, A. Weintraub, P. D. Meister, and H. M. Leigh, *J. Am. Chem. Soc.,* **74,** 5933 (1952).
71h. Peterson, D. H., A. H. Nathan, P. D. Meister, S. H. Eppstein, H. C. Murray, A. Weintraub, L. M. Reineke, and H. M. Leigh, *J. Am. Chem. Soc.,* **75,** 419 (1953).
72. Sanchez-Marroquin, A., and L. Arciniega, *Anales escuela nacl. cienc. biol.* (Mex.), **5,** 19 (1948).

73. Sauberlich, H. E., and C. A. Baumann, *J. Biol. Chem.,* **176**, 165 (1948).
74. Sauberlich, H. E., and C. A. Baumann, *J. Biol. Chem.,* **181**, 871 (1949).
75. Schatz, A., K. Savard, and I. J. Pintner, *J. Bact.,* **58**, 117 (1949).
76. Scheibler, C., quoted *J. Am. Chem. Soc.,* **28**, 460 (1906).
77. Schmidt, L. H., and H. B. Hughes, *J. Biol. Chem.,* **143**, 771 (1942); see also U. S. Patent 2,360,447 (1940).
78. Schmidt, L. H., H. B. Hughes, M. H. Green, and E. Cooper, *J. Biol. Chem.,* **145**, 229 (1942).
78a. Scully, N. J., H. E. Stavely, J. Skok, A. R. Stanley, J. K. Dale, J. T. Craig, E. B. Hodge, W. Chorney, R. Watanabe, and R. Baldwin, *Science,* **116**, 87 (1952).
78b. Shull, G. M., J. L. Sardinas, and R. C. Nubel, *Arch. Biochem. Biophys.,* **37**, 186 (1952).
78c. Smith, P. F., and D. Hendlin, *J. Bact.,* **65**, 440 (1953).
78d. Stacey, M., *Nature,* **149**, 639 (1942).
79. Stacey, M., and G. Swift, *J. Chem. Soc.,* **1948**, 1555.
80. Stacey, M., and F. R. Yound, *Biochem. J.,* **32**, 1943 (1939).
81. Stahly, G. L., U. S. Patent 2,310,263 (1943).
82. Stahly, G. L., and W. W. Carlson, U. S. Patent 2,203,702 (1940).
83. Stahly, G. L., and W. W. Carlson, U. S. Patent 2,203,703 (1940).
84. Stahly, G. L., and W. W. Carlson, U. S. Patent 2,203,704 (1940).
85. Stahly, G. L., and W. W. Carlson, U. S. Patent 2,203,705 (1940).
86. Stahly, G. L., and W. W. Carlson, U. S. Patent 2,229,941 (1941).
87. Stahly, G. L., and W. W. Carlson, U. S. Patent 2,236,386 (1941).
88. Stahly, G. L., and W. W. Carlson, U. S. Patent 2,239,980 (1941).
89. Stahly, G. L., and W. W. Carlson, U. S. Patent 2,249,544 (1941).
90. Stahly, G. L., and W. W. Carlson, U. S. Patent 2,328,036 (1941).
91. Stahly, G. L., and W. W. Carlson, U. S. Patent 2,344,179 (1944).
92. Stahly, G. L., and W. W. Carlson, U. S. Patent 2,344,180 (1944).

93. Stahly, G. L., and W. W. Carlson, U. S. Patent 2,380,879 (1945).
94. Stahly, G. L., and W. W. Carlson, U. S. Patent 2,385,553 (1945).
95. Stahly, G. L., and W. W. Carlson, British Patent 517,397 (1940).
96. Stahly, G. L., and W. W. Carlson, British Patent 517,398 (1940).
97. Stahly, G. L., and W. W. Carlson, British Patent 517,820 (1940).
98. Stahly, G. L., and W. W. Carlson, Canadian Patent 391,240 (1940).
99. Stahly, G. L., and W. W. Carlson, Canadian Patent 394,662 (1941).
100. Steele, B. F., H. E. Sauberlich, M. S. Reynolds, and C. A. Baumann, *J. Biol. Chem.*, **177**, 533 (1949).
101. Tarr, H. L. A., and H. Hibbert, *Can. J. Research*, **5**, 414 (1931).
102. Tiselius, A., and B. Ingelman, *Förh. Svenska Suckerfabriks-dirigenternas Fören. Sammanträden*, **1942**, II, 16 p.; *C. A.*, **38**, 4465 (1944).
103. Turfitt, G. E., *Biochem. J.*, **38**, 492 (1944).
104. Turfitt, G. E., *Biochem. J.*, **40**, 79 (1946).
105. Turfitt, G. E., *J. Bact.*, **54**, 557 (1947).
106. Turfitt, G. E., *Biochem. J.*, **42**, 376 (1948).
107. Van Tieghem, P., *J. Chem. Soc.*, **38**, 908 (1880).
108. Vašátko, J., and B. Jelinek, *Listy Cukrovar.*, **57**, 181 (1939); *Z. Zuckerind. Cechoslovak. Rep.*, **63**, 275 (1939); *C. A.*, **33**, 9699 (1939).
109. Vasseur, A., *Rev. pathol. comp. hyg. gén.*, **24**, 240 (1924); *Bull. soc. hyg. aliment.*, **12**, 441 (1924); *C. A.*, **19**, 374 (1925).
110. Vercellone, A., and L. Mamoli, *Z. physiol. Chem.*, **248**, 277 (1937).
111. Wadsworth, D. V., and M. F. Hughes, U. S. Patent 2,409,816 (1946).
112. Waldie, W. A., and J. E. Bersuder, U. S. Patent 2,344,190 (1944).
113. Waldie, W. A., and J. E. Bersuder, U. S. Patent 2,386,994 (1945).
114. Walten, C. F., Jr., and C. R. Fort, *Ind. Eng. Chem.*, **23**, 1295 (1931).
115. Whiteside-Carlson, V., and W. W. Carlson, *J. Bact.*, **58**, 135 (1949).

116. Whiteside-Carlson, V., and W. W. Carlson, *J. Bact.*, **58**, 143 (1949).

117. Wolff, I. A., and C. E. Rist, *J. Am. Chem. Soc.*, **70**, 2779 (1948).

117a. Zaffaroni, A., O. Hechtner, and G. Pincus, *J. Am. Chem. Soc.*, **73**, 1390 (1951).

118. Zettnow, E., *Z. Ver. Zuckerind.*, **57**, 971 (1908); *C. A.*, **2**, 594 (1908).

119. Zimmerman, W., and G. May, *Zentr. Bakt. Parasitenk., Abt. I.*, **151**, 462 (1944).

CHAPTER 11

LACTIC ACID FERMENTATION OF CUCUMBERS, SAUERKRAUT AND OLIVES

Reese H. Vaughn

INTRODUCTION

Preservation of foods by lactic acid fermentation (pickling) has been one of the important methods of food preservation for centuries. Until the development of canning and freezing processes during the past 150 years, pickling and drying constituted the major methods for preserving foods. However, with the development of canning and freezing techniques, food habits have changed, particularly in the United States. Now the pickled foods, cucumbers and olives especially, are used as food adjuncts which function as appetizers.

The most important pickled products still produced in whole or in part by lactic acid fermentation include cucumbers (pickles), cabbage (sauerkraut), and olives. That these pickled foods still constitute an important segment of the food industry of the United States is shown by production figures.

Cucumbers probably constitute the most important single crop produced for pickling. Michigan and Wisconsin lead in the production of cucumbers in the eastern United States, while California leads among the western states. In 1947, Michigan, Wisconsin, and

California harvested 35,600, 18,900, and 2,020 acres, with yields per acre of 1.5, 2.4, and 8.3 tons, respectively. Data showing the number of tons of salt stock and dill pickles produced in the United States are found in Table 38. It should be remembered in addition that the cucumber pickle is a popular and extensively home-prepared

TABLE 38. COMMERCIAL PRODUCTION OF CUCUMBER PICKLES[a] IN THE UNITED STATES

Year	Salt stock tons	Dills tons	Total tons	Ratio of salt stock to dills
1930	138,048	35,592	173,640	3.87
1931	110,208	23,184	133,392	4.75
1932	28,080	11,160	39,240	2.52
1933	61,440	16,680	78,120	3.68
1934	79,896	12,912	92,808	6.18
1935	93,648	14,616	108,264	6.41
1936	117,024	17,784	134,808	6.58
1937	138,144	25,944	164,088	5.32
1938	119,808	16,680	136,488	7.18
1939	70,752	13,968	84,720	5.06
1940	116,040	18,888	134,928	6.14
1941	151,776	18,912	170,688	8.03
1942	155,736	20,520	176,256	7.59

[a] Calculated from data in *Western Canner and Packer,* **35**, No. 5, 221 (1943) on the basis of 48 lb per bu of cucumbers.

TABLE 39. COMMERCIAL PRODUCTION OF SAUERKRAUT[a] IN THE UNITED STATES

	Total canned, cases, all sizes		
Year	Sauerkraut (thousands)	Sauerkraut juice (thousands)	Total bulk sauerkraut in barrels (45 gal per barrel) (thousands)
1925	2,395	—	—
1927	3,101	—	—
1929	4,224	—	—
1931	3,645	69	—
1933	3,345	—	—
1935	4,405	120	—
1937	4,583	234	—
1939	4,458	135	—
1940	6,480	—	—
1941	6,202	—	360
1942	3,062	—	460
1943	3,012	—	160
1944	2,537	—	160
1945	6,076	—	360
1946	5,904	—	460
1947	1,650	—	120

[a] Tin and glass containers.
Source: *Western Canner and Packer,* **40**, No. 5, 135 (1948).

TABLE 40. CALIFORNIA OLIVE PRODUCTION COMPARED WITH IMPORTS FROM SPAIN

Season[a]	Total California crop tons	California canned ripe olives tons	California canned ripe olives gallons[b] (thousands)	California green olives, Sicilian type tons	California green olives, Sicilian type gallons[b] (thousands)	California green olives, Spanish type tons	California green olives, Spanish type gallons[b] (thousands)	Total imported green olives year[c]	Total imported green olives tons[d]	Total imported green olives gallons[b] (thousands)
1940-41	69,000	16,200	5,890	4,200	1,527	5,100	1,854	1941	15,529	5,647
1941-42	55,000	16,700	6,072	1,600	582	4,500	1,636	1942	14,039	5,105
1942-43	57,000	11,100	4,036	1,900	691	5,500	2,000	1943	23,914	8,696
1943-44	57,000	15,500	5,636	1,300	473	5,200	1,891	1944	26,958	9,803
1944-45	43,000	13,800	5,018	500	182	3,600	1,309	1945	25,251	9,182
1945-46	30,000	13,500	4,909	500	182	3,000	1,091	1946	29,384	10,685
1946-47	48,000	27,500	9,999	500	182	3,300	1,200	1947	17,966	6,533
1947-48	40,000	13,700	4,982	1,100	400	2,600	945	1948	30,583	11,121
1948-49	58,000	13,500	4,909	1,200	436	5,700	2,073	1949	17,498	6,363
1949-50	35,000	20,400	7,418	800	291	2,600	945	1950	34,326	12,482
1950-51	42,000	24,900	9,055	1,200	436	2,600	945	1951	28,174	10,245
1951-52	64,000	33,300	12,109	1,300	473	3,300	1,200	1952	37,164	13,514
1952-53	57,000	24,400	8,873	900	327	4,500	1,636	—	—	—

[a] California olive season begins in September.
[b] A gallon is approximately equal to 5.5 lb of olives.
[c] Import data available on calendar year basis.
[d] Includes all imported green olives, whole, pitted, pitted and stuffed.

Sources: California Olive Association; United States Department of Commerce, Foreign Commerce and Navigation of the United States.

food, particularly in the rural areas throughout the United States.

Sauerkraut (fermented cabbage) is the only important pickled vegetable product other than cucumbers fermented in any significant quantity in the United States. Its commercial production is localized largely in New York, Wisconsin, Ohio, Indiana, and Michigan, other states bordering on the Great Lakes, and in Colorado and Washington. Sauerkraut is a very popular food among the people of northern European extraction, especially among the Germans, as its name suggests. Information concerning the amount of sauerkraut produced commercially is found in Table 39. Sauerkraut is esteemed for its taste as well as its economy. It is produced in quantity in the home.

California produces almost all of the olives grown in the United States. There are some 25,000 bearing acres which normally produce crops ranging from about 30,000 to nearly 70,000 tons per season. Although olives may be grown throughout the state, most of them are grown in the two large interior valleys, the San Joaquin Valley and the Sacramento Valley. Arizona is the only other state which produces a commercial quantity of olives. Data concerning recent California olive production and the chief type of pickled olives produced are shown in Table 40. These data do not include fresh shipments; the quantities used on the farms in California;

TABLE 41. VALUE OF FERMENTED PRODUCTS PROCESSED IN THE UNITED STATES

Product	Dollar value 1939	Dollar value 1947
Pickles (all kinds)	44,876,000	105,647,000
Sauerkraut		
Canned	6,439,075	8,705,000
Bulk	2,101,373	1,642,000
Juice	180,398	(not available)
Total value	8,720,846	10,347,000
Olives		
Ripe[a]	3,116,395	15,950,000
Bottled[b]	6,127,410	16,735,000
Total value	9,243,805	32,685,000
VALUE OF ALL PRODUCTS	62,840,651	148,679,000

[a] Produced in California.

[b] Produced in California and in Spain—final processing in the United States, green olives.

Sources: United States Bureau of Census, 16th Census of the United States, 1940, Manufacturers, 1939, Vol. 2, part 1, pages 114, 125-128, U. S. Government Printing Office, Washington, D. C., 1942.

United States Bureau of Census, Census of Manufacturers: 1947, Vol. 2, pages 97, 98, U. S. Government Printing Office, Washington, D. C., 1949.

those utilized to produce chopped ripe olives; oil; Greek-type olives and other processes; or the small commercial crops of Arizona.

The total value of pickles, sauerkraut, and olives processed in the United States in 1939 and 1947 is shown in Table 41. The chemical, physical, and microbiological aspects of the production and processing of each of these pickled foods will be discussed separately.

THE PICKLING OF CUCUMBERS

The cucumber (*Cucumis sativus*) is one of the oldest vegetables cultivated by man. It is supposed to have had its origin in Asia, perhaps in India, at least 3,000 years ago. It is popular both as a fresh and as a pickled vegetable and is grown widely in temperate climates although it was of semitropical origin. However, care must be taken to protect it against frost and drought and to control its microbial parasites and insect pests.

Pickling cucumbers must be grown from varieties known to have regular form, firm texture, and good pickling qualities. The common pickling varieties recommended by various authorities include the Chicago pickling, Boston pickling, Jersey pickling, Snow's perfection, National pickling, Heinz pickling, Fordhook pickling, Packer, Davis blend and various strains.[53,82]

Pickling cucumbers are harvested while still immature. Fully mature (ripe) cucumbers are undesirable for pickling, because they become too large, change color and shape, and are too soft for most commercial uses. Care must be taken in picking and transporting the cucumbers to avoid undue bruising and crushing. After picking, it is mandatory to deliver the cucumbers to the salting station or factory as soon as possible to prevent deterioration. Sorting to remove defective and distorted cucumbers (wilt, rot, nubbins, crooks, etc.) is done before salting or brining. In order to minimize spoilage during fermentation, it is very important to remove all unsound and decomposed cucumbers. Prior to fermentation, the cucumbers are generally graded for size with mechanical graders or by hand sorting. Final size grading is performed at the completion of the fermentation process. Some factories grade into four or more sizes; others remove only the large cucumbers for dills and ferment the rest field run.

There are two main types of pickled cucumbers made: salt-stock pickles from which the majority of the well-known sour, sweet, mustard, processed or manufactured dills, mixed pickles, etc., are made and fermented dill pickles.

Preparation of Salt-Stock Pickles

The cucumbers must undergo a preliminary fermentation and curing in brine before they are suitable for processing and finishing into sweet, sour, spiced, and other kinds of pickles. The fermentation takes place in circular tanks or vats of 8 to 14 or more feet diameter, and of 6 to 8 or more feet height. These fermentation tanks are usually made of cypress, redwood, fir, or pine, and sometimes reinforced concrete, and have capacities for 100 to 1,600 bu (2.4 to 38.4 tons) of cucumbers.

The salt-stock tanks may be located at the factory or near the center of the production area. If located apart from the processing and finishing factory, the collection of tanks is commonly called a salting station. Many of the salting stations are frame buildings which house, or partially house, the salting tanks and receiving platforms. Others are little more than a collection of tanks with a suitable receiving platform.

There are two general methods for making salt-stock pickles: dry salting and brining. The dry-salting technique is not used extensively because of the tendency to yield soft, flabby, shriveled pickles that do not fill out properly when processed.

Dry-Salting Procedure

Dry salting is accomplished by first adding 40° Salometer salt brine to the tank to a depth of 12 in. to prevent bruising, breaking, or crushing of the fresh cucumbers when they are dumped into the tank. The salt hydrometer (salometer, salimeter, or salinometer) is calibrated so that the Salometer degrees equal the percentage saturation of salt (NaCl) at 60°F. A saturated solution of pure sodium chloride (100° Salometer) contains 26.359 g salt at 60°F. Dry salt is added at the rate of about 50 lb for every 1,000 lb of small cucumbers, and 65 lb for every 1,000 lb of large cucumbers. When full, the tank is covered with a weighted, circular, slatted, wooden head until there is room for about 6 in. of brine above the cover. The weighted head is then secured by heavy cross timbers held at the ends with clamps. If the brine does not cover the pickles or wooden cover when the tank is closed, 40° Salometer brine is added to the desired level. The brine should be circulated a few days after filling the tank to equalize the concentration of salt in the brine. The strength of the brine should be increased slowly until it is about 60° Salometer.

Brining Technique

The majority of picklers use the brine-salting technique for fermenting cucumbers rather than the dry-salting process just described. A "low" or a "high" salt-brining procedure may be used. The "low" salt brine has a salt concentration of about 30° Salometer, whereas the "high" salt brine has a salt concentration of about 40° Salometer. To start this process, brine of the desired concentration is placed in the bottom of the tank to a depth of about 1 ft to prevent damage to the cucumbers dumped into the tank. Dry salt and sufficient water may then be added to the cucumbers to obtain the desired brine strength, or the tanks may be filled with cucumbers and covered with brine of the desired concentration. This method, unlike the dry-salting process, does not depend on osmosis for water to fill the tank with brine.

When filled, the tanks are closed in the same manner as described for the dry-salting process. Circulation of the brine is desirable because it insures an equalized concentration of salt. Furthermore, it aids in more accurate control of the salt concentration in the brine. There probably are as many minor variations in the brining procedure as there are picklers. Regardless of the variations, it is mandatory to increase the salt concentration of the brine to at least 60° Salometer (approximately 15% salt). More detailed discussion of the brining procedures are given by Campbell,[11] Fabian,[32,34] Fabian and Bryan,[37] Fabian and Fulde,[39] Etchells and Jones,[29] and Jones.[51]

Fermentation of Salt Stock

The brined cucumbers undergo a "spontaneous" lactic acid fermentation. No attempt is made to control artificially the microbial population of the brine. The natural controls of the microbial population of the fermenting cucumbers are the concentration of salt in the brine, the temperature of the brine, the availability of fermentable substances, and the types and relative numbers of microorganisms present in the brine and on the cucumbers at the start of the fermentation. The fermentation proceeds slowly at first, but after 3 or 4 days the cucumbers generally are in full fermentation (Figure 23).

FIGURE 23. *Salt-Stock Cucumbers Undergoing Active Fermentation*

Besides the evolution of gas, which is one of the first visual manifestations of fermentation, there are other evidences of change in the cucumbers and in the brine. The brine becomes cloudy, takes on a straw-yellow color, and is effervescent because of the activity of the microorganisms. The color of the cucumbers changes, as curing progresses, from the fresh pale-green color to a darker olive-green color. The white, opaque appearance of the flesh changes to a translucent green. This color change proceeds from the periphery to the center so that the pickler can examine cross-sections of the fermenting cucumbers and follow the rate of curing. When the cucumber has become completely translucent, curing is considered to be complete.

The chemical changes which occur in the brined cucumbers are typical of a mixed fermentation.[52] Bacteria, yeasts, and sometimes molds are responsible for the conversion of the fermentable substances present in the fresh cucumbers to gases, volatile and nonvolatile acids, alcohol, and traces of other end products. The desirable conversion of the fermentable matter is to the nonvolatile lactic acid which, together with the salt, preserves the pickles. Most of the lactic acid is formed by the lactic acid bacteria, although the coliform bacteria and others may also produce some lactic acid during the initial stage of fermentation. The lactic acid is formed in concentrations varying from about 0.5 to 1.0%.

Other products of fermentation which remain in some quantity in the salt stock are acetic acid and alcohol. The concentration of alcohol varies between 0.5 and 2.0% by volume. The acetic acid content is about 0.05 to 0.15%. Still other products, such as glycerol and mannitol, which may be formed generally are converted by the lactic acid bacteria and other microorganisms to other end products before completion of the fermentation. Therefore, if present, they are found only in traces, together with small quantities of acetylmethylcarbinol, esters, and other naturally occurring flavoring constituents that withstand the fermentation.

The gases have been identified as carbon dioxide and hydrogen. Hydrogen is formed only during the initial stage of fermentation, whereas carbon dioxide is produced as long as fermentable matter is present. Both gases are developed by the coliform bacteria, but only carbon dioxide is produced by the gas-forming lactic acid bacteria and yeasts.

As already stressed, the lactic acid bacteria are chiefly responsible for the desirable fermentation of cucumbers. However, there is a definite sequence of micoorganisms and microbial populations which occurs in both salt-stock and dill-pickle fermentations. In many respects, the population changes are similar to those for sauerkraut and olive fermentations.

The microbial sequence in normal fermentations of cucumbers (either salt stock or dills) may be divided into three stages: primary, intermediate, and final. During the primary or initial phase of fermentation, a great many unrelated bacteria, yeasts, and molds have been isolated. All of them are widely distributed in nature. At the beginning of this first stage of fermentation, the desirable lactic acid bacteria are far outnumbered by these other extraneous microorganisms. It is obvious, therefore, that the primary stage of fermentation is the most important phase of the pickling process. During this critical period, if for any reason the fermentation does not proceed in a normal fashion, any of the unessential microorganisms may predominate and contribute to undesirable deterioration of the cucumbers.

The primary stage normally requires 2 or 3 days, exceptionally as long as 7 days. During this time, the number of lactic acid bacteria increases rapidly, both fermenting and oxidizing types of yeasts become increasingly abundant, and the undesirable forms of bacteria decrease in number rapidly and may disappear entirely. At the same time, there is a steady increase in total acidity and a corresponding decrease in the pH of the brine.

A mixture of the low-acid-tolerant species of *Leuconostoc* and the high-acid-tolerant species of *Lactobacillus* are the predominating bacteria in the intermediate phase of the fermentation. The undesirable bacteria have completely disappeared by the end of 10 to 14 days if the fermentation is normal. Yeasts are still present in significant numbers. The total acidity increases and the pH of the brine further decreases. The intermediate phase of fermentation lasts for a variable length of time. During this period, however, the number of *Leuconostoc* types of bacteria increases and predominates, but at the end starts to decrease rapidly and is replaced by species of *Lactobacillus* as the acidity increases.

The final stage of fermentation is dominated by the acid-tolerant lactobacilli which have replaced the species of *Leuconostoc*. The titratable acidity increases to a maximum (0.5 to 1.0%) and the pH decreases to a minimum of 3.5 to 3.8. However, once the fermentable substances have been exhausted and the species of *Lactobacillus* can no longer produce lactic acid, the oxidative yeasts begin to dominate the microbial population. As a result, the titratable acidity decreases and the pH value increases as the yeasts oxidize more and more organic acids.

The common genera and species of bacteria found to predominate in the different stages of cucumber fermentations are shown in Table 42. It must be remembered that the salt concentration has a marked effect on the bacterial populations of cucumber fermentations. All of the bacteria listed will be found in cucumbers fermenting in 20° or 30° Salometer brine. However, as the salt concentration increases, the relative abundance of both species and total numbers will decrease until, at 60° Salometer, only the most salt-tolerant strains of *Lactobacillus* and the salt-tolerant yeasts will survive. This does not preclude the presence of halophilic microorganisms which, as far as is known, play no significant role in the fermentation.

TABLE 42. BACTERIA FOUND TO PREDOMINATE IN THE DIFFERENT STAGES OF CUCUMBER FERMENTATION

Stage of fermentation	Predominating species of bacteria
Primary	*Aerobacter aerogenes*[a,b]
	Aerobacter cloacae[a,b]
	Escherichia freundii[b]
	Escherichia intermedium[b]
	Bacillus mesentericus-Bacillus megatherium groups[c]
	Bacillus (Aerobacillus) polymyxa[d]
	Bacillus (Aerobacillus) macerans[d]
Intermediate	*Leuconostoc mesenteroides*[e]
	Lactobacillus plantarum[e,f]
	Lactobacillus brevis[e]
	Lactobacillus fermenti[e]
Final	*Lactobacillus plantarum*[e,f,g]
	Lactobacillus brevis[e]
	Lactobacillus fermenti[e]

[a] Etchells, Fabian, and Jones,[24] Etchells and Jones.[27]
[b] Unpublished data of Vaughn and Foda.
[c] Fabian and Johnson.[40]
[d] Unpublished data of Vaughn.
[e] Unpublished data of Vaughn and Tabachnick.
[f] Etchells and Jones.[30]
[g] Costilow and Fabian.[11a]

Yeasts are present throughout the fermentation and storage of salt-stock pickles. The presence and abundance of genera and species of these yeasts are influenced by the temperature, the relative amount of air, the salt concentration, and the presence or absence of sunlight. At low temperatures, the majority of the yeasts grows better than do most of the lactic acid bacteria found in the fermentation.

If the brine is frequently recirculated, it will carry enough occluded and absorbed oxygen from the air to allow yeasts to grow throughout the depths of the fermentation. Furthermore, since the salt-stock fermentations are conventionally conducted in open tanks, the yeasts very frequently dominate the fermentation, sometimes almost to the exclusion of the lactic acid bacteria.

The concentration of salt in the brine also influences the yeast population. At low concentrations (20° to 30° Salometer) more genera and species of yeasts are present. As the concentration of salt is increased, however, the number of yeast types able to grow in the brine is markedly reduced.

The presence of sunlight also considerably influences the yeast populations. Sunlight, because of its germicidal rays, prevents the growth of the oxidative, film-forming yeasts which grow on the surface of brines contained in sheltered or shaded tanks.

The genera of yeasts found in pickle brines in the United States include *Debaromyces, Pichia, Mycoderma, Torulopsis, Brettanomyces, Zygosaccharomyces, Hansenula, Torulaspora,* and *Kloeckera.*[11a,22,23,23a,23b,67] The chief film-forming types are *Debaromyces, Pichia, Hansenula,* and to a lesser extent *Mycoderma.* The data of Etchells and Bell[23] indicate a sequence of types of yeasts. It is not known with certainty whether any of these yeasts play an important role in the fermentation or are detrimental to the production of the best salt-stock pickles. The author takes the view that most of them are detrimental, because they consume the lactic acid which, together with the salt, preserves the pickles for future processing. It is certain that they are detrimental in the fermentation of dill pickles and green olives, as will be discussed later.

Storage

After fermentation, the salt-stock pickles commonly are allowed to cure for 3 to 6 months before being used. Some manufacturers will not use salt-stock pickles for making sweet pickles unless they are at least 1 year old. Salt-stock pickles may be stored in brine for several years if given adequate attention. Generally, the only controls consist of routine checking of the salt concentration and some effort to combat the film-forming yeasts.

The salt concentration must be increased to at least 60° Salometer for storage. The increase in salt concentration from 40° to 60° Salometer is usually accomplished in about 1 month after the tank is filled and headed.

If the salt-stock tanks are out of doors, the film-forming yeasts do not present a serious problem. In sheltered or shaded tanks, however, these yeasts remain active and destroy the lactic acid. They may reduce the acidity enough to allow the development of undesirable microorganisms. Many picklers prefer to remove the film or membrane of yeasts (also called "scum" and "mycoderma") by skimming the surface of the brine regularly. As the yeasts cannot grow without an adequate supply of air, some prefer to use a layer of neutral mineral oil which is floated over the surface of the brine. The membrane also may be effectively controlled by use of ultraviolet radiation. Recently, sorbic acid (0.1% in the brine) has been recommended by Phillips and Mundt[78] for control of membrane formation. It appears to be more effective than acetic acid or mustard oil emulsions which also have been advocated.[7,21]

The Processing of Salt-Stock Pickles

The removal of the excess salt from the pickles by leaching in water is termed "processing." Wooden tanks or vats equipped with steam pipes for heating the leaching water are commonly used for processing. The salt-stock pickles are placed in the processing tank, covered with cold water, and allowed to soak for 12 hours, or overnight. The leaching water is then changed and the pickles are allowed to soak for 6 to 8 hours more. The water is changed again and heated to 130°F (new, firm stock), or 150°F (old stock), and allowed to remain overnight. Alternative leaching methods include the use of four soakings in cold water; the use of three heated processing waters of 110°, 130°, and 140°F, respectively (the water being changed every 12 hours); or the use of water heated to 112°F, stirred at 10-minute intervals, and allowed to stand overnight. A steam siphon or pump is frequently used with the last method to keep the processing water in circulation. The following morning, the leaching water is removed, the pickles are flushed thoroughly with fresh water, covered with fresh, cold water, and sorted.

The same steps are taken in processing salt-stock pickles for sweet pickles as for sour pickles. Most picklers attempt to remove all of the salt during processing, so that on finishing, the concentration of salt can be more easily controlled.

Alum (potassium alum) and calcium chloride may be used during processing to firm and crisp the texture of the pickles. Alum is added at the rate of 1 lb per barrel (approximately 48 gal) of pickles. The alum is dissolved in warm water and added to the third processing water just before heating or, if the leaching process involves constant circulation of the water, it is added at the start.

Calcium chloride is added at the rate of 1 lb per barrel, either to the third processing water, just before heating, or to the final sweet liquor added to the pickles during the finishing process.

Sometimes when the salt-stock pickles are bleached near the blossom end, turmeric is used to restore some of the color. Turmeric gives the pickles a slightly yellowish appearance. It is added at the rate of 2 oz per barrel to the second processing water at the time of heating or at the start of processing, if the water is circulated. Turmeric, like alum and calcium chloride, must be listed on the label, if used.

After processing, the pickles may be sorted for size and quality directly into barrels or kegs. The sizes, based on the number of cucumbers per 45-gal cask, generally conform to those adopted by the National Pickle Packers Association.

The Finishing of Salt-Stock Pickles

After processing and sorting, the pickles are *finished,* i.e., prepared as sour, sweet, processed dills, mixed, mustard pickles, etc.

Sour Pickles. Preparation of sour pickles is very simple. The processed stock is placed in kegs or barrels and covered with distilled vinegar. The containers are then sealed and, as soon as osmotic equilibrium of the vinegar and the pickles has been reached, the pickles are ready for shipment. An alternative procedure is to "prime" the processed stock in distilled vinegar for a few days and then replace that vinegar with the final lot. At equilibrium the acidity of the pickle brine should vary between 2.0 and 3.5% acetic acid. Glass-packed sour pickles generally are less acid than those packed in bulk in barrels. Spiced sour pickles are made by adding suitable mixed spices to the containers of pickles, or by using spiced vinegar. Formulas for spicing sour and other kinds of pickles are to be found in Campbell[11] or Joslyn and Cruess.[58]

Sweet Pickles. The procedure selected for making sweet pickles depends on the quality desired, the plant capacity, and the time available for preparation. Sweet pickles are prone to shrink and wrinkle badly if sugar is added too rapidly. Consequently, every pickler has his own idea as to acidity, sugar content, spicing, and step-wise procedure for finishing sweet pickles.

The most rapid method for finishing sweet pickles makes use of a spiced, priming vinegar to cover the pickles. The pickles are soaked in this vinegar for a day or two. Sugar is then added directly to the priming mixture twice a day for 7 to 10 days (depending on the rate of addition and the desired finishing concentration) to complete the process. In this procedure, the liquid is mixed at frequent intervals to prevent too rapid osmosis which otherwise will make the pickles shrivel. This method is designed for use with finishing tanks which hold large quantities of pickles.

The slow methods for finishing sweet pickles involve the usual priming or soaking of the pickles in distilled vinegar, generally in kegs or barrels. Then, instead of adding sugar directly to the priming vinegar mixture, this is drawn off and replaced with a sweetened vinegar. The mixture is then allowed to stand until the osmotic equilibrium has been reached. The liquor is again drawn off and replaced with sweetened vinegar which has been spiced. The mixture is allowed to stand as before. Then, if the sugar content has increased to a satisfactory level, the pickles are ready for shipment in bulk or for packing in glass. The acidity of finished sweet pickles should not be less than 2.0% as acetic. The sugar content may vary between 20 and 40%, according to the quality.

Sweet pickles are subject to attack by yeasts of the genus *Zygosaccharomyces*. These yeasts are osmoduric (sugar tolerant) or osmophilic (sugar loving). They may attack the sugar during the finishing process or cause fermentation in the finished pickles packed in glass, kegs, or barrels. Scrupulous sanitation is necessary to control them. Because they are so difficult to control, once the plant has been contaminated, many picklers resort to the use of sodium benzoate or pasteurization to destroy them in the finished product.[26,28,55]

Process or Imitation Dill Pickles. Process or imitation dill pickles are made by flavoring processed salt stock with a brine containing 5 to 6% salt and 3 to 4% acetic acid (distilled vinegar). It is desirable to flavor the brine with genuine dill brine and dill emulsion. It is common practice to place about 3 lb of pickled dill weed in the bottom of the barrel, a similar quantity in the middle of the half-filled barrel of pickles, and more dill weed and dill spices on top of the pickles in the filled barrel. The barrel is then headed and filled with the dill brine. The barrels are closed and stored for proper aging. After the aging period which is necessary for proper flavoring of the pickles, they may be used in bulk or packed in glass.

Mixed Pickles. Pickles, cauliflower, onions, green beans, green tomatoes, carrots, and peppers are sometimes used together to make mixed pickles and mixed pickle relishes. In general, these vegetables may be prepared in much the same manner as already described for salt-stock pickles. For more detailed discussion of the variations used for pickling vegetables other than cucumbers as well as for pickling fresh cucumbers, other fresh vegetables, and fruits, the publications of Campbell,[11] Cruess,[17] Fabian and Blum,[36] Etchells and Jones,[29] and Joslyn and Cruess[58] should be consulted.

Fermented Dill Pickles

Genuine dill pickles differ from the other well-known cucumber pickles because they are the product of bacterial fermentation in spiced salt brine. They owe their distinctive flavor and aroma to the products of fermentation of the lactic acid bacteria and to the blending of dill herb and spices which are added to the brine.

The larger cucumbers are generally used for preparing fermented dill pickles. The fresh cucumbers are washed and placed in 45 to 50-gal barrels with pickled dill weed and dill spices as described for process dills. Some picklers add dry salt with the dill herb and other spices and then fill the headed barrel of cucumbers with water. Others prefer to use a salt brine (5 to 7% sodium chloride) which may or may not be spiced with dill emulsion. Many picklers also add 2 to 3 qt of vinegar per barrel of pickles to help retard the growth of undesirable microorganisms by decreasing the pH value of the brine.

One formula given by Campbell[11] for spicing dill pickles is as follows:

Drained dill weed (cured in vinegar and salt)	8 lb
Dill vinegar from barreled dill weed	3 qt
Mixed dill spices	3 pt
Salt	16 lb

Enough cucumbers to fill the barrel. Potable water to cover the contents and completely fill the barrel.

Some dill pickles are heavily seasoned with dried cayenne peppers; others may be flavored with garlic, caraway seed, anise, or fennel to make specially seasoned dill pickles for certain trade preferences. Sweetened dill pickles may also be produced.

The filled barrels may be bunged tightly to exclude air; or small holes may be drilled through the bungs to allow for the escape of fermentation gases. Some omit the bungs during the period of active fermentation. Whatever the procedure, it is mandatory to keep the barrels full of brine at all times because the exposed pickles will spoil rapidly. The salt concentration in the brine should be maintained at 20° to 25° Salometer. Optimum temperature for the fermentation is between 70° and 80°F. The barrels may be stacked indoors or allowed to ferment outdoors, as shown in Figure 24. The fermentation is usually active for 3 to 4 weeks after which an additional curing period of 3 to 4 weeks is considered essential. The flesh of dill pickles becomes entirely translucent when completely cured. The brine has about 0.5 to 1.2% total acidity calculated as lactic acid. In addition, there is a small amount of volatile acid (about 0.2% calculated as acetic) and alcohol (about 1.0% by volume) produced by fermentation.[56]

As already mentioned, the desirable fermentation of dill pickles is caused by the lactic acid bacteria. There is usually a definite

FIGURE 24. *Barrels of Fermenting Dill Pickles in a Fermentation Yard*

sequence of bacterial types. During the early stage of fermentation, the coliform bacteria and aerobic spore-forming bacteria may predominate. Soon, however, the Gram-positive cocci of the genus *Leuconostoc* gain the ascendancy. These in turn are eclipsed by the Gram-positive rods of the genus *Lactobacillus*. All of the bacteria listed in Table 42 have been recovered from fermenting dill pickles by the author and his students.

Yeasts may also be present, but if they are present in appreciable numbers they have an undesirable effect on the fermentation. The yeasts compete with the lactic acid bacteria for fermentable carbohydrates and besides may oxidize the lactic acid formed in the fermentation. The dissipation of both fermentable sugars and lactic acid by the yeasts decreases the total acidity and increases the pH of the brine. Thus the yeasts may contribute to spoilage of the pickles by making conditions favorable for the growth of other undesirable microorganisms.

OVERNIGHT DILL PICKLES

"Overnight" dill pickles are somewhat similar to genuine dill pickles, with the important exception that they are stored at 38°F where a slow lactic acid fermentation produces an acidity of only 0.3 to 0.6%, calculated as lactic at the end of 6 months. "Overnight" dill pickles retain some of the fresh cucumber flavor, but are so perishable that they are being replaced by unfermented, pasteurized dill pickles made from freshly harvested cucumbers.[26,32]

Deterioration of Pickles

Extensive study has been made of the deterioration of cucumbers during fermentation, curing, and storage.[31] For the most part, deterioration of pickles is caused by microorganisms. Chemical deterioration generally is confined to metallic contamination or unanticipated alteration of flavor and aroma by the use of specific chemicals or undefined congenerics used for spicing purposes.

The role of yeasts in the deterioration of pickles has already been discussed. The most important defect caused by both bacteria and molds is softening. Gaseous deterioration or "floater" spoilage is another common spoilage caused, for the most part, by bacteria. A comparatively rare deterioration, in California at least, is the formation of dark (black) brines and pickles caused by the activity of certain bacteria. Other spoilage of infrequent occurrence is the butyric fermentation and formation of "ropy" brine.

SOFTENING

Softening is a progressive spoilage which occurs most frequently soon after the brining of dill pickles or salt-stock pickles brined in less than 30° Salometer brine. The skin of the cucumber is attacked first, usually in spots at the blossom end. In a short time, the entire skin may be affected, become slippery, and be easily removed. This characteristic manifestation of softening has given rise to the terms "slips" or "slippery" pickles in the industry. "Mushy" pickles result when softening involves the deeper layers of cells in the pickles.

Most of the softening of commercially processed pickles results from the activity of microbial enzymes which attack the pectic substances present in the middle lamella separating the individual cells of the cucumber. Kossowicz[59] was probably the first to discover the importance of *Bacillus mesentericus* and closely related species of aerobic, spore-forming bacteria as agents of cucumber-pickle softening. His work has been confirmed by Rahn,[81] LeFevre,[60,61] Joslyn,[57] and more recently by Fabian and Johnson.[40] The latter investigators made a study of the purely chemical and physical types of softening as well as of softening caused by bacteria.

On the basis of all of these investigations, it may be concluded that most of the microbial softening is caused by members of the *Bacillus mesentericus-Bacillus megatherium* group of bacteria. According to the author's own unpublished experience, however, other microorganisms which may also cause softening of pickles include bacteria of the *Bacillus (Aerobacillus) macerans-Bacillus polymyxa* group, and molds of the genus *Penicillium*. Softening occurs when microorganisms are capable of elaborating pectin-destroying enzymes (see Phaff and Joslyn[77] and Kertész[58a] for detailed discussions of pectic enzymes) under the conditions of salinity, acidity, etc., which exist in pickle brines. It appears that the pectolytic enzymes produced by bacteria are most active in the alkaline range, i.e., above pH 7.0 and are rapidly inactivated in acid surroundings (see Wood,[94] Kraght and Starr,[59a] and Nortje and Vaughn[69a]). The reverse is true of the fungal pectolytic enzymes, according to the work of Ehrlich,[21a] Waksman and Allen,[90a] Phaff,[76a] Luh and Phaff,[64a] Pavgi, Pithawala, Savur, and Sreenivasan,[70a] and others.

Recently Bell and coworkers[4a,5a] have demonstrated the presence of the enzyme pectinesterase in the seeds, leaves, petioles, stems, flowers, and fruit of the pickling cucumber. The role this enzyme may play in softening of fermenting cucumbers is not clear. Thus it is possible that the pectic enzymes elaborated by bacteria, molds, yeasts, and perhaps the intrinsic enzymes of the cucumber itself may be involved in softening. At present, it is not possible to state clearly whether one or a combination of these agents causes most of the softening.

Softening may also be induced by purely chemical or physical changes not involving microorganisms. Thus, Lesley and Cruess[64] concluded that softening might, in certain instances, result from acid hydrolysis of pectic substances, particularly in immature cucumbers. Fabian and Johnson[40] confirmed this conclusion and at the same time showed that cooking also contributed to softening of pickles. It is to be stressed, however, that neither use of strong mineral acids (HCl to pH 2.2) nor cooking are integral steps in the production of salt-stock or dill pickles.

Faville and Fabian[44] have suggested that bacteriophages or antibiotic substances antagonistic to the lactic acid bacteria may contribute to softening of cucumbers by favoring the growth of the spoilage bacteria. It is unfortunate that their *in vitro* studies were not substantiated by comparable *in vivo* tests.

The recently developed polygalacturonase enzyme test of Bell, Etchells, and Jones,[5] if applied as suggested, should permit early detection of potentially undesirable softening of fermenting cucumbers.

Gaseous Deterioration

Cucumbers frequently undergo a type of deterioration which causes them to become so distended from gas pressure, that the tissue is permanently distorted and large gas-pockets form (Figure 25). Such pickles have an unnatural puffed or bloated appearance and, if the hollow centers are filled with gases they will float. Such pickles are most commonly called "bloaters" or "floaters."

Hollow pickles have been known to occur for a long time. Many possible explanations for their formation have been set forth at one time or another: the wrong variety of cucumbers; too rapid fermentation; insufficient weighting and too loose packing of the

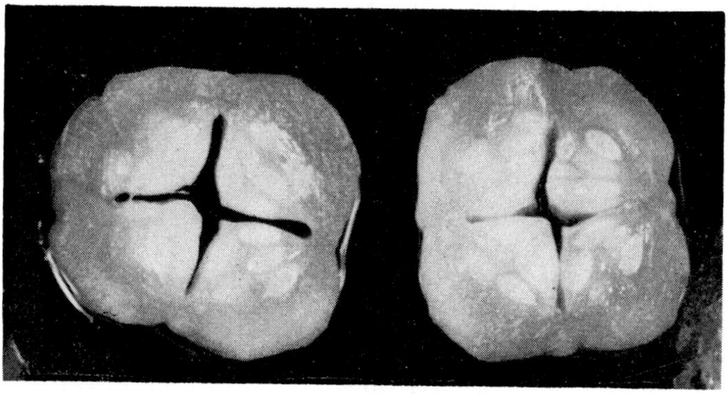

FIGURE 25. *Cross-Section of Salt-Stock Pickle to Show Rupture of Flesh Resulting from Gas Production*

cucumbers in the tanks; insufficient salt in brine; and too large cucumbers. LeFevre[61] was one of the first to observe that "floaters" were hollow and most frequently occurred in the larger cucumbers pickled in brines containing too little salt. However, until very recently, the cause of their formation was largely conjectural.

Although it was suspected earlier, the first substantial evidence that gaseous deterioration was caused by microorganisms was presented by Veldhuis and Etchells[90] in 1939. These investigators found that hydrogen was produced in significant quantities in the pickle fermentations in 60° Salometer brine and in some, but not all, fermentations at lower salt concentrations. They also found that gas from the hollow cucumbers from 60° Salometer brines contained essentially the same amount of hydrogen and about the same ratio of hydrogen to carbon dioxide as the gas collected from the surface of the fermentations. They also isolated, but did not identify, an organism which produced significant quantities of hydrogen. Somewhat later (1941) Jones, Etchells, Veerhoff, and Veldhuis,[54] and Etchells and Jones[25] suggested that gaseous fermentation by unidentified yeasts was the cause of "floater" formation. Still later (1945) Etchells, Fabian, and Jones[24] presented evidence that coliform bacteria of the genus *Aerobacter* were responsible for the formation of hydrogen in pickle fermentations.

"Floater" spoilage can be minimized under commercial conditions by elimination of the largest sizes of cucumbers; by the use of a needling machine to puncture the cucumbers and thus provide for escape of gases during the fermentation; by rigid control of the fermentation by trained personnel; and by scrupulous sanitation.

BLACK BRINE AND PICKLES

This form of deterioration is characterized by the formation of dark brine (deep brown to black) and the occurrence of darkened areas on the outside of the pickles. The blackening starts in the upper layers of the brine and extends downward. Hydrogen sulfide or a definite putrid odor is characteristic of this spoilage.

Rahn[81] was one of the first to investigate the problem of black pickles. He found that the blackening resulted from the production of iron sulfide. The bacteria present in the brine caused the reduction of gypsum ($CaSO_4$). The hydrogen sulfide thus formed reacted with iron present in the brine to give iron sulfide. Fabian, Bryan, and Etchells[38] confirmed and extended the work of Rahn to include a discussion of the possible liberation of hydrogen sulfide in the course of protein breakdown by bacteria. An additional source of blackening was found by Fabian and Nienhaus[42] to result from the production of a black, water-soluble pigment by *Bacillus nigrificans.* According to Fabian and his students, blackening can be prevented by controlled fermentation which insures a significant increase in acidity of the brine and elimination of iron and sulfate contamination.

In all probability, black brine in pickle fermentation, instead of always involving sulfate or protein reduction, frequently is the result of the reaction of iron with complex tannin compounds to form complex iron tannates which are black, bluish black, or greenish black. Iron tannates will not form unless the iron is oxidized to the ferric state. Iron sulfide, however, will also form from ferrous iron, so that the blackening of pickle brines should be nearly uniform throughout instead of starting at the surface and extending downward.

It is also possible that some of the darkening of pickles, particularly sweet ones, may result from nonenzymic chemical reactions not involving iron, tannins, or hydrogen sulfide (Stadtman[84]).

ROPY BRINE

Ropy brine is another defect which appears during the fermentation of pickles. Fabian and Nienhaus[42] have investigated the problem and advocate the control of slimy or ropy pickle brines by rapid increase in the salt content of the brine. However, unless the slimy or ropy brine has a deleterious effect on the quality of the finished pickles it should not be considered as a defect. Many of the bacteria known to produce polysaccharides (dextrans, levulans, etc.) are found in normally fermenting pickles; *Leuconostoc mesenteroides,* in particular, is usually present. Furthermore, other bacteria and yeasts which normally occur in the pickle fermentations will, in time, probably utilize the polysaccharides.

Chemical Deterioration

The chief cause of chemical deterioration of pickles is the direct addition (wittingly or unwittingly) of undesirable chemicals to the brines or to the pickles themselves. The changes which occur may affect the appearance or the flavor and aroma of the pickles.

Contamination with copper and iron is probably the most common type of chemical deterioration. Vinegar or lactic acid which may be used for preparation of brines or liquors for pickles may be heavily contaminated with copper and iron (see Chapter 17 of Volume 1). Other copper and iron contamination may result from contact of the comparatively corrosive brines used for pickling with corrodible equipment made with alloys which contain copper and iron.

Iron, as already described, is involved in the blackening of pickle brines and pickles. Copper replaces magnesium in chlorophyll contained in the pickles and causes them to turn an unnatural, artificial green color. Only 5 to 10 ppm of copper in the brine or liquor is enough to spoil the pickles. Both copper and iron may cause a metallic flavor.

Other possible sources of chemical deterioration include contamination of cucumbers with chemical sprays, etc., used to combat insects and microorganisms in the fields. Some of these compounds may remain on the cucumbers and cause off-flavors and off-odors in the pickles.

A comparatively new source of deterioration is the too generous use of "spice oils" for flavoring pickles. "Spice oils" first were recommended by Fabian[33,35,43] and others,[49] because many whole spices were too heavily contaminated with undesirable bacteria.[41,91] Methods were developed which insure the availability of commercially sterile spices and other flavoring agents. For reasons of economy, it is still thought desirable to increase the use of "spice oils." However, if these preparations are not manufactured correctly and are not used properly by the pickler, undesirable medicinal flavors and tastes result.

Packaging Finished Pickles

Formerly, almost all pickles were sold in bulk from wooden kegs, barrels, casks, etc., but now glass and sometimes tin or wooden containers are also used. Pickles do not keep very long in wooden containers. Therefore, care must be taken to provide cool storage (50°F) and to protect against drying out of the barrels or freezing the contents. The wooden containers should be inspected at weekly intervals and all broken or leaky ones should be repaired. All lost liquor must be replaced to prevent darkening and softening of exposed pickles.

Pickles may also be packed in tin cans. The high acidity of the pickle products, however, makes prevention of can corrosion a major problem. The tin plate must be heavily lacquered or coated with paraffin to retard can corrosion. Pickles have been most commonly canned in No. 2½, No. 10, and "full gallon" cans. For canning, the pickles are hand-packed into the cans and covered with dill brine, vinegar, or spiced sweet vinegar, according to the type of pack. Pickles canned in lacquered cans are exhausted to insure a suitable vacuum in the cooled, closed container. To prevent undue softening, canned pickles should be cooled immediately after heating. Exhausting and cooking are impossible if paraffined cans are used. Pickles packed in cans will keep somewhat longer than those packed in barrels or kegs, if corrosion is retarded.

Glass-packed pickles are packed by hand into jars which are filled with liquor of the appropriate type and exhausted in some manner before closing to insure a vacuum in the container, or are closed without a vacuum. Entirely mechanized lines, similar to those used with cans, are available for handling the jars, once they have been packed with pickles. Pickles packed in glass will keep much longer than either canned or bulk-packed pickles, if they have been properly packaged and stored.

THE PRODUCTION OF SAUERKRAUT

The use of cabbage (*Brassica oleracea*) as a food by man, as well as the use of cucumbers and olives, probably antedates recorded history. According to Magruder,[65] cabbage has been used rather commonly for about 4,000 years and, next to potatoes, is the most common vegetable in the diet of man.

Sauerkraut, a food resulting from the lactic acid fermentation of cabbage, to quote Prescott and Proctor,[80] "has been a popular food in Europe for at least 4,000 years, and probably has as its antecedent the somewhat similar preserved cabbage soaked in sour wine or vinegar which was popular even before the Christian era."

Originally sauerkraut was made only in the home because it provided a means of preserving fresh cabbage which might otherwise spoil before it could be used. However, the commercial production of sauerkraut has become an important food industry in the United States.[17,62,79,80] Because the major part of the industry is localized in those states bordering on the Great Lakes, sauerkraut manufacture is a seasonal industry which usually works from September through December. Cabbage varieties best suited for growth in these regions are used.

The varieties grown include Early Flat Dutch, Late Flat Dutch, Early Jersey Wakefield, Early Winningstadt, Glory of Enkhuizen, Copenhagen Market, Marion Market, Globe, Danish Ballhead, or special varieties and strains bred from them. When the cabbage is harvested, care is taken to select only the best heads and to prevent undue injury because only sound, firm, compact, uncolored cabbage should be used for sauerkraut. Immature and partially spoiled heads, if used, will result in an inferior product.

Preparation for Fermentation

The harvested heads are stored to allow the outer leaves to wilt before the sorting, trimming, and cutting operations are started. Storage may take place in piles outside the plant or in special bins or rooms where the temperature and air supply can be controlled to the best advantage. Controlled storage allows the maintenance of a uniform temperature of at least 70°F and the trimming and shredding of thus wilted cabbage with the least breaking and tearing of the leaves.

When the cabbage has been wilted sufficiently, it is passed by conveyor belts to workers who remove the wilted outer green leaves, sort and cut out any bad spots, and then core each head by machine. There are two types of coring machines. One removes the core entirely. The other, more common type, cuts the core into fine pieces without removing it from the head of cabbage. In the first case, the core is discarded; in the second case, the finely cut core is converted into sauerkraut. Alternatively, the wilted cabbage may be cored by machine and then passed through a trimming and washing machine before it is shredded.

The trimmed and cored cabbage then passes by conveyor to the cutters, where it is shredded. The cutters are power-driven, rotary, adjustable knives, capable of cutting the cabbage into long shreds as fine as 1/16 to 1/32 inch in thickness. In general, long, finely cut shreds are preferred, but the thickness is determined by the judgment of the manufacturer. The shredded cabbage (known as slaw) is then conveyed by belts or by hand-carts to the vats for salting and fermentation (Figure 26).

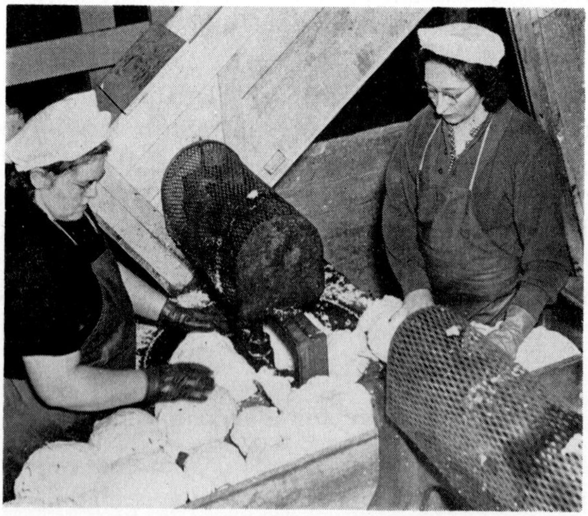

FIGURE 26. *Slicing Fresh Cabbage for Fermentation* (Courtesy—National Kraut Packers Association)

Salt plays a very important role in the making of sauerkraut and the concentrations used are rigidly controlled. According to legal definition,[86] the concentration of salt must not be less than

2%, nor more than 3%. As a result, most producers use a concentration of about 2.5% of salt (2.5 lb of NaCl per 100 lb of cabbage). Salt is required for several reasons. It extracts water from the shredded cabbage by osmosis, thus forming brine. It suppresses the growth of some undesirable bacteria which might cause deterioration of the product while making conditions favorable for the desirable lactic acid bacteria which cause fermentation. Salt also contributes to the flavor of the finished sauerkraut.

It is essential that the salt is uniformly distributed throughout the mass of shredded cabbage. For this reason, some producers prefer to salt the shredded cabbage in the hand-carts which convey it to the fermentation vats. Uniform mixing is much more certain under such conditions. Others salt the cabbage as it is packed into the vats.

The vats should be filled carefully, to make certain that the cabbage is evenly distributed and packed as solidly as possible. The vats are filled to about 1 ft above the top, the slaw is covered with a thick layer of clean, outer leaves, and then is fitted with a heavy cover which is heavily weighted. The additional weighting is necessary to facilitate formation of the brine and to assure that the brine will completely cover the cabbage at all times. Unless this is done, the top layers of sauerkraut will have to be discarded when the vat is emptied, because they may have become soft and discolored or otherwise deteriorated. In extreme cases, all of the kraut may have to be discarded.

Within a few hours, the brine has formed and the fermentation has started. The head is then fixed in position in much the same manner as with pickle tanks. Within 3 or 4 days, the fermentation has become very active and the acidity increases rapidly. Under normal conditions of temperature, the fermentation is usually complete within 3 or 4 weeks.

The Fermentation

The normal fermentation of sauerkraut is a spontaneous lactic acid fermentation caused by a natural mixture of lactic acid bacteria. The fermentation proceeds rapidly at temperatures between 60° and 75°F and generally some provision is made to keep the temperature within this range. Temperatures much below 60°F result in a slow fermentation which may not be completed until the following spring. Temperatures above 75°F may result in a very rapid fermentation, but opinion is divided concerning the quality of kraut produced at temperatures of 85°F and above. At temperatures of 75° to 85°F, the final acidity is usually attained within 2 to 3 weeks, whereas at lower temperatures, the fermentation may be prolonged to 4 or more weeks.

The sauerkraut fermentation normally proceeds more rapidly than either the pickle or olive fermentations. Fresh cabbage generally contains more sugar (about 2.9 to 6.4%, according to Atwater and Bryant[2] and Peterson et al[75,76]). The sugar is more readily available because the cabbage is shredded. Furthermore, the fermentation occurs in the presence of much less salt.

The chemical changes which occur in the sauerkraut fermentation are typical of a mixed fermentation caused by bacteria and yeasts. A variety of microorganisms is responsible for the conversion of the fermentable sugars present in the shredded cabbage to gases (carbon dioxide and hydrogen), volatile and nonvolatile acids, alcohol, and traces of other end products.[73] The major end products of the sauerkraut fermentation are lactic acid, acetic acid, alcohol, and carbon dioxide. Small quantities of hydrogen may be produced by the coliform bacteria during the early stages of fermentation. Glycerol and mannitol may also be formed, but they usually are found only in traces in the completely fermented kraut, together with quantities of flavoring constituents, which may be of natural occurrence or result from the fermentation.

Lactic acid bacteria are mainly responsible for the normal fermentation of sauerkraut. A definite sequence of microorganisms and microbial populations is known to occur (Pederson[71,72a,72b,74a]). The sauerkraut fermentation may also be divided into three stages: primary, intermediate, and final. At the beginning of the fermentation, the desirable lactic acid bacteria are outnumbered by a great many unrelated bacteria, yeasts, and molds. During the primary stage of fermentation, the desirable lactic acid bacteria begin to develop very rapidly and, by the end of 2 or 3 days, the gas-producing cocci dominate the population during the intermediate stage.

During this period, the total acidity has increased very rapidly to a 0.6 to 0.8% total acidity, calculated as grams of lactic acid per 100 ml of brine. After 4 to 6 days of fermentation, the rod-shaped lactic acid bacteria begin to dominate the population. During this final stage, the gas-forming (heterofermentative) and non-gas-forming (homofermentative) species of *Lactobacillus* are active. However, the homofermentative species probably complete the fermentation, because they dominate the *Lactobacillus* population, according to Pederson.[71]

The sequence of microorganisms observed in the fermentation of sauerkraut is essentially the same as that found in the fermentation of cucumbers in low-salt brines (see Table 42). However, the rate of acid production in cabbage is significantly faster than in either cucumbers or olives, because the low concentration of salt and readily available fermentable sugar favor the development of much larger total population levels of the desirable lactic acid bacteria in the sauerkraut. Furthermore, the sauerkraut fermentation is comparatively anaerobic. Therefore, fermentable substances are not dissipated by the yeasts to the extent they are in tanks of salt-stock pickles or olives.

Completely fermented sauerkraut must contain "not less than 1.5% of acid, expressed as lactic acid."[86] In reality, most of the completely fermented sauerkraut contains more than the required minimum of 1.5% of acid. The final acidity may range up to almost 2.0% as lactic acid. The total acidity is composed of lactic acid and acetic acid; the ratio of lactic acid to acetic acid varies from about 4:1 to 6:1. The finished kraut also contains alcohol in concentrations varying from about 0.2 to 0.8% by volume.

Storage, Processing and Packing

It is not customary to keep sauerkraut for more than one season, as is sometimes done with salt-stock pickles. As soon as the kraut has been completely fermented, it is ready for canning or shipment in bulk in wooden kegs or barrels. The first sauerkraut made in the fall is usually canned or shipped in bulk within a short time after the fermentation is completed. Frequently, a tank is used for fermentation two or more times in a season. However, it is necessary to store some of the sauerkraut for as long as 6 to 8 months under certain circumstances. It may be stored in kegs or barrels or in the original fermentation tanks. It must be covered with brine and kept from contact with air at all times. Storage in sealed wooden containers provides good anaerobic conditions, but is not economical. Tank storage is commonly used.

When the fermentation is completed, the top layer of kraut, protecting cabbage leaves, and brine is cleared of all foam and microbial film. If the brine is low, more is added to cover the shreds. Then the shreds and juice may be covered with a muslin cloth to aid in keeping the top layer of kraut moist. The cloth is then covered with the wooden tank cover which is set in place as before. An alternative method is to flood the top of the sauerkraut and brine with a layer of neutral mineral oil to provide anaerobic conditions and then reset the cover.

Processing

The vat of sauerkraut is opened by removing the weights and anchoring the timbers used to keep the wooden cover in place. All liquid above the cover is withdrawn. Then the cover is removed. If the sauerkraut has been kept well covered with brine, there is little loss, but if the brine has been allowed to recede below the surface of the shreds, the discolored and otherwise deteriorated product must be discarded until sound sauerkraut has been reached. Then, the plug in the bottom of the tank is opened to draw off the juice ahead of the removal of the sauerkraut for repacking in bulk or canning.

Packing in Bulk

Bulk distribution of sauerkraut is accomplished by repacking the kraut in kegs or barrels. The wooden container is filled with "dry pack" sauerkraut, consisting of about 90% solids and 10% brine. Because of the perishable nature of sauerkraut, bulk distribution is limited primarily to the cooler months.

Canning

Sauerkraut is relatively simple to can. It is a very acid food and thus does not require prolonged cooking under pressure to preserve it. It is easily canned with little equipment.

The kraut is removed from the tanks with forks and passed to the canning tables in large pans or other containers or by means of conveyor belts.

In the simplest canning operation, the compacted mass of shreds is thoroughly broken apart by hand, allowed to drain, hand packed into cans (Figure 27), brined, and then exhausted to insure heating the kraut in the center of the can to 150°F or more. In this type of operation, the filling of the cans may be expedited by use of the rotary hand-pack filling machine.

FIGURE 27. *Hand Packing Sauerkraut into Cans* (Courtesy—National Kraut Packers Association)

Some packers prefer to heat the sauerkraut before filling it into cans. In this method, the compacted kraut is first broken apart by hand and then heated batchwise in its own brine at 110° to 120°F. The warm sauerkraut is then passed to the canning line for the conventional hand filling, brining, exhausting, closing, cooking, and cooling. If the sauerkraut has been warmed in this manner, hot brine can be added and very little subsequent exhausting is necessary.

Much labor savings are possible if special kraut filling machines are used. By this method, the cans may be filled with sauerkraut and brine hot enough to make subsequent exhausting and cooking unnecessary. However, most prefer to use a short cook as an added protection.

Irrespective of the method, the center temperature of the can of sauerkraut with brine should be about 150°F at the time of closing for an adequate exhaust.

Although sauerkraut is a very acid food product, it is believed necessary to cook (process) the cans at 212°F in order to insure commercial sterility and prevent microbial spoilage of the canned product. Processing is necessary because of the very irregular heat penetration resulting from clumps of heavily compacted shreds which are not broken apart well before canning. According to Bitting,[6] the cooks for sauerkraut should be 10 minutes at 190°F for No. 2½ cans and 15 minutes for No. 10 cans in the agitating cooker, or 35 minutes at 212°F for No. 2½ cans and 45 to 50 minutes for No. 10 cans in stationary cookers. Campbell[11] points out, however, that if the filled cans are closed with a center temperature of over 160°F, it should not be necessary to cook the kraut. According to Cruess,[17] canned sauerkraut should be processed in steam long enough for the contents of the can to reach 180°F.

It is very desirable to cool the canned sauerkraut thoroughly after cooking. Judicious cooling preserves the color, prevents cooked tastes and undue softening of the shreds.

Defects of Sauerkraut

There are several rather common defects of sauerkraut which have been recognized. Some of these are of microbial origin; others apparently are chemical in nature. Discoloration of sauerkraut, one of the most common defects, may be caused by chemical reactions or by the activity of microorganisms. Softened, slimy, or rotted krauts, according to the literature, are largely of microbial origin. Off-flavored kraut results either from the activity of microorganisms or from direct chemical contamination.

Darkening of sauerkraut probably is the most common defect. If the fermenting kraut is not covered and weighted properly, the brine may fail to cover or moisten the upper layer of shreds. Darkening will result either by direct chemical oxidation or by the growth of aerobic microorganisms. Excessive amounts of iron may react with tannins from the wooden tanks or covers or sulfides may be produced by fermentation to cause black discoloration. Uneven distribution of salt or use of too much salt may disrupt the normal fermentation and allow undesirable microorganisms to produce discoloration.

Pink sauerkraut is caused by the activity of aerobic yeasts. Since the yeasts grow only in air, they usually develop in the brine at the surface of the tank of sauerkraut, on the exposed shreds, or in the air pockets formed between the sides of the vat and the shredded product. The yeasts produce a pigment that varies from a light pink to an intense red color. They are undesirable because they utilize sugar and acid, in addition to causing discoloration. According to the literature, any inhibition of normal fermentation may favor the development of the pink yeasts. (See Brunkow, Peterson, and Fred,[9] Fred and Peterson,[46] Marten, Peterson, Fred, and Vaughn,[66] Pederson,[72] Pederson and Kelley,[74] Peterson, Parmele, and Fred.[76])

Soft kraut is usually associated with the presence of air pockets, faulty salting, high temperatures, or abnormal fermentation (Fabian[32]). Nevertheless, it should be stressed that although it is thought that softening is caused by microorganisms, their identity is unknown.

It is claimed in the literature that slimy or ropy sauerkraut is the result of the growth of *Lactobacillus cucumeris* and *Lactobacillus plantarum.* (The name *L. cucumeris* is regarded by most authorities as a synonym of *L. plantarum.*) It is not likely that these homofermentative lactobacilli are involved in the production of slimy or ropy kraut. Neither of them is known to produce polysaccharides. Furthermore, they occur in the normal fermentation of sauerkraut.

Rotting of sauerkraut is claimed to be caused by the presence of certain types of bacteria, yeasts, molds, and even fruit flies. Apparently, cases of rotting occur only when the sauerkraut is seriously neglected. Adequate protection of the surface layers of the sauerkraut will prevent rotting. However, it should be stressed that if rotting has progressed to such an extent that insects have infested the sauerkraut, better plant sanitation is essential.

Off-flavors in sauerkraut have been demonstrated by Pederson[71] and others to result from changes in the sequence of types of desirable lactic acid bacteria in the fermentation. For this reason, starters of pure cultures of some of the lactobacilli are not recommended. Off-flavors may also result from the activity of undesirable microorganisms which cause softening or rotting of sauerkraut. In some cases, chemical contamination resulting from improper use of sterilizing agents, insecticides, etc., may produce off-flavors.

Spoilage of canned sauerkraut results from understerilization or from can corrosion. Either type of spoilage is due to development of gas which causes the cans to swell. Growth of lactic acid bacteria or yeasts in understerilized canned sauerkraut will yield carbon dioxide. Can corrosion may cause the liberation of hydrogen.

OLIVES

The first olive trees grown in California are said to have originated from seeds planted at the Mission San Diego in 1769. According to Lelong,[63] the seeds were brought from San Blas, Mexico, by Don Joseph de Galvez, while on an expedition to rediscover the port of Monterey. The seedling trees thus obtained are thought to be the source of the present Mission variety of olive (*Olea europaea*).

Although olives were utilized for oil production in the California Missions as early as 1780, they were of little commercial importance until about 1900. The first commercial production of olive oil outside of the missions did not occur until 1871. Prior to 1890, little attention was given to the utilization of olives for commercial pickling, although pickling had been practiced for some time on the farms and in the homes.

Until the development of the California ripe olive, olive oil was the chief product of the industry. The first attempts at commercial production of California ripe olives were made between 1890 and 1895. By 1900, there were several commercial producers of ripe olives. The development of California ripe olives was successful and, with modern technology, achieved worldwide prestige. During the same period, attempts were also made to produce commercial quantities of Spanish-type, fermented green olives.[93] In contrast, however, the pickling of green olives has been of no great importance in California until comparatively recently, despite the fact that, according to Dunn,[20] the Spanish Queen olive, or Sevillano variety, was introduced into the state some time before 1866.

That the canned ripe olive originally developed and even now produced only in California is the most important product of the California industry has already been shown in Table 40.

Varieties and Harvesting of Olives

Commercial Varieties

There are five main varieties of olives grown in bearing acreage in California. The Mission variety (of California origin), which constitutes a little over half of the total bearing acreage of olives in the state, is followed in commercial importance by the Manzanillo, Sevillano, Ascolano, and Barouni varieties in the order named. The Manzanillo and Sevillano varieties were imported from Spain. The Ascolano variety originated in Italy. The Barouni variety was obtained from Tunisia. Each of these varieties is quite distinctive in size, shape, and general utility.

The different species vary markedly in size of fruits. The Sevillano olives are the largest, followed in order by the Ascolano, Barouni, Manzanillo and Mission varieties. The Manzanillo variety most nearly approaches the requirements of an all-purpose olive, since it may be used for the production of California ripe olives, green fermented olives, or olive oil. The other varieties have characteristics which limit their utility in one way or another. The Sevillano or Queen olive is desirable for production of the largest sizes of ripe or green fermented olives, but bears too little oil to be of great value for the production of olive oil. The Mission variety is best utilized for the production of ripe olives or oil and is not well suited for making green fermented olives. The Ascolano variety is used primarily for the production of canned ripe olives because its oil content is too low for economical recovery and its color is too light a green for good green fermented olives. The Barouni olive may be used to a certain extent for the production of green fermented olives but is not acceptable for the preparation of good quality California ripe olives. Consequently, a significant amount of the Barouni crop is shipped fresh from California to the eastern United States where the fruit is purchased for home pickling.

All of these varieties have been used for the production of other olive products, including Sicilian-type green olives and brined or dry salt-cured Greek-type olives, products not well known to the general public.

Harvesting

Most of the olives destined for ripe or green pickling are harvested between October 1 and November 25, the time depending on the variety of fruit, the locality, the growing season, and other factors. Olives which are too green do not process well. Those picked near the end of the season often become undesirably softened or discolored and are prone to spoil more easily during the early stages of the pickling process.

Unfortunately, fresh olives are quite easily bruised. Care must be exercised in picking the fruit as well as in transportation and handling during preparation for pickling. To avoid unnecessary bruising, the olives are picked into padded buckets. The buckets of fruit are then emptied into boxes known as fruit lugs. These boxes are fitted with cleats to protect the olives against bruising and crushing when they are stacked and transported. The lugs vary in size, but the majority contains about 36 lb of olives (Figure 28).

FIGURE 28. *Harvesting Olives*

Because the canned ripe olive is the chief product of the California olive industry, harvesting practices naturally favor the selection of fruit most desirable for production of this commodity. However, high labor cost prevents the use of as much hand labor in California as is customarily used in other olive-growing areas of the world.[10,12,48] When the majority of the fruit in an orchard has reached the desired maturity, the harvest is started and continued until all the olives destined for pickling have been gathered. The fruit in the top of the tree generally is more mature and is picked first. The industry is fully cognizant of the value of careful harvesting and selection of the fruit for maturity, but with prevailing economic conditions, it is impossible to maintain the desired high standards for picking.

Mature (ripe) olives do not have the final dark purple to jet-black color at the time of harvest for pickling, but are at the color-turning stage, from green to straw yellow or, at most, light cherry red. More highly colored fruit is overripe, deteriorates rapidly during the pickling processes, and consequently is used for the production of oil or preparation of such specialties as dry salt-cured or brine-cured Greek-type olives. Unfortunately, there are no chemical or physical methods for determining maturity and it is left to the field men to estimate maturity by empirical means.

California ripe olives (brown to black in color after canning) are made from fruit ranging in color from green to cherry red at the time of picking. The less well-known California green-ripe or "home-cured" olives (yellowish-green to light brown, frequently mottled color after canning) are made from olives having a green to not more than cherry-red color. California green olives are produced from fruit ranging from green to straw yellow.

The harvested fruit is transported to the pickling plants by truck or other motorized equipment. Use of cleated lug boxes (Figure 28) allows the fruit to be hauled without undue bruising and crushing.

Preparation and Processing

The only olives produced in significant commercial quantity without lactic acid fermentation are small quantities of California ripe olives pickled directly from the trees; California green-ripe olives which of necessity must be prepared immediately after harvest; and salt-cured Greek-type olives which are dehydrated and preserved in large quantities of dry salt. Since these olive products are not allowed to ferment, they will not be discussed here. For descriptions of their preparation the publications of Cruess[17] and Vaughn[87,88] should be consulted. All other types of olives undergo at least some lactic acid fermentation in salt brines.

Sorting and Grading

On arrival at the factories, the olives generally are carefully graded for defects, color, and size. All defective and damaged fruit is removed by hand as the olives are passed along sorting tables on endless belts. Color grading is accomplished in the same manner and it is customary to sort the olives for color into three main grades, green, cherry-red, and black fruit. The green fruit is then used for green fermented or ripe olives. The cherry-red fruit is used for ripe pickling and the black (overripe) fruit is used for oil or other by-products.

Whereas sorting and grading for defects and color is largely a hand operation, size grading is entirely mechanical. Size-grading machines consisting of diverging steel cables are almost exclusively used (Figure 29). The olives are graded into sizes according to the number of olives per pound, as shown in Table 43.

FIGURE 29. *Size Grading of Olives* (Courtesy—Pacific Olive Co.)

Three blends of olive sizes—family, king, and royal—have been standardized recently in an attempt to reduce the number of sizes of olives now canned.[10a] The family blend, a mixture of the medium, large, and extra-large sizes, ranges between 91 and 105 olives per pound. The king size, a blend of the giant, jumbo, and smaller colossal olives, ranges between 45 and 53 olives per pound. The royal size is a blend of the largest olives not to exceed 34 olives per pound. If these new sizes are accepted in retail channels the number of sizes eventually will be reduced from nine to five and will include standard (a very competitive size), family, mammoth (a size that cannot be blended in all varieties), king and royal.

Rapid handling of the fresh fruit is necessary to avoid "sweating" which may predispose the olives to softening and discoloration. Therefore, when fresh fruit is being received in great abundance at the height of the harvest, it is sometimes necessary to forego size grading and to handle the fruit "orchard-run" until it is pickled and ready for final packaging.

TABLE 43. SIZE GRADES OF CALIFORNIA OLIVES[a]

California standard size grades	Count per pound (average)	Count per kilogram (average)
Super-colossal	32[b]	60-70
Colossal	36-40	80-90
Jumbo	46-50	100-110
Giant	53-60	120-130
Mammoth	70	140-160
Extra-large	82	170-190
Large	98	200-230
Medium	113	240-260
Small, select, or standard	135	280-310

[a] Based on the California Ripe Olive Standardization Act as amended in 1933 and 1935.
[b] Maximum number of olives allowed.

CALIFORNIA RIPE OLIVES

As already stressed, the California canned ripe olive may be prepared as soon as the fruit is harvested. However, most of the factories do not have enough pickling vats to handle all of the olives at once. Therefore, a large majority of the fruit destined for production of canned ripe olives is held in salt brine (holding solution) for some time until the pickling vats are available.

The strength of the brine used for holding olives prior to ripe pickling varies according to the type of fruit. The large varieties

(Sevillano and Ascolano) are very susceptible to salt shrivel. Consequently, it is customary to use a brine of 2.5 to 5% salt content (calculated as grams NaCl per 100 ml of brine) to cover the large varieties. Since the smaller Manzanillo and Mission varieties are comparatively immune to salt shrivel, they are generally placed in holding brines which contain 5 to about 10% salt.

The salt content of the original brine decreases because the olives take up salt from the brine by osmosis. It is necessary to in-

FIGURE 30. *Concrete and Redwood Storage Tanks for Olives* (Note customary method for increasing salt content by dissolving sacks of salt in surface layer of brine)

crease gradually the salt content of the holding solutions. Experience has shown that olives stored in brine keep best in the presence of 7 to 8 or even 10% sodium chloride.

It is customary to store the olives in the pickling vats or regular storage tanks, both of which may be constructed of reinforced concrete or redwood staves (Figure 30). The filled vats or tanks are closed with slatted covers which keep the olives submerged. If the brining is properly controlled, the olives cured in holding solutions can be stored safely for 6 to 8 months, or even longer.

Processing of Ripe Olives

The characteristic dark-brown to black color of the California canned ripe olive is obtained by oxidation of unknown constituents in the fruit by treatment with dilute lye (sodium hydroxide) solutions alternated with exposure to air or to highly aerated water. After lye treatment and oxidation, which fix the color of the olives, the excess caustic is removed by leaching with water. Finally, the olives are soaked in dilute salt brine (2.5 to 3.0% sodium chloride) to stabilize the salt content of the fruit just prior to canning.

The number of lye applications varies from 3 to 6 or 8. The strength of the caustic solution is between 0.5 and 2.0% sodium hydroxide, depending on the variety of fruit, the views and experience of the processor, and other factors. The first lye is allowed to penetrate just through the skin of the majority of the olives in the pickling vat. The first lye is thought to intensify the color change caused by the oxidation. Subsequent applications of sodium hydroxide also aid color fixation and hydrolyze the bitter glucoside *oleuropein,*[8,18] which, if not destroyed, leaves the olives so bitter as to be almost inedible. The last lye solution must penetrate the pit to destroy the last traces of the glucoside.

Each application of sodium hydroxide is followed by a period of oxidation. During this time, the olives must be stirred frequently to prevent uneven development of color when contact surfaces of the fruit are not exposed to the air. The unexposed surfaces do not darken to the same intensity, thus giving a mottled appearance to the fruit. For stirring, the olives are covered with water and the liquid and olives are agitated for several minutes with large wooden paddles or rods through which air is blown. Alternatively, oxidation is accomplished by covering the olives with water which then is continuously aerated by blowing air through a distributing device placed in the bottom of the pickling vats.

Excess lye is removed by leaching the olives with water until all traces of the sodium hydroxide have disappeared. The water is changed frequently and stirred often to facilitate leaching. Acidulated water is sometimes used to neutralize a portion of the sodium hydroxide. Removal of the sodium hydroxide from the fruit, as well as progress of its penetration, generally is followed by placing phenolphthalein on the cut surfaces of the olives.

After removal of the sodium hydroxide, the olives are stabilized with dilute salt brines prior to canning. Three changes of brine are normally used for stabilization. The first contains 2.0, the second 2.5, and the third 3.0% sodium chloride. This stabilization insures a uniform concentration of salt in the brine and in the olives after canning.

When the olives have been subjected to a final sorting and grading, they are canned in a salt brine of about 3.0% sodium chloride content. All canning is under the supervision of the California State Department of Health. It is required that all canned olives are heat processed at 240°F for 1 hour.

Brined Greek-Type Olives

Brined or wet-process Greek-type olives are usually made from firm yet highly colored purple to jet-black fruit. The color may also be intensified by lye treatment and oxidation as described before. However, since this product is bitter, only one or at most two lye applications can be used.

The olives are sorted and graded for size, placed in 16, 32, or 50-gal barrels, and covered with brine containing 7 to 10% sodium chloride. The salt content of the brine is then increased until it is about 15% or more. The olives cure out slowly, lose some of their natural bitterness, and take on a purple color. Little is known about the microorganisms involved in the fermentation. It is obvious that the high salt content of the brine suppresses acid formation. Since yeasts abound, it is assumed that they utilize the fermentable carbohydrates present in the olives.

Only small quantities of these olives are produced. Most suitable olives are harvested before they have reached the extreme maturity required for making brined olives of this type. Furthermore, the natural bitterness and saltiness of the brined Greek-type olive do not appeal to the normal American taste.

Green Fermented Olives

There are two characteristic kinds of green fermented olives produced in California: Spanish- and Sicilian-type olives. Formerly, Sicilian olives were the principal fermented product in California. Since about 1938, however, Spanish-type olives have been produced in increasing amounts.

Spanish-Type Olives

Fresh olives of green to straw-yellow color are treated with lye to destroy most of the bitter glucoside. The lye solutions which contain 1.25 to 2.0% sodium hydroxide are allowed to penetrate to about three-fourths or more of the way to the pits. A small quantity of untreated flesh is allowed to remain, for a slight bitterness is essential to give the fermented fruit part of its flavor.

After lye treatment, the residual sodium hydroxide is removed from the olives by leaching with water. During the leaching process, care is taken to avoid undue exposure of the olives to air or to aerated water; otherwise, they darken undesirably. After removing the excess lye, the olives are placed in 50-gal barrels, covered with salt brine, rolled into the fermentation yard, and allowed to undergo a lactic acid fermentation (Figure 31). Experimental bulk fermentations of green olives have been made in redwood tanks which hold about 15,000 lb of olives. These studies indicate that there is no significant gross chemical or bacteriological difference between the barrel and the bulk processes. The bulk fermentation, if used under the direct supervision of trained personnel, is economical of labor and time and insures a greater uniformity of product than does the conventional barrel process.[4]

The concentration of salt in the brines varies widely. For the Sevillano fruit, it is customary to use a brine containing not over 5% salt at the time of brining. With Manzanillo and Barouni olives, it is customary to use a brine of 7 to 8% salt. The fermentation takes place, as a rule, in the presence of about 4 to 7% salt.

The fermentation, which generally requires 6 months or longer, depending on the amount of laboratory control, results in the

FIGURE 31. *Barrels of Olives in a Fermentation Yard* (Courtesy—Lindsay Ripe Olive Co.)

production of 0.7 to over 1.0% total acid (calculated as grams lactic acid per 100 ml of brine). After the fermentation, the olives are ready for packing in glass or for shipment to repacking centers.

SICILIAN-TYPE OLIVES

Sicilian-type olives are made from green to straw-colored and even light-pink-colored Sevillano fruit and are fermented without lye treatment to remove the bitter glucoside *oleuropein*. For this process, suitable olives, after grading and sorting, are immediately packed in 32-gal paraffin-lined barrels. The barrels of olives are then rolled into the fermentation yard, placed in an upright position and filled with water or a weak salt brine until the entire head of the barrel is covered. Salt is then added until the concentration in the brine has increased to 7 or 8% (calculated as grams NaCl per 100 ml of brine). The olives undergo a lactic acid fermentation similar to that observed in holding-solution (storage) olives except that conditions in the barrels suppress the excessive populations of yeasts generally found in storage brines.

Formerly, it was the practice to add spices when the olives were barreled. Now most of the Sicilian-type olives are fermented in unspiced brines and, after fermentation, look and taste almost identical with storage fruit.

Fermentation

The important, normal floral and chemical changes occurring in the brines of fermenting olives have been found to be similar to those reported for the fermentation of sauerkraut by Pederson[71] and for pickles by Etchells and Jones.[27,29] There are, however, important differences in the rates of fermentation. Olives, under the conditions which exist in the California industry, do not ferment nearly as rapidly as either cabbage or cucumbers.

Seasonal temperatures play a major role in the rate of fermentation of olives. Even during the earlier part of the season (October 1 to 15), the temperatures of the fermenting brines (storage, Sicilian or Spanish-type), in most cases, do not increase significantly above the temperatures of the water supplies with which the brines were made. Temperatures averaging 68°F are normal for brines during this period. Later, the temperatures decrease markedly and, during the winter months, average between 40° and 45°F.

The concentration of salt, availability of fermentable carbohydrates, and the presence or absence of desirable bacteria, as will be shown, also affect the rate of fermentation. It may also be assumed that there are differences in naturally occurring bacterial growth factors and antibiotic substances in olives as compared with either cabbage or cucumbers.

Fermentation in Holding Solutions

Although detailed studies have not been published, it is known from the work of the author and his students that olives held in salt brines undergo a typical lactic acid fermentation caused by the same types of bacteria as those found in the sauerkraut and cucumber fermentations.

The acidity developed is somewhat variable, but usually ranges between 0.4 and 0.6% (total acidity calculated as grams lactic acid per 100 ml of brine). The development of acidity is usually not very pronounced in holding brines, although the sugar content of the different varieties of olives is sufficient to allow it. The high salt concentration of some of the brines; an abundance of yeasts in the brines; and the low temperatures of the brines during the late fall and winter months all tend to prevent, suppress, or reduce the acidity.

The major varieties of olives differ appreciably in their fermentability in holding brines. The Ascolano and Sevillano varieties ferment very readily, whereas the Manzanillo and Mission varieties ferment slowly. The apparent varietal differences are correlated with the content of fermentable sugars. The Sevillano and Ascolano varieties contain the most, the Manzanillo and Mission varieties the least sugar. Their order in oleuropein content (which may eventually be shown to suppress the activity of the lactic acid bacteria) is exactly reversed. It has been stressed previously that the Sevillano and Ascolano varieties are quite sensitive to salt shrivel and must be stored in brines containing less salt than the other varieties. It is known that the rate of fermentation is decreased as the salt content of the brine is increased. It is obvious, therefore, that the Sevillano and Ascolano varieties should ferment more rapidly under commercial conditions.

Sicilian-Type Fermentation

The gross chemical and bacteriological changes that occur in the brines of Sicilian-type olives are similar in every respect to those found in holding-brine fermentations. Normally, all of the Sicilian-type olives are made from Sevillano fruit. The total acid production in the Sicilian-type fermentation is somewhat higher than in storage brines. The fermentation takes place in barrels and consequently is more anaerobic (Figure 32). The barrels are stored in the open where the surface brines are irradiated by the sun. These two factors combine to markedly reduce the large populations of acid-destroying yeasts normally present in holding brines.

Spanish-Type Fermentations

It is generally acknowledged that successful, large-scale production of Spanish-type green olives depends in large measure on a thorough knowledge of the changes which occur during preparation and fermentation of the fruit. Consequently, these changes are better known than those which occur in other fermented olives.

Although early attempts were made to produce Spanish-type, green fermented olives in California, there was no significant commercial production until comparatively recently. The first extensive

FIGURE 32. *Barrel of Fermenting Sicilian-Type Olives*

commercial-scale experiments made to study this fermentation were reported by Cruess.[13] Since about 1935, there has been a gradual increase in the interest in production of this olive. The increased production is the result of conscientious efforts of the whole industry to adapt the fermentation process to conditions which prevail in California. It is now recognized throughout the industry that the green-olive fermentation must be rigorously controlled. There are several reasons why control of the fermentation is mandatory if it is to be successful.

To insure the production of lactic acid which, together with salt, preserves the fermented olives, it is necessary to have a sufficient quantity of fermentable matter present in the fruit. Also, the brine and olives must have a concentration of salt that will not be harmful to any desirable lactic acid bacteria which may be present. Finally, the brined olives must be held at a temperature which will allow the desirable bacteria to dominate the fermentation.

Fresh olives of the varieties desirable for green pickling do, according to Nichols,[69] contain enough fermentable substances (glucose, fructose, sucrose, and mannitol). However, as Cruess[13] has shown, up to 65% of the total fermentables may be lost by lye treatment and washing to destroy the bitter glucoside. Therefore, failure of the olives to ferment properly may result from lack of fermentable sugars.

According to data accumulated by the author and his students, lye treatment and use of too concentrated salt brines can deplete the population of desirable lactic acid bacteria to such an extent that they cannot compete with undesirable bacteria and yeasts even if sufficient sugar is present. Therefore, to insure fermentation of green olives, it is necessary to add supplementary sugar and to insure the presence of desirable lactic acid bacteria. However, unless the temperature is controlled by artificial means during the winter months, fermentation may cease, for it is well known that the lactic acid bacteria have minimum temperature ranges for growth and acid production. A more detailed discussion of some aspects of the control of the fermentation is given by Vaughn, Douglas, and Gililland.[89] It is desirable to control the fermentation until the pH of the brine has decreased to a value of 3.8 or less. Under such conditions, the total acidity calculated as lactic acid will be in the range of 0.7 to 1.0% (calculated as grams lactic acid per 100 ml of brine). It is possible, by proper manipulation, to complete the fermentation in 3 weeks to 1 month under commercial conditions. To accomplish this, 2 to 5 lb of supplementary sugar must be added to the brines; desirable lactic acid bacteria must be present in the brines; the salt concentration must be at 5 to 6% (calculated as grams NaCl per 100 ml of brine); and a temperature of at least 75°F must be maintained. *These control measures must be carefully integrated.* Inoculation with a starter will not accelerate acid production unless there is sufficient fermentable sugar present in the brines. It is unreasonable to expect appreciable acid production in the presence of 10% salt or in brines at 45°F, regardless of other favorable conditions.

Crystallized corn sugar (glucose) is commonly used as the source of supplementary sugar, although cane or beet sugar (sucrose) may also be used.

To insure desirable populations of lactic acid bacteria, pure-culture starters of various strains of *Lactobacillus plantarum* may be used.[14] In the absence of suitable facilities for developing and maintaining pure-culture starters, it is common practice to use quantities of normal, actively fermenting brine. In many cases, for reasons as yet imperfectly understood, use of some kind of starter is necessary.

Incubation of the olives is accomplished by storing the barrels in special incubating rooms where the temperature is maintained at 75° to 85°F. For obvious economic reasons, only a portion of the crop is incubated.

The sequence of bacteria observed in the fermentation of olives is essentially the same as that described in Table 42 for the fermentation of cucumbers in low-salt brines. However, because the olives ferment more slowly, the changes in microbial populations are more clear-cut and it is easier to delimit the different stages of fermentation.

As already indicated, the salt concentration of the brines has a marked effect on the total numbers and types of bacteria and other microorganisms found in them. Thus, the fermentations of the Manzanillo variety, which have been customarily conducted in the presence of 7 to 8% salt, have a different bacterial population than the Sevillano olives, which are fermented in the presence of from 3 to about 5% salt. Whereas both homofermentative and heterofermentative species of *Lactobacillus* have always been recovered from the brines of Sevillano olives, only the homofermentative species *Lactobacillus plantarum* has been recovered from Manzanillo brines containing 7 to 8% salt. *Leuconostoc mesenteroides,* being salt tolerant, is found in both fermentations.

This difference in bacterial populations is a manifestation of the differences in salt tolerance of the lactobacilli. When Manzanillo olives are fermented in the presence of 3 to 5% salt, the heterofermentative as well as the homofermentative lactobacilli are recovered. When Sevillano olives are fermented in brines containing 7 to 8% salt, only homofrmentative *Lactobacillus plantarum* is found. When either variety is fermented in the presence of 10% salt, it is difficult to recover any lactobacilli at all and there is no significant acid production in the brines even if supplementary sugar is constantly supplied for as long as a year.

Determination of the salt tolerance of pure cultures of lactic acid bacteria is difficult. However, these more recent observations concerning the salt tolerance of lactic acid bacteria from green-olive brines, together with those previously reported by Vaughn, Douglas, and Gililland[89] amply support the conclusions of Orla-Jensen[70] that heterofermentative species of *Lactobacillus* are less resistant to salt than the species of *Leuconostoc* or homofermentative species of *Lactobacillus.*

Spoilage Problems

Bacteria, yeasts, and molds may cause deterioration of olives at any time between the harvest and the final packaging of the olive product. Bacteria generally are responsible for the most serious losses, although, at times, yeasts and molds may also be involved. Care in handling the freshly harvested olives and control of the course of biological, chemical, and physical events taking place during the primary stage of fermentation determine, in a large measure, the extent of deterioration of the fruit. However, losses may result from general lack of adequate sanitation or from neglect of the fruit during any subsequent step in the process of preparing the various olive products.

The commonest and, consequently, best-known types of spoilage caused by the activity of microorganisms are softening, gassy deterioration, and butyric fermentation. Of these, gassy spoilage occurs most frequently and under the widest variety of conditions.

Gassy, "Floater," or "Fish-eye" Spoilage

This deterioration (Figure 33) is characterized by the formation of blisters resulting from accumulation of gases which separate the skin from the flesh of the olives (not to be confused with blisters caused by too concentrated or too warm lye solutions) or by the formation of fissures or gas pockets which may extend to the pits of the fruits. Since this type of spoilage has been extensively studied by Cruess and Guthier,[19] Alvarez,[1] Tracy,[85] and Vaughn and his students,[45,89,92] it is well established that the coliform bacteria are chiefly responsible for gassy spoilage.

The coliform species most frequently encountered are *Aerobacter aerogenes* and *A. cloacae. Escherichia freundii* and *E. intermedium* are also found, but no true *E. coli* cultures have been reported. Cultures of *Bacillus* (*Aerobacillus*) *polymyxa* and *B. macerans* also cause gas-pocket formation in olives. Various species of the saccharolytic anaerobes of the genus *Clostridium* may also produce gas pockets in olives, but these butyric acid bacteria are more important for the malodorous fermentations they produce.

The bacteria mentioned all produce hydrogen gas as an end product of the decomposition of glucose, fructose, and mannitol. It is the contention of the author that only those bacteria which produce hydrogen gas are dangerous. Carbon dioxide is much more

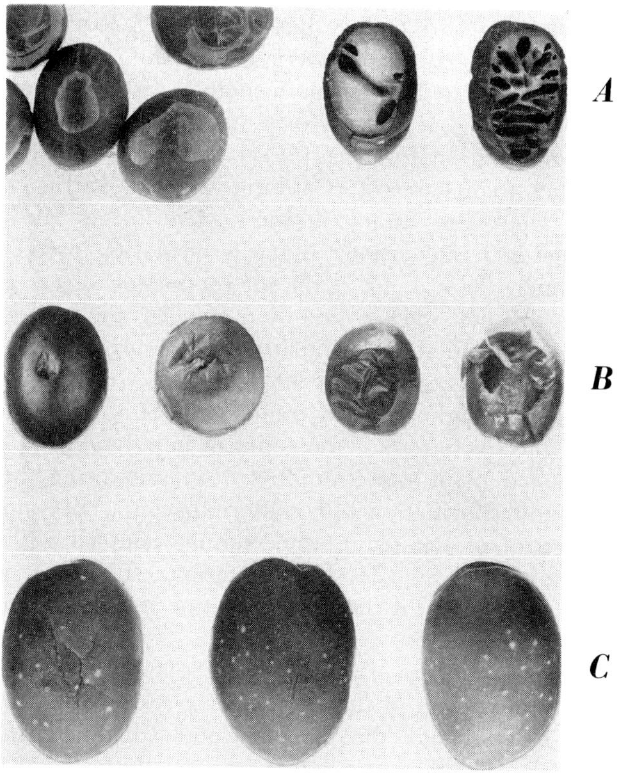

FIGURE 33. *Spoilage of Olives:* ***A*** *Manifestations of gas formation;* ***B*** *Softening, left to right, "nail-head," soft stem-end, mushiness and sloughing, respectively;* ***C*** *"Yeast spots"*

soluble than hydrogen. Yeasts and heterofermentative lactic acid bacteria (*Leuconostoc mesenteroides, Lactobacillus brevis,* and others) produce large quantities of carbon dioxide and abound in all olive fermentations; yet, pure-culture studies with them have failed to substantiate that either the yeasts or lactic acid bacteria cause gassy spoilage of olives.

Gas-pocket formation may occur in olives in storage brines, Sicilian-type fermentations, Spanish-type fermentations, or during the ripe pickling process, particularly in those olives pickled without preliminary storage in brine.

Conditions which favor the gaseous spoilage are to be found

in an environment containing not more than 5% salt and having pH values of 4.8 to 8.5, a temperature of 75° to 85°F, and no significant population of lactic acid bacteria to compete with the coliform bacteria for an abundance of fermentable carbohydrates. It is obvious, therefore, that control measures to effectively eliminate serious losses resulting from gas-pocket formation should be preventive rather than curative. Coliform bacteria, particularly the *Aerobacter* types, are found in most olive brines during the primary stage of fermentation. If the fermentation proceeds in a normal manner, there is no significant production of gas pockets. If, however, this first stage is unduly prolonged, the coliform bacteria dominate the microbial populations in the brines and spoilage results. The prevalence of the coliform bacteria in nature makes it impossible to eliminate them completely. However, much can be done to effectively reduce the coliform populations by control of contaminated plant water supplies, thus preventing the plant from continuous inoculation with coliform bacteria. In commercial practice, control of the plant water supply, coupled with maintenance of a general high level of sanitation, will reduce gaseous spoilage to a minimum, if the usual attention is given to the olive fermentations themselves.

Nevertheless, in cases of acute spoilage where there is potential danger of complete loss of the fruit, speedy curative measures are dictated. These include the use of strongly acidified brines or pasteurization. Heating softens fresh or partially fermented olives and, therefore, speedy rebrining with acidified brine is recommended to check spoilage in such fruit. Pasteurization may be used particularly well in the ripe pickling of olives. However, commercial pasteurization will not kill the spores of the *Bacillus polymyxa-macerans* group under all conditions.

There is an extremely salt-tolerant form of *Aerobacter aerogenes,* which was recently described by Foda and Vaughn.[45] This salt-tolerant form will grow and produce gas in the presence of 14% salt (calculated as grams NaCl per 100 ml of brine). Obviously, salt alone will have no effect on the control of this particular type of coliform bacterium, which often occurs in olive brines.

Malodorous Fermentations

The butyric acid bacteria have been associated with the deterioration of olives since Hayne and Colby,[50] in 1895, first recorded their spoilage by the "butyric ferment." This abnormal fermentation is characterized by its butyric acid or rancid odor. In the early stages of the fermentation, the odor is distinctly that of butyric acid or rancid butter; but, as the spoilage progresses, the odor becomes increasingly obnoxious and finally results in a very malodorous, fecal stench.

Pure cultures of anaerobic, spore-forming butyric acid bacteria were first isolated from a large number of samples of spoiled olives by Gililland and Vaughn.[47] All of the cultures studied were sugar-decomposing types not capable of digesting protein. Most of the cultures were closely related to or identical with *Clostridium butyricum*. All of the isolates caused malodorous fermentation of olives if the chemical and physical conditions, such as salt concentration, pH, and sugar content in the brines, were favorable for their growth.

Butyric acid spoilage is less widespread than the abnormal gas-pocket fermentation commonly caused by the coliform bacteria. Nevertheless, losses may be extensive, because olives affected by butyric fermentation cannot be salvaged. Control, therefore, must be designed to prevent the activity of the butyric acid bacteria rather than to check the fermentation once it is started. Scrupulous sanitation and rigid control of the primary phase of the fermentation are the only feasible methods for preventing the butyric acid fermentation. If the fermentation has been carefully controlled until acid production has decreased the pH value of the brines to a value of 3.8 to 4.0, there is little danger that the butyric acid bacteria will spoil the olives.

"Zapatera" spoilage is another malodorous fermentation which occurs in holding brines, Sicilian-type fermentations, and Spanish-type fermentations of olives in California. This abnormality is characterized by the development of a very penetrating, unpleasant, "sagey" odor in olives undergoing fermentation. Under California conditions, zapatera spoilage, unlike the butyric fermentation, occurs when the desirable lactic acid fermentation is allowed to cease before the pH of the brine has decreased below 4.5. At the onset of deterioration, the pH of the affected brines increases and the titratable acidity decreases as spoilage progresses. Smyth,[83] who made a study of the bacteriology of the Spanish green olive, concluded that spoilage was due "to one or more of a group of spore-forming, proteolytic, facultative rods normally present in the soils of Andalusia."

The author has isolated large numbers of viable spores of *Clostridium butyricum* from the brines of certified samples of imported Spanish green olives affected with zapatera. Attempts to isolate other bacteria which produce evil-smelling compounds in olives have been unsuccessful. It is known that lactic and acetic acids disappear during development of the spoilage and are replaced by propionic, butyric, valeric, caproic and caprylic acids.[19a] Although causative organisms have not been proven, the author and his students have isolated a number of *Cl. butyricum* and related cultures that ferment lactic acid if acetic acid is also present.

Control of zapatera spoilage, to be successful, must provide the same measures as have been used to prevent the butyric acid fermentation.

Softening

The softening of olives manifests itself in one of three characteristic changes in the appearance of the fruit. These changes are known in the industry by the very descriptive terms "soft stem end," "nail head" and "sloughing," respectively. Examples of these three types of softening are shown in Figure 33. These types of softening have been observed in all kinds of pickled olives. The changes are so characteristic and the common names are so indicative of the types of change that further description is superfluous.

Histological studies made by the author and his students have shown that softening of olive tissue results from destruction of the pectic substances which hold the individual cells together and give firmness and shape to the fruit.[69a] It was also reported that strains of *Aerobacillus* (the *Bacillus polymyxa-macerans* group of bacteria) capable of elaborating pectolytic enzymes could cause softening of olives. Cruess[16] used a commercial pectolytic-enzyme preparation to demonstrate softening of olives after having observed softening on laboratory samples of olives caused by an unidentified species of *Pencillium*. It is known from the literature that other bacteria and fungi also elaborate pectolytic enzymes. However, the form and texture of fresh olives may also be altered by use of too concentrated sodium hydroxide solutions, by frosting, or by heating. Thus, it is not always easy to determine whether the softening is of microbial origin or whether it results from chemical or physical mistreatment of the fruit.

Detection of softening resulting from microbial activity has been somewhat simplified recently. Baier and Manchester[3] have developed a pectate-gel medium which may be used for the direct isolation of microorganisms which elaborate pectolytic enzymes. Bell, Etchells, and Jones[5] have also devised a qualitative test for the enzyme polygalacturonase in cucumber brines. The author has used the Baier-Manchester gel medium for the isolation of aerobic pectolytic microorganisms (strains of *Bacillus megatherium, Bacillus macerans, Penicilli,* and *Aspergilli*) from olive brines. It is hoped that further applications of the polygalacturonase-enzyme test will be made so that softening may be forecast in time to prevent severe losses in olives.

Although practical control at present is limited by lack of knowledge, softening is curtailed or eliminated by a normal lactic acid fermentation. Other preventive measures include careful handling of the freshly harvested olives to prevent "sweating." Storage of the fresh fruit for more than 3 days before brining or lye treatment greatly increases the amount of possible softening. It is also mandatory to maintain a high level of plant sanitation.

Yeast Spots

One of the very common defects associated with fermented olives in California is the abnormality known throughout the industry as "yeast spots." This defect is characterized by the development of raised white spots or pimples which form between the epidermis and the underlying flesh of the olive (Figure 33). Although commonly called "yeast spots," most, if not all of the pimples contain lactobacilli.[89a] These bacteria have been isolated and studied, and they have been identified as *Lactobacillus plantarum* and *Lactobacillus brevis* or other heterofermentative species closely related to *L. brevis.*

This defect is also found in green olives certified to have been imported from Spain and has been observed in cucumber pickles and pickled green tomatoes. Evidently, any factor which influences the entrance of the lactobacilli through the pores (stomata) or other breaks in the skin of the fruit may contribute to the development of the colonies of bacteria. Since such olives are considered unsightly, some loss of value of the pickled fruit is experienced although the olives are perfectly normal and healthful in other respects. No control measures are known.

Refermentation

Unsightly sediments and gas formation frequently occur in glass-packed fermented green olives.[68,89] The principal cause for this refermentation is a reservoir of fermentable sugars left in the olives or unwitting addition of fermentable substances contained in various spices and flavoring constituents used when the fruit is packed in glass. If such substances are present, refermentation, with gas formation and development of turbidity and a heavy sediment in the brine, appears quickly unless control measures are used.

Most refermentation is caused by lactic acid bacteria. However, if the containers are not completely filled with brine or vacuum-sealed to exclude air, yeasts may form much of the sediment and both yeasts and molds form a pellicle in the head space between the brine and the jar cap.

To control refermentation effectively, it must be carefully determined that most or all of the fermentable matter in the olives, brine, and other constituents has been decomposed. It is also desirable to wash the olives to remove lactic acid bacteria and yeasts adhering to the surface of the fruit. The brines should be made with high-quality salt and edible lactic acid. It is also necessary to exclude as much air as possible either by complete filling of the jar with brine or by use of vacuum-type jar closures. Pasteurization may also be used to destroy yeasts and bacteria which might otherwise cause refermentation. The use of any chemical preservative per se is not practiced in the California industry.

Other Abnormalities

Heavy yeast and mold growths on the surface of brines covering olives, whether in barrels or storage tanks, may also impart off-flavors and aromas to the fruit. Aside from the off-flavors and aromas developed through such neglect, it is known that the yeasts and molds rapidly destroy the lactic acid contained in the brines. Loss of the lactic acid by oxidation in this manner makes the olives susceptible to other spoilage. Control of yeast and mold growth is best accomplished by routine rebrining of the barrels with fresh brine during active fermentation and keeping them tightly closed during storage to eliminate air. Storage brines are easily kept free of film-forming fungi if the tanks are exposed to sunlight, but present a problem if the tanks are roofed over. With recently developed pumping techniques for handling storage olives, however, closed tanks could be used, thus eliminating the need for ultraviolet lamps, layers of oil, or other control measures.

Purely chemical deterioration of olives is caused by the same contaminating metals (copper and iron) as have already been discussed with cucumber pickles. The problems are nearly identical and the same preventive measures are required for both products.

BIBLIOGRAPHY

1. Alvarez, R. S., *J. Bact.*, **12**, 359 (1926).
2. Atwater, W. O., and A. P. Bryant, "The Chemical Composition of American Food Materials," *U. S. Off. Expt. Sta. Bull., 28* (revised ed.) (1906).
3. Baier, W. E., and T. C. Manchester, *Food Industries*, **15**, 94 (1943).
4. Ball, R. N., Ed. Van Dellen, J. B. Jaquith, R. H. Vaughn, J. Tabachnick, and G. T. Wedding, *Food Technology*, **4**, 30 (1950).
4a. Bell, T. A., *Botan. Gaz.*, **113**, 216 (1951).
5. Bell, T. A., J. L. Etchells, and I. D. Jones, *Food Technology*, **4**, 157 (1950).
5a. Bell, T. A., J. L. Etchells, and I. D. Jones, *Arch. Biochem., Biophys.*, **31**, 431 (1951).
6. Bitting, A. W., *Appetizing or The Art of Canning; Its History and Development*, San Francisco, The Trade Pressroom, 1937.
7. Blum, H. B., and F .W. Fabian, *Fruit Prod. J. and Amer. Vinegar Ind.*, **22**, 326 (1943).
8. Bourquelot, E., and J. Ventilesco, *Compt. rend.*, **147**, 533 (1908).
9. Brunkow, O. R., W. H. Peterson, and E. B. Fred, *J. Am. Chem. Soc.*, **43**, 2244 (1921).
10. Bull, W. E., *Econ. Geog.*, **12**, 136 (1936).
10a. California Agricultural Code, Chapter 272, 870.5, Stats. 1953.
11. Campbell, C. H., *Campbell's Book—Canning, Preserving and Pickling*, (revised ed.), Chicago, Vance Publishing, 1937.
11a. Costilow, R. N., and F. W. Fabian, *Appl. Microbiol.*, **1**, 314 (1953).

12. Cruess, W. V., "Olive Pickling in Mediterranean Countries," *Calif. Agr. Sta. Cir., 278* (1924).
13. Cruess, W. V., "Pickling Green Olives," *Calif. Agr. Exp. Sta. Bull., 498* (1931).
14. Cruess, W. V., *Fruit Prod. J.,* **17**, No. 1, 1 (1937).
15. Cruess, W. V., *Ind. Eng. Chem.,* **33**, 300 (1941).
16. Cruess, W. V., *The Role of Enzymes in Olive Processing,* 26th Ann. Tech. Rept., Calif. Olive Assoc., San Francisco (Mimeo), 1947.
17. Cruess, W. V., *Commercial Fruit and Vegetable Products,* 3rd. ed., New York, McGraw-Hill, 1948.
18. Cruess, W. V., and C. L. Alsberg, *J. Am. Chem. Soc.,* **56**, 2115 (1934).
19. Cruess, W. V., and E. H. Guthier, "Bacterial Decomposition of Olives during Pickling," *Calif. Agr. Exp. Sta. Bull., 368* (1923).
19a. Delmouzos, J. G., F. H. Stadtman, and R. H. Vaughn, *Agr. and Food Chem.,* **1**, 333 (1953).
20. Dunn, H. D., *California–Her Agricultural Resources,* Trans. Calif. State Agr. Soc., **1866-1867**, 507-542.
21. Erickson, F. J., and F. W. Fabian, *Food Research,* **7**, 68 (1942).
21a. Erlich, F., *Biochem. Z.,* **250**, 525 (1932).
22. Etchells, J. L., *Food Research,* **6**, 95 (1941).
23. Etchells, J. L., and T. A. Bell, *Farlowia,* **4**, 87 (1950).
23a. Etchells, J. L., and T. A. Bell, *Food Technol.,* **4**, 77 (1950).
23b. Etchells, J. L., R. N. Costilow, and T. A. Bell, *Farlowia,* **4**, 249 (1952).
24. Etchells, J. L., F. W. Fabian, and I. D. Jones, "The Aerobacter Fermentation of Cucumbers during Salting," *Mich. State Coll., Agr. Exp. Sta. Tech. Bull., 200,* 56 p. (1945).
25. Etchells, J. L., and I. D. Jones, *Fruit Prod. J.,* **20**, 370 (1941).
26. Etchells, J. L., and I. D. Jones, *Fruit Prod. J.,* **21**, No. 11, 330 (1942).
27. Etchells, J. L., and I. D. Jones, *Food Industries,* **15**, No. 2, 54 (1943).
28. Etchells, J. L., and I. D. Jones, *Food Research,* **8**, 33 (1943).
29. Etchells, J. L., and I. D. Jones, *Fruit Prod. J.,* **22**, 242 (1943).
30. Etchells, J. L., and I. D. Jones, *J. Bact.,* **52**, 593 (1946).
31. Etchells, J. L., and I. D. Jones, *Am. J. Pub. Health,* **36**, 1112 (1946).
32. Fabian, F. W., *The Chemistry and Technology of Food and Food Products,* Vol. **2**, p. 362-393, New York, Interscience, 1944.

33. Fabian, F. W., *Fruit Prod. J.,* **26**, No. 8, 234 (1949).
34. Fabian, F. W., *Food Packer,* **31**, No. 9, 23 (1950).
35. Fabian, F. W., and H. J. Bertraw, *Fruit Prod. J.,* **23**, 196 (1944).
36. Fabian, F. W., and H. B. Blum, *Fruit Prod. J.,* **22**, 228 (1943).
37. Fabian, F. W., and C. S. Bryan, "Experimental Work on Cucumber Fermentation." I. "The Influence of Sodium Chloride on the Biochemical and Bacterial Activities in Cucumber Fermentation," *Mich. Agr. Exp. Sta. Tech. Bull. 126,* 1 (1932).
38. Fabian, F. W., C. S. Bryan, and J. L. Etchells, "Experimental Work on Cucumber Fermentation." V. "Studies on Cucumber Pickle Blackening," *Mich. Agr. Exp. Sta. Tech. Bull. 126,* 49 (1932).
39. Fabian, F. W., and R. C. Fulde, *Food Packer,* **31**, No. 11, 28 (1950).
40. Fabian, F. W., and E. A. Johnson, "Experimental Work on Cucumber Fermentation." IX. "A Bacteriological Study of the Cause of Soft Pickles"; X. "Zymological Studies of the Cause of Soft Cucumbers"; XI. "Histological Changes Produced in Pickles by Bacteria, Acids and Heat"; XII. "Chemical Changes Produced in the Pectin Substances of Pickles by Bacterial Enzymes," *Mich. Agr. Exp. Sta. Tech. Bull. 157,* 1 (1938).
41. Fabian, F. W., C. K. Krehl, and N. W. Little, *Food Research,* **4**, 269 (1939).
42. Fabian, F. W., and A. L. Nienhaus, "Experimental Work on Cucumber Fermentation." VII. *"Bacillus Nigrificans* n. sp. as a Cause of Pickle Blackening," *Mich. Agr. Sta. Tech. Bull. 140,* 23 (1934).
43. Fabian, F. W., and M. C. Van Wormser, *Food Research,* **8**, 95 (1943).
44. Faville, L. W., and F. W. Fabian, "The Influence of Bacteriophage, Antibiotics, and Eh on the Lactic Fermentation of Cucumbers," *Mich. Agr. Exp. Sta. Tech. Bull., 217,* 1 (1949).
45. Foda, I. O., and R. H. Vaughn, *Food Technology,* **4**, 182 (1950).
46. Fred, E. B., and W. H. Peterson, *J. Bact.,* **7**, 257 (1923).
47. Gililland, J. R., and R. H. Vaughn, *J. Bact.,* **46**, 315 (1943).
48. Gracey, W. T., "Olive Growing in Spain," *Dept. Commerce, Bur. Foreign Domestic Commerce, Special Consular Rept.,* **79**, 1 (1918).
49. Hamann, E. H., *Food Packer,* **28**, No. 5, 38 (1947).
50. Hayne, A. P., and G. E. Colby, *Calif. Agr. Exp. Sta. Reports,* **1895**, 1-37 (Appendix to report for 1894-95).

51. Jones, I. D., *Ind. Eng. Chem.*, **32**, 858 (1940).
52. Jones, I. D., and J. L. Etchells, *Food Industries*, **15**, No. 1, 62 (1943).
53. Jones, I. D., and J. L. Etchells, *The Canner*, **110**, No. 1, 34 (1950).
54. Jones, I. D., J. L. Etchells, O. Veerhoff, and M. K. Veldhuis, *Fruit Prod. J.*, **20**, 202 (1941).
55. Jones, I. D., J. L. Etchells, M. K. Veldhuis, and O. Veerhoff, *Fruit Prod. J.*, **20**, No. 10, 304 (1941).
56. Jones, I. D., M. K. Veldhuis, J. L. Etchells, and O. Veerhoff, *Food Research*, **5**, 533 (1940).
57. Joslyn, M. A., *Fruit Prod. J.*, **8**, No. 8, 19; No. 9, 16 (1928).
58. Joslyn, M. A., and W. V. Cruess, "Home and Farm Preparation of Pickles," *Calif. Agr. Exp. Sta. Cir.*, *37*, 1 (1943).
58a. Kertesz, Z. I., *The Pectic Substances*, New York Interscience, 1951.
59. Kossowicz, A., *Z. L. Landwirtschaft. Versuchswesen Oesterreich*, **11**, 191 (1908).
59a. Kraght, A. J., and M. P. Starr, *Arch. Biochem. Biophys.*, **42**, 2 (1953).
60. Le Fevre, E., *The Canner*, **48**, 205 (1919).
61. Le Fevre, E., *The Canner*, **52**, 207 (1921).
62. Le Fevre, E., "The Commercial Production of Sauerkraut," *U. S. Dept. Agr. Cir.*, *35* (1928).
63. Lelong, B. M., *Calif. State Bd. Hort., Ann. Rept.*, **1890**, 185.
64. Lesley, B. E., and W. V. Cruess, *Fruit Prod. J.*, **7**, No. 10, 12 (1928).
64a. Luh, B. S., and H. J. Phaff, *Arch. Biochem. Biophys.*, **33**, 212 (1951).
65. Magruder, R., "Improvement in the Leafy Cruciferous Vegetables," *Yearbook of Agriculture*, U. S. Dept. Agr., Washington, U. S. Govt. Printing Office, 1937.
66. Marten, E. A., W. H. Peterson, E. B. Fred, and W. E. Vaughn, *J. Agr. Research*, **39**, 285 (1929).
67. Mrak, E. M., and L. Bonar, *Zentr. Bakt. Parasitenk., Abt. II*, **100**, 289 (1939).
68. Mullen, G. W., *Glass Packer*, **2**, No. 12, 497 (1929).
69. Nichols, P. F., *J. Agr. Research*, **41**, 89 (1930).
69a. Nortje, B. K., and R. H. Vaughn, *Food Research*, **18**, 57 (1953).
70. Orla-Jensen, S., *Mem. Acad. Roy. Sci. Lettres Danemark, Sec. d, Sci. 8 ser.*, **5**, No. 2, 81 (1919).

70a. Pavgi, M. S., H. P. Pithawala, G. R. Savur, and A. Screenivasan, *Proc. Indian Acad. Sci.,* **34**, (1) Sec. B, 33 (1951).

71. Pederson, C. S., "Floral Changes in the Fermentation of Sauerkraut," *N. Y. Agr. Exp. Sta. Tech. Bull., 168* (1930).

72. Pederson, C. S., "Sauerkraut," *N. Y. Agr. Exp. Sta. Bull., 595,* (1931).

72a. Pederson, C. S., and M. N. Albury, *New York Agr. Exp. Sta. Bull. 744* (1950).

72b. Pederson, C. S., and M. N. Albury, *Food Research,* **18**, 290 (1953).

73. Pederson, C. S., and C. D. Kelly, *Food Research,* **1**, 277 (1936).

74. Pederson, C. S., and C. D. Kelly, *Food Research,* **3**, 583 (1938).

74a. Pederson, C. S., and L. Ward, *New York Agr. Exp. Sta. Tech. Bull. 288* (1949).

75. Peterson, W. H., E. B. Fred, and J. A. Viljoen, *The Canner,* **61**, No. 4, 19 (1925).

76. Peterson, W. H., H. B. Parmele, and E. B. Fred, *Soil Science,* **24**, 299 (1927).

76a. Phaff, H. J., "The Biochemistry of the Exocellular Pectinase of *Penicillium chrysogenum,*" Ph.D. Thesis. Library, University of California, Berkeley, Calif., 1943.

77. Phaff, H. J., and M. A. Joslyn, *Wallerstein Labs. Commun.,* **10**, 133 (1947).

78. Phillips, G. F., and J. O. Mundt, *Food Technology,* **4**, 291 (1950).

79. Prescott, S. C., and C. G. Dunn, *Industrial Microbiology,* 2nd ed., p. 429-461, New York, McGraw-Hill, 1949.

80. Prescott, S. C., and B. E. Proctor, *Food Technology,* 1st. ed., p. 515-519, New York, McGraw-Hill, 1937.

81. Rahn, O., *The Canner and Dried Fruit Packer,* **37**, No. 20, 44; No. 21, 43 (1913).

82. Seaton, H. L., R. Hutson, and J. H. Muncie, "The Production of Cucumbers for Pickling Purposes," *Mich. Agr. Exp. Sta. Special Bull., 273,* 1 (1936).

83. Smyth, H. F., *J. Bact.* **13**, 56 (1927).

84. Stadtman, E. R., *Advances in Food Research,* **1**, 325 (1948).

85. Tracy, R. L., *J. Bact.,* **28**, 249 (1934).

86. U. S. Dept. Agr., Services and Regulatory Announcements, Food and Drug No. 2, 5th revision, November 1936.

87. Vaughn, R. H., "Olives and Pickles," Quartermaster Food and Container Institute for the Armed Forces **4**, Part 2, 1st ed. (1944).

88. Vaughn, R. H., *Food Packer,* **28**, No. 5, 36 (1947).

89. Vaughn, R. H., H. C. Douglas, and J. R. Gililland, "Production of Spanish-Type Green Olives," *Calif. Agr. Exp. Sta. Bull., 678* (1943).

89a. Vaughn, R. H., W. D. Won, F. B. Spencer, D. Pappagianis, I. O. Foda, and P. H. Krumperman, "*Lactobacillus plantarum*, the Cause of 'Yeasts Spots' on Olives," *Applied Microbiol.*, **1**, 82 (1953).

90. Veldhuis, M. K., and J. L. Etchells, *Food Research*, **4**, 621 (1939).

90a. Waksman, S. A., and M. C. Allen, *J. Am. Chem. Soc.*, **55**, 3408 (1933).

91. Wenzel, F. W., and F. W. Fabian, "Experimental Work on Cucumber Fermentation." XIII. "Influence of Garlic on the Softening of Genuine Kosher Dill Pickles," *Mich. Agr. Exp. Sta. Tech. Bull. 199*, 1 (1945).

92. West, N. S., J. R. Gililland, and R. H. Vaughn, *J. Bact.*, **41**, 341 (1941).

93. Wetmore, C. A., *The Fig and Olive J.*, **4**, No. 3, 14; No. 4, 15, (1919).

94. Wood, R. K. S., *Nature*, **167**, 771 (1951).

CHAPTER 12

SELECTION AND MAINTENANCE OF CULTURES

Elizabeth F. McCoy

It has been said that the working capital of a bacteriologist is his set of stock cultures. Certainly, they are valuable and he must handle them so as to preserve that value and to profit by it in their fermentation use. In fact, the culture is the first of the limiting factors in a successful commercial fermentation.

SELECTION OF CULTURES

The original selection of suitable microbial cultures for commercial use is of considerable importance. In every group of fermentation organisms, there is a variety of strains producing the desired product, and it is important to select those (a) capable of highest yield under the commercial conditions to be used and (b) stable enough to give that yield consistently. Often, stability is more important, because in the plant, there is greater need for a vigorous, dependable starter culture than for one giving the highest, but erratic yields in laboratory tests. It may even be necessary to settle for the use of a good, but not high-yielding strain for this reason.

It is difficult to generalize on the original selection of the cultures. There are some, like the lactobacilli, which can be freshly isolated relatively easily and such fresh isolates from nature are likely to be superior to old laboratory cultures. For example, the following procedure for isolation of *Lactobacillus delbruckii* (or the recently recognized spore-forming lactic organisms, such as *Bacillus dextrolacticus*) is likely to yield a high acid-producing strain.

It is well to start with several potential source materials, such as grains, malt sprouts, and soil, because, while *L. delbruckii* is widespread, one cannot assume it present in every sample. Tubes of 5% glucose-yeast water or malt broth (deep broth to favor the microaerophilic lactobacilli should be inoculated). Incubation should be carried out at 45° to 55°C for 24 to 36 hours, at which time high acidity should have developed with resultant killing of the less acid-tolerant lactobacilli. Transfer should be made one or more times to 10% glucose broth containing excess calcium carbonate added as a sterile slurry. Growth under these conditions, at high temperature in high-sugar medium, results in further selection in favor of *L. delbruckii*. One may even use 12 or 15% glucose broth at this stage of culture provided that rapid transfers, e.g., at 12 to 24 hours, are made so as not to lose the vigorous *L. delbruckii* enrichment culture. When *L. delbruckii* is in the desired state of dominance as shown by microscopic evidence, the culture may be plated and a pure culture picked. Incubation of plates under anaerobic conditions is desirable for the lactobacillus in question.

In our experience, this procedure is very suitable for the isolation of *Lactobacillus delbruckii* from the various lactobacilli probably present in the original source inoculum; if any interference is encountered, it is generally due to the spore-forming lactic organisms, *Bacillus dextrolacticus,* and related forms which will develop under the same enrichment conditions. Such a lactic culture, if isolated, may be useful, since it is a strong lactic fermenter. But, if one wishes to avoid the spore former, one may detect it (and not use the enrichment for plating) by microscopic examination. The spore-forming organism in vegetative state is a slender Gram-negative rod in contrast to the Gram-positive, but variable rods which are the lactobacilli. In addition, the spore formers are strongly catalase positive and a catalase test with hydrogen peroxide on the enrichment culture will be positive, whereas it should be negative for a lactobacillius enrichment.

In other cases, the finding of a desirable strain for a commercial fermentation is much more difficult, but can still be done with certainty provided the microbiologist sets up proper selective enrichments and is persistent until he isolates a desirable strain for his purpose. Let us consider, for example, the problem presented by the butyl-butyric group. They are members of the genus *Clostridium,* which is a large and heterogeneous group containing pathogens and putrefactive members as well as the desired butyl-butyric fermentative types. Enrichment of the butyl-butyric anaerobes is relatively simple, if the primary cultures are started in starch or sugar media to select in favor of the saccharolytic anaerobes. It is important also to eliminate non-spore-forming contaminants at the outset, because lactobacilli, if present, would seriously interfere with the desired enrichment of butyl anaerobes. They may be eliminated by heating the inoculum for the original culture, e.g., by a treatment for 1 to 5 minutes at 100°C. Even at best, it is necessary to start many enrichments to get a few desirable ones. If one wishes to isolate the butyl types, one must expect them to be present in only 25 to 30% of the original source materials; butyrics, or at least low-producing butyl types, are contained in many more natural materials and it is thus necessary to distinguish the enrichments in which the butyl types appear in sufficient numbers to be successfully isolated. This may be done, to some extent, by noting those tubes in which the odors of butyl alcohol and acetone are pronounced and plating from such rather than from tubes with cheesy, butyric odors. One might think that to get the butyl organisms in dominant numbers, subculture in series would be attempted, but this is not done. In fact, the butyl types do not compete well in serial transfer and may even be lost by such a procedure before plating. The better plan is to allow sporulation of the primary or secondary culture for 5 to 7 days and then to plate from spores for isolation. Even so, it is necessary to persist until the desired type of butyl anaerobe is obtained, e.g., types for corn versus molasses (based on good utilization of starch versus high sugar); butyl-acetone versus butyl-isopropyl alcohol; and especially the highest ratio of butyl alcohol to other solvents. The chances of getting an outstanding culture of the desired type are certainly small, but it is possible to obtain them within reasonable time. It appears that the butyl patent situation reflects the culture situation. When an outstanding culture is found, it is likely to be characterized as a new species and its use in a fermentation process made the basis of a new butyl patent. Such a situation is justifiable only in view of the rarity of outstanding butyl cultures and by the fact that diversity of species does exist in the group (in contrast to the situation in the *Lactobacillus* group in which the single species, *L. delbruckii,* in the broad sense, covers the type of industrial interest).

Lastly, there are industrial starter cultures, which represent purely chance findings. The new user would be well advised to obtain his starter culture from someone who has it, rather than to attempt isolation. The penicillin cultures are good examples. From the first *Penicillium notatum* of Fleming to the currently used strains of *Penicillium chrysogenum,* there has been a long program of isolation and selection of better strains and, interestingly, in the latter period of development, the greatest progress was made by selecting substrains within the parent cultures. This was done by a combination of plating and selection of colonies, which by chance were better than their parent (Raper, Alexander, and Coghill,[22] and Raper and Fennell[23]), by induced variation (mutation) by X-ray (Demerec[5]), and by ultraviolet radiations (Backus, Stauffer, and Johnson[2]). The development of the present strains represents an immense amount of research effort and time by government, university, and industrial research workers before efficient starter cultures were ready for plant use. In such cases, it is wise to make arrangements for use of the best available culture rather than to attempt new isolations.

MAINTENANCE OF CULTURES

There is no single best method of maintenance of stock cultures. The two objectives are maintenance of viability and preservation of the original desirable characteristics of the cultures. Obviously, the first must be achieved or all is lost, but it is not always realized that the second objective does not necessarily accompany the first. Degeneration of the old stock cultures is very common and is well recognized by specialists.

There is one principle that can be highly recommended, i.e., the choice of culture medium and/or conditions of storage that will allow the least number of culture transfers, so as to avoid slow degeneration and variation (mutational) loss of the desirable characters of the cultures. Everyone who carries stock cultures is (or should be) aware of this. As a result, there is very extensive literature on stock-culture methods. Much of that literature does not apply to industrial stocks, because it deals with pathogens and the effort to maintain virulence as well as viability. But some of the principles and claims can well be considered here.

Paraffin-Oil Method

The use of paraffin-oil seal, although an old technique, has recently been favorably reported on (Morton and Pulaski,[18] Hac,[12] Simmons,[25] and Gordon and Smith[10]). Actually, this is the method of Ungerman (Michael[17]) and consists merely of culture of the organisms on a suitable medium until they are well grown and then of layering with sterile mineral oil to a depth of perhaps 1 cm. The oil seal checks water evaporation, which is desirable, since water loss, with consequent increase in osmotic pressure and salt effects, is in part the cause of death in old cultures, as they approach air dryness. At the same time, the oil seal allows slow diffusion of oxygen into the culture, as is evident from slow growth of obligate aerobes under the seal. It also allows transfer by loop of culture material to a new culture medium without destroying the parent stock. From our own experience, such an open stock is very convenient and time saving.

In the report of Simmons,[25] the paraffin-oil method was recommended for a variety of pathogens, such as staphylococci, streptococci, and diphtheria bacteria, which survived 15 to 30 months in storage at room temperature in the dark. Of particular interest is his report that *Neisseria meningitidis* survived 12 months even at 37°C in the paraffined sealed stock; this result is apparently not a chance observation, since Hac[12] also reported paraffin-oil storage favorable for *Neisseria gonorrheae.* He even found it superior to lyophilizing *N. gonorrheae.* Another notoriously difficult culture to keep in stock is *Phytomonas sepedonica,* which Sherf[24] succeeded in keeping 18 months without transfer under oil seal. Perhaps because of such recent recommendations of the oil method, Gordon and Smith[10] report that tests are underway at the American Type Culture Collection with species of some forty-two genera of bacteria. In a preliminary report, they stated that tests were being made at 6-month intervals and that the preliminary indications were of success with: *Achromobacter, Actinomyces, Agrobacterium, Bacterium, Brucella, Caseococcus, Cellulomonas, Chromobacterium. Coccobacillus, Corynebacterium, Eberthella, Erysipelothrix, Escherichia, Flavobacterium, Gaffkya, Klebsiella, Kurthia, Listerella, Microbacterium, Micrococcus, Mycobacterium, Mycoplana, Neisseria, Pasteurella, Rhizobium, Sarcina, Serratia, Shigella, Spirillum,* and *Xanthomonas.* They added that "The method was unsatisfactory for some cultures of the following genera: *Aerobacter, Alcaligenes, Cytophaga, Erwinia, Leuconostoc, Pseudomonas, Staphylococcus,* and *Vibro.*" The reason why such organisms would not survive is not clear. It would seem that the paraffin-oil storage method is generally useful, but that it should be tested out for the particular group of stocks concerned. Little or no mention is made of the culture medium to be used, but it might have a bearing on survival. If one assumes slow growth during storage, the medium should be so chosen as not to be exhausted of nutrient and not to accumulate excessive acidic products; in general, a well-buffered medium, such as a rich protein digest with minimal sugar content, would be indicated.

Soil-Stock Method

A second, rather simple method for preservation of cultures may be considered for those organisms for which it is advantageous, i.e., dry spore stock on sterile soil. It is difficult to trace the origin of this method for stock cultures. It was reported by Bredemann[3] that passage of *Clostridium pasteurianum* on sterile soil would stimulate its nitrogen-fixing ability. Thaysen[28] applied the principle to stock cultures of the Weizmann butyl organism, but used acid-washed sand instead of soil, apparently to standardize the substrate. However, the use of neutral garden soil has become general in the handling of butyl anaerobes industrially. A small proportion of sand, such as 5 to 10%, is often added to prevent caking of the soil in the tube and a small amount of calcium carbonate, such as 1 or 2%, may be added with the culture suspension to neutralize acids. A word of caution is offered on sterilization of the soil. It cannot be done in a dry oven without charring the organic matter. Autoclaving is, therefore, utilized, but dry soil is very difficult to sterilize. It wets over the surface; air is trapped in the spaces; and then sterilization must proceed by conduction of heat through a material of very poor conductivity. It is not uncommon to require 8 to 15 hours at 15-lb steam pressure to sterilize a basket packed with tubes of soil or a large flask of soil. The sterilization can be speeded up and made much more certain by moistening (but not water logging) the soil with a few milliliters of tap water, added after the soil is placed in tubes. Steam from this internal water drives out the air from the soil and materially hastens sterilization. However, it is always wise to sterilize at least 2 or 3 hours, at 15-lb steam pressure, with the tubes loosely held in the baskets and to test random tubes for sterility before a batch of soil is accepted for stock-culture use.

In making stocks on soil, one has to take into account whether the cultures are to grow in the soil or merely to be carried on it in dormant state. The soil for cultures of the butyl-butyric anaerobes serves only as carrier of spores. Thus a good spore stock, ripened in some suitable fermentation medium, must first be prepared. When the spore count is high, as judged by number of free spores, and when the spores are properly ripened, perhaps 2 or 3 weeks of age, the spore stock can be made. Enough volume of the spore suspension is added by pipette to saturate the soil. The soil tube is then dried spontaneously in air or in a vacuum desiccator, if convenient. Storage may be in air at room temperature for the anaerobes, even the thermophilic anaerobes (McClung[15]). In our experience, the butyl-butyric anaerobes may remain viable in soil stocks as long as 21 years; one of the cultures, *Clostridium pasteurianum,* was tested physiologically for nitrogen fixation and fermentation balance on glucose-yeast medium and was found unchanged from laboratory records on the same culture 12 years earlier.

The soil-stock method lends itself well to the preservation of mold cultures (Greene and Fred[11]) and of streptomycetes (Jones[15]). Both molds and streptomycetes grow in the soil, as can be seen from the whitish mycelium filling the pores of the soil. In the Jones' technique, moist soil is employed which would surely favor massive formation of spores in the culture. No doubt, the very high numbers of spores are a factor in the survival of such cultures in storage. Jones reported "at least 4.5 years without transfer" for *Streptomyces* species. Our own experience confirms at least that duration in observations on an extensive set of *Streptomyces* species stocks; we have also the casual note of viability of some soil *Streptomyces* species at 11 years of age. Our experience with a large mold collection has been also very good. *Aspergillus* and *Penicillium* species are viable for years; certain of the cultures of Dr. Greene, prepared in soil in 1935, were tested in 1949 and found alive. Some eight or nine other genera in the stocks are also successfully preserved, but *Rhizopus* species have occasionally been lost, so that it is our present practice to keep the *Rhizopus* species in double set, one on soil and one on malt agar, the second with yearly transfer. At each transfer of the malt cultures, we test the corresponding soil stock and, if viable, hold it longer. In case one of these *Rhizopus* cultures is needed for experiment, it is taken from the soil stock which is considered most likely not to have degenerated.

The Lyophilization Method

A third general procedure, that of lyophilization of cultures, is to be highly recommended from the data in recent literature. Lyophilization involves desiccation in vacuum from the frozen state to give a product which is readily hydrated for subculture. The process has much to recommend it in principle. Protein denaturation, including destruction of enzyme systems, is minimized by the dry state which is rapidly approached. The harmful concentration of salts and other solutes resulting from loss of water is avoided by the prefreezing, and oxygen is excluded by the high vacuum involved in the procedure. Obviously, the culture must be sealed in vials under vacuum to maintain these conditions. The very fact of sealing in glass vials is considered troublesome by some, but it need not be. It is possible to use manifolds so that a group of cultures can be made at one time and the sealing itself is simple. Subculture involves breaking the seal and thus the individual culture is destroyed at the time of subculture.

There are several modifications of the lyophile process. The apparatus may be of the Cryochem or Flosdorff-Mudd type[7,8] or that of Wickerham and Andreasen,[30] or various improvisations of the principle.[9,14] Photographs of Wickerham's apparatus are shown in Figures 34 and 35. Success appears to depend on rapid attainment of the frozen state, complete dehydration, and maintenance of vacuum or inert atmosphere[20] at all times during production and storage. Even so, the lethal effects on the culture are appreciable and studies have been made to discover the cause of death of a very large percentage of cells during or soon after

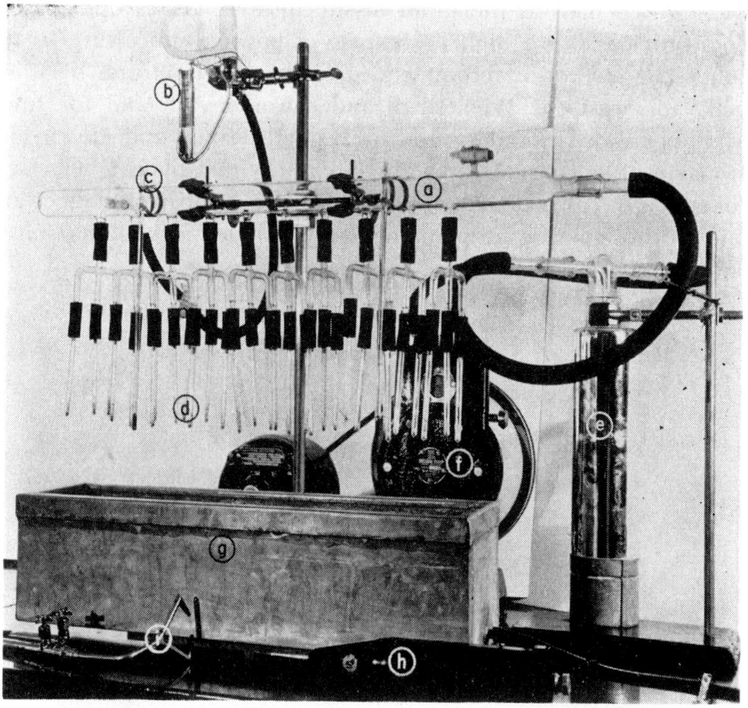

FIGURE 34. *Stationary Table Lyophil Apparatus* (Courtesy—Northern Utilization Research Branch)

lyophilization.[9,20,26,29] Based on the observation that such natural colloids as milk, blood serum, and mucin have a protective effect on dried cells, Heller[13] generalized that death of cells from the desiccated state was greater in a menstruum containing crystalline compound of high solubility, which the organism could metabolize. Conversely the percentage survival was greatest in presence of protective colloids of low gold number and high hydrophilic property. Peptone, normal rabbit serum, and "broth" served very well in this capacity. Actually for stock-culture purposes, one is not necessarily interested in the highest percentage survival during lyophilization and subsequent storage, but it may be a factor in the successful subculture from the stock at later time. There need be only sufficient numbers surviving *in all cultures* to assure 100% success when the transfer is made. It is general practice to use the whole dried mass as inoculum in subculture. That lyophilization does provide such a high percentage of positive subcultures is the rule. The method is recommended for bacterial cultures by Bushnell,[4] for yeasts by Wickerham and coworkers,[30,31] and for fungi by Raper and Alexander[21] and by Fennell, Raper, and Flickinger.[6] It may well be considered by any laboratory that plans a very large collection of cultures or has need of a large number of vials of any one culture for use as permanent and stable stock of its starter culture.

FIGURE 35. *Completely Contained Mobile Lyophil Unit* (Courtesy —Northern Utilization Research Branch)

The stability of physiological characteristics is generally considered good for lyophilized cultures,[6] but there are a few reports of change. Nymon, Gunsalus, and Gortner,[19] after trying lyophilization of microbiological assay strains of *Lactobacillus arabinosus* and *L. casei,* reported that the first maintained its original acid-producing capacity whereas that of the second decreased somewhat, but remained stable afterward. Stanier, Adams, and Ledingham[27] advocated lyophilization to stabilize the highly variable *Bacillus polymyxa* in the desired phase and in high-yielding state for 2,3-butanediol fermentation. However, marked changes may follow lyophilization in certain cases. Atkin, Moses, and Gray[1] found that certain brewers' yeasts underwent changes in vitamin requirements, suggestive of mutation or segregation of characters. It should be noted that a very large proportion (99.98%) of the cells of their cultures failed to survive lyophilization. One might speculate that the "changed" cultures might represent minor types within the parent population, which for reasons unknown proved more resistant to lyophilization. At least there is no apparent reason why lyophilization should act as a mutagenic process and this case involving yeasts is the only such report.

In conclusion, it may be said that there is available a variety of methods for maintenance of stock cultures, no one of which is recommended to the exclusion of the others. The choice may be determined by available apparatus or, if all are available, by experience with them for the cultures to be stocked.

BIBLIOGRAPHY

1. Atkin, L., W. Moses, and P. P. Gray, *J. Bact.,* **57**, 575 (1949).
2. Backus, M. P., J. P. Stauffer, and M. J. Johnson, *J. Am. Chem. Soc.,* **68**, 152 (1946).
3. Bredemann, G., *Zentr. Bakt. Parasitenk., Abt. II,* **23**, 385 (1909).
4. Bushnell, O. A., *J. Bact.,* **42**, 152 (1941).
5. Demerec, M., U. S. Patent 2,445,748 (1948).
6. Fennell, D. I., K. B. Raper, and M. H. Flickinger, *Mycologia,* **42**, 135 (1950).
7. Flosdorff, E. W., and S. Mudd, *J. Immunol.,* **29**, 389 (1935).
8. Flosdorff, E. W., and S. Mudd, *J. Immunol.,* **34**, 469 (1938).
9. Frobisher, M., Jr., E. I. Parsons, S. E. Pal, and S. Hakim, *J. Lab. Clin. Med.,* **32**, 1008 (1947).
10. Gordon, R. E., and N. R. Smith, *J. Bact.,* **53**, 669 (1947).
11. Greene, H. C., and E. B. Fred, *Ind. Eng. Chem.,* **26**, 1297 (1934).

12. Hac, L. R., *Proc. Soc. Exptl. Biol. Med.*, **45**, 381 (1940).
13. Heller, G., *J. Bact.*, **41**, 109 (1941).
14. Hornibrook, J. W., *J. Lab. Clin. Med.*, **34**, 1315 (1949).
15. Jones, K. L., *J. Bact.*, **51**, 211 (1946).
16. McClung, L. S., *J. Bact.*, **29**, 173 (1935).
17. Michael, M., *Zentr. Bakt. Parasitenk., Orig.*, **86**, 507 (1921).
18. Morton, H. E., and E. J. Pulaski, *J. Bact.*, **35**, 163 (1938).
19. Nymon, M. C., I. C. Gunsalus, and W. A. Gortner, *Science*, **102**, 125 (1945).
20. Proom, H., and L. M. Hemmons, *J. Gen. Microbiol.*, **3**, 7 (1949).
21. Raper, K. B., and D. F. Alexander, *Mycologia*, **37**, 499 (1945).
22. Raper, K. B., D. F. Alexander, and R. D. Coghill, *J. Bact.*, **48**, 639 (1944).
23. Raper, K. B., and D. I. Fennell, *J. Bact.*, **51**, 761 (1946).
24. Sherf, A. F., *Phytopathology*, **33**, 330 (1943).
25. Simmons, R. T., *Med. J. Australia*, **29**, 283 (1942).
26. Stamp, L., *J. Gen. Microbiol.*, **1**, 251 (1946).
27. Stanier, R. Y., G. A. Adams, and G. A. Ledingham, *Can. J. Research*, **23F**, 72 (1945).
28. Thaysen, A. C., *J. Inst. Brewing*, **30**, 349 (1924).
29. Weiser, R. S., and L. A. Hennum, *J. Bact.*, **54**, 17 (1947).
30. Wickerham, L. J., and A. A. Andreasen, *Wallerstein Labs. Commun.*, **5**, 165 (1942).
31. Wickerham, L. J., and M. H. Flickinger, *Brewers Digest*, **21**, 55 (1946).

CHAPTER 13

CULTURAL VARIATION AND GENETICS

Carl C. Lindegren

Cultural variation in microorganisms occurs so abundantly that it is debatable if any such thing as a "pure" culture exists or can be maintained. This is not merely an academic question, but one of great importance to practical workers interested in reproducing fermentations or biological syntheses. Frequent single-cell isolation is often proposed as a means of maintaining pure cultures. Actually, this may have exactly the opposite result by selecting variants that may differ radically from the population which makes up the culture. Therefore, the maintenance of "pure" cultures will not be discussed in this chapter, but rather procedures which will maximize or minimize cultural variation.

GENERAL PRINCIPLES

Minimizing Cultural Variation

There are two empirical rules governing cultural variation of microorganisms: (1) Variation can be minimized by frequent transfer (just before full growth is attained), using large inocula on a uniform good medium (or simply on a uniform medium). (2) Variation can be maximized by infrequent transfer, using small inocula on a variety of deficient media.

There is, however, no technique for eliminating variability. It can only be minimized by providing nutrients which will insure the *continuity* of cell structure in the culture resembling as closely as possible the cells from which they were produced. *Continuity*, however, does not mean *identity*. No new cell is exactly like its progenitor and the operator can only insure continuity of the culture. The similarity of organisms of the same species which we isolate from nature is not evidence of the stability of the hereditary apparatus of the species. It is merely evidence of the response of the organism to its environment and of the selectivity of the natural environment for the specific type of organisms. When we transfer the culture to artificial environment, the inherent variability (which was concealed by the operation of natural selection under natural conditions) becomes evident.

The principal bases for the validity of these rules are:

(1) In organisms with complex life cycles (especially if two or more mating types are involved), *large inocula* are essential to avoid separating different mating types or other essential components of the life cycle.

(2) *Frequent transfers* are essential to avoid selecting specialized cells, such as spores which survive after the death of most of the vegetative population.

(3) A good medium is necessary to insure the continued growth of all the forms present in the original culture.

(4) A uniform medium is necessary to prevent selection of specialized types different from the original culture. This uniform medium need not necessarily be a *good medium* if only one specialized type of cell is to be propagated. Then a medium capable of favoring this type and unfavorable for other variants can serve to maintain the type culture. However, frequent transfers of large inocula are still required to avoid the selection of other variants.

(5) Variation either due to gene mutation or transmissible extrachromosomal change (Dauermodifikation[15]) may occur in the vegetative phase and large inocula on uniform good medium are necessary to avoid selecting and separating variants of this type from the mixed population which makes up the culture. Large inocula are also necessary to preserve the *mutational or adaptive potential* of the population. For example, Luria and Delbruck[37] showed that by inoculating about five hundred cells of colon bacillus into 0.5 ml of medium, one could obtain a crop of cells which

contained either very few or a great many cells capable of resisting bacteriophage lysis. Shapiro[41] showed, however, by calculation from these results that an ordinary large inoculation would insure that a statistically uniform fraction of the population would be capable of resisting bacteriophage lysis. Large inocula insure the transfer of a sufficient number of those mutants which occur with a regular frequency to maintain an adequate population of the mutants and if the culture is carried for the *mutant* (rather than the cell type from which the mutant arises), it will be present in large inocula.

Life Cycles

Understanding of cultural variation requires a knowledge of the life cycle of the organism. Life cycles are divided into two principal phases with regard to chromosome number: (1) the haplophase, with a single set of chromosomes in each nucleus and (2) the diplophase, with pairs of chromosomes present in each nucleus. Diplophase nuclei are produced by the fusion of two *haplophase* nuclei. The haplophase is produced by the "reduction" of a *diplophase* nucleus by two consecutive nuclear divisions which finally results in the production of four haplophase nuclei. In "reduced" haplophase nuclei, the gene complexes have been reorganized (1) by random assortment of the different chromosomes and (2) by crossing-over, interchange between pairs of homologous chromosomes. In Figure 36 the single lines represent the haplophase and the double lines, the diplophase. Life cycles are sometimes diagramed as circles with the implication that after a series of operations, the same kind of individual that was originally present is reconstituted. However, no two individuals are exactly alike and all biological systems are changing with time from a former condition to a future one. Stability in a biological system is based on continuity in time, but this does not imply identity of the individual organisms.

Figure 36 shows some of the possibilities when a medium is inoculated with a diploid yeast cell. The sidetrack indicates the reduction of one of the diploid cells. Each of the crossbars indicates a cell division. The continued vegetative multiplication of the original diplophase culture does not mean, however, that all of the offspring are identical since the environment is being changed constantly by the continued growth of the cells themselves. These new diploid cells may have nearly the same total *genetical potentials* as their forbears, but this potential has been reassorted and the cells have undergone differentiation (Dauermodification) by the change

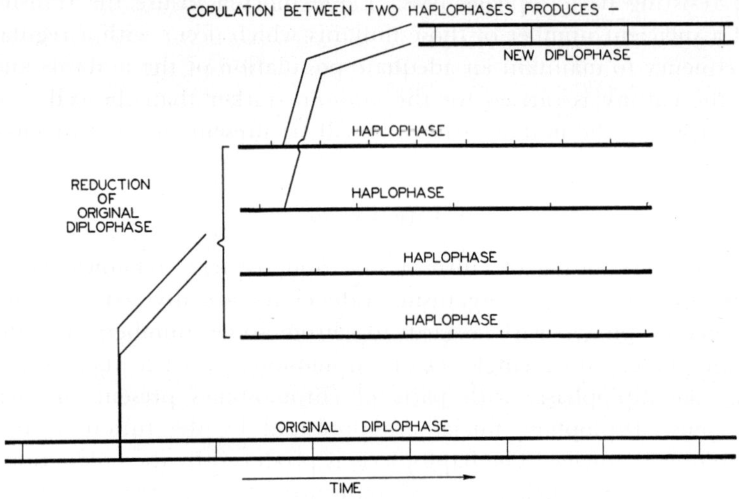

FIGURE 36. *Life-Cycle Diagram Illustrating the Effects of Reduction and Copulation in Increasing the Complexity of a Population*

in the environment. That close genetical relationship does not assure identity is obvious when we consider that the highly differentiated cells in a metazoan are all descended from a single original diploid cell.

Competition

In the culture indicated in Figure 36 the clone of differentiated diplophase cells competes with the four haplophase cultures descended from one of the diploid cells, as well as with the other diplophase and haplophase clones. Each cell has its own inherent rate of division or "biotic potential," but it is influenced by and itself affects the rates of division of all its neighbors by the effect it has on the food supply and accumulated waste products. Competition is the most potent factor controlling the characteristics of the culture. A newly constituted diplophase of high vigor may tend to overgrow and crowd out its weaker relatives. However, conditions are changing so rapidly and are so nonuniform, even in a small volume, that no *one clone* can be the best adapted and thus a variety of genotypes is maintained.

Functions of the Sexual Mechanism

There are two different phases in the life cycle of a micro-organism, the sexual phase and the asexual phase.[6,8,23] The survival of the sexual mechanism lies in the opportunity which it provides the organism for reassortment and recombination of hereditary characteristics. Recombinations usually result in the reconstitution of the wild type and the mutants usually fail to survive in competition since the wild type is adapted to the natural environment. Lederberg and Tatum's[17] work on *Escherichia coli* describes a mating of two mutants, each characterized by different vitamin deficiencies, which produces a small number of "prototrophs" in which the original nondeficient stock is reconstituted. Presumably, all the other possible recombinations are also made, but the prototrophs are most easily detected for they do not have any of the vitamin deficiencies and can grow on unenriched medium. In nature, the sexual mechanism often reconstitutes a *type of hybrid* better adapted to changed conditions. In wheat rust, the sexual cycle is confined to the barberry plant. Whenever plant breeders develop a new type of wheat capable of resisting the current rust, new combinations are produced on the barberry capable of attacking the resistant wheat.

The continuance of the sexual mechanism depends on the maintenance of fertility.[28] The sexual reaction is so complex that any one of an enormous number of defects can prevent its continued operation. When this occurs, the organism reproduces asexually and, therefore, loses the ability to recombine its transmissible genetical characters with those of its relatives. This does not mean that it has lost the capacity for variation, since the asexual phase is still capable of a wide spectrum of genetical change, but the latitude of the variability is now restricted and may proceed in a direction quite different from what would occur if the stabilizing effect of sexual reproduction intervened occasionally.

In the haplophase, the single set of chromosomes is present in the cells, while in the diplophase, the double number of chromosomes is present.[22,25] The diplophase is formed by the fusion of two similar or dissimilar haplophases. The function of mating type, or sex, is to insure that the two fusing haplophases will not be closely related and thus to encourage fusions between widely dissimilar haplophases. The survival value of this mechanism depends on the fact that vigor often results from matings between dissimilar individuals. When haplophases are again produced from the diplophase, the hereditary factors are reshuffled and the haplophases produced are often extremely different from the two parents which produced the original diplophase. This mechanism, involving the recombination of hereditary factors, has an enormous survival value and has been demonstrated to exist in the lowest forms of life even including bacteriophages and bacteria. The opportunity for adaptation by recombination which is available to organisms capable of recombining their hereditary factors by a sexual mechanism suggests that they would outgrow closely related individuals whose reproductive devices were limited to an asexual apparatus. The diplophase is often an inconspicuous fraction of the total life cycle and this fact has sometimes led to the conclusion that it is absent. *Neurospora*[8] and the colon bacillus[17] were studied for nearly 100 years before the discovery of their sexual stages, but in both of these organisms, the inconspicuous diplophase has been the principle factor in producing the recombinations required in their evolutionary development.

In bacteriophage, bacteria, fungi (except yeast), and some algae, the haplophase is the predominant phase of the life cycle. In higher plants and animals, the diplophase is the predominant phase of the life cycle. The evolution from predominant haplophase to predominant diplophase is the result of the requirement of organisms with complex structures for a maximum stability of the cells which act as building blocks for the complex structures. The absence of complex structures in lower organisms makes possible the extensive growth in the haplophase with the accompanying extraordinary variability. The variability may lead to sterility of a small fraction of the haplophase and prevent the sterile forms from recombining their abilities, thus draining the sterile haplophases off into a dead end, so far as the continued perpetuation of the species is concerned.

Variability in the Haplophase

The best understood type of variation which occurs in the haplophase is gene mutation which can be recognized by the regular segregation of the mutated gene at the reduction division. Such a change may be either spontaneous or induced. X-rays, ultraviolet light, and nitrogen mustards are all potent mutation-inducing agents. Gene mutations often involve decreases in the ability of the cell to synthesize specific substances, generally enzymes, or vitamins, or other essential nutrilites.[4,5] Evolution has developed an array of synthetic mechanisms in each species essential to the continued maintenance of the organism in nature. Frequent recourse to sexual reproduction restores the integrity of these synthetic mechanisms and maintains the wild-type characteristics which enable the organism to compete successfully in its natural environment. The combination of abilities which are essential to the normal wild organism are not necessarily essential, nor even desirable, in a cultivated plant or animal. A desirable cultivated plant like maize, or a domesticated animal, like the milch cow might be incapable of survival under natural conditions, because the genetical changes which have been selected operate against the organism in a state of nature. In microorganisms, the loss or imbalance of a function may also prove desirable for a cultivated organism, although deleterious to the organism in its natural habitat. For example, the normal amount of riboflavin produced by yeast may be enormously increased by selection, but this imbalance might not be helpful to the organism in its natural environment. An increase in the amount of a given substance produced by an organism need not mean a total increase in synthetic ability, but may involve a block in some synthetic process which comprises many individual steps. Whenever any single step is blocked, the precursor accumulates and the increased amount of the precursor is the result of a stoppage of its utilization rather than a fundamental increase in its production. The acquisition of a new synthetic ability by an organism may involve an extraordinarily long evolutionary development and probably is encountered seldom in the laboratory. There are many reports of mutations by which the ability to synthesize various vitamins is achieved. These may be back mutations; a relatively complete system which has been dormant may have been reactivated by some rather simple process.

Adaptation to an Undesirable Environment

Many organisms show the ability to adapt to an otherwise undesirable environment. This may result when large inocula are introduced into the unfavorable medium or may occur during continued cultivation in a medium containing small amounts of some harmful substance.[24,35] Mutation usually involves a change in a very small fraction of the population and the overgrowth of the single original mutant cell.

Adaptive Enzymes

A cell may also become adapted by the production of an adaptive enzyme.[36] Adaptation to the fermentation of sugars has been most extensively studied. The ability to adapt may be under the control of a single Mendelian gene and an entire population containing this gene may adapt *en masse* to the presence of the substrate. Other stocks apparently incapable of fermenting the substrate may achieve the ability after long exposure due to the back mutation of the recessive gene for this ability and the selection of the individual mutant in the presence of a substrate. In the second case, a *de novo* mutation has not occurred, but a latent (recessive) gene has been reactivated.

LIFE CYCLES OF SACCHAROMYCES

Saccharomyces may serve as an example of the different types of variation which one may find in all fungi and bacteria. It differs principally from other fungi in having a predominant diplophase. The cells of *Saccharomyces* exist in both haplophase and diplophase (Figure 37). There are two mating types in the haplophase and the haploid cells of different mating type copulate to produce diploid cells. Kruis and Satava,[16] in Prague, and Winge,[43] in Copenhagen, showed that the standard vegetative cells are diploid, produced by copulation of two spores or gametes derived from spores. The diploid nuclei undergo reduction at spore formation to produce four haploid ascospores. The large, ellipsoidal vegetative yeast cell is produced by the fusion of two round haploid gametes derived from ascospores. Winge established the basic facts of this life cycle by a classical series of observations on the germination of ascospores and fusion of haploid cells. Winge and Laustsen,[44,45,46] in a series of notable papers, showed that colonial characteristics, fermentative ability, and cell shape are under the control of genes which segregate at the reduction division. They hybridized yeasts by placing a haploid ascospore from one strain in close proximity to an ascospore of a second strain by use of the micromanipulator. When all conditions are favorable, the two spores fuse to produce a diploid hybrid cell.

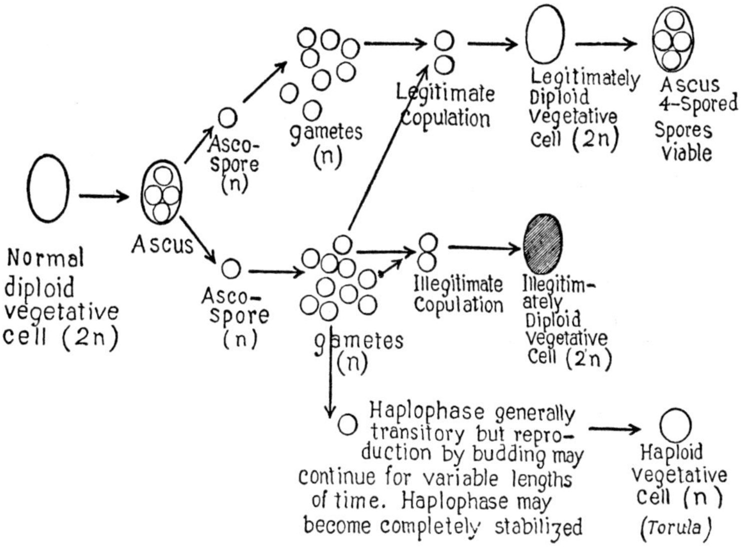

FIGURE 37. *Life Cycle of* Saccharomyces (Reprinted by permission from *Ann. Rev. Microbiol.,* **1948**, 58)

Lindegren and Lindegren[30] developed a method for hybridizing yeasts based on the fact that single ascospores of *S. cerevisiae* usually produce persistently haploid cultures. It is possible to hybridize these with other similarly derived haploid cultures by mixing the cells together in an appropriate medium. The mixtures result in copulation if each culture is paired with a complementary type. One parent culture can be mated to a large number of other clones. The parents can be classified for biochemical and other characteristics prior to the matings and the haploid progeny can be classified subsequently. Lindegren and Lindegren[30] have developed stocks heterozygous for genes controlling the fermentation of galactose, sucrose, maltose, melibiose, and raffinose. They have also developed stocks differing in their abilities to synthesize the various vitamins, amino acids, and nucleic acid components. These cultures are all interfertile and over two thousand hybrid asci have been analyzed. Several cases of linkage have been discovered and maps have been constructed.

Lindegren and Lindegren[34] discovered that haplophase cultures of *S. cerevisiae* fall into two mating types which they have designated *a* and α. Each haploid culture is made up of potential gametes which can be mated with other cells of appropriate genetic composition from another culture. As the pure haplophase cultures age, illegitimate copulations occur in some cultures between two cells of the same reaction in the same culture. Diploid cells produced by illegitimate copulations are generally distinguished by diminished ability to produce four-spored asci containing viable spores. Legitimate matings between *a* and α haplophase cultures derived from a variety of strains of *S. cerevisiae* usually produce diploid cells which sporulate to form four-spored asci containing viable spores. The procedure is as follows: 1 ml of broth is placed in a test tube. The broth is first inoculated with a large loop of cells from an agar slant of the tester culture and shortly after with a large loop from the haploid culture to be tested. Copulations sometimes appear within 6 hours. The tube is kept overnight in an incubator at 19°C, and inspected under the microscope for the presence or absence of copulations.

Some haplophase mutants obtained by continued selection involving numerous platings and transfers over a prolonged period (usually more than 1 year) were found to be incapable of copulation. Prolonged competition and selection probably result in loss of fertility because the genes which insure fertility do not have a high survival value in competition with other rapidly growing mutants.

Mendelian Inheritance

Figure 38 summarizes data on the inheritance of a Mendelian gene controlling the adaptive fermentation of galactose. No exceptions to standard Mendelian inheritance were found. Thirteen asci were analyzed from a heterozygous hybrid made by mating a galatose fermenter (*G*) by a nonfermenter (*g*); two spores in each of these asci carried the dominant gene controlling fermentation of galactose and two carried the recessive allele. A backcross of fermenter to the fermenter parent produced thirteen asci; all four spores in each of these asci carried the fermenting gene. A backcross of the nonfermenter to the nonfermenting parent produced seven asci, each of which contained four nonfermenting spores. A heterozygous zygote was produced by backcrossing a nonfermenter to the fermenting parent; six asci were analyzed and each contained two fermenting and two nonfermenting spores. This analysis shows quite convincingly that in some pedigrees the genes controlling fermentation of galactose may behave in a regular Mendelian manner.

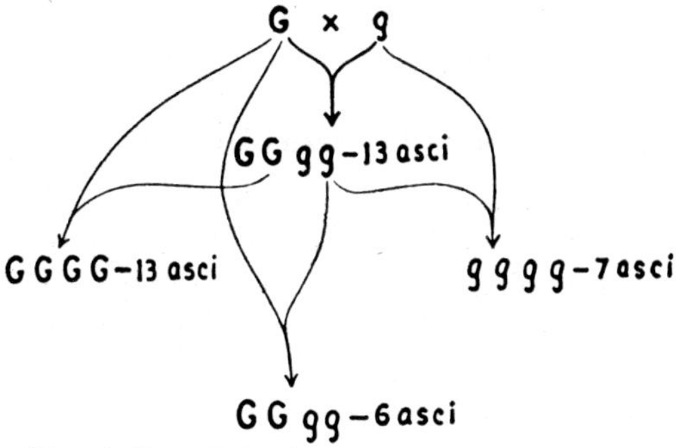

FIGURE 38. *Mendelian Inheritance of Galactose Fermentation in* Saccharomyces (Reprinted by permission from *Ann. Rev. Microbiol.,* **1948**, 60)

Non-Mendelian Inheritance

In contrast to the regular Mendelian behavior of genes controlling galactose fermentation described before, irregular segregations have also been discovered in yeast.[26,33] Some instances of irregular segregations comprise *bona fide* examples of non-Mendelian inheritance due to gene conversion. On this view, a chromosome may carry a synthesizing mechanism which can be transferred from an active to an inactive chromosome or can be lost from an active chromosome. Some irregular segregations are due to changes in ploidy. Heterozygous hybrids (*G/g*) have produced ascospores which are diploid and heterozygous for the character in question. Outcrossing the diploid ascospores has produced triploid and tetraploid zygotes.

Yeast cultures produce some variants which are not gene mutations, since their characteristics cannot be transmitted through the sexual cycle. We have thus confirmed Jollos'[15] concept of the Dauermodifikation. The variants derived from vegetative cultures are stable when propagated vegetatively but when the variant is mated with a normal culture only normal offspring are produced. The variant strains[40] are usually characterized by smaller colony size, loss of red pigment (if the parent culture was red), absence of certain cytochrome oxidase enzymes, or inability to utilize alcohol. These defects may all be the result of the same basic extra chromosomal deficiency which can be restored to normal by the outcross. This study has suggested that a considerable degree of the variation encountered in microorganisms is not due to gene mutation but is the result of a defect in some structure other than the chromosome.

Haplophase Variability versus Diplophase Stability

Nearly all haplophase yeast cultures, when plated on agar, produce numerous relatively stable colonial "mutants," while the corresponding diplophase cultures only rarely produce colonial variants. Hybrids are produced by mixing the cells of two apparently unstable haplophase cultures together in a small amount of broth. After copulations occur, diploid cells are produced which subsequently sporulate. Genetical analyses of the hybrid are made by dissecting four-spored asci at random after sporulation has occurred. One of the puzzling things about this process was the fact that in spite of the considerable potential of the haplophase cultures for variation, analyses of the offspring from matings usually yielded surprisingly regular results. This was supposed at first to be the result of statistical sampling; in the enormous population of cells, the chances that a given zygote had been derived from the two preponderant genotypes is presumably very great. However, the fact that Dauermodifikationen occur so frequently suggests that many of the variants which appear on the plates cannot be transmitted through the zygote. The reliability of the method of mass mating is greater than the variability of the haplophase would lead one to believe and numerous trials have confirmed this conclusion.

Segregation

Segregation of genes occurs when the chromosomes of the diploid cells are segregated at the reduction division just prior to sporulation. The haplophase originates by the reduction of the diplophase at spore formation and the segregation of a heterozygote produces segregants of different genotypes. Each of the four spores formed in a single ascus is usually genetically different. The haploid segregants are usually rough colonied. The segregant cultures also vary in fermentative ability, in ability to synthesize amino acids, vitamins, and nucleic acid components, in color, and in the size and shape of the haploid cells. The type of cell aggregation is also characteristically different. Haplophase clones generally tend to produce aggregated or agglutinated cells much more frequently than diploid clones.

Haploid yeast cells are much smaller and more variable than diploid cells, varying more than diploid cells both from culture to culture and within a single culture. These differences are also reflected in the colonies. Haplophase yeasts are nearly always inferior in synthetic ability when compared quantitatively, or qualitatively, to the diploid parent from which they originated; many of them have lost certain specific characteristics.

Variations in the haplophase enormously increase the number of colonial forms, but the original segregant can generally be distinguished from the secondary variants when the culture is plated. At first, the variants are usually slow growing and produce small, round colonies, but on transfer, they become adapted and stabilized and their specific colonial character becomes apparent, distinguishing them from the original segregant. Whether there is a process involving several genetical steps has not been determined. Variations occurring in the haplophase can be selected and propagated. In spite of the wide variety of types that occur, the existing genotype of any cell limits its potentialities and the range of its possible variations. This fact has been especially brought out in experiments aimed at adapting haplophase by selection. *S. cerevisiae* is unable to ferment melibiose and prolonged exposure of haplophase cultures of *S. cerevisiae* to melibiose failed to produce any mutants capable of fermenting this sugar. A haplophase variant of *S. cerevisiae* incapable of fermenting galactose could not be induced through a 4-month period to produce mutants capable of fermenting galactose, although this strain produced an abundance of colonial variants during the same period. Although "losses" occur easily, "gains," as in the case of specific fermentative or synthetic abilities, apparently do not occur at all under certain conditions.

Recombination

Segregation and mutation produce a great variety of haploid gametes. Copulations between these gametes selected at random produce new recombinations which may differ in a number of ways from the diploid from which they originated. Genetical variations in yeasts are, therefore, produced by (1) segregation, (2) vegetative variation (gene mutation and Dauermodifikation), and (3) recombination.

LIFE CYCLE OF NEUROSPORA

Neurospora crassa may be studied as a type specimen for the higher ascomycetes since its life cycle is so well known. Shear and Dodge[42] named the genus *Neurospora,* which contains both eight- and four-spored species. Dodge[8] discovered that the eight-spored species were heterothallic, while the four-spored species produced homothallic spores, due to the inclusion of one nucleus of each mating type in the spore. The life cycle shown in Figure 39 is practically the same as that of *Penicillium, Aspergillus,* and similar forms.

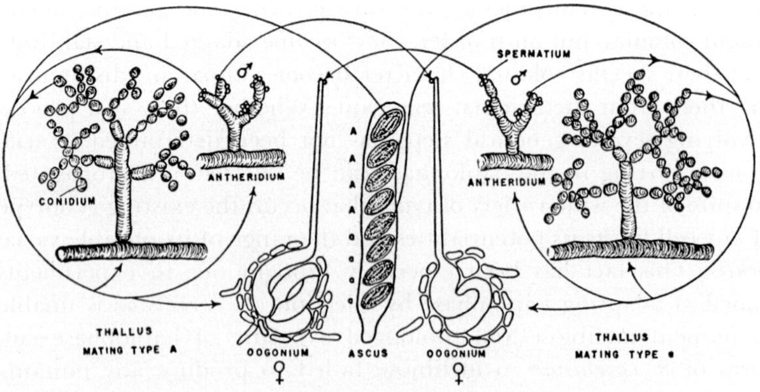

FIGURE 39. *Life Cycle of* Neurospora crassa (Reprinted by permission from *Ann. Rev. Microbiol.,* **1948**, 64)

The perithecium of *Neurospora crassa* produces a large number of asci, each containing eight spores. The spores fall into two categories with regard to mating type. Each spore gives rise to a thallus which produces a mycelium, on which male and female sex organs, as well as asexual conidia, are developed. The female sex organ is called a bulbil and contains an oogonium and trichogyne. The male sex organ produces sperm called spermatia. The spermatia from an *A* type thallus are incapable of fertilizing the oogonium developed on the thallus of the same type, but will cross-fertilize an oogonium of a complementary mating type. The superimposition of sex organs and mating type shows that mating-type genes are self-sterility alleles and bear no relation to either sex organs or the evolution of sex.

Single ascospore cultures[29] are characterized either by morphological or biochemical properties and hybrids are made either by transferring spermatia from a tube of one mating type to a tube of the complementary mating type or by inoculating a single tube with conidia from each culture and allowing them to grow together.

COENOCYTIC FUNGI

Although the hyphal threads of the *Ascomycetes* are divided into chambers of cross walls, these septations are practically always perforated and function to keep the tubes from collapsing rather than to separate the organism into cells. The genetic mixture of haploid nuclei suspended in the common cytoplasm passes freely through the septations as the cytoplasm streams along the hyphal threads. If one of these haploid nuclei mutates, the mutant may continue to increase by division and become distributed throughout the plant by cytoplasmic streaming. The genetical heterogeneity of the nuclei prevents such an organism from producing structures of great complexity; in fact, in these organisms, genetical complexity is maintained at the expense of the ability to produce complex structures. In some species, there are elaborate adaptations which assure that the sexual spores (which contain only a single type of nucleus in most species) will contain two genetically different kinds of nuclei.

Hyphal Fusion

Hyphal fusion or anastomosis is a device which functions to preserve and increase the genetical complexity of the fungi. When two hyphal tips come into contact, their contiguous walls dissolve and their contents intermingle. When a hyphal wall becomes old and thick, it is incapable of fusion. These fusions have no sexual significance and are not followed by karyogamy. They occur between mycelia of the same mating type as frequently as between mycelia of opposite mating types. They sometimes occur between genetically different hyphae and may even occur between hyphae of different species or genera.

Heterokaryosis

The facility of nucleo-cytoplasmic interchange produces a mixture of genetically different haploid nuclei. This condition is called heterokaryosis and the heterokaryotic thallus is the biological equivalent of the diplophase in other forms. When masses of spores are transferred, the young germ tubes fuse with one another as soon as they emerge from the spores and when the spores are genetically different, the new mycelium contains a mixture of different kinds of haploid nuclei. An interesting result of this capacity to produce nuclear mixtures is the fact that lethal genes in haploid nuclei can be carried in culture because the normal nuclei in the cytoplasm provide a sufficient excess of the substances necessary for subsistence. The fact that the "weaker" pseudo-parasitic or even lethal mutant nuclei can be supported by the more vigorous normal or wild-type nuclei means that practically all new mutations are preserved and live in the cytoplasm elaborated by the more vigorous "normal" hosts. Since a gene is only "poor" or "good" in certain combinations, the low vigor of a genotype does not mean that the gene itself is not valuable for the species. The new genotype multiplying passively in the host cytoplasm recombines with other mutants in the sexual cycle. If a genotype of exceptional vigor or specific adaptability is eventually produced, it may attain supremacy in the heterokaryon and finally produce sectors of new growth. Mutations in diploid cellular plants rarely attain supremacy in the plant in which they occur. Mutations which occur in somatic cells, even if they are dominant, are walled off from the rest of the plant and fail to affect its character unless they occur early in a bud (and cause a bud sport). An advantageous mutation in a fungus thallus, however, because of the coenocytic character of the plant, has a good chance of dominating the thallus, although it can rarely become completely separated under natural conditions from the mixture of nuclei from which it arose.

Some of the diverse nuclei found in fungus thalli are capable of performing specific functions. For example, in *Neurospora,* forms found in nature may be heterokaryotic for normal conidial and nonconidial mutant nuclei. These nonconidial mutants are incapable of producing the asexual spores (conidia) by which the fungus is disseminated vegetatively. Since they do not expend energy in the formation of asexual spores, they grow more rapidly than the normal conidial forms. The heterokaryon has the rapid growth character of the nonconidial mutant and the conidium-forming capacity of the conidial form and is, therefore, more vigorous than either pure form. It is conceivable that mixtures of nuclei are important in other ways not so easy to demonstrate. A heterokaryon carrying mutants capable of elaborating different kinds and amounts of enzymes would probably have a higher survival value than a more restricted and less flexible homokaryotic form.

Organisms, such as fungi and bacteria, in which single thalli or colonies contain mixtures of genetically differentiated nuclei, can possess only a minium of differentiated structures and in place of structures capable of performing a variety of functions, genetically different nuclei may be specifically adapted for the performance of different functions. In higher plants, which are genetically uniform and protected against variation by diploidy, different functions are performed by structures differentiated morphologically from the other cellular tissue of the organism. Fungi, such as the *Hymenomycetes* (mushrooms and toadstools) which produce relatively complex structures, are not coenocytes, but are built of cells each of which contains a stable dikaryon which is duplicated by synchronous "conjugate division" of each of the nuclear mates. In mushrooms, each cell contains two nuclei and this pairing insures stability against the possibility that variation might produce a new phenotype and distort the structure. Stability is, therefore, assured by dikaryons in *Hymenomycetes* just as it is assured by diploidy in the cells of higher plants. However, a *Hymenomycete* dikaryon is an extremely flexible structure and one of the members of a dikaryon is often replaced by an "invading" nucleus which is genetically different. Smut mycelia capable of invading host tissue and causing disease are dikaryotic. Monokaryotic strains of smuts are usually incapable of parasitism, while the dikaryotic bisexual mycelium easily gains entrance to the host tissue.

Asexual Cycle

Asexual spores (conidia) of *Neurospora* are usually multinucleate, thus insuring the vegetative propagation and dissemination of the heterokaryon existing in the thallus. When a conidium is finally separated from the conidiophore by a solid wall, it may contain a few to a dozen or more haploid nuclei. In nature, the variety of mutant nuclei present in a single conidium usually contains representatives of each of the two mating types. A conidium planted on agar produces a germ tube which branches rapidly and grows radially from the point of inoculation, with simultaneous multiplication of the various nuclei.

The mycelium is made up of branching tubes divided into cylindrical compartments by septa (cross walls), but these cross walls do not separate the threads into cells in the regular sense. Two partitions may enclose a mass of cytoplasm containing any number of nuclei, from a few dozen to many hundreds, each of which is haploid. Furthermore, each septum has a large perforation at the center and nuclei migrate freely through the perforations. When the hyphae have attained nearly full growth, an abundant deposit of nuclei and cytoplasm has accumulated. During the early stages of growth, only a few conidia are produced, but finally, when the growth of the hyphal mat nears its limit, conidia are formed in abundance.

The large hyphal tubes in the center of the mat are completely emptied of their nucleo-cytoplasmic mixture and, as the emptying progresses, a plug is formed at each septal perforation. (Sometimes a thin hyphal thread will grow back into the large, emptied hypha through the perforation.) The abundance of already elaborated cytoplasm and nuclei is stuffed like the filling of a sausage into the terminal hyphae, which now take on a predominantly aerial growth, probably stimulated by the staling of the substrate. The aerial hyphae branch dichotomously and the conidia are cut off by walls as the streaming into the tips continues.

Sexual Cycle

The sexual spores of *N. crassa* are black, lemon-shaped ascospores with longitudinal markings.[22] After a heat treatment at 60°C for an hour or at 90°C for a shorter time, the ascospores put out germ tubes from pores at each end of the spore. In nature, the spores lying dormant in the ground usually germinate after a fire. Before any green vegetation gets started, the aerial hyphae covered with salmon-pink conidia (asexual spores) cover the burnt-over region. The mold sometimes completely obscures the object on which it is growing. The growth is so thick that conidia can be picked up by handfuls from the surface of bread or sugar-cane bagasse. Slight disturbances of the air release clouds of conidia to spread the infection. Finally ascospores are produced as a result of the sexual cycle and the ascospores may lie dormant in the soil for several years until the next fire occurs. The asexual spores usually die in a few months.

In the laboratory, it is possible to obtain pure cultures from the ascospores of *N. crassa* by planting them on agar and subjecting them to a heat treatment. Heat treatment kills the vegetative mycelia and asexual spores and stimulates the ascospores to germinate. After the ascospores germinate, a small block of agar containing a single spore is cut out and transferred to a culture tube. Such cultures contain an abundance of vegetative mycelia and asexual spores, but no ascospores appear in a culture grown from a single ascospore.

The fungus is heterothallic, producing **A** and **a** ascospores. Cultures from either **A** or **a** spores look exactly alike to the naked eye and under the microscope. When **A** and **a** mycelia are planted together in a tube, perithecia filled with ascospores are produced.

The male gametes are called spermatia. They are produced on both **A** and **a** mycelia. The spermatia are small (3 to 4 μ), thin-walled, uninucleate cells which act as sperm. They are extruded from a pore at the tip of a bell-shaped hypha or from pores in the sides of short, stubby hyphae (spermatiophores). **A** spermatia only fertilize **a** cultures, and **a** spermatia can fertilize only **A** cultures.

The bubils are the female sex organs. Each contains an oogonium tightly wrapped in a dense ball of thick-walled hyphae. From the oogonium, a branched trichogyne extends sometimes for a long distance. **A** bubils are transformed into perithecia containing asci full of ascospores if they are spermatized by spermatia from an **a** culture. Bubils which are indistinguishable from each other are found in both **A** and **a** cultures.

Segregation in the Ascus

The ascospores are contained in the slender tubular asci inside the perithecium. From fifty to one hundred are normally found in each perithecium and each ascus contains eight spores.[18,21]

Each ascospore is binucleate, but the ascospore at its origin contains only a single nucleus; the paired nuclei are genetically identical and the culture produced by growing a single spore is homokaryotic. Four of the spores from each ascus are **A** and four are **a**. The zygote is produced by a nuclear fusion occurring in the young ascus. One haploid **A** gametic nucleus and one haploid **a** gametic nucleus always participate in the fusion. The zygote immediately undergoes reduction and the distribution of the two types of spores in the ascus is due to the reduction of an **A**/**a** gene pair in the zygote. The fact that (1) the zygote is always heterozygous for the **A**/**a** gene pair, (2) that no zygote can be produced unless it is heterozygous for these genes, and (3) that every member of the species (and probably the genus) falls into either the **A** or **a** category with no alternative class, differentiates these mating type alleles from other sterility factors. Sterility factors are well known in *Neurospora* and have been extensively studied, but cannot possibly be confused with the sex factors. We have, therefore, two mechanisms involved in producing the zygote: (1) the sex organs, spermatia (male) and bulbils (female), and (2) the mating type factors, **A** and **a**, for which the zygote is always heterozygous. The two kinds of spores are arranged in the ascus in six ways:

Arrangement	1	2	3	4	5	6	7	8
(1)	**A**	**A**	**A**	**A**	**a**	**a**	**a**	**a**
(2)	**a**	**a**	**a**	**a**	**A**	**A**	**A**	**A**
(3)	**A**	**A**	**a**	**a**	**A**	**A**	**a**	**a**
(4)	**a**	**a**	**A**	**A**	**a**	**a**	**A**	**A**
(5)	**A**	**A**	**a**	**a**	**a**	**a**	**A**	**A**
(6)	**a**	**a**	**A**	**A**	**A**	**A**	**a**	**a**

The first two arrangements (four and four) are the result of Mendelian segregation of the mating type factors at meiosis I. The last four arrangements (two and two) are the result of Mendelian segregation of the factors at meiosis II.

The third division is equational, so that each spore pair (1 and 2, 3 and 4, 5 and 6, 7 and 8) is genetically identical. A fourth division occurs in each spore which is also equational and results in each spore containing a pair of genetically identical nuclei.

PENICILLIUM

The discovery of penicillin has resulted in considerable interest in variations among species of *Penicillium* in penicillin production.

Derx[7] proved by mating single ascospore cultures isolated from *Penicillium luteum* that this species is heterothallic. *P. notatum* (and *P. chrysogenum*), from which penicillin is currently produced, is a form species rather than a true species. It probably originated from *Penicillium* ascospores and is perpetuated in nature asexually. The existence of heterothallic species in *Penicillium* suggests that the sexual mechanism may be exploited to obtain penicillin-producing strains. It may be significant that Derx mated the cultures which he obtained from single ascospores very shortly after having isolated them. If the same cultures had been tested a few months later, they might have been incapable of producing perithecia. A parallel situation is found in yeasts. Lindegren and Lindegren[31] have shown that haploid yeast cultures capable of copulating vigorously may lose their mating-type specificity and become completely sterile when carried in culture. Lindegren, Beanfield, and Barber[28] described variations in the fertility of *Neurospora* cultures.

Heterokaryons in Penicillium

Foster, Woodruff, and McDaniel[12] have shown that *P. notatum* becomes heterokaryotic by mutation, with a rapid loss of penicillin-producing ability. They found that the new mutants which occurred in the mat seemed to decrease its capacity to produce penicillin.

Baker[2] showed that the septations in *P. notatum* are perforated and that fusions between germinating spores and between different hyphae occur frequently. Her figures suggest that conditions in *Penicillium* may approximate those in *Neurospora*, in which heterokaryosis is extremely important, especially for survival in nature.

Hansen[13] has described the "dual phenomenon" in a wide variety of fungi and Hansen and Snyder[14] have made observations on *P. notatum*, which, they conclude, also exhibits the "dual phenomenon." What Hansen referred to is obviously simple heterokaryosis. Previously to Hansen's work, both Dodge[9] and Lindegren[19] had shown that heterokaryosis is the rule in *Neurospora* and also that the condition is not "dual" but multiple. The commonest type of mutation involves loss of the ability to produce conidia. Lindegren found at least ten different nonconidial mutants in *Neurospora*, all of which were nonallelic. To speak of a heterokaryon of conidial and nonconidial forms as a "dual phenomenon" is to oversimplify the situation. The nonconidial mutants are easy to detect by observation. This is especially true in *P. notatum* and may be due to the fact, recently discovered by Pontecorvo and Gemmel,[39] that the nonconidial form may be dominant in the heterokaryon. However, in addition to nonconidial forms, natural heterokaryons also contain a number of other types of mutants.

Lindegren and Andrews[27] made heterokaryons between different strains of penicillin-producing *Penicillia.* The first step in the formation of heterokaryons is hyphal fusion and our first experiments were designed to test the readiness with which fusions occurred between different cultures of *P. notatum* and to select strains which were able to anastomose readily. The fact that hyphae fuse has no significance from the standpoint of mating type or sexual difference. It is characteristic that some cultures are able to fuse more readily than others. In one mycelium, fusions may occur readily between adjacent hyphae, while in another strain of the same species, practically no fusions will occur.

The experimental procedure was as follows: Sterile slides were coated with a thin layer of nutrient agar and parallel streaks of conidia from a single *P. notatum* culture were made on each slide with the inoculating needles adjusted so that the points were approximately 2 mm apart. These slide cultures were allowed to incubate at room temperature (about 80°F). Conditions were optimum for observation at 48 hours. Fusions between the hyphae could be observed in the region between the parallel streaks at 48 hours, but after the second day, overgrowth generally obscured precise observations.

Three of twenty-five NRRL strains of *P. notatum* showed a marked tendency to fuse, seven others did so with somewhat less vigor, and the remainder either demonstrated a pronounced reluctance to fuse or the mycelium failed to grow across the intervening space. In very few cases did fusions occur with any of the strains where the parallel streaks were more than 3 mm apart. When the streaks were 3 to 5 mm apart, the mycelia would advance to within about 1 mm of each other and growth on the inner sides would cease, although it continued on the outside. The three NRRL strains, No. 49, No. 72, and No. 75, each of which had demonstrated a strong tendency to produce hyphal fusions with themselves, were matched in pairs on agar slides. The three possible pairings all exhibited abundant fusions after 2 days. This is exceptional, for in most other instances where the parallel streaks originated from two different strains, a pronounced barrage occurred.

When fusions occurred between two different strains, a transfer was made to an agar slant and the resulting culture was studied carefully for uniformity. In no case were sectors observed and the mycelium was always intermediate in roughness and other observable characters. Since the previous tests had shown that the mixed mycelia were compatible, we concluded that the combinations produced heterokaryons rather than fungal mixtures.

Hyphal fusion is not a mating-type or a sexual phenomenon and there is no apparent limit to the number of genotypes which can be built into one heterokaryon. A triple heterokaryon was made by mixing No. 72, 75, and 49.

Culture No. 72 (which was most compatible from the standpoint of penicillin production) was paired with two cultures which we had been using in the laboratory for penicillin production. Mat cultures of all the original strains and the heterokaryons were grown in penicillin medium in 2-qt milk bottles. All tests were run in triplicate and many in quadruplicate. The results were averaged and the agreement in all cases was rather close. Before seeding the heterokaryons, the mixed cultures were grown on agar slants. The bottles were inoculated and kept 1 day at room temperature and 7 days at 23°C. The filtrate from the bottles was assayed by the Oxford method, using glass cylinders on an agar plate sown with staphylococci. The concentrations of penicillin in Oxford units per ml are shown in Table 44.

TABLE 44. PENICILLIN PRODUCTION BY DIFFERENT STRAINS OF *PENICILLUM NOTATUM*

Strain	Penicillin units/ml	Strain	Penicillin units/ml	Strain	Penicillin units/ml
No. 49	20	No. 49	18	No. 20.4	50
72	47	72	90	40.1	5
75	25	75	22	72	33
49 & 72	12	49 & 72 & 75	2	20.4 & 72	35
49 & 75	7			20.4 & 40.1	5
72 & 75	47			40.1 & 72	27

Culture No. 49 has a definitely destructive effect on the penicillin production of cultures with which it is associated. No. 75 and

No. 72 seem to be compatible, if not complementary. No. 72 and No. 20.4 also seem compatible.

Heterokaryons in Nature and in the Laboratory

Under natural conditions, many conidia usually germinate together and if the strains are compatible, anastomoses immediately produce heterokaryons. Therefore, in nature, heterokaryosis is an especially favored condition and opportunity is afforded for continued increase in complexity of the thallus. However, laboratory manipulation of fungal cultures leads to a different end. On agar plates, single conidia germinate separately and cultures are selected which are of monoconidial origin. Many of these cultures are homokaryotic, and this is especially true of fungi, such as *Penicillium* which has uninuclear conidia. The high lability of penicillin production, as indicated by our experiments on heterokaryons, suggests that under natural conditions, where mixtures are the rule, little or no penicillin is produced. Only in the laboratory where homokaryons are handled, or when a single conidium falls in a plate (as in Fleming's original observation), can appreciable amounts of penicillin be expected. The possibility still exists, however, that the yield of penicillin can be increased by making mixtures. Dodge[11] and Beadle and Coonradt[3] showed that mutants deficient for certain physiological characteristics may mutually supply each other's deficiencies in a heterokaryon.

Because of our limited knowledge of sexuality in the *Penicillia* and *Aspergilli,* and the general conviction that with few exceptions these species are imperfect, attempts to improve penicillin production have been confined principally to the selection of supposedly imperfect cultures isolated from nature. But this program is very inflexible, for even if the isolation of a large number of different forms of *P. notatum* should yield exceptionally effective penicillin producers, improvement of the culture is limited by the potentialities of the original isolate because this so-called species apparently is incapable of mating and our own attempts at improvement by cytoplasmic mixtures do not show too much promise.

The following procedure might yield improved forms: It should be possible to find an ascogenous heterothallic *Penicillium* or *Aspergillus* with the ability to produce penicillin. Once this organism has been discovered, its sexual mechanism can be exploited to produce different segregants and those which are best from the point of view of penicillin production can be mated with each other, as well as with widely different forms. This method has been used successfully with yeasts.[32]

In *Neurospora* many, and probably most, of the isolates from nature are heterokaryotic for both the **A/a** alleles,[19] but sterility factors prevent copulation, so that the mycelium appears to be of one mating type. The most experienced observers may fail to detect such self-sterile heterokaryons. Lindegren[21] has reported obtaining five cultures of *Neurospora* from Dr. B. O. Dodge, which were labelled Mexico A, Cuba A, Japan A, Japan B (= a), Panama B (= a). An elaborate series of experiments was required to show that all three of the A strains carried two kinds of nuclei (A and a), but sterility factors prevented the production of perithecia.

Moreau and Moruzi[38] performed experiments in which a culture of *Neurospora,* freshly isolated from nature, was planted in one side of a U-tube filled with agar and a laboratory culture isolated from a single *Neurospora* ascospore (and, therefore, homokaryotic as far as mating type is concerned) was planted in the other arm of the U-tube. Under such conditions, the natural culture produced perithecia and ascospores. They proved conclusively that mycelium did not grow through the U-tube and thus only the diffusion of some stimulating substance through the agar could account for the appearance of the perithecia. The culture which had been isolated from nature was incapable of producing perithecia except under these conditions. They concluded that this experiment proved that A/a genes did not control the production of perithecia in *Neurospora.* Dodge[10] and Aronescu[1] criticized these experiments adversely, but Lindegren[20] interpreted them as indicating that the culture which Moreau and Moruzi had isolated from nature was a self-sterile heterokaryon which was made fertile by the diffusion of a substance from the homokaryotic culture. Lindegren[19] suggested that self-sterile heterokaryons occur frequently in nature among the filamentous ascomycetes because of their obvious high survival value. They enable the thallus to preempt a large area due to its rapid vegetative growth, but when vegetative growth has ceased, the conidia can fertilize other thalli with which they come in contact. The practice of plating out *Penicillia* with their uninuclear conidia breaks up any heterokaryons which may be collected. In *Neurospora* the multinucleate conidia facilitate the maintenance of the heterokaryon.

BIBLIOGRAPHY

1. Aronescu, A., *Mycologia,* **26**, 244 (1934).
2. Baker, G. E., *Bull. Torrey Bot. Club,* **71**, 367 (1944).
3. Beadle, G. W., and V. L. Coonradt, *Genetics,* **29**, 291 (1944).
4. Beadle, G. W., and E. L. Tatum, *Proc. Nat. Acad. Sci.,* **27**, 499 (1941).
5. Bonner, D., E. L. Tatum, and G. W. Beadle, *Arch. Biochem.,* **3**, 71 (1943).
6. Buller, A. H. R., *Bot. Rev.,* **7**, 335 (1941).
7. Derx, H. G., *Bull. Soc. Mycol. France,* **41**, 375 (1925).
8. Dodge, B. O., *J. Agr. Research,* **35**, 289 (1927).
9. Dodge, B. O., *Mycologia,* **28**, 226 (1928).
10. Dodge, B. O., *Bull. Torrey Bot. Club,* **58**, 517 (1932).
11. Dodge, B. O., *Bull. Torrey Bot. Club,* **69**, 75 (1942).
12. Foster, J. W., H. B. Woodruff, and L. E. McDaniel, *J. Bact.,* **46**, 421 (1943).
13. Hansen, H. N., *Mycologia,* **30**, 442 (1938).
14. Hansen, H. N., and W. C. Snyder, *Science,* **99**, 264 (1944).
15. Jollos, V., *Arch. Protistenkunde,* **83**, 197 (1934).
16. Kruis, K., and J. Satava, *Ovyvoji a kliceni spor jakoz i sexualite kvasinek,* Nakl. C., Akad. Praha, 67 pp. (1918).
17. Lederberg, J., and E. L. Tatum, C. S. H. Symposia on Quantitative Biology, **11**, 113 (1946).
18. Lindegren, C. C., *Bull. Torrey Bot. Club,* **59**, 119 (1932).
19. Lindegren, C. C., *J. Genetics,* **28**, 425 (1934).
20. Lindegren, C. C., *Am. Nat.,* **70**, 404 (1936).
21. Lindegren, C. C., *J. Genetics,* **32**, 243 (1936).
22. Lindegren, C. C., *Iowa State Coll. J. Sci.,* **16**, 271 (1942).
23. Lindegren, C. C., *Ann. Rev. Microbiol.,* **2**, 184 (1948).
24. Lindegren, C. C., *VII Congress Internat. Industries Agricoles,* p. 1-8 (1948).
25. Lindegren, C. C., *The Yeast Cell,* St. Louis, Educational Publishers, 1949.
26. Lindegren, C. C., Genetics and Cytology of Saccharomyces, pp. 256-266 (Chapt. 16) in *Heterosis,* edited by J. W. Gowen, Ames, Iowa, Iowa State College Press, 1952.
27. Lindegren, C. C., and H. N. Andrews, *Bull. Torrey Bot. Club,* **72**, 361 (1945).
28. Lindegren, C. C., V. Beanfield, and R. Barber, *Bot. Gaz.,* **100**, 592 (1939).
29. Lindegren, C. C., and G. Lindegren, *J. Heredity,* **32**, 404 (1941).

30. Lindegren, C. C., and G. Lindegren, *Proc. Nat. Acad. Sci. U. S.,* **29**, 306 (1943).
31. Lindegren, C. C., and G. Lindegren, *Ann. Mo. Bot. Gardens,* **31**, 203 (1944).
32. Lindegren, C. C., and G. Lindegren, *Science,* **102**, 33 (1945).
33. Lindegren, C. C., and G. Lindegren, *Ann. Mo. Bot. Gardens,* **34**, 95 (1947).
34. Lindegren, C. C., and G. Lindegren, *Proc. Nat. Acad. Sci. U. S.,* **33**, 11 (1947).
35. Lindegren, C. C., and C. Raut, *Ann. Mo. Bot. Gardens,* **34**, 85 (1947).
36. Lindegren, C. C., S. Spiegelman, and G. Lindegren, *Proc. Nat. Acad. Sci. U. S.,* **30**, 346 (1944).
37. Luria, S. E., and M. Delbruck, *Genetics,* **28**, 491 (1943).
38. Moreau, F., and M. C. Moruzi, *Compt. rend.,* **192**, 1476 (1931).
39. Pontecorvo, G., and A. R. Gemmell, *Nature,* **154**, 514 (1944).
40. Reaume, S. E., and E. L. Tatum, *Arch. Biochem.,* **22**, 331 (1949).
41. Shapiro, A., *C. S. H. Symp. Quant. Biol.,* **11**, 228 (1946).
42. Shear, C. L., and B. O. Dodge, *J. Agr. Research,* **34**, 1019 (1927).
43. Winge, O., *Compt. rend. trav. lab Carlsberg, Ser. physiol.,* **21**, 77 (1935).
44. Winge, O., and O. Laustsen, *Compt. rend. trav. lab. Carlsberg, Ser. physiol.,* **22**, 235 (1939).
45. Winge, O., and O. Laustsen, *Compt. rend. trav. lab. Carlsberg, Ser. physiol.,* **22**, 337 (1939).
46. Winge, O., and O. Laustsen, *Compt. rend. trav. lab. Carlsberg, Ser. physiol.,* **23**, 17 (1940).

CHAPTER 14

FERMENTATIONS IN WASTE TREATMENTS

Arthur M. Buswell

Efficient methods for disposal of wastes constitute a problem which faces almost every municipality and many industries. Since such wastes represent large volumes of liquids, they usually find their way into nearby streams or other bodies of water. However, it is generally not permissible to discharge raw wastes into natural waters, since these wastes contain organic matter which causes depletion of the oxygen below the requirements of normal fauna and flora and thus constitutes a nuisance.

Various methods have been devised by different industrial concerns and municipalities for the disposal of wastes. Waste treatment may be practiced for the recovery of valuable products and frequently, the monetary savings thus effected more than pay for the cost of treatment. In other cases, waste treatment is mandatory to diminish stream pollution. The most important processes for waste treatment involve the microbiological destruction of organic matter. Both aerobic and anaerobic processes are extensively employed and are important, but perhaps the greatest interest attaches to the anaerobic degradation of organic matter. Many of these degradation reactions, occurring under anaerobic conditions, result in the formation of methane accompanied by carbon dioxide and, in some cases, hydrogen. In this manner, nature has contrived to stabilize the carbon, hydrogen, and oxygen of which organic matter is primarily composed by the formation of these three stable products.

METHANE FERMENTATION

The formation of methane in nature has interested bacteriologists and chemists for more than 60 years. The studies in this field, commencing with a search for an explanation of the "will-o'-the-wisp," have progressed to a point where methane, enough for fuel for power plants of several thousand horsepower, is produced by anaerobic fermentation.

The literature in this field was summarized some years ago by McBeth and Scales[50] and by Stephenson.[78] Briefly, it has been known that moist organic matter, when allowed to decompose under restricted oxygen conditions, yields hydrogen, carbon dioxide, methane, and a variety of organic acids in greater or lesser amounts. Little or no quantitative data on the yield of the various products were available and little had been done with pure compounds, although Söhngen[73] had shown that lower fatty acids with an even number of carbon atoms could be decomposed by mixed cultures giving methane and carbon dioxide. Whether or not hydrogen and fatty acids were necessary intermediates in the process of methane formation was not known. The information on the susceptibility of various natural products to methane fermentation was conflicting. Cellulose was known to ferment, to some extent,[58,61] but so-called "ligno-cellulose" was said to be resistant.[37] One author held that grease would decompose anaerobically,[60] while another stated that it would not ferment to any substantial extent.[85] The addition of lime to favor methane production was preferred by one group of workers and opposed by another.

Of the earlier systematic studies of this fermentation, those of Söhngen were the most extensive.[73] More recent studies are those of Fowler and Joshi,[37] Sen, Pal, and Ghosh,[69] Fischer, Lieske, and Winzer,[34] and Buswell, et al.[18]

Characteristics of Methane Fermentation

Mixed Cultures

The anaerobic fermentations, as carried out for the production of methane, differ in many respects from other types of fermentations. The most important difference is perhaps the fact that it is not necessary to use a pure culture of organisms nor is it required to maintain "purified" cultures for inoculation or reinoculation. The bacteria which are capable of producing methane are found almost universally in nature, although in preponderant numbers in mud and decaying matter. Under proper conditions, these bacteria can be cultivated to a high degree of activity within a few days. The culture can then be maintained at this activity level indefinitely, if a few simple rules, concerning chemical and physical environment, are followed. In fact, microscopic and subculture studies of this fermentation reveal such a mixed and variable flora that it appears at present to be environment rather than flora which determines the results obtained.

Continuous Process

This situation makes possible the continuous operation of the fermentation, a procedure which is quite unusual in fermentology. It is possible to carry out this process in apparatus arranged to allow the substrate to enter continuously at one point and the exhausted or inert residue to be discharged continuously at another, while the products, methane and carbon dioxide, are given off at a steady rate. There is apparently no limit to the size of apparatus which can be used. Large tanks, yielding several hundred thousand cubic feet of gas a day, operate as smoothly as laboratory-size flasks.

Independent of Substrate

Another characteristic of these fermentations is that practically any sort or kind of organic matter may be used as a substrate. Nearly a hundred different pure substances[1,80,81] and some thirty or forty natural plant and animal products,[6,7,16] such as cornstalks, milk whey, etc., have been used successfully as fermentation material. There is apparently no decomposition of mineral oils and lignin, when isolated, is not attacked at all or with difficulty.[9,49,57]

Quantitative Yields

The nearly quantitative yields of the two simple products, carbon dioxide and methane, are somewhat unique. It is true that in the various commercial fermentations of grains, the starch is practically quantitatively recovered in the products, but the fats, proteins, and fiber are not attacked at all. The methane fermentation converts the entire grain, with the possible exception of a small amount of fiber, to carbon dioxide and methane within 24 to 72 hours. The reaction is an oxidation-reduction involving water. If the composition of the substrate is known, the yield of gas can be calculated from the following equation:

$$C_nH_aO_b + \left(n - \frac{a}{4} - \frac{b}{2}\right) H_2O \longrightarrow \left(\frac{n}{2} - \frac{a}{8} + \frac{b}{4}\right) CO_2 + \left(\frac{n}{2} + \frac{a}{8} - \frac{b}{4}\right) CH_4$$

With a little care, it is possible to get 95 to 100% agreement with this equation. The uniformity of the end products, carbon dioxide and methane, was at first puzzling, but a study of the energy[80] of the reaction of various compounds with water showed the maximum free energy when the final products were carbon dioxide and methane.

In the experience of the author, neither carbon monoxide nor higher hydrocarbons have been observed among the products. A few earlier reports in the literature stating that ethane was formed have not been verified and the amounts reported are within the limits of analytical error in the experiments cited. However, in a private communication from Dr. John B. Davis, Magnolia Petroleum Company, Dallas, Texas, Dr. Davis has stated that very recent work, involving special techniques in concentration, and analysis by means of the mass spectrograph has detected minute quantities of ethane, ethylene, and acetylene, and even smaller quantities of propane and propylene, in the fermentation gas. The range of concentration is 0.1 to 7 ppm by volume for these various hydrocarbons.

Wide Temperature Range

In this fermentation, there is no narrow optimum temperature range. The rate of fermentation increases with increase in temperature from about 0°C to 55°C or a little higher. Maxima have been reported at 26°, 37°, and 50° to 55°C.[30] These maxima are not very pronounced and have not been noticed by some observers.

Although this fermentation can be carried out over a wide range of temperature, it may be upset by sudden temperature changes. A drop of 5°C from the temperature at which the culture has been working will frequently arrest the production of gas without materially affecting the production of acids. This may result in a dangerous accumulation of acids.

Control of Methane Fermentation

Conditions

The high yields, uniformity of products, and successful inoculation and cultivation are dependent on four conditions.

1. Large Surface. The organisms seem to require a certain amount of surface for their propagation. When material of a fibrous or granular nature is fermented, this requirement is satisfied by grinding or shredding the substrate, but when dissolved substrates are employed, some sort of inert surface must be supplied. Breden and Buswell[10] used 25 g shredded, washed asbestos per l of effluent. Fischer, Lieske, and Winzer[34] found ferric hydroxide or ferrous sulfide useful for the same purpose. With such techniques, many soluble substances have been used as substrates and over considerable periods of time. Successful inoculation can be made by transferring single fibers of asbestos[10] from an active culture to a flask or test tube containing sterile asbestos and culture medium.

2. Volatile Acids. The volatile acids,[21] which are intermediates in the decomposition of higher compounds, must not exceed a predetermined value, usually 2,000 to 3,000 ppm (calculated as acetic). If the volatile-acid value is allowed to rise much above 2,000 ppm (as acetic) gas formation drops off, the quantity of acids increases rapidly, and, usually within 24 to 48 hours, all fermentation ceases.

Volatile-acid levels are followed in the fermentations by daily determinations by the method of Duclaux,[30] after acidification of the sample with phosphoric or sulfuric acid. This procedure gives a value which includes both the free acids and their salts and is independent of the pH value. It is somewhat surprising to encounter such a situation, especially since in many fermentations pH is an almost all-important factor in fermentation control. Since it is the total of volatile acids, plus their salts which controls the methane fermentation, the addition of alkali is of little use. In fact, the production of acid is stimulated in some cases by the addition of lime.

There is only one way to limit the accumulation of volatile acids. That is to limit the rate at which the substrate is added to the fermentation vessel, so that the acids will be fermented to methane and carbon dioxide as rapidly as they are formed from the raw substrate. Many of the early failures were due to the "batch" type of fermentation in which too much substrate was present.

If a culture has developed too much acid, the only remedy is dilution. This has been adapted to a series or stage digestion for certain industrial wastes.[13]

3. Scum. The formation of considerable amounts of scum must also be avoided. This is sometimes encountered when fermenting greasy material.[53] The objection to scum is that it constitutes a zone of high substrate concentration in which acids are likely to accumulate. In laboratory fermentations, this can be avoided by mechanically breaking up the scum. In plant-scale fermentations, the power required for a scum breaker is too great. Moistening the scum with liquid pumped from beneath it is a very effective and inexpensive remedy.[11] Unpublished studies show that recirculating the gas from the collecting dome, discharging it at several points beneath the scum, has distinct advantages in scum control.

4. Fibrous Material. Fibrous materials, such as paper, shredded cornstalks, etc., form a tough mat at the top of the fermentation vessel.[64] The objection to this mat is, again, that it favors the accumulation of large amounts of acid. A mat cannot be broken up mechanically nor by the circulation of liquor with any success. It is necessary to provide the fermentation vessel with suitable connections so that it can be operated alternately in an upright and an inverted position. In this manner, the mat is broken up after each inversion by the fermentative action itself. A fermentor containing a slowly rotating drum has been developed for large-scale operation.[15]

Effects of Salts

Rudolfs,[63] and Rudolfs and Zeller[66,67] have reported a retarding effect of sulfates and chlorides in concentrations of 500 to 5,000 ppm, in batch experiments with sludge digestion. However, it was stated that, if sufficient time was allowed, digestion proceeded to approximately the same extent as in the controls. Recently Buswell,

Pagano, and Sollo[22] have shown that sodium sulfate, in concentrations of 4,000 ppm or more, makes the methane fermentation of glucose at 55°C impossible. Acetic acid, however, can be readily fermented in the presence of as much as 10,000 ppm of either sodium chloride or sodium sulfate at 37°C.

Recently, Buswell and Pagano[21a] have found that nitrates in the amount of 50 ppm (calculated as nitrogen) will completely inhibit methane fermentation.

Loadings

It should be pointed out that the difficulties mentioned before are not met with in the ordinary plants for the reduction of municipal wastes, except at times of starting new plants or if the supervision is inexcusably bad. The digestion or fermentation tanks are so large in proportion to the load that volatile acids seldom accumulate and any scum which may form may be allowed to accumulate for a year or more without seriously interfering with the operation of the tank. It is common practice to design municipal plants with 50 cu ft of digester capacity for each pound (3 cu m per kg) of organic matter which is to be fermented a day.

The writer uses loadings of 0.2 lb per cu ft (3 kg per cu m) per day to 1.0 lb per cu ft (16 kg per cu m) per day quite successfully and continuously. It is necessary, however, to observe the rules laid down previously.

Quantitative Research Techniques

The apparatus and experimental procedure for quantitative studies of the methane fermentation were described by Buswell et al.[18] as follows:

Inverted bottles or large filter flasks were used for these experiments. The setup shown in Figure 40 was connected to a gas collector and gasometer. Inoculation for these tanks was, at first, liquor from anaerobic digestion tanks; later, liquor from one experiment was used as the inoculum for another.

The amount of substrate fed was regulated so that the weight of gas produced between feedings was equivalent to not less than 85% of the theoretical yield. At the same time that the substrate was fed, inorganic nitrogen, in the form of ammonium hydroxide, was added at the rate of 6 mg of nitrogen per gram of substrate.

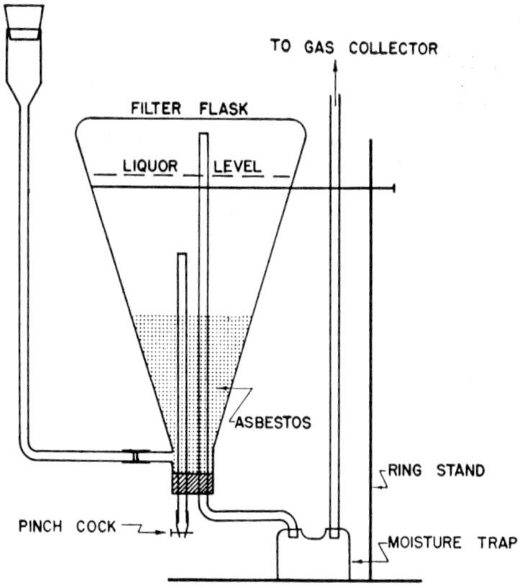

FIGURE 40. *Apparatus for Fermentation Studies*

The concentration of ammonia nitrogen in the liquor was maintained at 400 to 600 mg per liter.

Sludge containing high amounts of organic matter introduced a large possible error into the carbon balances. To obviate this difficulty and still provide a resting place for the bacteria, finely chopped asbestos[10] (previously ignited to remove volatile matter) was added to the flasks in amounts of 25 g per 1 of solution. To each liter of inoculating solution, 100 mg of each of the following salts was added: K_2HPO_4, $Mg_3(PO_4)_2$, $(NH_4)_2SO_4$.

For all inoculating media, the following were determined: residue, volatile matter, total carbon dioxide, ammonia nitrogen, organic (Kjeldahl) nitrogen, volatile acids (reported as acetic acid), lactic acid, chlorides, and alkalinity. At the end of an experiment, the same determinations were made on the culture medium. Analysis of the liquor was also made for unfermented substrate and qualitative tests were carried out for ethyl alcohol and formic acid. The asbestos was analyzed to determine total increase in dry weight, total increase in volatile matter, and gain in total organic nitrogen. The gas was analyzed every few days. From these data, it was possible to strike a balance and account for the amount of substrate fed but not converted to carbon dioxide and methane and the amount of carbon dioxide that was dissolved in the solution, as well as for the gas formed.

The usual laboratory incubator will house only two fermentations if carried out in apparatus of this size, even though the gasometer is placed outside with connections running through the wall of the incubator.

A smaller digester[22] (Figure 41), consisting of a 250-ml extraction flask with a brine-displacement gasometer fastened above the digester, has been designed recently. A glass tube of 14-mm outside diameter and 10 mm long is sealed to the side of the flask. This is closed with a serum stopper, through which, by means of a hypodermic syringe, samples are taken and additions are made. Fitted with stoppers and a tube for gas flow, a calibrated tube of 48-mm outside diameter is fastened above the flask. Above this is set another reservoir for the displaced brine.

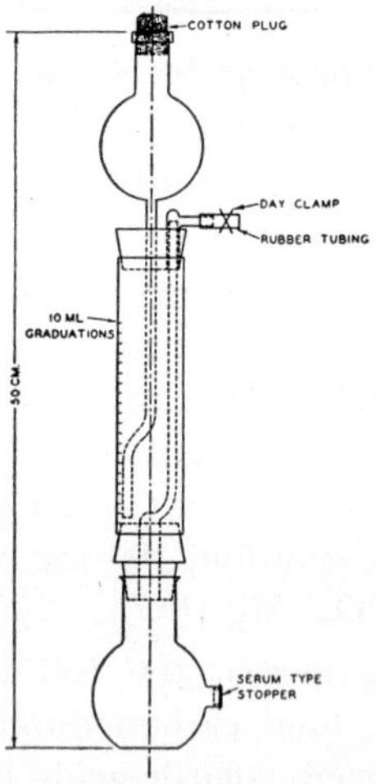

FIGURE 41. *Small Fermentation Vessel*

Gas production is observed directly by the position of the brine level in the calibrated tube and this level is easily reset to zero by opening a Day clamp at the top of the tube and releasing the gas to the air. The digester is operated with an effective volume of 200 ml and the capacity of the gasometer is roughly also 200 ml. The only disadvantage of the apparatus observed in several months' operation is that it is top heavy, and care must be taken to avoid accidents during manipulation.

These digesters could, of course, be constructed on an even smaller scale if the fermentations were not carried out on a quantitative basis. The minimum volume compatible with accuracy in gas measurement and analysis of the medium was considered to be 200 ml. Accuracy in gas measurement with this design is ±2 ml and errors are less than 2% when the daily gas production is over 100 ml. Up to 50 ml of the fermentation medium may be withdrawn daily for analysis without disturbing the fermentation.

Products of Methane Fermentation

Gas Yields

The actual gas yields[14] range from 12 cu ft per pound decomposed (700 ml per g) for protein to about 20 cu ft per pound decomposed for fats (1,250 ml per g). From the loadings listed before, it is seen that the ratio of the volume of gas yield to the volume of fermentor varies from 1:1 to 4:1 in most cases. In laboratory experiments, as much as 10 l of gas was obtained per day per liter of fermentor volume.

Hydrogen Formation

Apparently, hydrogen is the only combustible gas other than methane which is formed by fermentation. Occasional reports of ethane may be found in the literature and carbon monoxide has also been mentioned. The amounts have been so small that when the limits of accuracy in even the most careful gas analyses are considered, it is very doubtful whether the data have justified the conclusions in regard to the presence of the last two gases.

Hydrogen is not formed in detectable amounts during the methane fermentation of the lower fatty acids, although it may well be an intermediate. When carbohydrates are used as substrates, the hydrogen yield may be as high as 30% by volume,[81] although with proper control it can be held to about 1%.

There appear to be only two types of fermentations yielding hydrogen which have been subjected to quantitative investigation. One is the fermentation of sugars by the coli group of organisms and the other the butanol-acetone fermentation.

Extensive quantitative data on the coli fermentation have been given by Rogers, Clark, and Adams.[62] Their work indicated that 30 to 50% of the carbohydrate was decomposed and probably not more than a few per cent occurred as gas. Three types of fermentation reactions were encountered. The one giving the highest yield of gas produced carbon dioxide and hydrogen in the ratio of about 2:1. The second type produced about half as much gas with a CO_2:H_2 ratio of 1:1. The third type gave carbon dioxide only in a relatively low yield. The second type has been investigated by Scheffer (reported by Kluyver[45]) who found about 10% gas having a CO_2:H_2 ratio of 1:1.

A discussion of the butanol-acetone fermentation is given in Chapter 11 of Volume 1. In this fermentation, as carried out commercially, a yield of 1.5 lb of mixed carbon dioxide and hydrogen is obtained for each pound of mixed solvents produced.[38] These gases may be utilized to produce methanol by catalytic synthesis.

Mechanism of Methane Fermentation

References on the mechanism of the bacteriological degradation of cellulose and cellulose derivatives, and especially those dealing with the anaerobic fermentation of these substances, are scarce and far from enlightening. Some information may be found in the periodical literature and in texts, such as Stephenson's[78] *Bacterial Metabolism.* For a review of the earlier literature which covers not only the bacterial degradation of cellulose and cellulose materials but also the decompositions attributed to other organisms, the reader is referred to Thaysen and Bunker's book, *The Microbiology of Cellulose, Hemicelluloses, Pectin, and Gums,*[83] and to Thaysen and Galloway's *The Microbiology of Starch and Sugars.*[84]

Since acetic acid is always found during methane fermentation, one is inclined to regard it as an important, if not universal, precursor of methane. Three possible mechanisms for the methane fermentation of acetic acid are suggested by the findings and reasoning of previous workers in this field. From the work of Omelianski[58,59] and Söhngen,[72,73] a preliminary decomposition of the acetic acid to hydrogen and carbon dioxide would be expected, with subsequent reduction of carbon dioxide to methane by the hydrogen. Barker's work[2] indicates that carbon dioxide would be directly reduced to methane and the acetic acid oxidized to carbon dioxide. The reasoning of Buswell and Neave[21] leads to simple decarboxylation as the mechanism.

The first mechanism was based on the similarity of the hydrogen and methane fermentations, but the low concentration of hydrogen found in the gas from the methane fermentation of acetic acid is evidence against this mechanism. The data of Symons and Buswell[80] show that only 3.5 l of methane was formed over a period of 100 days with regular circulation of hydrogen and carbon dioxide through a culture of 2.0 l total volume. In the same time, such a culture fermenting acetic acid could be expected to form 75 l of methane.[18,81]

The second mechanism avoids this weakness by implying a direct reduction of the carbon dioxide, without the intermediate stage of free hydrogen. This mechanism seems rather indirect and involved, but parallels that found by Barker in the fermentation of alcohols.

The last mechanism appears to be the simplest and most direct. A similar reaction *in vitro* is the chemical decarboxylation of sodium acetate with sodium hydroxide. The internal oxidation-reduction could be effected by the transfer of the hydrogen atom from the carboxyl to the methyl group. Evidence against the decarboxylation mechanism is found in Thayer's work.[82] He reasoned that if the reaction was a decarboxylation the fermentation of propionic and butyric acids should yield ethane and propane. However, in the fermentation of these acids, he found no hydrocarbon other than methane. This has been confirmed in all work where the gases have been analyzed. Therefore, if this mechanism of simple decarboxylation is to be accepted for acetic acid, it must be considered a special case, not applicable to the higher fatty acids.

Buswell, Fina, Mueller, and Yahiro[19] further investigated the action of enrichment cultures on propionates having carbon atoms 1, 2, and 3 labeled, respectively. The distribution of the radioactivity in the evolved gases, carbon dioxide and methane, indicated that propionate is decomposed in such a manner that both gases may arise from all three carbons. Carbon dioxide is, in part at least, a precursor of the methane formed. A preliminary hydrolysis of propionic acid to ethyl alcohol and formic acid is consistent with the results obtained. This was partly confirmed by simultaneous adaptation experiments. When labeled carbon dioxide was administered to cultures stabilized on ethyl alcohol and formic acid, respectively, labeled methane was found in both cases. Stadtman and Barker[76] have extended their studies on the fermentation of various lower fatty acids with C^{14} labeled substrates. Using highly purified, methane-producing cultures, capable of carrying out the incomplete oxidation of fatty acids, they were able to show that a beta oxidation is involved in the decomposition of butyric, caproic, and valeric acids. Propionic acid was shown to be oxidized in such a way that carbon dioxide is derived from the carboxyl carbon, and that acetate is formed from the alpha and beta carbons. In each of these anaerobic fermentations, methane was derived entirely from the carbon dioxide.

Stadtman and Barker,[77] using C^{14} labeled compounds, were also able to show that in the fermentation of methyl alcohol, methane, as in the case of the acetate fermentation, is derived chiefly from the organic substrate and that the reduction of carbon dioxide is of little importance.

Barker, Ruben, and Kamen[4] found evidence for the reduction of carbon dioxide in the fermentation of acetic acid through the use of $C^{11}O_2$, but they stated that the radioactive methane found might have been due to the presence of methanol carried over with the inoculum. It should also be mentioned that these workers were using a pure culture of *Methanosarcina methanica* and that even if methane should be formed by reduction of carbon dioxide in that case, this might not be the predominant mechanism in the usual methane fermentation where a mixed culture is used.

This question of the mechanism of fermentation of acetic acid is of more than academic interest. Culture failure is almost invariably accompanied by, or preceded by, the accumulation of high concentrations of volatile organic acids, largely acetic. Thus, any information concerning this mechanism might lead to methods of treatment or operation which would alleviate or prevent this accumulation of acids and possibly the failure of many cultures.

It may be seen at once that if the carbon of the carbon dioxide were marked isotopically, this question of mechanism could be settled. If either of the mechanisms involving reduction of carbon dioxide were involved, the methane produced should be similarly marked. If the reaction were a simple decarboxylation, the methane should not be so marked.

Buswell and Sollo[24] have reported that by comparing the activity of the methane and the carbon dioxide, when using C^{14} labeled carbon dioxide with acetic acid for methane production, it was shown that only a very small portion of the methane was derived from the carbon dioxide. By studying the relation of the amount of methane formed by reduction of carbon dioxide to the time of incubation, a correlation was found which was interpreted as indicating a slow reduction unassociated with the general fermentation, such as the formation of cell substance and subsequent autolysis. From this, it was concluded that the acetic acid was fermented entirely, or very nearly so, without reduction of carbon dioxide and that the methane is predominantly derived from the acetic acid and not from carbon dioxide.

Stadtman and Barker,[75] in a series of parallel experiments using C^{14} carboxyl- and methyl-labeled acetate and labeled carbon dioxide, confirmed this observation and showed that the methane is derived mainly from the methyl group of the acetic acid, whereas the carbon dioxide is derived mainly from the carboxyl group. But when using ethanol as the substrate, together with labeled carbon dioxide, these authors reported that 82 to 100% of the methane came from the carbon dioxide.

Bacteriology of Methane Fermentation

Methane, as a product of a wide variety of bacterial fermentation processes, has been recognized for many years.[18,50] A partial list of fermentations yielding methane would include composting of straw, leaves, and manure; retting of flax; the decomposition of fertilizer in the soil; the liquefaction of cellulose which causes losses in various textile industries, such as paper manufacture, etc.; and the production of various fatty acids from starch, sugars, and cellulose. Methane is one of the early products in spontaneous combustion.

Although attempts were made to isolate the bacteria responsible for this reaction very early in the history of bacteriology, or some 50 years ago, none was successful in producing actual pure cultures until within the last 15 years. Many experimenters, including the writer, had been successful in producing cultures of very high activity, but microscopic examination always revealed a mixed bacterial growth.

Most experimenters reported two somewhat different rod-shaped forms and two distinctly different spherical or coccus forms. The two coccus forms were of different size; one was usually seen in groups of four or eight, resembling *Sarcina*. These organisms are rather large, i.e., 4 or 5 μ in diameter. The other coccus form is very small, being something less than 1 μ in diameter, and is referred to as a *Micrococcus*. The rod forms are similar, being rather long and thin and slightly bent. The distinguishing characteristic is a tendency of one of them to form long filaments.

Two new species of the methane-producing bacteria were recently obtained by Stadtman and Barker.[76] These organisms differ from the methane-producing bacteria previously described in that they are able to carry out the incomplete anaerobic oxidation of fatty acids.

The methane organisms, in addition to producing methane, also have the following common characteristics: They are non-motile, non spore bearing, and Gram negative. They are obligate anaerobes and are capable of using ammonia as a source of nitrogen. They all develop at a relatively slow rate, requiring several weeks after inoculation for the development of an active culture.

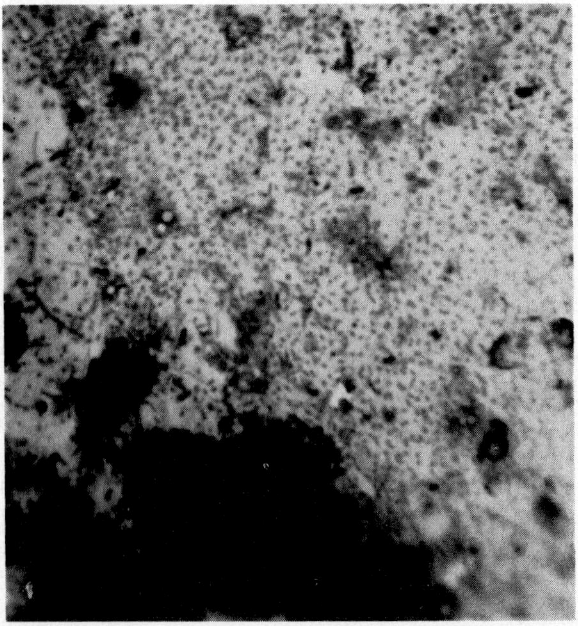

FIGURE 42. *Bacterial Flora from Jefferson Junction Plant; numerous* Micrococcus *forms; fermentation rapid*

The earlier workers found that many of the organisms could produce methane from ethanol or calcium acetate. Barker,[2,3,4] using a medium in which the only source of food was ethanol, was successful in isolating a pure culture of one of these four types of organisms. This was the *Methanobacterium omelianski*. Recently, in the author's laboratory, there was obtained what appears to be a pure culture of the large *Methanosarcina* form, but this is incapable of producing methane from calcium acetate.

The relation of bacterial flora to successful operation is well illustrated by Figure 42. This sample was taken from a plant after it had been in successful operation for a little over a year and when it was producing gas at a very high rate. It will be noticed that this figure shows a very nearly pure culture of a *Micrococcus,* one of the organisms observed previously by practically all workers in methane-producing fermentations.

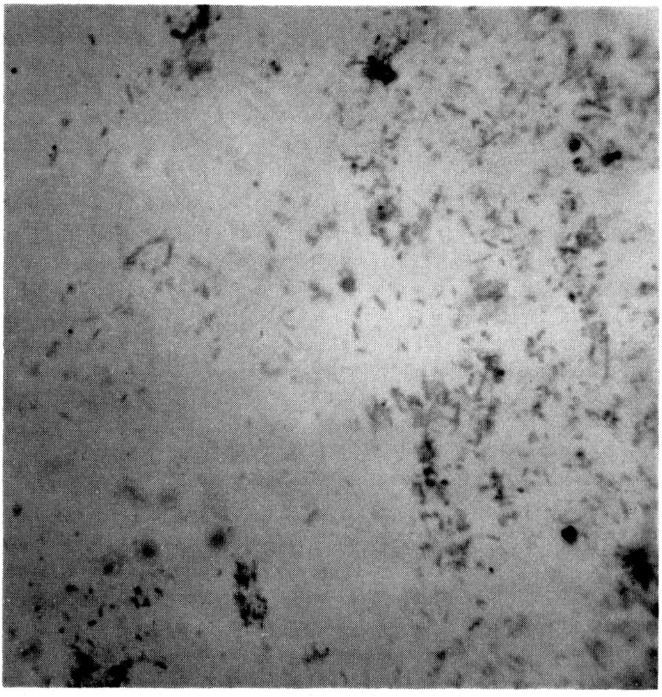

FIGURE 43. *Bacterial Flora from Crystal Lake Plant; mixed flora; fermentation poor*

Unfortunately, a comparable figure at the time of poor operation of this plant is not available. Figures 43 and 44, however, made from contents of a fermentation tank at Crystal Lake, Illinois, reflect operating conditions very strikingly.

Figure 43 shows relatively few organisms and no particular predominance of any one methane-producing form. At the time this sample was taken, the fermentation was very slow, producing only about 100 cu ft of gas per hour.

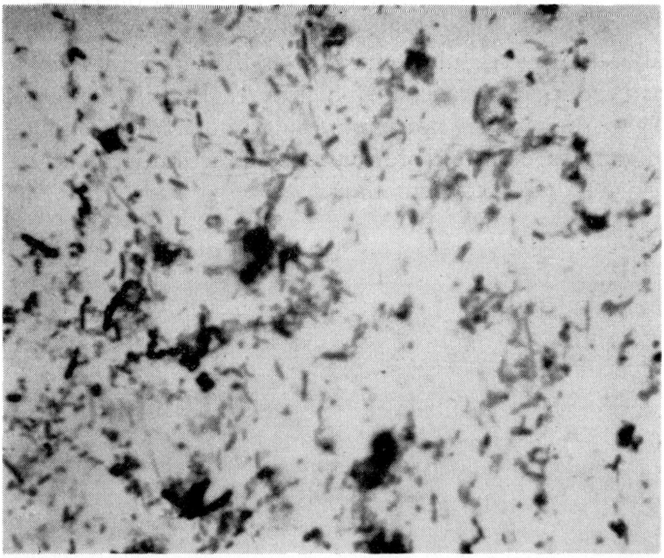

FIGURE 44. *Bacterial Flora from Crystal Lake Plant; mixed flora, methane producers predominating; fermentation good*

Figure 44 shows a luxuriant growth of typical methane-producing organisms, particularly the *Micrococcus,* some of the rods, and some of the large *Sarcina.* Careful control of temperature, feed rate, and circulation had increased the gas yield to about 1,000 cu ft per hour.

Figure 45 shows a greatly enriched culture grown in laboratory flasks with calcium acetate as a source of food. This photograph shows a large number of the large *Sarcina* forms and a considerable number of the small *Micrococci.* The long threads are asbestos fibers which were added to the culture to provide the mechanical support which has been found necessary for rapid growth.

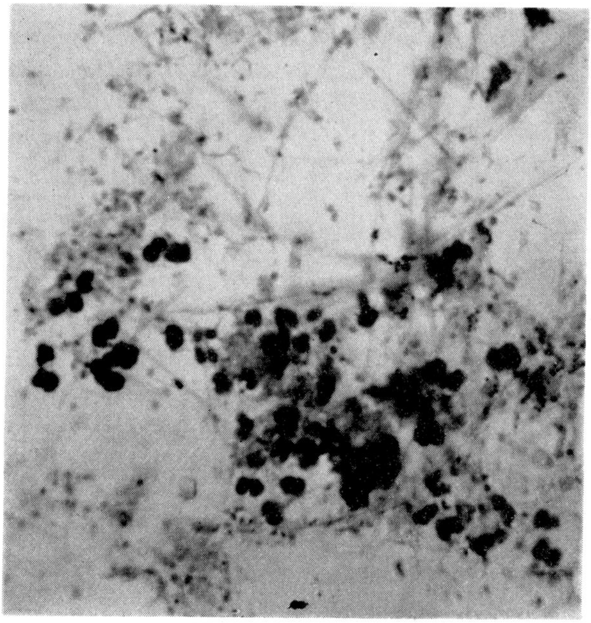

FIGURE 45. *Laboratory Methane-Producing Culture; filaments are asbestos fibers*

MICROBIOLOGICAL TREATMENT OF MUNICIPAL AND TRADE WASTES

General Considerations

The term "municipal wastes" as used here applies to sewage and the term "trade wastes" applies to the water-carried wastes discharged by manufacturing plants. Certain biological processes will be discussed, which, broadly speaking, are applicable to the stabilization of all organic wastes before they are discharged into a stream. A few of these waste liquors carry mineral salts or acids which may be responsible for more or less damage to the stream. A great majority of these wastes contains organic matter of biological origin which, on decomposing, causes putrefactive odors and a depletion of the oxygen of the stream below the requirements of normal fauna and flora. It is obviously this second class of wastes with which we are primarily concerned in the discussion of biological processes in industrial-waste treatment.

The primary object of treatment is to prevent oxygen deficiency in streams. Oxygen deficiency is the basic cause of nuisance and destruction of fish and other aquatic life, including normal vegetable life. The elements carbon and hydrogen and, to a small extent, nitrogen are the chemical factors responsible for consumption of dissolved oxygen. Therefore, the problem is to remove substances containing unoxidized compounds of these elements.

Various bacteria are capable of accomplishing this by forming either (a) relatively insoluble gases or (b) insoluble solids, from the carbon in the organic matter. These processes might be classified as fermentative or gas producing and precipitating (bioprecipitation), respectively. A more common classification of the processes for biological stabilization is made on the basis of the classes of bacteria and other microorganisms which are responsible for the results produced. This classification divides the stabilizing processes into the aerobic and anaerobic groups; that is, those produced by organisms requiring air or oxygen and those carried on in the absence of oxygen.

Except in the case of very simple substances, like acetic acid or glucose, aerobic and anaerobic actions both yield solid and gaseous end products under practicable conditions. When carried to an extreme, the solids may be almost, if not quite completely, gasified by both processes.

The character of the solids and gases is distinctly different in the two cases. The aerobic process naturally yields carbon dioxide, the hydrogen forming water. The solids produced are largely microbial protoplasm, together with colloidal and suspended matter[17,23,26,42] caught by the biological jelly. The solids are 60 to 70% of the original material, the remainder going to carbon dioxide and water.

The anaerobic process is essentially a methane fermentation and yields a humuslike solid, while the gas evolved is composed of methane and carbon dioxide, with small quantities of hydrogen. The methane and carbon dioxide together usually amount to 95% of the total gas collected, and the ratio of $CH_4:CO_2$ varies from 1:1 to 3:1, depending on the composition of the waste decomposed and, to a lesser extent, on the conditions under which the fermentation is carried out.[6,7,9,54,80,81] The humuslike solid remaining at the end of active anaerobic fermentation may amount to 40% of the original when woody material is decomposed, while relatively simple and less resistant substances, such as sugars, fats, and certain proteins are completely gasified.[7,20,21,27,47]

Data in the literature cited indicate that, with the exception of hydrocarbons, practically all carbonaceous substances can be quantitatively converted to methane and carbon dioxide if fermented long enough and under proper conditions. Even lignin yields slowly to this action.[9]

Both the aerobic and anaerobic processes effectively accomplish the stabilization of putrescible waste liquors. The choice between the two will depend on the particular problems involved. Where conditions permit its use, the anaerobic process is much cheaper per pound of organic matter stabilized. A good example is the calculation made in the author's laboratory[20] of the relative cost of treating dairy wastes by anaerobic fermentation and on trickling filters.

The anaerobic process requires simple covered tanks of suitable design. Assuming an average dosage of 1/25 of a volume of milk waste (undiluted basis) per day, per unit of tank volume, it would require a tank, or tanks if operated as a two-stage process, of 5.72 cu ft capacity for the anaerobic fermentation of milk waste containing 1 lb dry weight of milk waste solids. At 50¢ per cu ft, this amounts to only $2.86 per pound of milk solids treated. This fermentation would remove at least 95% of the pollution load. The remaining 5% contained in the overflow liquor could be stabilized readily on filters. Assuming that this final treatment could be made at a cost similar to that given by Kimberly[44] for filter treatment of milk wastes, the total investment for complete treatment would be $8.70 per pound of solids if trickling filters were used following the anaerobic digestion, or $17.46 per pound if sand filters were used, as compared with $116.80 per pound if trickling filters were used alone, or $292 per pound if sand filters were used alone. These figures are not given to show actual costs, but rather the relative costs of the two processes.

While the anaerobic process is much less expensive, it has two serious limitations. First, for reasons as yet obscure, it is impossible to produce a final effluent comparable to that yielded by aerobic treatment. The discharge from anaerobic fermentation is usually dark colored and rather foul with a b.o.d. (biological oxygen demand) of several hundred. Where a high dilution is available, this is not a serious handicap and where heavy untreated wastes are being discharged, the 75 to 90% reduction in the b.o.d. load effected by anaerobic treatment is such a great improvement that it may be adopted as a preliminary measure. As will be described in more detail later, the anaerobic process may be used as a preliminary step followed by the more expensive aerobic process where necessary. The second limitation is that the cost of anaerobic treatment increases as the concentration of the waste decreases. This is due to the large tankage required to accommodate a given number of pounds of organic matter when dilute liquors are to be handled. It has usually been found that the economic advantages of the process disappear when wastes containing less than 1% dry weight of organic matter are to be treated. There is no upper limit of concentration to this process. However, when the concentration of the waste reaches 3%, it can usually be evaporated and dried at a cost which permits its profitable sale as a stock feed or as a fertilizer.

Commercial Development of Anaerobic Methane Fermentation

The commercial application of this fermentation dates well back into the last century when it was employed merely to stabilize and humify organic wastes (i.e., in septic tanks).

It was not until 1897 that a waste-disposal tank serving a leper colony in Matunga, Bombay, was equipped with gas collectors and the gas used to drive gas engines.[36] At about the same time, the waste-disposal tanks at Exeter, England, were partially equipped with gas collectors and the gas was used for heating and lighting at the disposal works. In 1911, a company was formed in Australia for producing and using fuel gases which resulted from the biological decomposition of municipal wastes. Kessener of the Hague, Netherlands, in 1914, described seeding and culture methods for optimum gas yields from straw-board wastes. In the United States, in 1915, Hommon[40] equipped waste-treatment tanks with gas collectors and used the gas. In 1920, Watson[89] of Birmingham, England, reported a study of methane production from sludge digestion and called attention to the fact that a considerable amount of methane can be produced in this way. Following his suggestion, the new disposal plant which was put into operation in 1927 by his successor, Whitehead, was equipped with gas engines that are operated on the gases produced from sludge digestion. This use of the gas cuts down very materially the operating cost of the disposal works. In the meantime (1925), Imhoff in Germany had equipped the sludge-reduction tank in Essen with gas collectors and connected them to the city mains. The gas was found satisfactory for general municipal use and sold to the city. In the same year, Buswell and Strickhouser[25] observed that the sludge-reduction tanks at Decatur, Illinois, were producing about 200,000 cu ft of gas a day. This large yield is due to a considerable amount of wastes from a starch works which are discharged into the city drainage system. The average yield at Decatur was about 125,000 cu ft of gas a day.

At present, the use of gas for power or heat at municipal disposal plants is almost universal in many countries. Fischer[33] reported in 1946 that fuel and fertilizer from sewage was the aim of German treatment plants. Digester gas was developed as a motor-vehicle fuel to replace gasoline. The purified gas contained 90 to 94% methane and about 135 cu ft of this gas at atmospheric pressure was equivalent to 1 U.S. gal gasoline. Its cost was about 10¢, not considering fixed charges on equipment and facilities. Isham[41] in 1948 stated that in a conventional-type sewage-treatment plant, with separate sludge digestion, the digester gas is used in a gas engine to supply pumping power, saving the plant $100 a month on fuel bills. The dried sludge is sold for orchard fertilizer and the chlorinated plant effluent is used for irrigation. A private communication from the superintendent, C. T. Wilson, disclosed that in the calendar year of 1949, the disposal plant of the city of Waterloo, Iowa, received $47,090.20 for gas, besides providing all the heat used about the plant. A recent communication from the manager, J. W. Pray, stated that the city of Fort Dodge, Iowa, has a 250-hp gas engine driving a 175-kw generator, the fuel for which is methane gas produced at the sewage-treatment plant. Disposal of surplus electric power is made to the local power company under an interchange of power contract.

In addition to the general use as fuel, gases from sludge-digestion plants can be also utilized for chemical manufacture. Wilson[91] described the use of digester gas as raw material for the production of formalin in an English plant. In this process, all sulfur compounds are removed from the gas, water vapor is added, and the methane-carbon dioxide mixture is passed over a granular nickel catalyst at 950°C to produce carbon monoxide and hydrogen. Treatment of these gases at 2,500 psi pressure over a zinc chromite catalyst converts them to methanol, which is then dehydrogenated to formaldehyde.

An interesting possible application of this fermentation is that proposed by Fischer, Lieske, and Winzer,[35] namely, the detoxication of coal gas by means of the reactions,

$$CO + 3H_2 \longrightarrow CH_4 + H_2O$$
$$CO + H_2O \longrightarrow CO_2 + H_2$$
$$CO_2 + 4H_2 \longrightarrow CH_4 + 2H_2O$$

Söhngen[73] found that hydrogen was formed in the methane fermentation and suggested that it was intermediate in the formation of methane. He showed that in the presence of the methane-forming culture, the reaction

$$4H_2 + CO_2 \longrightarrow CH_4 + 2H_2O$$

took place quite readily. This observation of Söhngen has been confirmed in the author's laboratory.[79] The volume efficiencies have not been developed to date, but this phase of the process holds interesting possibilities.

TREATMENT OF INDUSTRIAL WASTES BY ANAEROBIC METHANE FERMENTATION

When an industrial waste is very concentrated (containing 1% or more solids) as in the case of distillery, dairy, packing-house, and other wastes, methane fermentation offers an attractive method of treatment.[5,8,12,16] The operating cost is small, although the original investment is rather large. Altogether, it appears that gas can be produced from these wastes at 5 to 10¢ per 1,000 cu ft.

Design of the Digester Fermentation Plant

The preferred design (Figure 46) for an anaerobic fermentation plant consists of two, or preferably three, tanks with floating covers arranged to operate in series. A heat exchanger is necessary to maintain the desired temperature (about 95°F for cold wastes and 130°F for hot wastes). The floating covers are conical and carry domes for gas collection. These are protected by flame traps. The bottoms of the tanks are conical, sloping toward the center. The

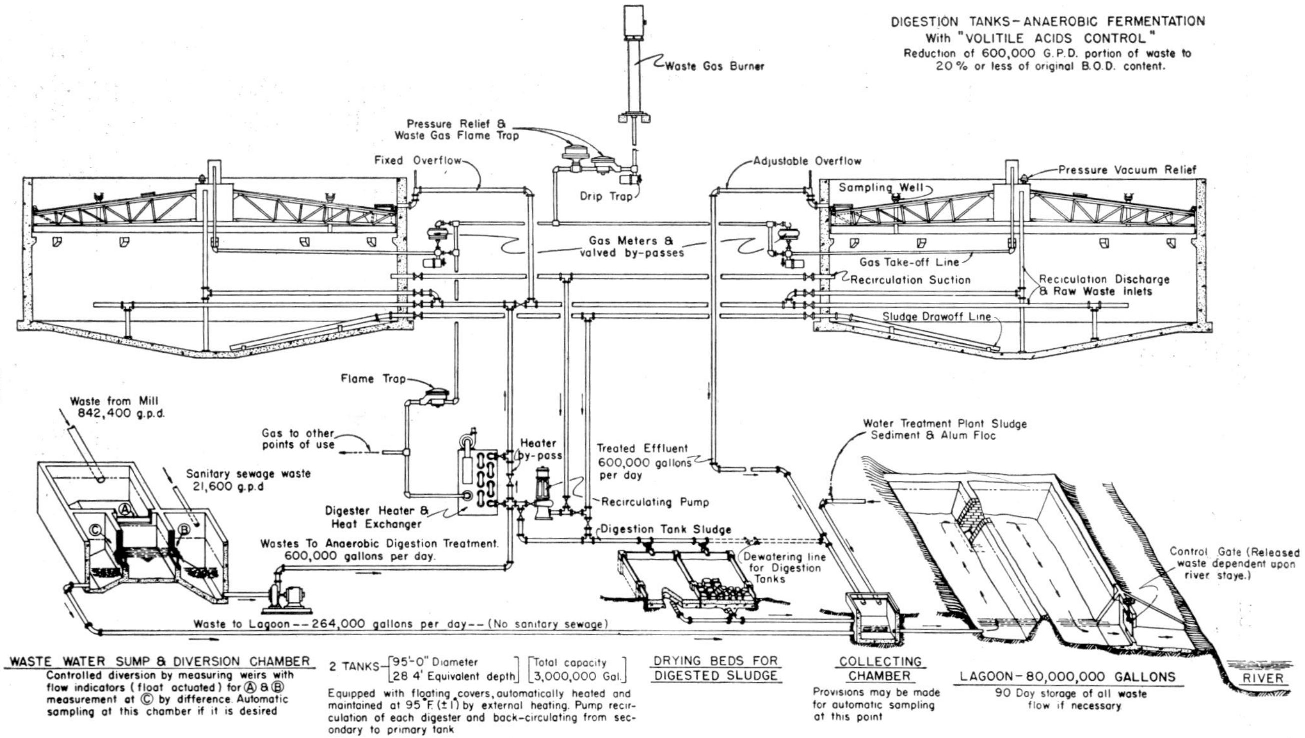

FIGURE 46. *Design for an Anaerobic Fermentation Plant*

circulating pumps and piping system must be laid out to provide the following variations in a three-tank system: (1) use of the series in any order, i.e., 1, 2, 3; 2, 3, 1; 3, 1, 2; or 3, 2, 1; etc.; (2) circulation of contents of each tank from bottom or center to center or top; (3) circulation of the contents of any pair of tanks or all three tanks in a similar manner.

Performance in Digester Fermentation Plants

The largest plant constructed to date is at Pekin, Ill. In a private communication, the late Mr. Almon Fales furnished the following design data:

> The weekly average volatile solids content of the four major wastes to be treated by anaerobic digestion ranged from 0.79% to 1.36% averaging 1.05% for the 93 weeks ending January 1, 1938.
>
> The average quantity of these wastes during this period was 222,300 gal per day, 6 days a week, the 6-day average for the week of maximum flow being approximately 320,000 gal per day.
>
> The digestion capacity of the six digestion tanks to be built will be approximately 285,000 cu ft not including the space reserved for gas but including an allowance of 10% for accumulation of sludge. The effective digestion capacity is accordingly approximately 259,000 cu ft.
>
> The average 6-day loading for the 93-week period, with an average of 222,300 gal per working day and an average of 1.05% volatile solids, would be 0.075 lb per cu ft per working day. An additional digestion period of 1 day per week with little or no wastes added is accordingly afforded.
>
> The loading based on the maximum average flow per working day for the week of maximum flow and the maximum weekly average of 1.36% of volatile solids, which combination would rarely if ever occur, would be 0.140 lb per cu ft per working day.
>
> I should like to point out that in the anaerobic digestion tests, we obtained excellent results with loadings as high as 0.3 lb per cu ft of digestion flask capacity after favorable biological action had become established. As a matter of fact, we originally proposed four digestion tanks having an effective total digestion capacity of 170,000 cu ft which allowed a considerable factor of safety for the loading of the large digestion tanks in practice. In order to make it possible to maintain high efficiency in case of continuous operation of the factory at capacity production, it was decided to provide six tanks instead of four, thus increasing the digestion capacity by 50% and also increasing the flexibility of operation.

Table 45. ANAEROBIC DIGESTION OF TRADE WASTES

Location	Pekin, Ill.	Peoria, Ill.	Crystal Lake, Ill.	Jefferson Junction, Wis.	Carthage, Ohio
Waste	Yeast (Molasses)	Butanol (Grain)	Yeast (Molasses)	Malt steep water trickling filter sludge	Distillery (Grain)
Volume flow, gal/day	222,300	300,000	160,000	2,000	500
Tank, gal	2,300,000	3,000,000	623,000	40,000	7,200
Solids					
volatile, %	1.05	3.0	0.7	3.5	3
B.O.D.					
Raw	10,000	17,000	5,000		16,000
Effluent	2,000	2,420	1,500		1,600
Removal, %	80	69.8	70		90
Loadings					
lb/cu ft/day	0.108	0.114	0.104	0.10	0.143
Gas					
cu ft/lb	5.0	9.3	4.25	9.2	11.0
vol. per tank vol.	0.525	1.06	0.43	1.0	1.5

In Table 45 are summarized extensive performance data based on several years of operating experience in five trade waste-digestion plants.

The second column shows the usual performance of the Standard Brands yeast plant at Pekin, Illinois. The purification amounts to 80% at a loading of 0.10 lb per cu ft per day. Since the third-stage tanks show practically no improvement over the second stage, the effective load is 0.15 lb per cu ft per day. This plant has been in continuous operation for 10 years.

The third column shows the performance of digesters operated by the Peoria Sanitary District while digesting the "slops" or "still bottoms" from the Commercial Solvents butanol-acetone plant. We are indebted to Mr. L. S. Kraus for these data. The digesters, which have been in operation for about 15 years, except for a few periods when the butanol plant was shut down, are accomplishing a 69.8% reduction in b.o.d. at a loading of 0.114 lb per cu ft per day.

The fourth column shows the normal operating results of a plant at Crystal Lake, Illinois, put into operation in 1944. This plant consists of a two-stage digester for handling wastes from the manufacture of yeast. The waste is slightly more dilute than that at Pekin and the loading is correspondingly less, as is the gas yield. The removal of b.o.d. is 70%. The primary digester is equipped with a floating cover and has a capacity of 366,500 gal. The secondary digester has a fixed cover and a capacity of 264,300 gal. We are indebted to Col. M. W. Tatlock of Ralph L. Wolpert Company, Dayton, Ohio, for performance data on this plant.

The data in the fifth column were obtained on an installation treating the waste steep water from a malt-manufacturing plant. The steep liquor is free from suspended solids and is discharged directly to a trickling filter. The filter "unloads" continuously and the load passes to a settling tank which is provided with mechanical sludge removal. This sludge is digested in two 20,000-gal digesters equipped with floating covers. The effluent from the digesters is returned to the influent to the trickling filter. The sludge is digested until it drains well on sand beds.

Although results in the sixth column for grain-distillery wastes were obtained on a two-stage pilot plant operated for about 4 months, they are believed to be representative of what can be accomplished under carefully controlled conditions. A purification of 90% was obtained with a loading of 0.143 lb per cu ft. This pilot-plant experiment has been described previously in detail.[19]

Summarizing, it may be said that: (1) Anaerobic fermentation plants for the reduction of b.o.d. of organic wastes are capable of continuous operation for indefinite periods; (2) loadings of 0.14 lb per cu ft per day of organic wastes have been found feasible in many pilot-plant studies; (3) existing plants built with a large safety factor are loaded to 0.1 lb per cu ft per day, and regularly produce a 70 to 80% improvement in the quality of the wastes.

Recent Investigations on Anaerobic Digestion

Numerous laboratory experiments indicate the possibility of much higher loadings than have been employed in plant installations. Because of the newness of the process, commercial installations have been designed with a large safety factor which experience is showing to be unnecessary. Table 46 lists some of the laboratory results with heavier loadings.

Recent literature gives information on the stabilization of other industrial wastes by anaerobic digestion, a few examples of which are worthy of brief mention. Sawyer[68] gave the results of some experiments conducted at the University of Wisconsin on the feasibility of treating waste sulfite liquors by the standard biological methods. Good and uniform b.o.d. removals, varying from 93 to 95%, were obtained with sulfite liquor concentrations as high as 10%, even though very little of the original color was removed. Singleton[70] reported that pilot-plant studies with anaerobic digestion of wool-scouring wastes gave a b.o.d. reduction of 82%, volatile matter 65%, and grease 90%. Morgan[52] reported pilot-plant treatment of deinking wastes. Primary sedimentation for 96 hours reduced suspended solids from 2441 to 1326 ppm and b.o.d. from 530 to 399 ppm. Activated-sludge treatment reduced the suspended solids to 885 ppm and the b.o.d. to 169 ppm. Trickling-filter treatment was not satisfactory. Rudolfs and Trubnick[65] found on the pilot-plant scale that two-stage digestion gave average removals of 54% total solids, 68% volatile matter, 85% b.o.d., and 74% organic nitrogen with compressed-yeast wastes.

Lamb[45] reported that when preliminary study is made to determine the proper retention time and the amount of air necessary per gallon of sewage, the activated-sludge process can be satisfactorily

Table 46. FEEDING AND GAS–PRODUCTION RATES OBTAINED DURING THERMOPHILIC DIGESTION OF VARIOUS INDUSTRIAL WASTES

Waste	Rate of Feed		Gas Recovery	
	Volume basis, volume per day per tank volume	Solids basis, lb total solids per day per cu ft of tank	Volume basis, volume of gas per day per unit of tank volume	Solids basis, cu ft of gas per lb of total solids
Commercial Solvents' beer-slop (butyl-acetonic) While fermenting rye, 3.3% total solids, 91% volatile	0.500 0.250 0.167	0.930 0.620 0.310	7.00 4.20 3.00	7.50 8.60 9.70
While fermenting corn, 2.0% total solids, 92% volatile	0.250	0.330	3.70	11.00
Heinz distillery waste 7.1% total solids, 96% volatile, pH 4.1	0.036	0.160	2.00	12.50
Buttermilk and whey waste 7.0% total solids, 91% volatile, pH 5.0, high sugars	0.035	0.150	1.60	10.70
Artichoke waste		0.095	0.81	8.35
Extracted chicory		1.560	1.64	10.50
Sugar-beet waste		0.078	0.96	12.30
Packing-house paunch manure		0.370	3.10	8.95

adapted to wastes from canneries, packing plants, milk-processing plants, and acid-pickling processes, either alone or when mixed with domestic sewage. Uhlmann[87] found that activated-sludge treatment of rendering wastes, followed by lime settling, aeration, secondary settling, and sand filtration, gave a b.o.d. reduction of more than 95%.

BACTERIOLOGICAL ASPECTS OF USES OF SEWAGE

In order to complete the picture and make a more rounded presentation of the whole problem of waste disposal, some of the uses of waste waters not covered elsewhere, and perhaps lying somewhat outside of the principal purpose of these discussions, will be reviewed briefly.

Sewage Farms

One of the oldest attempts at the use of wastes is the recovery of fertilizer. As is well known, excreta contains considerable amounts of available nitrogen, which is a very important plant food. The Chinese probably have the most effective method for the recovery of nitrogenous matter. This consists, as most of us know, in the collection of excreta in dry latrines with tight containers which makes possible the transportation of the undiluted material directly to agricultural land. This method introduces health problems, since the consumption of fresh vegetables produced on lands so fertilized is a well-known method of transmitting intestinal diseases, including both the amoebic and bacillary types of dysentery.

The more modern method of removal of wastes by water carriage presents an entirely different picture. In these systems, 1 gal waste is usually diluted with 50 to 150 gal of water, so that even strong domestic sewage contains less than 0.05% of organic matter. In other words, strong domestic sewage is 99.95% pure, which exceeds the purity of a certain well-advertised soap! With this high dilution, the cost of concentration of the nitrogenous material exceeds the cash value of the fertilizer. An early attempt to utilize sewage was the development of sewage farms. In this case, the sewage, after removal of coarse fibrous and granular material, is distributed over the land through irrigation ditches supplying both moisture and some plant food. In arid regions, as for example, the area near

Denver, Colorado, the value of the moisture is of prime importance and the process is, in general, successful.

On the older sewage farms, such as those operated by Paris, France, and Berlin, Germany, the amount of moisture was sometimes greater than the soil required and sour and soggy land resulted.

Another example of the use of sewage as a fertilizer is the development of fish ponds in the vicinity of Munich, Germany. Here carp, which depend on aquatic vegetation for their food supply, are successfully raised. Sewage, after the removal of coarser particles, serves to fertilize and stimulate the growth of aquatic vegetation and, of course, as the organic material is removed from the water, purification is accomplished.

Recovery of Fertilizer

The next development in the recovery of fertilizer from sewage was the introduction of purification processes which removed the organic matter in the form of sludge. This sludge, after digestion, is drawn out to sand beds to be dried. The digestive process liquefies a considerable portion of the nitrogenous compounds so that the sludge is relatively poor in nitrogen (1 to 3% on a dry basis). This sludge contains 1 to 2% of available phosphoric acid and a small amount of potash. The principal value of this sludge is, therefore, as a humus material. It is sometimes sold for a fair price where soil in the vicinity requires humus, but in many cases, it is impossible even to give it away.

The introduction of the activated-sludge process resulted in a much higher recovery of nitrogen, since by this process, the soluble and colloidal nitrogen compounds are taken up by biological growth which later settles. This process is best described as a bioprecipitation. Activated sludge is chemically treated and artificially dried in considerable quantities at Milwaukee and Chicago as well as several other large cities. This sludge contains 6% of nitrogen, with 3% of phosphoric acid, and is, of course, of greater value. The total amount of activated sludge sold annually in the United States is in the neighborhood of 100,000 tons.[28]

Interesting bacteriological problems exist in connection with both the digestion of settled sludge and the production of activated sludge by bioprecipitation. These are strictly bacteriological processes and advances in their operation have depended on studies of their flora.

Reuse of Wastes

The handling of industrial wastes and the recovery of valuable products has been handicapped by the large volume of water to be handled. Notable strides in decreasing the volume of industrial wastes have been made in the last 25 years. Starch manufacture and the pulp and paper industry are examples.

In starch manufacture, corn is steeped in water before final grinding for the removal of the starch grains. According to Greenfield, Cornell, and Hatfield,[39] the wastes originally consisted of all the corn kernel, except the starch. Early in the history of the industry, the germ was separated for the oil, and the grits and gluten meal were added to the oil cake for cattle feed. The remaining soluble sugars and proteins, dissolved from the corn in the steeping and starch-washing process, were 2 to 4% of the corn. The liquor from a single steep is too dilute to permit its evaporation and the recovery of the gluten. Introduction of sulfur dioxide prevents the growth of putrefactive bacteria, so that this weak process water can be used in the steeps instead of fresh water and the heavy steep water can be evaporated and added to the cattle feed. This procedure closed the cycle and bottled up the process; fresh water only enters the process for washing the starch on vacuum filters.

In paper making, the fiber, as it goes to the paper machine or pulp drums, is suspended in 95 to 99 parts of water. A certain amount of soluble matter is taken up and the water, after the removal of the fiber (so called "white water") serves as a medium for the growth of slime-forming bacteria which eventually attack the paper stock. In the last 25 or 30 years, the practice of disinfecting the white water with chlorine and treating it with some newer bacteriostatic compound (e.g., certain chlorinated phenols) has made it possible to reuse this white water indefinitely. According to Skinner,[71] many paper mills now operate on this closed system where all, or practically all, of the water is reused. This has resulted in the lessening of stream pollution by white water and also in financial savings to the mills. A saving of a very considerable amount of paper fiber is made since some fiber always passes through the paper machines. Savings of fiber easily amount to $100 per day per machine.

Southgate[74] reported a substantial saving in flax-retting establishments by reusing the retting liquor continuously. The waste water from the anaerobic retting of flax is very polluting. Although chemical treatment of these wastes has been beneficial, the necessity for large-scale waste-water treatment was avoided by reusing the retting liquor and continuously aerating it in the retting tank.

The extremely high demand for water for industrial purposes in the last few years has led to some very interesting developments in the reuse of sewage. This was a subject of a symposium held by the joint meeting of the American Water Works Association and the Federation of Sewage Works Associations in San Francisco in 1948. Veatch,[88] in this symposium, cited nine large installations where the purified effluent of a sewage-treatment plant was used for such industrial purposes as steam raising, cooling, and quenching of steel and coke. Probably the largest such installation is that at Baltimore and described by Wolman, where some 20 million gallons of reclaimed sewage are used a day as an industrial water supply. This development is of interest to bacteriologists, since it presents a very important problem of control of the disinfection of such a large amount of heavily polluted liquor.

FERMENTATIVE UTILIZATION OF INDUSTRIAL WASTES

Certain large industries produce tremendous quantities of wastes having high organic content, particularly of carbohydrates, which present a considerable disposal problem. Prohibition of stream pollution requires treatment of such wastes before they can be discharged into streams. Activated-sludge or anaerobic fermentation plants have been installed by some of these industries to assume the disposal load, as has been discussed before. Other industries have attempted to realize some practical benefit from such wastes by other fermentation procedures. Space does not permit a detailed treatment, but a few recent papers related to this approach to the disposal problem may be reviewed briefly.

Nolte, von Loesecke, and Pulley[56] investigated the production of feed yeast and industrial alcohol from citrus-waste press juice. *Torula utilis* was used in the yeast production. Four different yeasts were used in the production of alcohol. About 25 gal of press juice were required to yield 1 gal of 190° proof alcohol, with a production cost of 11.3¢ per gallon of alcohol. McNary[51] reported that the disposal problem of citrus wastes in the Florida canning industry was solved by pressing, drying the press cake as carbohydrate cattle feed, and concentrating the press juice to citrus molasses for addition to animal feeds. Fenstel and Thompson[32] have investigated the possibility of using pear-canning wastes for yeast culture.

The disposal of vast quantities of sulfite waste liquor from the wood pulping industry has become an increasingly important problem. The carbohydrate content of the waste liquor may be utilized for fermentation to yield useful products. For example, sulfite waste liquor is now used commercially in a few plants for the production of industrial alcohol and of food and feed yeast (see Chapters 4 and 10 of Volume 1). Tyler[86] reported that fodder-yeast production reduced the sugar content of sulfite waste liquor by 68%, and reduced the b.o.d. by 32 to 56%. The alcohol fermentation reduced the b.o.d. of the sulfite waste liquor by 43%. Joseph[43] reported the same reduction in b.o.d. in the industrial-alcohol plant operated by the Ontario Paper Co. Other fermentations which have been proposed for utilization of sulfite waste liquor are the acetone-butyl alcohol fermentation investigated by Wiley, Johnson, McKay, and Peterson,[90] the butyric acid fermentation investigated by Daniels and McCarthy,[29] and the lactic acid fermentation investigated by Leonard, Peterson, and Johnson.[47] Although these have not achieved commercial exploitation, sulfite waste liquor is receiving increasing attention as a potential raw material for fermentation industries.

BIBLIOGRAPHY

1. Ardern, E., Ann. Rept. Revenue Dept. City of Manchester (England), year ending March 25, 1931.
2. Barker, H. A., *Arch. Mikrobiol.*, **7**, 404 (1936).
3. Barker, H. A., *Antonie von Leeuwenhoek, J. Microbiol. Serol.*, **6**, 201 (1939-1940).
4. Barker, H. A., S. Ruben, and M. D. Kamen, *Proc. Nat. Acad. Sci.*, **26**, 426 (1940).
5. Boruff, C. S., *Ind. Eng. Chem.*, **25**, 703 (1933).
6. Boruff, C. S., and A. M. Buswell, *Ind. Eng. Chem.*, **21**, 1181 (1929).
7. Boruff, C. S., and A. M. Buswell, *Ind. Eng. Chem.*, **22**, 931 (1930).

8. Boruff, C. S., and A. M. Buswell, *Ind. Eng. Chem.*, **24**, 33 (1932).
9. Boruff, C. S., and A. M. Buswell, *J. Am. Chem. Soc.*, **56**, 886 (1934).
10. Breden, C. R., and A. M. Buswell, *J. Bact.*, **26**, 379 (1933).
11. Buswell, A. M., *Ind. Eng. Chem.*, **21**, 322 (1929).
12. Buswell, A. M., *Ind. Eng. Chem.*, **22**, 1168 (1930).
13. Buswell, A. M., *Water Works and Sewage*, **82**, 135 (1935).
14. Buswell, A. M., and C. S. Boruff, *Sewage Works J.*, **4**, 454 (1932).
15. Buswell, A. M., and C. S. Boruff, *Ind. Eng. Chem.*, **25**, 147 (1933).
16. Buswell, A. M., C. S. Boruff, and C. K. Weisman, *Ind. Eng. Chem.*, **24**, 1423 (1932).
17. Buswell, A. M., A. L. Brensky, et al., *Ill. State Water Survey Bull. 18,* 7 (1923).
18. Buswell, A. M., et al., *Ill. State Water Survey Bull. 32* (1936).
19. Buswell, A. M., L. Fina, H. F. Mueller, and A. Yahiro, *J. Am. Chem. Soc.*, **73**, 1809 (1951).
20. Buswell, A. M., and M. LeBosquet, *Ind. Eng. Chem.*, **28**, 795 (1936).
21. Buswell, A. M., and S. L. Neave, *Ill. State Water Survey Bull. 30* (1930).
21a. Buswell, A. M., and J. F. Pagano, *Sewage and Ind. Wastes*, **24**, 897 (1952).
22. Buswell, A. M., J. F. Pagano, and F. W. Sollo, *Ind. Eng. Chem.*, **41**, 596 (1949).
23. Buswell, A. M., R. A. Shive, and S. L. Neave, *Ill. State Water Survey Bull. 26* (1928).
24. Buswell, A. M., and F. W. Sollo, *J. Am. Chem. Soc.*, **70**, 1778 (1948).
25. Buswell, A. M., and S. I. Strickhouser, *Ind. Eng. Chem.*, **18**, 407 (1926).
26. Buswell, A. M., S. I. Strickhouser, et al., *Ill. State Water Survey Bull. 26* (1928).
27. Buswell, A. M., H. L. White, G. E. Symons, et al., *Ill. State Water Survey Bull. 29* (1929).
28. Committee of the Federation of Sewage Works Association, *Manual of Practice No. 2* (1946).
29. Daniels, H. S., and J. L. McCarthy, *Paper Trade J.*, **126**, Tappi Sec. 53 (1948).
30. Duclaux, E., *Ann. chim.*, **2**, 289 (1874); *Traité microbiologie*, **3**, 385 (1900).

31. Fair, G. M., and E. W. Moore, *Sewage Works J.*, **6**, 3 (1934).
32. Fenstel, I. C., and J. H. Thompson, *Western Canner and Packer*, **38**, No. 4, 60 (1946).
33. Fischer, A. J., *Civil Eng.*, **16**, 448 (1948).
34. Fischer, A. J., R. Lieske, and K. Winzer, *Biochem. Z.*, **236**, 247 (1931).
35. Fischer, A. J., R. Lieske, and K. Winzer, *Brennstoff-Chem.*, **14**, 301, 328 (1933).
36. Fowler, G. J., *An Introduction to the Biochemistry of Nitrogen Conservation*, p. 132, London, Edward Arnold, 1934.
37. Fowler, G. J., and G. V. Joshi, *J. Indian Inst. Sci.*, **3**, Part IV, 39 (1920).
38. Gabriel, C. L., and F. M. Crawford, *Ind. Eng. Chem.*, **22**, 1163 (1930).
39. Greenfield, R. E., G. N. Cornell, and W. D. Hatfield, *Ind. Eng. Chem.*, **39**, 583 (1947).
40. Hommon, C. C., *Eng. Record*, **73**, 182 (1916).
41. Isham, O. L., *Am. City*, **43**, No. 9, 135 (1948).
42. Johnson, J. W. H., *J. Econ. Biol.*, **9**, 105, 127 (1914).
43. Joseph, H. G., *Sewage Works J.*, **19**, 60 (1947).
44. Kimberly, A. E., *Water Works and Sewage*, **78**, 48 (1931).
45. Kluyver, A. J., *Chemical Activities of Microorganisms*, p. 55, London, University of London Press, 1931.
46. Lamb, M., *Public Works*, **80**, No. 8, 24 (1949).
47. Larson, T. E., C. S. Boruff, and A. M. Buswell, *Sewage Works J.*, **6**, 24 (1934).
48. Leonard, R. H., W. H. Peterson, and M. J. Johnson, *Ind. Eng. Chem.*, **40**, 57 (1948).
49. Levine, M., G. H. Nelson, D. G. Anderson, and P. B. Jacobs, *Ind. Eng. Chem.*, **27**, 195 (1935).
50. McBeth, I. G., and F. M. Scales, *U. S. Dept. Agr., Bur. Plant Ind., Bull. 266* (1913).
51. McNary, R. R., *Ind. Eng. Chem.*, **39**, 625 (1947).
52. Morgan, P. F., *Sewage Works J.*, **21**, 512 (1949).
53. Neave, S. L., and A. M. Buswell, *Ind. Eng. Chem.*, **19**, 1012 (1927).
54. Neave, S. L., and A. M. Buswell, *J. Am. Chem. Soc.*, **52**, 3308 (1930).
55. Neave, S. L., and A. M. Buswell, *Ill. State Water Survey Circular 8* (1930).
56. Nolte, A. J., H. W. von Loesecke, and G. N. Pulley, *Ind. Eng. Chem.*, **34**, 670 (1942).
57. Norman, A. G., *Science Progress*, **30**, 442 (1936).

58. Omelianski, W., *Zentr. Bakt. Parasitenk., II Abt.*, **8**, 193, 225, 257, 289, 321, 353, 385 (1902).
59. Omelianski, W., *Zentr. Bakt. Parasitenk., II Abt.*, **11**, 369 (1905).
60. O'Shaughnessy, F. R., *J. Soc. Chem. Ind.*, **33**, 3 (1914).
61. Popoff, M. A., *Arch. ges. Physiol.* (Pflügers), **10**, 113 (1875).
62. Rogers, L. A., W. M. Clark, and A. C. Evans, *J. Infect. Diseases*, **17**, 137 (1915).
63. Rudolfs, W., *U. S. Pub. Health Repts.*, **43**, 874 (1928).
64. Rudolfs, W., and H. M. Heisig, *Sewage Works J.*, **1**, 519 (1929).
65. Rudolfs, W., and E. H. Trubnick, *Sewage Works J.*, **21**, 294 (1949).
66. Rudolfs, W., and P. J. A. Zeller, *Ind. Eng. Chem.*, **20**, 48 (1928).
67. Rudolfs, W., and P. J. A. Zeller, *Sewage Works J.*, **4**, 771 (1932).
68. Sawyer, C. N., *Ind. Eng. Chem.*, **32**, 1469 (1940).
69. Sen, H. K., P. P. Pal, and S. B. Ghosh, *J. Indian Chem. Soc.*, **6**, 673 (1929).
70. Singleton, M. T., *Sewage Works J.*, **21**, 286 (1949).
71. Skinner, H. J., *Ind. Eng. Chem.*, **31**, 1334 (1939).
72. Söhngen, N. L., *Proc. Roy. Acad. Amsterdam*, **8**, 327 (1905).
73. Söhngen, N. L., Proefschrift, Delft (1906); *Rec. Trav. Chim.*, **29**, 238 (1910).
74. Southgate, B. A., "Water Pollution Research," 146 p., *Dept. Sci. Ind. Research (British), Tech. Paper No. 10* (1948).
75. Stadtman, T. C., and H. A. Barker, *Arch. Biochem.*, **21**, 256 (1949).
76. Stadtman, T. C., and H. A. Barker, *J. Bact.*, **61**, 67 (1951).
77. Stadtman, T. C., and H. A. Barker, *J. Bact.*, **61**, 81 (1951).
78. Stephenson, M., *Bacterial Metabolism*, 3rd ed., New York, Longmans, Green, 1949.
79. Symons, G. E., Ph.D. Thesis, University of Illinois, 1932.
80. Symons, G. E., and A. M. Buswell, *J. Am. Chem. Soc.*, **55**, 2028 (1933).
81. Tarvin, D., and A. M. Buswell, *J. Am. Chem. Soc.*, **56**, 1751 (1934).
82. Thayer, L. A., *Bull. Am. Assoc. Petroleum Geol.*, **15**, 441 (1931).
83. Thaysen, A. C., and H. J. Bunker, *The Microbiology of Celluloses, Hemicelluloses, Pectin and Gums*, London, Oxford University Press, 1927.

84. Thaysen, A. C., and L. D. Galloway, *The Microbiology of Starch and Sugars,* London, Oxford University Press, 1930.
85. Thumm, K., *Vierteljahrsschr. gerichtl, Med.,* **48**, Suppl. 2, 73 (1914).
86. Tyler, R. D., *Sewage Works J.,* **19**, 70 (1947).
87. Uhlmann, P. A., *Sewage Works Eng.,* **20**, 330 (1949).
88. Veatch, N. T., *Sewage Works J.,* **20**, 3 (1948).
89. Watson, J., *Engineering,* **112**, 456 (1921).
90. Wiley, A. J., M. J. Johnson, E. McCoy, and W. H. Peterson, *Ind. Eng. Chem.,* **33**, 606 (1941).
91. Wilson, H., *Surveyor,* **105**, 27 (1946).

INDEX

B

C

D

G

H

I

M

P

S